THE DICTIONARY OF
SPACE TECHNOLOGY

DEDICATION

To Joseph, James, Jennifer, and other children of Earth
who will soon inherit the stars.

THE DICTIONARY OF
SPACE
TECHNOLOGY

by

Joseph A. Angelo, Jr.

FACTS ON FILE, Inc.
460 Park Avenue South
New York, N.Y. 10016

Dictionary of Space Technology

by
Joseph A. Angelo, Jr.

Library of Congress Cataloging in Publication Data

Angelo, Joseph A.
   Dictionary of space technology

   1. Astronautics—Dictionaries.   I. Title
TL788.A53   629.4′03′21   81-3144
ISBN 0-87196-583-6   AACR2

Printed in the United States of America
10 9 8 7 6 5 4 3 2 1

Foreword

Intense pride in the spirit and abilities of the American people—this was the irrepressible emotion that came upon us during our 84 days in space on Skylab III. Our nation had made the necessary investment of time, intelligence, and effort to gain entrance into a world which could only be imagined from the ground. The transition to a people who could walk on the moon had opened up a whole new realm of opportunities and challenges. Skylab, our country's first space station, was our first application of this new strength.

In subsequent years, however, we have seen our foresight give way to a fixation on immediate problems and our aggressive technical growth give way to a reluctance to new commitment. Let us go back and look at some of those national characteristics that gave us such great pride and see how they have evolved.

We felt proud not only of our fellow countrymen, whose spirit and abilities made our space station possible, but also of previous generations who had the wisdom to significantly support the research and development that assured our success. We saw our nation use the focal point of a lunar landing to help develop a strong broad-based technology that gave us a natural position of world leadership. However, since then we have witnessed a tendency to turn more and more inward to our immediate needs and comforts. Initiation of growth stimulating bold technological programs has brought us to the level of technical achievement that we enjoy today. As much as in any previous age, we now have the responsibility to adequately use our inherited technology and, most importantly, to develop it further for our future generations. We must become the leaders in producing the space products of tomorrow and not allow them also to become like an ever increasing portion of our technology of today—imported.

Fortunately, the Shuttle Program has survived the post Apollo era. However, it has done so at a scale and pace that is significantly lower than that which would bring optimum return to our Nation. Like the DC-3 of the space era, it will make possible frequent economical trucking to and from earth orbit and replace many of the wide variety of unmanned boosters currently in use. The next step, which could have been taken nearly in parallel, the permanent space station, remains to be initiated. Skylab has more than adequately demonstrated that a space station provides a platform for a wide variety of beneficial activities and has also provided the required knowledge of how to proceed. We do not wait for justification or technology, only the national will to act.

As we looked down on the homelands of almost all of the people living today, we took pride in the strength of the American economy which could support our growth into space. We also realized that this support is a two-way street. It has been shown that money spent on programs that provide a focal point for the stimulation and growth of new technologies is not really lost but is an investment which continues to pay off handsomely in the future. Chase Econometric, for example, has shown that the spending of one billion dollars per year between 1974 and 1984 on NASA research and development results in a $22 billion growth in the real gross national product, over a 1 percent increase in labor productivity, the addition of over 1 million new jobs, and a decrease of 2 percent in the rate of inflation. Our long-term economic viability is aided rather than hindered by our priming the economic pump in this manner. The 1980 NASA budget is less than 0.9 percent of our total national expenditure while the amount of immediate direct payments to individuals is 54 times larger. Unlike these direct payments, however, the economic benefits of NASA-type spending bear fruit in 5 to 10 years downstream. It should be clear that the economic vitality of our next generation is very dependent on our increased investment in research and development of new technology. The Shuttle should be a key factor in this investment.

During our flight, we also took pride in the immediate focussing of our new space technology on practical return, both direct and indirect. Vastly improved communications, navigation, and knowledge of our Earth's weather, agriculture, water resources, ocean state, and geology are now possible because of the direct use of space platforms. The spinoff of new technologies into other fields has led to improved medical care, computers, and materials to name only a few. Truly, space technology is like the goose that continues to lay golden eggs, as long as it continues to be fed.

We have finally reached that point where the Shuttle is ready to fly. This book examines the near-term mechanics of flight as well as several of the many possible long-term directions for human growth off of our planet. Dr. Angelo speaks with both knowledge and enthusiasm about what lies before us in a style that makes it absorbing reading for the layman and professional alike.

The possibilities for future flight are as extensive as the universe in which we live. Clearly, space flight is but an infant, and proper development depends on informed intelligent decisions. This book provides an excellent background for all those who will be directly involved in these decisions. It is also well suited to those who wish to watch and to appreciate our movement into space as one of the most rewarding and noble adventures yet made by man.

Edward Gibson

Introduction

The first flight of the Space Shuttle "Columbia" on April 12, 1981 opened an exciting new era of U.S. space achievement—the Shuttle Era. This book introduces you to the world of tomorrow, a world in which space, its resources and its distinctive advantages will enhance the quality of life for all the people on Earth.

Throughout history human development has been forged and molded in the crucible of challenge and adversity. If the human race is to continue to grow, develop and prosper, it must now acquire a sufficient resource base—or else, truly primitive situations will develop in which the very struggle for existence all but displaces any capability for higher order development. It appears to many who study and predict the future that the Earth itself cannot supply a sufficient resource base for centuries, or even decades, of exponential planetary development. Space, the ultimate physical frontier, must be exploited.

However, if man turns his back on the Universe, a static, essentially "closed world" civilization will evolve. As demonstrated by the laws of natural selection, once a life form has become fully adapted it establishes an intimate, quasi-equilibrium with its environment. No further evolutionary changes can occur unless the nature and characteristics of the environment change.

Exploitation of the "vertical" or space frontier, on the other hand, provides man an infinite environment in which to continue to develop and grow without physical limit. We have now acquired the technical foundation (through space technology) for initiating the second phase of planetary development, expansion of the human resource base throughout the Solar System. This "open world" approach to planetary development, also called "the humanization of space" may be viewed as the socio-technical evolutionary process in which man learns to use space to better life on Earth, and also learns (a selective few at first) to live in space. The first phase of planetary development began with the origins of life on Earth and will culminate with the full use and eventual depletion of the terrestrial resource base. The third, and perhaps final, phase of planetary civilization involves man's eventual migration to the stars.

Space technology enables man to accept his cosmic challenge. The Space Transportation System, or Space Shuttle, is a critical next step on this technological pathway. Through its clever use and application, we will have routine access to space and will engage in a variety of activities that directly and indirectly improve the conditions of life on Earth. With its versatility and reusability, Shuttle opens the age of space exploitation!

Space no longer belongs to just the scientists and engineers; it is the common heritage of all mankind. In the Shuttle Era and beyond, man's cultural and social activities will experience as dynamic a transition as man's technical and economic advancements. Space, in your lifetime, will continue to exert a significant pull on all areas of human activity, eventually triggering a terrestrial renaissance.

This book opens up the language of space to all who wish to join man's march to the stars. Linked heavily to the Space Transportation System and many of its companion programs, it was written neither as an historic compendium of past space achievements nor an exhaustive treatment of all possible future space alternatives without regard to engineering sensibilities. Rather, it was written as a well-illustrated guide to an exciting assortment of U.S. and foreign space programs that will occur, or at least have the potential of occurring, in the next few decades. Many complementary terms from physics, chemistry, astronomy and a host of other physical sciences and engineering disciplines accompany the body of space technology terms—but in necessarily abbreviated form. (For additional details in these areas, you are referred to the many fine source documents and dictionaries that already exist.) Soviet space program entries, because of their government's penchant for secrecy are very limited to avoid speculative entries, or ones which lack the quantitative details that accompany so many of the other entries found throughout the book.

Another, not obvious, but unavoidable limitation also deserves mention. The field of space technology itself is rapidly changing. Some new topics, not now listed as entries, will emerge almost overnight; while other, seemingly very significant present programs, will suffer premature "fiscal" demise. A sufficiently wide topic base has been chosen from on-going and projected U.S., European Space Agency and Japanese space programs to compensate for such oscillations.

Finally, I wish to express my most sincere appreciation to the National Aeronautics and Space Administration and the United States Air Force, who generously provided much of the material that made this book possible. Several individuals have also been extremely helpful in the creation of this work and they deserve special acknowledgement here: Captain Chester (Chet) Lee [USN-retired], Lt. Col Gotthard Janson III [USAF], Mrs. Donna Miller, Mr. Les Gaver, Mr. Donald Zylstra, Mr. Charles Redmond, Mr. Jesco von Puttkamer, and Mr. Michael Collins—all presently at NASA Headquarters (except Mr. Collins now a fulltime student); Dr. Charles Hollinshead and Mr. Dick Young—Kennedy Space Center; Mr. Donald Engel—Eastern Space and Missile Center; Mrs. Alice Price—Air Force Art Collection; Mr. Edward (Larry) Noon and Mr. Ben Casados—Jet Propulsion Laboratory; Mr. Dick Barton—Rockwell International; Mr. William Rice—Boeing Aerospace Co.; Mr. David Buden—Los Alamos National Laboratory and Dr. Jay Burns—Florida Institute of Technology. Dr. Krafft Ehricke and Astronaut (Dr.) Edward Gibson deserve a very special acknowledgement for their conversations and stimulating thoughts on space and its potential for mankind. Last, but certainly not by any means least, I wish also to acknowledge four individuals without whose personal sacrifices this manuscript would never have been completed. They are my wife, Joan, and our three children: Joseph, James and Jennifer. Over the past year they have very patiently endured weekend after weekend of manuscript preparation, helping to sort, edit, file and review a myriad of material that eventually was distilled into a final product. Whatever success this book enjoys is also very much their success, whatever limitations appear are solely my own.

Joseph A. Angelo, Jr.
Cape Canaveral
January, 1982

A

A *ampere.*

Å *Angstrom* unit.

aberration 1. In astronomy aberration of starlight is the apparent displacement of the direction of light from *celestial* bodies. It is caused by the combination of the velocity of the observer—e.g., a terrestrial observer would have the Earth's orbital velocity (v_e) around the Sun, which is approximately 30 km/sec—and the *velocity of light* (c), which is approximately 300,000 km/sec. This phenomenon was first observed by the English astronomer James Bradley in the 1720s. A familiar analogy illustrating this phenomenon is the oblique traces made by raindrops on the side window of a moving automobile, train or airplane. In the course of a year the light from a *"fixed" star* appears to move in a small *ellipse* around its mean position on the *celestial sphere*. This principal aberration effect, called the "annual aberration" of fixed stars, is due to the Earth's orbital motion around the Sun. The maximum radius of this ellipse in radians (i.e., the semimajor axis of the ellipse) is approximately 20.5 seconds of arc. A terrestrial observer must incline his telescope forward by an amount α, called the angle of aberration, in order to receive light from a particular star. The mathemati-

cal relationship, which is described in Fig. 1, is

$$\sin \alpha = (v_e/c) \sin \beta$$

See: parallax

2. In optics aberration is a specific deviation from perfect imagery, as for example, spherical aberration, astigmatism, coma, curvature of field and distortion. In addition if light of more than one color is involved, false color effects may also appear in the image. This particular defect is called chromatic aberration. Spherical aberration results in a point's image appearing as a circular disk. The distortion of the lens- or mirror-produced image is due to different light rays from any one point on the object making different angles with the line joining that point to the optical center and coming to a focus in slightly different positions. The optical center for a thin lens is essentially a point at its geometric center through which a light ray passes without being deviated. For thicker lenses the optical center may be defined as that axial point through which all undeviated light rays pass. Astigmatism is a defect of a lens or mirror in which a point off the axis of a simple lens creates two images, each in the form of a line. Coma is an optical aberration in which the image of a point off the axis formed by the various lens zones will be made up of a ring of light. Because the light rings are of different diameters and are displaced from each other, the composite image presents a blurred, comet-like appearance. Curvature of field is an aberration of a lens or mirror

in which a plane (i.e., flat) object perpendicular to the axis of an optical system does not give rise to a plane image. Rather (in the absence of astigmatism) the image will lie on a paraboloidal surface. The effects of astigmatism, if also present, would be superimposed on those of curvature of field. Distortion is an aberration in which the magnification produced by a lens or mirror varies over the field of view it covers. Thus, a plane object gives rise to an image of a different geometrical shape. Fig. 2 depicts two common forms of distortion: pincushion distortion, in which magnification increases at the edge of the field, and barrel distortion, in which magnification decreases at the edge of the field. Finally, chromatic aberration is the formation by a lens (but not by a mirror) of an image containing colored fringes. This is due to the fact that the *refractive index* of all transparent substances varies with the wavelength of light. For example, blue light is bent (refracted) more than red light. Thus, false colors will appear in the image.

ablating material   A material designed to provide *thermal protection* to a body traveling at *hypersonic* speed in a planetary atmosphere. Ablating materials are used on the surfaces of some *re-entry vehicles* to absorb thermal energy by removing mass. This blocks the transfer of heat to the rest of the vehicle and maintains the temperatures of the vehicle's structure and interior (especially a manned compartment) at an acceptable level. Ablating materials absorb thermal energy by increasing in temperature and by undergoing changes in their chemical or physical state. This absorbed thermal energy is then dissipated from the surface by a loss of the ablating material's mass (in either the vapor or liquid state). The departing mass also blocks part of the *convective heat transfer* to the remaining ablating material in a manner similar to *transpiration cooling*.
See: ablation, mass transfer cooling, aerodynamic heating, heat shield

ablating nose cone   A *nose cone* designed to reduce thermal (heat) energy transfer to the interior of a *re-entry vehicle* through the use of an *ablating material*.

ablation   A form of *mass transfer cooling*

that involves the removal (frequently intentional) of surface material from a body, such as a *re-entry vehicle* or space probe, by vaporization, melting or other process due to the aerodynamic effects of moving at high speed through a planetary *atmosphere*. The rate at which ablation occurs is a function of the aerothermal environment through which the body passes. In general the ablation temperature remains relatively constant, thereby minimizing conduction of heat to the body's interior.

The *effective heat of ablation* (symbol Q_A^*) describes the efficiency of an *ablating material*. It can be defined by the following relationship:

$$Q_A^* = \dot{q}_o / \dot{m}_A$$

where: $\dot{q}_o$ - is the thermal energy flux to a non-ablating body which has the same dimensions as the ablating body (assuming the surface temperature is equal to the ablation temperature)
$\dot{m}_A$ - is the rate of mass loss of the ablating material

A high value for the effective heat of ablation (Q_A^*) is desired in designing re-entry bodies to minimize the mass loss needed to ensure an appropriate level of *thermal protection*. Furthermore, to provide maximum thermal protection, the ablating material should not conduct thermal energy into the body. Ablation is a function of many factors, including (1) the thermal energy needed to raise the ablating material to the ablation temperature, including *phase change;* (2) the heat-blocking action created by *boundary layer* thickening due to injection of mass; and (3) the thermal energy dissipated by the body at the ablation temperature. Finally, ablation systems can be classified as (1) those that involve melting materials, (2) those that involve *subliming* materials and (3) those that involve a combination of subliming and melting materials.
See: ablating material, thermal protection, mass transfer cooling

abort   1. To cancel, cut short or break off an action, operation or procedure with an aircraft, space vehicle or the like, especially because of equipment failure (for example, "the launch was aborted," or "to abort a mission").

2. An aircraft, space vehicle, etc., that aborts.

3. An act or instance of aborting.

4. To cancel or cut short a flight after it has been launched.

The *Space Shuttle* system provides *intact abort* throughout all mission phases. There are four basic Shuttle abort modes: (1) Return-to-Launch-Site, (2) Abort-Once-Around, (3) Abort-to-Orbit and (4) Abort-from-Orbit. The Shuttle *Orbiter* is designed to de-orbit and land with a 14,515 kg (32,000 lb$_m$) maximum cargo weight. However, the overall Orbiter flight capability will permit cargo weighing more than the nominal 14,515 kg—but not more than 29,484 kg (65,000 lb$_m$)—to be returned under abort conditions. The circumstances under which this may occur include (1) the Return-to-Launch-Site abort, (2) the Abort-Once-Around and (3) aborts from orbit due to a *payload* malfunction. However, no Shuttle mission with a landing cargo weight of more than 14,515 kg should be planned.

Abort-Once-Around [abbr. AOA]
See: abort

Abort-to-Orbit [abbr. ATO]
See: abort

absolute   1. Complete, as in "absolute vacuum."

2. Pertaining to measurements or units of measurements relative to a universal constant, natural datum or fundamental relationships of space, time and *mass*, as in "absolute temperature."

absolute altimeter   An instrument which gives accurate, direct indications of *absolute altitude* (for example, a *laser* altimeter).

absolute altitude   The distance along the local vertical between an aircraft or a *spacecraft* and the point where the local vertical intersects the actual surface (either land or water) of a *planet* or a natural *satellite.*

absolute coordinate system   An *inertial coordinate system* that is fixed with respect to the *stars.* In theory an absolute coordinate system cannot be established because the reference stars themselves are in motion.

However, in practice such a system can be established to meet the demands of a particular navigation problem by selecting appropriate reference stars.

absolute delay   The time interval between the transmission of sequential signals. Also called "delay."

absolute temperature   A *temperature* expressed in terms of degrees Kelvin (K) or degrees Rankine (R).
See: absolute temperature scale

absolute temperature scale   The absolute temperature scale in the SI unit system is the Kelvin scale (symbol K). By international agreement since 1954 a reference value of 273.16 K has been assigned to the triple point of water. The Celsius (formerly centigrade) scale (symbol C or °C) is a relative temperature scale that is related to the absolute Kelvin scale by the formula:

$$T_C = T_K - 273.16$$

A second absolute temperature scale also in common use is the Rankine scale (symbol R). In this scale the triple point temperature for water is 491.69 R. The Fahrenheit scale (symbol F) is a relative temperature scale that is related to the absolute Rankine scale by the formula:

$$T_F = T_R - 459.67$$

It should be noted that the absolute and relative temperature scales are related by the formulas:

$$1 \text{ K} = 1 \text{ C}$$

and

$$1 \text{ R} = 1 \text{ F}$$

The two absolute temperature scales are related by the equation:

$$1 \text{ K} = 9/5 \text{ R}$$

It should also be noted that the absolute temperature scales (i.e., K and R) are useful for expressing levels of temperature above an absolute datum and temperature intervals or differentials. The relative temperature scales (i.e., C and F) are convenient and are commonly used to express temperature intervals. In thermodynamic applications the symbol T should always be taken as implying the absolute temperature scales **K** or **R**, unless it is specifically stated as being in terms of the relative temperature scales.

absolute vacuum   A (hypothetical) void completely empty of matter. Also called a "perfect vacuum."

absorber   Any material that absorbs or diminishes the intensity of *ionizing radiation*. For example, neutron absorbers, like boron, cadmium and hafnium, are in *control rods* for nuclear reactors, while steel and concrete in *reactor shields* absorb both *gamma rays* and *neutrons*. It is also interesting to note that a thin sheet of paper or metal foil will absorb or attenuate *alpha particles* and all but the most energetic *beta particles*.

absorption spectrum   The array of absorption lines and absorption bands which results from the passage of *electromagnetic energy* from a continuous source through a selectively absorbing medium cooler than the source.

The absorption spectrum is characteristic of the absorbing medium, just as an emission spectrum is a characteristic of a radiator or source.

An absorption spectrum formed by a monatomic gas exhibits discrete dark lines, whereas that formed by a polyatomic gas exhibits ordered arrays (bands) of dark lines, which appear to overlap. This type of absorption is often referred to as line absorption. The spectrum formed by a selectively absorbing liquid or solid is typically continuous in nature (continuous absorption).

See: electromagnetic spectrum

abundance   The number of *atoms* of a particular *isotope* expressed as a percentage of the total number of atoms of the *element*. The *natural abundance* is the abundance as found in a naturally occurring isotopic mixture of an element. Consider, for example, the element uranium (U). Its naturally occurring isotopes have the following abundance values:

U-isotope	abundance
^{234}U	0.005%
^{235}U	0.719
^{238}U	99.276

The **cosmic abundance** of an atom or element can be expressed as its percentage of the total mass found in the universe. Using the abundances found in the *Solar System* as a "cosmic standard" and incorporating other astrophysical data (e.g., stellar spectra), the estimated cosmic abundance of the most common materials in the universe is:

element	abundance
	(% of total mass)
hydrogen (H)	73 %
helium (He)	25
oxygen (O)	0.8
carbon (C)	0.3
neon (Ne)	0.1
nitrogen (N)	0.1

Analysis of solar and stellar spectra have indicated that *hydrogen* and helium, the two lightest elements, are by far the most abundant in the universe. All the other elements taken together account for only some two percent of the mass in the universe.

acceleration   1. Linear acceleration (symbol a) is the rate of change in velocity with time, expressed as:

$$a \equiv \frac{dv}{dt}$$

Typical units of linear acceleration are meters per second per second (m/sec^2). Negative acceleration is commonly called "deceleration."

2. Angular acceleration (symbol α) is the time rate of change of angular velocity, with typical units being radians per second per second (rad/sec^2).

accelerometer   An instrument, usually a transducer, that measures change in *velocity*, or measures the gravitational forces capable of imparting a change in velocity.

acceptor charge   Charge on the internal side of an *igniter* that is detonated by the *shock wave* transmitted through the internal *diaphragm* in a through-*bulkhead* initiator.

acoustic absorber   An array of acoustic resonators distributed along the wall of a *combustion chamber*, designed to prevent oscillatory combustion by increasing *damping* in the engine system.

acoustic velocity   The speed at which sound

waves spread. Also called "the speed of sound" (in the particular medium).

acoustic vibration *Vibrations* transmitted through a gas as opposed to those transmitted through other operational *environments*. These vibrations may be *subsonic, sonic* and *ultrasonic.*

acquisition 1. The process of locating the *orbit* of a *satellite* or the *trajectory* of a *space probe* so that *tracking* or *telemetry* data can be gathered.
2. The process of pointing a telescope or antenna so that it is properly oriented to gather tracking or telemetry data from a satellite or space probe.

acquisition and tracking radar A *radar* system that locks onto a strong signal and tracks the object reflecting the signal.

actinic light The portion of the visible light region of the electromagnetic spectrum that causes chemical changes to occur in light-sensitive photographic emulsions. Actinic light creates images on light-sensitive material. The blue or violet portion of the visible spectrum is generally considered the most actinic band of light.

actinic radiation Electromagnetic radiation capable of causing photochemical reactions, as for example, in photography or in pigment fading.
See: actinic light

actinide series The series of elements beginning with actinium (symbol: Ac, element number 89) and continuing through lawrencium (symbol: Lw, element number 103) that together occupy one position in the *periodic table.* The series, also referred to as the "actinides," includes uranium and all the man-made *transuranic elements.*

action time Interval from attainment of a specified initial fraction of maximum *thrust* or pressure level to attainment of a specified final fraction of maximum thrust or pressure level.

active 1. Transmitting a signal, as for example, an active *satellite.*
2. Fissionable, referring to an active material.
3. Radioactive, as for example, an active sample.
4. Receiving *energy* from some source other than a signal, as for example, an active element.

active satellite A *satellite* that transmits a signal, in contrast to a *passive satellite.*

active tracking system A system that requires the addition of a *transponder* or transmitter on board the *vehicle* to repeat, transmit or retransmit information to the tracking apparatus.

actuation time Elapsed time from receipt of signal to first motion of the device being actuated.

actuator A *servomechanism* that supplies and transmits energy for the operation of other mechanisms or systems.

actuator struts Two actuator struts bolted to the *main combustion chamber* are used, in conjunction with *hydraulic actuators*, to *gimbal* the *Space Shuttle Main Engines* during flight when it is necessary to change the direction of the *thrust.*
See: Space Shuttle Main Engine

adapter skirt A flange or extension of a *launch vehicle stage* or section that provides a means of fitting on another stage or section.

adhesive Material applied between components to bond the components together structurally.

adhesive force The strength of the attraction between two different substances at the boundary between them.

adiabatic Term applied to a *thermodynamic* process in which heat is neither added to nor removed from the system involved; when the process is irreversible, it is called *isentropic.*

adiabatic process A thermodynamic change of state (condition) of a system in which there is no heat energy transfer across the boundaries of the system. In an adiabatic process, for example, compression always results in "warming" (that is, raising the level of *internal energy* of the *working fluid*), while expansion results in "cooling" (that is, lowering the level of internal energy of the working fluid).

adiabatic wall In a *thermodynamic* system, a boundary that is a perfect heat insulator (neither emits nor absorbs heat); in a *boundary-layer* theory, the wall condition in which the temperature gradient at the wall is zero.

adiabatic wall temperature Steady-state temperature of the wall of a *combustion* *chamber* or *nozzle* when there is no heat transfer between the wall and the exhaust gas; also called recovery temperature.

Advanced X-Ray Astrophysics Facility [abbr. AXAF] An astrophysics x-ray facility that NASA has planned to complement visual and radio observations made from the ground and from space observatories such as the Space Telescope. The AXAF will study stellar structure and evolution, large-scale galactic phenomenon, active galaxies, clusters of galaxies, quasars and cosmology. The basic objectives are to determine the positions of X-ray sources, their physical properties and the processes involved in X-ray photon production. The facility is a grazing incidence x-ray telescope with nested pairs of mirrors and a 9 to 12 m focal length. The 10,000 to 12,000 kg facility will have an overall length of approximately 13 to 16 m. It will be placed in a 28.5 degree inclination, approximately 500 km altitude orbit. The facility is expected to last 10 years.

aeroballistics The study of the interaction of high-velocity projectiles or vehicles with the *atmosphere*. The problem of the effect of atmospheric entry on the *Space Shuttle Orbiter's trajectory* is a problem in aeroballistics.

aerodynamic force The *force* exerted by a moving gas upon a body completely immersed in it. This aerodynamic force is proportional to the expression:

$$\rho \; u^2 \; L^2 \; (Re)^n$$

where: ρ is the fluid density

u is the velocity of the undisturbed stream relative to the body (i.e., the "free stream" velocity)

L is a characteristic linear dimension of the body

$(Re)^n$ is the Reynolds number raised to the power n, an experimentally determined constant

This expression for aerodynamic force is sometimes called the "Rayleigh formula." The component of the aeordynamic force parallel to and opposite the direction of vehicle motion is called *drag*.

aerodynamic heating The surface heating of a body produced by the passage of air or other gases over the body. It is caused by friction and by compression processes and is significant mainly at high velocities.

aerodynamic performance Portion of the *nozzle* performance due to nozzle divergence efficiency (the degree of perfection of the nozzle contour).

aerodynamics The science that deals with the motion of *gases* (especially air) and with the *forces* acting on solid bodies when they move through gases or when gases move against or around the solid bodies.

aerodynamic skip An atmospheric entry *abort* caused by entering the atmosphere at an excessively shallow angle. This results in a trajectory back out into space rather than downward toward the planet. The process is analogous to skipping a stone across the surface of a pond.

aerodynamic throat area Effective flow area of the *nozzle* throat; the effective flow area is less than the geometric flow because the flow is not uniform.

aerodynamic vehicle A vehicle, such as a glider or an airplane, that is capable of flight only within an atmosphere and relies on aerodynamic forces to stay aloft. (See Fig. 1.) Compare this term to "aerospace vehicle" and "space vehicle."

aeroelasticity The study of the response of structurally elastic bodies to *aerodynamic forces*.

aeroembolism An *aerospace medicine* term describing the formation or liberation of gases (mostly nitrogen) in the blood vessels of the body, brought on by an excessively rapid change from an *environment* of high atmospheric *pressure* to one of lower pressure (e.g., the upper atmosphere). The condition, which is also called decompression sickness, is characterized mainly by neuralgic pains, cramps and swelling and sometimes results in death.

aeronautics The study of atmospheric flight, including *aircraft* design and *aerial* navigation.

aeronomy The study of the upper regions of the *Earth*'s atmosphere where *ionization*, *dissociation* and chemical reactions occur due to *solar radiation*. The term has also been extended to include the study of other planetary atmospheres.

aerosol Very small particles of dust or

droplets of liquid suspended in the *Earth*'s *atmosphere.* The term includes smoke, dust, fog, clouds, haze, fumes and mist.

aerospace   A term, derived from *aero*nautics and *space*, meaning of or pertaining to the *Earth*'s atmospheric envelope and the space beyond it considered as a single realm, as in for example, an "aerospace vehicle," such as the *Space Shuttle.*

aerospace ground equipment   All equipment required on the ground to make an *aerospace system* operational in its intended environment.

aerospace medicine   That branch of medical science dealing with the effects of flight through the *atmosphere* or in space upon the human body and with the prevention or cure of physiological or psychological illnesses arising from these effects.

aerospace technology transfer   The generation of innovations in virtually all fields of science and technology as a result of the pursuit of *aerospace* technology goals and objectives. Aerospace technology is accomplishing things that otherwise could not be done economically—or perhaps could not be done at all. This is occurring in global communications, navigation, oceanography, meteorology, geology, astronomy and, of course, exploration of the *Solar System.*

So thoroughly have aerospace technology transfers pervaded our lives that the many benefits attributable to the stimulus of the space program are difficult to measure in their entirety. However, an economic study was performed tracing the "spin-offs" from four NASA projects in order to evaluate the benefits to the national economy from secondary applications of aerospace technology. This study (completed in 1977) estimated that the secondary benefits in four areas alone would return almost $7 billion to the economy by the early 1980s. The results of the study and the four areas examined are briefly summarized below:

integrated circuits
Developed for *satellites,* communications and other space uses, these circuits are now used in TV sets, automobiles and hundreds of industrial and household products (including electronic and video games). It is estimated that the improved technology will return over $5 billion between 1963 and 1982.

insulation for cryogenic uses
The *cryogenics* industry erupted as a direct result of the development of the liquid fuels (for example, *liquid hydrogen*) needed for *rocket propulsion* and *life support systems* in space. Today, hospitals and steel mills are among the dozens of beneficiaries that store and use liquid oxygen, helium, nitrogen and other frozen gases. The estimated benefits to the U.S. economy: over $1 billion by 1983.

structural analysis computer program
Developed originally to help design more efficient *space vehicles,* this NASA computer program is used today to help design railroad tracks and cars, automobiles, bridges, skyscrapers and other structures. Use of the computer program yields about a 60 percent improvement in predicting the behavior of stressed parts and a two-thirds reduction in calculation time. The program is expected to return more than $700 million in cost savings from 1971 to 1984.

gas turbines
Initially developed for *jet-engine* aircraft, but widely spun-off to electric power generation plants, these turbines will effect cost savings of an estimated $111 million between 1969 and 1982.

These are just four of the thousands of transfers of aerospace technology that have directly impacted—and will continue to impact—terrestrial living conditions. As the Shuttle era matures many more people will begin to realize that space is not just a "money sink" or a place to pursue some obscure scientific objectives, but rather space is part of the mainstream of the high-technology development of our global civilization.

aerospace vehicle   A vehicle, like the *Space Shuttle*, that is capable of operating both within the atmosphere and in outer space.

aerospike nozzle   *Annular nozzle* that allows the gas to expand from one surface— a centerbody spike—to *ambient pressure.*

afterbody   1. A *companion body* that trails a *satellite* or a *spacecraft.*

2. A portion or section of a *launch vehicle, rocket* or *missile* that enters the *atmosphere* unprotected behind the *nose cone* or other component that is protected for atmospheric entry.

3. The aft section of a vehicle.

afterburner   A device used for increasing the *thrust* of a *jet engine* by burning additional fuel in the uncombined oxygen present in the *turbine* exhaust gases.

aftercooling   The cooling of a *nuclear reactor* after it has been shut down. It is accomplished by pumping a liquid or gas through the *reactor core* to remove the excess heat (*afterheat*) generated by the continuing *radioactive decay* of the *fission products* within the core.
See: nuclear reactor, nuclear rocket, space nuclear power

afterglow   1. A broad, high arch of radiance, or glow, that is seen on occasion in the western sky above the highest clouds in the deepening twilight. Afterglow is caused when sunlight is scattered by the very fine particles of dust suspended in the upper *atmosphere*.

2. The transient decay of a *plasma* after the power has been turned off. The plasma decay time is a direct function of charged particle loss mechanisms, such as *diffusion* and *recombination*.

afterheat   The heat produced by the continuing *radioactive decay* of *fission products* and *transuranic nuclides* found in a *reactor core* after the *fission* process (*chain reaction*) has stopped. The vast majority of this afterheat is due to the decay of the fission products.
See: nuclear reactor, nuclear rocket, space nuclear power

aft flight deck [abbr. AFD]   A portion of the crew station module of the *Space Shuttle Orbiter* on the upper deck where *payload* controls can be located.
See: Orbiter, Orbiter structure

aft fuselage [abbr. AF]   A component of the *Orbiter* structure that carries and interfaces with the *orbital maneuvering system/reaction control system* pod (left and right sides), wing aft *spar*, mid fuselage, *Space Shuttle Main Engines*, heat shield, body flap and vertical stabilizer.
See: Orbiter structure

aft skirt   The component of the *Space Shuttle Solid Rocket Booster* used for mounting the *thrust vector control* system and for supporting the aft cluster of four *booster separation motors*. The weight of the entire Space Shuttle is borne by holddown post assemblies that provide rigid physical links between the *mobile launch platform* and the two aft skirts.
See: Solid Rocket Booster

agravic   Of or pertaining to a condition in which there is no *gravitation*.

airbreather   An *aerodynamic vehicle* propelled by the combustion of *fuel* that is *oxidized* by air taken in from the *atmosphere*; an airbreathing vehicle. A *cruise missile* is an example of an airbreather, while a *rocket* carries its own supply of oxidizer and operates independently of the surrounding atmospheric environment.

airflow   A flow or stream of air. Airflow can occur in a wind tunnel and in the intake portion of a *jet engine*; relative airflow occurs past the wing or other component of a moving *aircraft*. Airflow is usually measured in terms of mass or volume per unit of time.

Air Force Satellite Communications Systems [AFSATCOM]   The Air Force Satellite Communications (AFSATCOM) System was developed by the *Air Force Systems Command's Space Division*. It satisfies high priority Department of Defense communications requirements necessary to command and control U.S. nuclear forces around the world. AFSATCOM operational capability was achieved on May 19, 1979. (See Fig 1) The AFSATCOM space segment consists of a series of communications *transponders* carried by various host satellites belonging to several systems. One of its hosts is the *Fleet Satellite Communications System*. Another is the *Satellite Data System*. Since FLTSATCOM satellites have *equatorial orbits*, (see: orbits) and the SDS satellites have *polar* type *orbits*, AFSATCOM transponders make up a world-wide communications network that relays data between airborne and ground teletype terminals. The ground segment of *AFSATCOM* consists of mobile and fixed terminals. The terminals are modular in design and allow for easy replacement or upgrade. *Antennas* and power supplies vary in size and type, depending on the requirements of the terminals they service.

AFSATCOM message traffic usually takes the form of short, preformatted messages. These transmissions require a minimum amount of power and permit reliable and secure two-way communications between ground-based and air-borne forces and command posts. A conferencing capability is also provided between command posts. The Air Force *Space Division*, at Los Angeles AFS, Calif., and the *Electronic Systems Division*, Hanscom AFB, Mass., have played major roles in developing and fielding the AFSATCOM system for operation use. Space Divison has been responsible for the system's design, procurement and integration into the force. The space segment was also a prime responsibility for SD. ESD developed the airborne and ground terminals for the *AFSATCOM* system.

See: U.S. Air Force role in space.

airframe The assembled structural and aerodynamic components of an *aircraft* or *rocket* vehicle that support the different systems and subsystems integral to the vehicle. The term "airframe," a carry-over from aviation technology, remains appropriate for rocket vehicles, since a major function of the airframe is performed during flight within the *atmosphere*.

See: Orbiter structure

airglow A faint general luminosity in the *Earth*'s sky, most apparent at night. As a relatively steady (visible) radiant emission from the upper *atmosphere*, it is distinguished from the sporadic emission of *aurorae*. Airglow is a *chemiluminescence* that is due primarily to the emission of the *molecules* O_2 (oxygen molecule) and N_2 (nitrogen molecule), the *radical* OH (hy-

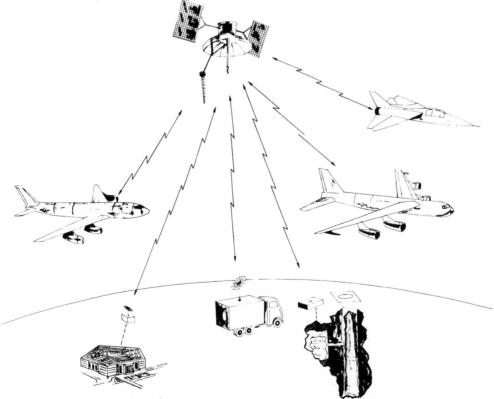

Fig. 1   Nuclear capable forces will transmit and receive communications from military satellites through the AFSATCOM network. Users of the AFSATCOM system include intercontinental ballistic missile control centers, strategic aircraft, submarines and nuclear weapons storage sites.   (Drawing courtesy of the U.S. Air Force.)

droxyl radical) and the *atoms* O (oxygen atom) and Na (sodium atom).

airlock   1. A small chamber capable of providing a passage between two places of different *pressure* (for example, between a *spacecraft* cabin and outer space).
2. A compartment capable of being depressurized without depressurizing the *Orbiter* cabin or the *Spacelab module*. It is used to transfer crew members and equipment. A similar compartment in the Spacelab module (the experiment airlock), is used to expose *experiments* to space. The airlock/ airlock mounting kit is a standard *Spacelab flight kit,* while the experiment airlock is an item of *Spacelab mission-dependent equipment.*
See: Orbiter, Spacelab, Orbiter Structure

airlock hatch   The *Orbiter* vehicle *airlock* hatch is used primarily for *extravehicular activity* but will also be used for stowage or retrieval of gear in the airlock and for transfer of materials or personnel to and from the *Spacelab* or a docked space vehicle.
See: Orbiter structure, docking

air sounding   The act of measuring atmospheric phenomena or determining atmospheric conditions at altitude, especially by means of apparatus carried by *balloons* or *rockets.*

airspace   Specifically, the *atmosphere* above a particular portion of the *Earth,* usually defined by the boundaries of an area on the surface projected upward.

airstart   An act or instance of starting an *aircraft*'s engine while in flight, especially a *jet engine* after flameout.

air-to-air-missile [abbr. AAM]   A *missile* launched from an *aircraft* at a target in the air.

air-to-surface missile [abbr. ASM]   A *missile* launched from an *aircraft* at a target on the surface of the *Earth.*

air-to-underwater missile [abbr. AUM]   A *missile* launched by an *aircraft* toward an underwater target.

alkali metal   A metal in group IA of the *periodic table*; namely, lithium, sodium, potassium, rubidium, cesium and francium. Alkali metals are frequently considered for use as coolants (in the liquid state) for *space nuclear power* systems.

all-fire limit   Minimum energy (e.g., *voltage*) required to consistently fire an *electroexplosive device.*

all-inertial guidance   1. The *guidance* of a *rocket* vehicle entirely by use of *inertial* devices.
2. the equipment used for this.

allowable load (or stress)   Load that, if exceeded, produces failure of the structural element under consideration. Failure may be defined as buckling, yielding, ultimate, or fatigue failure, whichever condition prevents the component from performing its intended function. Allowable load is sometimes referred to as criterion load or stress; allowable stress is equivalent to material strength.

alpha particle [Symbol: α or $^4_2He^{++}$]   A positively charged particle emitted by certain radioactive *nuclides*. It is made up of two *neutrons* and two *protons* bound together and thus is identical to the *nucleus* of a helium *atom* that has been stripped of its orbital *electrons*. Alpha particles are the least penetrating of the three common types of *radiation* (alpha, *beta* and *gamma*); they are stopped by materials such as a thin sheet of paper or even by a few centimeters of air.

altimeter   An instrument for measuring height (*altitude*) above a reference point (for example, sea level).

altitude   1. The height above a given reference point, for example, sea level or a planet's surface. In space navigation "altitude" generally designates the distance from (height above) the mean surface of the reference *celestial body* as contrasted to "distance," which usually means the distance from the center of the reference celestial body.
2. In astronomy the angular displacement above the *horizon;* the arc of a vertical circle between the horizon and a point on the *celestrial sphere,* measured upward from the horizon. Angular displacement below the horizon is called "negative altitude" or "dip."

altitude acclimatization   A physiological adaption to reduced atmospheric and oxygen pressure.

altitude chamber   A chamber within which the air *pressure,* temperature, etc., can be adjusted to simulate conditions at different *altitudes;* used for experimentation and testing.

2. In astronomy the angular displacement above the *horizon;* the arc of a vertical circle between the horizon and a point on the *celestrial sphere,* measured upward from the horizon. Angular displacement below the horizon is called "negative altitude" or "dip."

altitude engine *Rocket engine* that is designed to operate at high-altitude conditions.

ambient Surrounding, especially of or pertaining to the *environment* that is around an *aircraft* in flight or other body (e.g., a *nose cone*) but is unaffected by the body, as "ambient temperature" or "ambient pressure."

ampere [Symbol: A or amp] The SI unit of electric current, defined as the constant current that—if maintained in two straight parallel conductors of infinite length, of negligible circular cross-sections and placed one meter apart in a vacuum—will produce between those conductors a *force* equal to 2×10^{-7} newtons per meter of length. The unit is named after André-Marie Ampere (1775–1836), French physicist and pioneer in electrodynamics.

ancient astronaut theory A theory or hypothesis that has emerged in recent times as a complement to the traditional ("creationist") and Darwinian ("evolutionist") theories of man's origins. This hypothesis is based on a "suspicion" (as yet scientifically unfounded) that someone or something disturbed the normal biological evolution of the apes of ancient Earth.

There are several general trends of thought regarding the ancient astronaut hypothesis. The first is that modern man (i.e., Homo sapiens) arrived on Earth as colonists from another star system. After establishing an advanced civilization, perhaps even incorporating the primitive creatures then living on Earth, the "extraterrestrial" society was destroyed by some cosmic catastrophe. Modern man, the hypothesis reasons, is now rebuilding civilization on the wreckage of that ancient society. A second version of the ancient astronaut theory is that a migratory wave of intelligent creatures from another star system visited Earth in the distant past, planted the seeds of civilization among the primitive creatures found on the planet and then continued on to the edges of the *Galaxy.* A third variation of the ancient astronaut hypothesis proposes that extraterrestrial visitors performed some form of genetic engineering on apes, or even on primitive men, the by-product of which was the evolution of human intelligence and the emergence of Homo sapiens.

Although **unproven** in the context of traditional scientific methods, the ancient astronaut hypothesis has given rise to many popular books, films and stories—all speculations that seek to link such phenomena as ancient civilizations, unresolved mysteries from the past and *unidentified flying objects* with extraterrestrial sources.

android A term originating in the science fiction literature that describes a *robot* or an *automaton.*

angels A colloquial aerospace term describing a *radar* echo caused by a physical phenomenon not discernible to the eye.

angle of attack The angle between the chord line (i.e., the straight line joining the centers of curvature of the leading and trailing edges) on an *airfoil* and the relative *airflow,* normally the immediate flight path of the *aircraft.* It is also called the *"angle of incidence."*

angle of incidence The angle at which a ray of *energy* impinges upon a surface. It is usually measured between the direction of propagation of the energy and a perpendicular to the surface at the point of incidence.

angle of reflection The angle at which a reflected ray of *energy* leaves a reflecting surface. It is usually measured between the direction of the outgoing ray and a perpendicular to the surface at the point of reflection. Compare *angle of incidence.*

angstrom [Symbol Å] A unit of length commonly used to measure *wavelengths.* It is described by the following formula:

$$1 \overset{\circ}{A} = 10^{-10} \ meters = 0.1 \ nanometer$$

This unit is named in honor of Anders Jonas Ångstrom (1814–74), a Swedish physicist.

angular acceleration [Symbol α] The rate of change of *angular velocity.*

angular frequency   The *frequency* of a periodic quantity expressed in radians per second. It is equal to the frequency (in *hertz*) times 2π.

angular measure   Units of angle expressed in terms of degrees (°), arc minutes (′) and *arc seconds* (″); where $1° = 60′$ and $1′ = 60″$, while $360° = 2\pi$ (a full circle). In the case of *right ascension,* angular measure is expressed in terms of hours, minutes and seconds.

angular resolution   The angle between two points that can just be distinguished by a *detector* and a *collimator*. The normal human eye has an angular resolution of one arc minute.

See: angular measure

angular velocity [Symbol ω]   The change of angle per unit of time; the rate of change of angular displacement. Typical units of angular velocity are radians per second.

anistropy   Condition in a material in which properties are not the same in all directions; observed or measured properties change when the *axis* of observation or test is changed.

annealing   The application of *thermal energy* (heat) to a material to relieve stresses, change certain properties, improve machinability, etc., or to realign *atoms* in a distorted crystal lattice as caused, for example, by *radiation damage*. The heated material is cooled at a suitable rate to achieve the desired effect.

annihilation   The conversion of a *particle* and its corresponding *anti-particle* into *electromagnetic energy* (annihilation radiation) upon collision. This annihilation radiation has a minimum energy equivalent to the *rest mass* (m_o) of the two colliding particles. When a *positron* and an *electron* collide, for example, the minimum annihilation radiation consists of a pair of *gamma rays,* each of 0.511 MeV energy. The energy of annihilation is derived from the *mass* of the annihilated particles according to the famous Einstein mass-energy equation, $E = mc^2$.

annular   Pertaining to an *annulus* or ring; ring-shaped.

annular nozzle   *Nozzle* with an annular

throat formed by an outer wall and a centerbody wall.

anode   The positive *electrode* in a system (e.g., discharge tube or electrolytic cell) whereby *electrons* leave the system.

anodize   To produce a protective oxide film on a metal by electrochemical means.

antenna field   A group of *antennas* placed in a geometric configuration.

antenna gain
See: gain

antenna pair   Two antennas located on a *base line* of accurately surveyed length.

anti-cyclonic   Having a direction of rotation about the vertical opposite to the rotation of the *Earth;* that is, clockwise in the Northern Hemisphere, counterclockwise in the Southern Hemisphere and undefined at the equator.

anti-extrusion ring   Ring installed on the low-pressure side of a seal or packing to prevent extrusion of the sealing material; also called a backup ring.

anti-matter   *Matter* in which the ordinary nuclear particles (e.g., *neutrons, protons, electrons*) are conceived of as being replaced by their corresponding *anti-particles,* that is, *anti-neutrons, anti-protons, positrons,* etc. For example an "anti-hydrogen" atom would consist of a negatively charged anti-proton with an orbital positron. Normal matter and anti-matter would mutually annihilate each other upon contact, being converted totally into *energy.* Figure 1 is an artist's drawing showing a stylized conception of the symmetry of matter and anti-matter in the *Universe.* Depicted in this hypothetical figure are two similar enormous regions of the Universe containing clusters of *galaxies*—one consisting of matter and the other of anti-matter—along with adjacent matter–anti-matter regions. According to one theory at the boundaries where matter and anti-matter come together, nuclear *annihilation* occurs, resulting ultimately in the production of *gamma rays* (symbolized in Figure 1 by wiggling arrows), a by-product of the annihilation process. Although individual anti-particles have been discovered, bulk quantities of anti-matter have yet to be found in the Universe. Compare this term with "matter."
See: annihilation

Fig. 1   Artist's conception of a hypothetical matter–anti-matter universe.   (Drawing courtesy of NASA.)

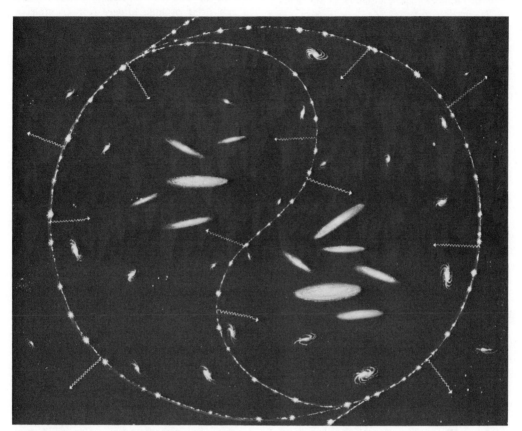

anti-neutrino   the *antiparticle* of the *neutrino*.
See: neutrino

antinode   The position of maximum disturbance in a *standing wave*.

anti-particle   Every *elementary particle* has a corresponding real (or hypothetical) anti-particle, which has equal *mass* but opposite *electric charge* (or other property, as in the case of the *neutron* and *anti-neutron*). The anti-particle of the *proton* in the anti-proton; that of the *electron* is the *positron;* that of the *neutrino* is the *anti-neutrino,* etc. However, the *photon* is its own anti-particle. When a particle and its corresponding *anti-particle* collide, they are converted into *energy* in a process called *annihilation.*
See: anti-matter, annihilation

anti-proton   the negatively charged *antiparticle* of the *proton*.
See: antimatter; proton

anti-rotation device   Mechanical device (e.g., a key) used in rotating machinery to prevent rotation of one component relative to an adjacent component.

anti-slosh baffle   A device provided in a tank to *damp liquid* motion. It can take many forms including flat ring, truncated cone, and vane.

anti-vortex baffles   Assemblies installed in both propellant tanks (i.e., *liquid oxygen* and *liquid hydrogen*) of the *External Tank* to prevent gas from entering the *Space Shuttle Main Engines*. The baffles minimize the rotating action as the propellants flow out the bottom of the tanks. Without these baffles the propellants would create a vortex similar to a whirlpool in a bathtub drain.
See: External Tank

apareon   That point on a *Mars*-centered *orbit* at which a *satellite* or *spacecraft* is at its greatest distance from the planet. The term "apareon" is analogous to the term "apogee."

apastron 1. That point in a body's *orbit* around a *star* at which it is at a maximum distance from the star.

2. Also, that point in the orbit of one member of a *binary star* system at which the stars are farthest apart. (That point at which they are closest together is called the *"periastron."*)

aperture 1. The diameter of the *objective* of a *telescope* or other optical instrument.

2. An opening, especially that opening in the front of a camera or optical instrument through which light passes.

3. Concerning a unidirectional *antenna,* that portion of a plane surface near the antenna, perpendicular to the direction of maximum radiation, through which the major portion of the radiation passes.

aperture ratio The ratio of the useful diameter (d) of a lens to its focal length (f). Expressed mathematically as

$$\text{aperture ratio} = d/f$$

It is the reciprocal of the *f-number.*

aphelion The point in a *celestial body's orbit* around the *Sun* that is most distant from the Sun. The orbital point nearest the Sun is called the *"perihelion."*

apoapsis That point in an *orbit* farthest from the orbited central body or center of attraction.

See: orbital elements

apocynthion That point in the lunar *orbit* of a *spacecraft* or *satellite* launched from the *Earth* that is farthest from the *Moon.* Compare this term with *apolune.*

apogee 1. The point that is the farthest from the Earth in a *geocentric orbit.* The term is now applied to both the *orbit* of the *Moon* and the orbits of *artificial satellites* around the Earth in a *geocentric orbit.* The term is now applied to both the *orbit* of the *Moon* and the orbits of *artificial satellites* around the Earth. At apogee, the orbiting body's velocity is at a minimum. To enlarge or "circularize" the orbit, a *spacecraft's thruster* (i.e., the *apogee motor*) is turned on at apogee, giving the vehicle and its *payload* increased velocity. Opposite of *perigee.*

apogee motor A device attached to a *satellite* or *spacecraft* that is used for boosting

the *payload* from an initial *parking orbit* to an orbit of high *apogee* (for example, a *geosynchronous orbit*). The apogee motor is designed to fire when the space vehicle is at apogee, the orbital point farthest from Earth. This *burn* of the apogee motor establishes a new orbit farther from Earth or permits the spacecraft to achieve *escape velocity.* Also called "apogee kick motor" and "apogee rocket."

Apollo Lunar Surface Experiments Package [abbr. ALSEP] Scientific materials set on the Moon by the Apollo astronauts and left there to transmit data back to Earth. Experiments included the study of meteorite impacts, lunar surface and seismic characteristics and analysis of the atmosphere.

Apollo Project NASA project, instituted in 1961, to put man on the Moon before the end of the decade. The first four Apollo flights, Apollo 7-10, launched between October 1968 and May 1969, orbited the Moon. Apollo 11 landed on the lunar surface on July 20, 1969. Before the completion of the Apollo program in 1973, 12 astronauts landed and explored the lunar surface during six missions. Experiments performed in the program included cosmic-ray and neutron detection, seismometry, laser ranging, magnetometry, solar wind and lunar atmosphere probing.

Apollo-Soyuz Test Project Joint U.S.-Soviet space mission (July 1975) centering on the docking of the Apollo 18-Soyuz 19 spacecrafts. The project resulted in the development of a standard docking system for possible space rescue.

apolune That point in the lunar *orbit* of a *spacecraft* or *satellite* launched from the *Moon* that is farthest from the Moon.

apparent In astronomy, observed.

True values are reduced from apparent (observer) values by eliminating those factors such as *refraction* and light time, etc., which affected the observation.

approach The maneuvers of a spacecraft such as the *Space Shuttle Orbiter* from a *stationkeeping* position toward an orbiting *payload* for the purpose of *capture.*

Approach and Landing Test(s) [ALT] The 9-month-long *Space Shuttle* approach and landing test (ALT) program was conducted at the *Dryden Flight Research Center* at *Edwards Air Force Base* in California. The Orbiter was carried aloft by a modified 747. (See: Fig. 1) There were 13 flights in the ALT program. Five of these were unmanned captive flights with the inert Orbiter mated on top of the 747 Shuttle carrier aircraft (SCA). Three manned captive flights followed with an *astronaut* crew onboard the Orbiter operating the flight control systems while the Orbiter remained on top of the SCA. Following these eight captive flights were five "free flights" in which the astronaut crew separated the Orbiter from the SCA and flew it back to a landing. For all eight captive flights and for the first three free flights, the Orbiter was outfitted with a tail cone covering its aft section to reduce aerodynamic *drag* and *turbulence.* The tail come was removed for the final two free flights and the simulated engines—three main engines and two orbital maneuvering engines—were *aerodynamically* exposed.

The approach and landing tests began on February 15, 1977, with three taxi tests conducted to determine structural loads and responses and ground-handling and control characteristics up to flight takeoff speed. The taxi tests also validated the steering and braking of the SCA.

The manned captive-active flights exercised and evaluated all systems in the flight environment in preparation for the Orbiter free flights. These flights included flutter tests of the mated craft at low and high speeds, separation *trajectory* tests, and a rehearsal for the first Orbiter free flight.

The free flights verified the Orbiter pilot-guided approach and landing capability; demonstrated the Orbiter subsonic automatic terminal area energy management (TAEM) and automatic landing approach capability; and verified the Orbiter *subsonic* airworthiness, integrated system operation, and selected subsystem operation in preparation for the first manned orbital flight. [*See:* STS-1] The free flights also demonstrated the Orbiter's capability to safely approach and land with a minimum gross weight using several *center-of-gravity* configurations within the operational envelope.

The approach and landing tests were successfully completed with the fifth and final free flight of Orbiter 101, "Enterprise," on October 22, 1977. Astronauts Fred Haise, Gordon Fullerton, Joe Engle, and Richard Truly comprised the two-man crews for these flights. (See Table 1)

Table 1 ALT Free Flights
August 12, 1977, Free Flight 1—Haise and Fullerton
September 13, 1977, Free Flight 2—Engle and Truly

Fig. 1   The Orbiter "Enterprise" and its Boeing 747 carrier aircraft over Edwards AFB, California during the ALT program. Two F-5 chase aircraft are also in view (February, 1977).   (Photograph courtesy of the U.S. Air Force.)

September 23, 1977, Free Flight 3—Haise and Fullerton
October 12, 1977, Free Flight 4—Engle and Truly
October 22, 1977, Free Flight 5—Haise and Fullerton

The final phase of the ALT program prepared the Orbiter for four "ferry" flight tests. The *fluid* systems were drained and purged, the tail cone was reinstalled, and the *elevon* locks were installed. The forward attach strut was replaced, lowering the Orbiter's cant from 6° to 3° to reduce drag during the ferry flights. After the ferry flight tests, Orbiter 101 underwent minor modifications for the vertical ground vibration test program at the Marshall Space Flight Center.

appulse 1. The near approach of one *celestial body* to another on the *celestial sphere,* as in *occultation, conjunction,* etc.
2. A *penumbral eclipse* of the *Moon.*

arc 1. A part of a curved line, such as a circle.
2. A luminous glow which appears when an electric current passes through *ionized air* or gas.
3. An auroral arc

arc-jet engine A type of electrical *rocket engine* in which the *propellant* gas is heated by passing through an electric *arc.*
See: electric propulsion.

arc-minute A unit of angle: 1° = 60 arc-minutes, 1 arc-minute = 60 arc-seconds. 1 arc-minute = 0.016667 degrees.

arc second An angle equal to $1/3,600$ (.02778 percent) of a degree or $1/202,265$ (.00048 percent) of a radian.

area ratio Ratio of the geometric flow area of the *nozzle* exit to the geometric flow area of the nozzle throat; also called expansion area or simply *expansion ratio.*

areocentric With *Mars* as a center. Areocentric is analogous to geocentric.

areography The study of the surface features of *Mars;* the geography of Mars.

areographic Referring to positions on Mars measured in latitude from *Mars'* equator and in longitude from the reference meridian.

Ares *Mars.* Ares is seldom used except in combining forms as areocentric, apareon.

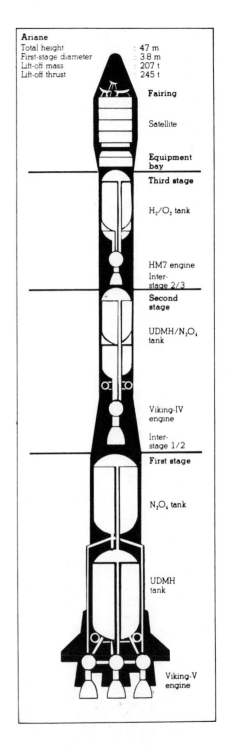

The three stage ARIANE launch vehicle. (Drawing courtesy of the European Space Agency.)

Ariane   A heavy launcher designed by the European Space Agency to give Europe an independent launching capability for its own satellites (previously launched by NASA). Ariane is a three-stage launch vehicle with a total height of 47.4 m and weighing 208 m tons at lift-off. 90% of the mass consists of propellant. The structures and payload account for about 9% and 1% of the total weight respectively. The first stage is equipped with four Viking-V engines mounted on a thrust frame. They can be swivelled in pairs to provide three-axis control. The second stage is equipped with a Viking-IV engine attached to the frame by a gimbal with two degrees of freedom for pitch and yaw control. Roll is controlled by auxiliary jets. The third stage is the first cryogenic stage developed in Europe. It is attached to a tapered thrust frame, gimbal-mounted for pitch and yaw control. Roll is controlled by auxiliary nozzles which eject hydrogen. Ariane can place a payload at 1,700 kg into geostationary orbit, 4,800 kg into low Earth orbit, 2,500 kg into sun-synchronous circular Earth orbit and 400 kg into a hyperbolic trajectory.

articulated   Segmented or jointed and thereby able to accommodate motion.

artificial asteroid   A man-made object placed in *orbit* around the *Sun*.
See: asteroid

artificial gravity   A simulated gravity established within a space vehicle, as by rotating a cabin about an axis of a *spacecraft*, the *centrifugal force* generated being similar to the *force of gravity*.

artificial intelligence   Commonly understood to refer to the study of thinking and perceiving as general information processing functions. These functions can be done by computer as well as man. All artificial intelligence involves elements of planning and problem solving; perception (the process of obtaining data and analyzing it), cognition and learning, and communications between man and machine or machine and machine. NASA uses three types of artificial intelligence:
1. automatic devices—those operating without direct human control.
2. teleoperational devices—those with a human operator in remote control of a mechanical system.
3. robotic devices—those with the capacity to manipulate or control other devices.

artificial satellite   An early aerospace term used to describe an object (e.g., a

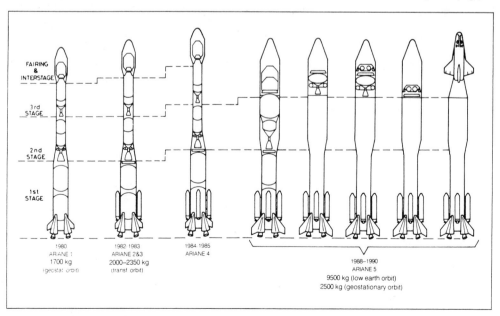

Projected development scenario for the ESA ARIANE launch vehicle (1980-1990). ARIANE 5 configuration is highly speculative at present.   (Drawing courtesy of the European Space Agency.)

spacecraft) built by man and placed in *orbit* around the *Earth* or another *celestial body*. "*Sputnik I*," launched by the Soviet Union on 4 October 1957, was the first artificial *satellite* to be placed in orbit around the Earth.

ascending node   That point at which a *planet*, planetoid, or *comet* crosses to the north side of the *ecliptic*; that point at which a *satellite* crosses to the north side of the *equatorial* plane of its *primary*. Also called northbound node. The opposite is descending node or southbound node.

aspect ratio   1. The ratio of width to height in a rectangular flow passage.
2. The ratio of *blade* height to chord length.
3. The ratio of width to depth in a rectangular *combustion chamber*.

asperities   Minute irregularities or roughnesses on the surface of a component, usually the result of the surface-refinishing process.

assembly   A number of parts and/or subassemblies joined together to perform a specific function and capable of disassembly. The distinction between an assembly and a subassembly is determined by the particular application. A grouping used in one application as a complete assembly may also be used in another application as a subassembly to form part of a larger assembly.

asteroid   A small, solid body orbiting the *Sun* independently of a *planet*. The majority of the asteroids or minor planets have *orbits* between those of *Mars* and *Jupiter*, with periods lying between three and six years. [*See:* asteroid belt] As depicted in Fig. 1 the asteroids are being considered as targets of future space missions. The name "minor planet" is preferred by astronomers; but asteroid is used extensively in the aerospace literature.
See: asteroid mining

asteroid belt   The region between the orbits of *Mars* and *Jupiter* that contains a large majority of the *asteroids* or minor planets. These *planetesimals* or planetoids have periods lying between three and six years and orbit at distances of 2.2 to 3.3 *astronomical units*.

asteroid mining   *Asteroids* contain many of

the major elements that provide the basis for industry and life on *Earth*. Fig. 1 shows a *satellite power system* (SPS) surrounded by the activity of asteroid retrieval. Creation of a new economy in space, the process of *space industrialization,* is the objective of retrieving an asteroid and placing it in high *Earth orbit*. Such asteroids could then be mined for materials needed for spacebased industries, the construction of satellite power systems and the development of large space habitats.

An artist concept of an asteroid retrieval mission.   (Illustration courtesy of NASA.)

astro   A prefix meaning *star* or stars and, by extension, sometimes used as the equivalent of *celestial*, as in astronautics.

astroballistics   The study of the phenomena arising out of the motion of a solid through a gas at speeds high enough to cause *ablation*; for example, the interaction of a *meteoroid* with the atmosphere.
    Astroballistics uses the data and methods of astronomy, aerodynamics, ballistics, and physical chemistry.

astrobiology   The study of living organisms on *celestial bodies* other than the *Earth*.
See: exobiology

astrodynamics   The practical application of *celestial mechanics, astroballistics,* propulsion theory, and allied fields to the problem of planning and directing the *trajectories* of space vehicles.

astrogeology   A term used to describe the geological areas of *lunar* and planetary science; the space-related aspects of geology; also called space geology; planetary geoscience.

astrolabe   An ancient instrument designed to measure the *altitudes* of *celestial bodies*. Modern versions of the astrolabe are still being used to determine stellar positions and consequently latitude and local time.

astronomical unit [Symbol: AU]   A unit of distance in *astronomy* and space science defined as the "mean" distance between the center of the *Earth* and the center of the *Sun*, i.e., the semimajor axis of the Earth's *orbit*. One AU is equal to 149.6×10^6 km (approximately 92.9×10^6 mi) or 499.01 *light seconds*.

astronomy   The science that treats of the location, magnitudes, motions and constitution of celestial bodies and structures.
See: Space telescope

astrophysics   The study of the nature and physics of stars and star systems.

asymetric separation   Separation of the exhaust jet from the *nozzle* wall nonuniformly or at localized regions not in the same plane.

"Atlantis"   The name given to *Space Shuttle Orbiter* vehicle (OV) 104.

Atlas   A family of launch vehicles used as primary intermediate boosters. They served as launch vehicles in a number of programs, including Mercury and Ranger. Atlas boosters have a cluster of three mainstage liquid propellant rocket engines and two small vernier engines. All engines are ignited on the ground and brought up to approximately full thrust before vehicle launch. The Atlas has a thrust of approx. 1,921,550 new tons at sea level. It can lift a variety of payloads up to approx. 5440 kg. (12,000 lb.) into low Earth orbit.

Atlas-Centaur Class   Payloads weighing approximately 1,800 kg to 2,000 kg (4,000 lb_m to 4,400 lb_m).

atmosphere   1. The gases and suspended solid and liquid materials that are gravitationally bound to the region around a planet or satellite.
2. The breathable environment inside a space capsule, spacecraft, or space station. The cabin atmosphere.

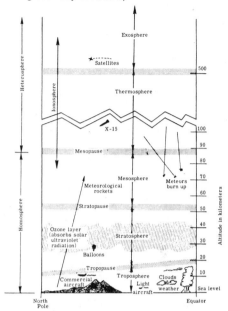

Basic structure of the Earth's atmosphere. (Drawing courtesy of NASA.)

3. A unit of (gas) pressure; 1 atmosphere = 1.01×10^5 $N/^2$.

4. The gaseous envelope surrounding the Earth. The normal composition of clean, dry atmospheric air near sea level is: nitrogen 78%, oxygen 21%, argon .9% and carbon dioxide .03%. In addition it contains water vapor, dust particles, bacteria, etc. The atmosphere is divided into two fundamental layers; below 90 km the homosphere; above 90 km the heterosphere. The heterosphere is divided into various layers based on temperature. In ascending order they are: the troposphere, stratosphere, mesosphere, thermosphere and exosphere. Fig. 1 shows both the composition and temperature of the shells.

atmospheric braking   The action of slowing down any object entering the *atmosphere* of the *Earth* or other *planet* from space, by using the *drag* exerted by air or other *gas* particles in the atmosphere.

atmospheric drag   The retarding *force* produced on a *spacecraft* or *satellite* by its passage through upper *atmosphere*. For the planet Earth it drops off exponentially with *altitude*, and has a small effect on satellites whose perigee is higher than a few hundred kilometers.

atmospheric entry   The penetration of any

planetary *atmosphere* by any object from outer space; specifically, the penetration of the *Earth*'s atmosphere by a manned or unmanned capsule, aerospace vehicle, or *spacecraft*. Also called "entry" and sometimes "re-entry" (although typically the object is making its initial return or entry after a space flight).

atmospheric pressure   The *pressure* at any point in an *atmosphere* due solely to the *weight* of the atmospheric gases above that point.
See: pressure

atom   A particle of matter indivisible by chemical means. It is the fundamental building block of the chemical elements. The *elements*, such as iron, lead, and sulfur, differ from each other because they contain different types of atoms. There are about six textillion (6 followed by 21 zeros, or 6×10^{21}) atoms in an ordinary drop of water. According to present-day theory, an atom contains a dense inner core (the *nucleus*) and a much less dense outer domain consisting of electrons in motion around the nucleus. Atoms are electrically neutral.

atomic bomb   A bomb whose energy comes from the *fission* of heavy elements, such as *uranium* or *plutonium*. Compare this term with *hydrogen bomb*.

atomic clock   A device that uses the extremely fast vibrations of *molecules* or atomic nuclei to measure time. These vibrations remain constant with time, consequently short intervals can be measured with much higher precision than by mechanical or electric clocks.

atomic cloud   The cloud of hot gases, smoke, dust and other matter that is carried aloft after the explosion of a nuclear weapon in the air or near the surface. The cloud frequently has a mushroom shape.
See: fireball, radioactive cloud

atomic energy   Nuclear energy.

atomic mass   The *mass* of a neutral atom of a nuclide usually expressed in atomic mass units. *See:* atomic mass unit, atomic weight, mass number

atomic mass unit [Symbol: amu]   One-twelfth the *mass* of a neutral atom of the most abundant *isotope* of carbon, ^{12}C.

atomic number [Symbol: Z]   The number of *protons* in the *nucleus* of an atom, and also its positive charge. Each chemical element has its characteristic atomic number, and the atomic numbers of the known elements form a complete series from 1 (*hydrogen*) to 103 (lawrencium).
See: Table (y).

atomic particle   One of the particles of which an *atom* is constituted, as an *electron*, *neutron* or a positively charged nuclear *particle*.

atomic reactor   *See:* nuclear reactor

atomic weapon   An explosive weapon in which the energy is produced by nuclear *fisson* or *fusion*.

atomic weight   The mass of an *atom* relative to other atoms. The present-day basis of the scale of atomic weights is carbon; the commonest isotope of this element has arbitrarily been assigned an atomic weight of 12. The unit of the scale is $1/12$ the weight of the carbon-12 atom, or roughly the mass of one proton or one neutron. The atomic weight of any element is approximately equal to the total number of *protons* and *neutrons* in its *nucleus*.

att   *attitude.*

attached shock wave: An oblique or conical *shock wave* that appears to be in contact with the leading edge of an *airfoil* or the nose of a body in a *supersonic* flow field. Also called "attached shock."

attenuation   In general, a reduction in radiation intensity basically through scattering and absorption caused by the action of a transmitting medium. Specifically, the reduction of radiation quantity, such as *energy flux density, intensity, particle flux density*, etc., upon passage of the radiation through a medium.

attitude   The position or orientation of a vehicle (e.g., the *Shuttle*, an *airplane*, a *spacecraft*), either in motion or at rest, as determined by the relationship between its axes and a reference line or plane (such as the horizon) or a fixed system of reference axes.

attitude control   1. The regulation of the *attitude* of an *aircraft*, *spacecraft*, or *aerospace vehicle*.
2. The device or system that automatically regulates and corrects attitude, especially of a pilotless vehicle.

attitude direction indicator An instrument that displays both vehicle *attitude* in all three axes and compass heading.

audible sound Sound containing frequency components lying between about 15 and 20,000 *hertz.*

audio Pertaining to *audiofrequency range.* The word audio may be used as a modifier to indicate a device or system intended to operate at audiofrequencies, e.g., audio-amplifier.

audiofrequency Any frequency corresponding to a normally audible *sound wave. See:* audiofrequency range.

audiofrequency range The range of *frequency* to which the human ear is sensitive, approximately 15 to 20,000 *hertz.* Also called audiorange.

augmenter tube A tube or pipe, usually one of several, through which the exhaust gases from an *aircraft reciprocating engine* are directed especially to provide additional *thrust.*

autoigniting propellant Any *propellant* that ignites by itself without external stimulation.

autoignition self-ignition or spontaneous combustion of *propellant.*

autoignition temperature The *temperature* at which combustible materials ignite spontaneously in *air.*

automated payloads *Payloads* that are supported by an unmanned *spacecraft* capable of operating independently of the *Space Shuttle.* Automated payloads are detached from the *Orbiter* during the operational phase of their flights. *See:* Shuttle mission categories

automatic control Control of devices and equipment, including *aerospace vehicles,* by automatic means.

automatic pilot Equipment which automatically stabilizes the *attitude* of a vehicle about its *pitch, roll,* and *yaw* axes. Also called autopilot.

auxiliary power unit [abbr. APU] 1. A power unit carried on a *spacecraft* or aircraft that can be used in addition to the main sources of power on the craft.
2. A component of the *Space Shuttle Solid Rocket Booster. See:* Solid Rocket Booster

auxiliary stage 1. A small propulsion unit that may be used with a *payload.* One or more of these units may be used to provide the additional velocity required to place a payload in the desired *orbit* or trajectory.
2. A propulsion system used to provide mid-course trajectory corrections, braking maneuvers and/or orbital adjustments.

avionics The contraction of "*avi*ation" and "elect*ronics*." The term refers to the application of electronics to systems and equipment used in *aeronautics* and *astronautics.*

Avogadro's hypothesis The volume occupied by a *mole* of *gas* at a given *pressure* and *temperature* is the same for all gases. This hypothesis can also be expressed as: Equal volumes of all gases measured at the same pressure and temperature contain the same number of *molecules.* This hypothesis was originally developed in 1811 by Amedeo Avogadro, an Italian physicist.

Avogadro's number (symbol N_A) The number of *molecules* per *mole* of any substance, a universal constant. This can be expressed as:

$$N_A = 6.022 \times 10^{23} \text{ molecules/mole}$$

See: ideal gas

axial deflection Elongation or compression along the longitudinal axis.

axial flow compressor A rotary *compressor* having interdigitated rows or stages of rotary and of stationary blades through which the flow of *fluid* is substantially parallel to the rotor's *axis* of rotation. Compare *centrifugal compressor.*

axis (plural: axes) 1. A straight line about which a body rotates, or along which its *center of gravity* moves *(axis of rotation).*
2. A straight line around which a plane figure may rotate to produce a solid; a line of symmetry.
3. One of a set of reference lines for a *coordinate system.*

axis of freedom Of a *gyro,* an axis about which a *gimbal* provides a *degree of freedom.*

axis of rotation The straight line, passing through a rotating body, about which the body rotates.

az azimuth

azimuth 1. Horizontal direction or *bearing.* Compare *azimuth angle.*
2. In navigation, the horizontal direction of

a celestial point from a terrestrial point, expressed as the angular distance from a reference direction, usually measured clockwise through 360°.

An azimuth is often designated as true, magnetic, compass, grid, or relative as the reference direction is true, magnetic, grid north or heading, respectively. Unless otherwise specified, the term is generally understood to apply to true azimuth, which may be further defined as the arc of the horizon, or the angle at the zenith, between the north part of the celestial meridian or principal vertical circle and a vertical circle, measured from 0° at the north part of the principal vertical circle clockwise through 360°.

3. In astronomy, the direction of a celestial point from a terrestrial point measured clockwise from the north or the south point of the meridian plane.

4. In surveying, the horizontal direction of an object measured clockwise from the south point of the meridian plane.

In surveying, an azimuth of a celestial body is called an astronomic azimuth.

B

background counts   In *radiation* counting or monitoring, the responses of a detection system or counting system caused by radiation originating from sources other than the one being measured.

background radiation   The radiation occurring in the natural environment, including *cosmic rays* and radiation from naturally *radioactive* elements. It is also called natural radiation. The term may also mean radiation that is unrelated to a specific experiment or series of measurements.
See: background counts

backlash   1. Dead space or unwanted movement that results from fabrication and assembly tolerance in linkages; excessive backlash produces poor positional control of the controlled element or errors in position instrumentation.

2. The clearance between the tooth spacing of a gear and the tooth thickness of a meshing gear.

backout   To undo things or events that have already been done during a *countdown*, usually in reverse order.

backup crew   A crew of *astronauts* trained to replace the prime crew, if necessary, on a manned space *mission.*

baffle   In a *combustion chamber,* an obstruction used to prevent *combustion instability* by maintaining uniform *propellant* mixtures and equalizing *pressures.* In a fuel tank, an obstruction used to prevent sloshing of propellant.
See: Space Shuttle Main Engine, External Tank

balance piston   The special mechanical device used to balance axial *thrust* in a *turbopump* and thereby reduce axial loads imposed on the bearings.

balance ribs   Blades on the back of the pump *impeller* that reduce the *pressure* on the impeller back side and help control the axial *thrust* loads.

ball flow tube   A tube inserted into a hollow ball to reduce *pressure* drop through a *ball valve.*

ballistic body   A body free to move, behave and be modified in appearance, contour or texture by ambient conditions, substances or forces, as by the pressure of *gases* in a gun, by the rifling or grooves in a gun barrel, by *gravity,* by *temperature* or by air particles. It should be noted that a *guided missile* or *rocket* with a self-contained propulsion unit is **not** considered a ballistic body during the period of its *guidance* or *propulsion.*

ballistic coefficient   An aerospace design parameter that indicates the relative magnitude of *inertial* and aerodynamic effects. This parameter is used in performance analysis of objects moving through the atmosphere.

ballistic flyby   Unpowered flight similar to a bullet's trajectory, governed by gravity and by the body's previously acquired *velocity.*

ballistic missile   A missile that is guided and propelled only during the initial phase of its flight. During the non-powered, non-guided portion of its flight, it describes a *trajectory* similar to that of an artillery shell and operates primarily in accordance with the laws of *ballistics.*

ballistics The science that deals with the motion, behavior and effects of projectiles, such as bullets, aerial bombs and missiles.

ballistic trajectory The path followed by a body being acted upon only by gravitational forces and the resistance of the medium through which it passes. A *rocket* without lifting surfaces describes a ballistic trajectory after its propulsive unit(s) stops operating.

ballistic vehicle A non-lifting vehicle that follows a *ballistic trajectory.*

balloon-type rocket A liquid-fuel *rocket,* such as the *Atlas,* that requires the *pressure* of its *propellants* (or other gases) within to give it structural integrity.
See: launch vehicle

ball valve A valve in which a ball regulates an opening by rising and falling due to fluid *pressure,* a spring or its own *weight.*

band 1. In computer technology a group of tracks on a magnetic drum that is used for storing data.
2. In *spectroscopy* a closely spaced set of spectral lines associated with atomic or molecular energy levels (e.g., an *absorption band*).
3. In communications a *frequency* band or range of frequencies.

band-pass filter A *wave filter* that has a single transmission band extending from a lower cutoff *frequency* greater than zero to a finite upper cutoff frequency.

bandwidth 1. The number of hertz (cycles per second) between the limits of a *frequency band.*
2. The frequency range over which a sensor or instrument can respond appreciably to input signals.
3. In information theory the data-carrying capacity of a communications channel.
4. In an *antenna* the range of frequencies within which its performance, with respect to some characteristic, conforms to a specified standard.

barbecue mode A slow roll of the *Space Shuttle Orbiter* for thermal conditioning. In outer space solar radiation unattenuated by any *atmosphere* is intense on the side of a *spacecraft* or satellite facing the Sun, while the side opposite the Sun is extremely cold.

The rolling, or barbecue, maneuver helps equalize the external temperature of the spacecraft.
See: attenuation

barn [Symbol: b] A unit of area used in expressing the *cross sections* of *atoms, nuclei, electrons* and other particles. One barn is equal to 10^{-24} square centimeter.

barrier cooling The use of a controlled mixture ratio near the wall of a *combustion chamber* to provide a film of low-temperature gases to reduce the severity of gas-side heating of the chamber.

barycenter The center of mass of a system of masses, such as the *Earth-Moon* system. For the Earth and Moon, the ratio of masses is approximately 81-to-1. Consequently, the barycenter is found at a distance from the Earth's center equal to $1/82$ of the distance from the Earth to the Moon. This is a distance of 4,700 km (2,919 mi)—a point actually located within the Earth.

baryon One of a class of heavy *elementary particles* that includes *hyperons, neutrons* and *protons.* All baryons have a mass equal to or greater than that of the proton. When several baryons are associated together, the stable form is a mixture of protons and neutrons, creating an atomic *nucleus.*

base line Any line that serves as the basis for measurement of other lines, as in a surveying triangulation or the measurement of auroral heights. The term is frequently used in the aerospace literature to mean a reference case, as in "base line design" or "base line schedule."

baud A unit of signaling speed. The speed in bauds is the number of *code elements* transmitted per second.

beam A narrow stream of particles (e.g., *electrons*) or *electromagnetic radiation* (e.g., a *photon* of *visible light*) going in a single direction.

beam splitter A partially reflecting (semi-transparent) mirror that permits some incident (arriving) light to pass through and reflects the remainder, thereby dividing the incident (arriving) light beam into two separate beams.

bearing The horizontal direction of an object or point, usually measured clockwise from a reference line or direction through 360°.

beat frequency   A *frequency* that is the sum of or the difference between two frequencies. For example, when two sources are emitting sound waves of a different frequency (f_1 and f_2), the combined sound swells and falls in intensity, producing "beats." The beat frequency then equals $f_1 - f_2$.

bell nozzle   A *nozzle* with a circular opening for a *throat* and an axisymmetric contoured wall downstream of the throat that gives the nozzle a characteristic bell shape.

bellows   The thin-wall, circumferentially corrugated cylinder that can be elongated or compressed longitudinally and, when integrated into a *duct* assembly, can accomodate duct movements by deflection of the corrugations (convolutions); also used in tanks to provide positive expulsion of fluid and in fluid systems to isolate a *regulator* or valve or similar component.

bends   An *aerospace medicine* term describing the acute pains and discomforts in the extremities (arms and legs), abdomen, joints and chest caused by the formation of nitrogen bubbles (together with other biological gases) in body tissues and fluids as a result of the reduction of air *pressure*.
See: aeroembolism

bent-pipe communication   A colloquial expression that describes the use of relay stations to achieve non-line-of-sight transmission links.

Bernoulli's theorem (or law)   Named after the Swiss mathematician, Daniel Bernoulli (1700–1782), this equation is a mathematical statement of the *conservation of energy* principle for a *nonviscous* fluid in steady motion (steady state flow). The specific energy (energy per unit mass) is composed of the kinetic energy ($u^2/2$), the potential energy (gz) and the work done by the *pressure* forces of a *compressible fluid* ($\int v\ dp$—the integration is always with respect to values of p and v on the same fluid parcel). Therefore, the equation

$$\frac{u^2}{2} + gz + \int v\ dp = \text{Bernoulli's constant (C)}$$

along a *streamline*

is valid for a compressible fluid in steady state flow, since the streamline is also the path. Here:

u is the fluid velocity (along a streamline)
g is the acceleration of gravity
z is the height above an arbitrary reference level
p is the pressure
v is the *specific volume* ($= 1/\text{density}$)

Bernoulli's theorem further states that in the flow of an incompressible fluid (a fluid for which the density is constant), the sum of the *static pressure* and the *dynamic pressure* along a streamline is constant if gravity and frictional effects are negligible. That is,

$$udu + vdp = 0$$

This differential form of Bernoulli's law indicates that where there is a *velocity* increase (that is, du > 0), there must be a corresponding pressure decrease (that is, dp < 0) for a flowing liquid.

berthing   The process of using the *remote manipulator system* to gently bring together an orbital element and the *Orbiter vehicle,* that is, the positioning of a *payload* on the repair/*maintenance* support fixture in the *cargo bay.*

beta angle   The minimum angle between the Earth-Sun line and the plane of an *orbit*.

beta decay   The *radioactive* transformation of a nucleus associated with the emission of a *beta particle*. When a negative beta particle ($_{-1}^{0}\beta$) is emitted, a *neutron* has in effect been converted to a *proton* in the nucleus, thereby increasing the *atomic number* (Z) by one and leaving the *atomic mass number* (A) unchanged. The decay of phosphorus-32 ($_{15}^{32}P$) into sulfur-32 ($_{16}^{32}S$) is an example of negative beta decay. The decay process is described as:

$$_{15}^{32}P \rightarrow\ _{16}^{32}S +\ _{-1}^{0}\beta + \bar{\nu}$$

where $\bar{\nu}$ is the *anti-neutrino*.

In the case of positive beta particle decay, or positron decay ($_{+1}^{0}\beta$), a positron is emitted from the radioactive nucleus, reflecting a change of a proton into a neutron. In the process the atomic number of the nucleus is decreased by one, while the atomic mass number remains unchanged. The decay of nitrogen-13 ($_{7}^{13}N$) into carbon-13 ($_{6}^{13}C$) is an example of positron, or positive beta, decay. The process is described as:

$$^{13}_{7}N \rightarrow {}^{13}_{6}C + {}^{0}_{+1}\beta + \nu$$

where ν is the *neutrino.*

It should be noted that the neutrino is associated with positron emission, while its *anti-particle,* the anti-neutrino, is emitted with a negative beta particle. They are similar in all respects, except that for the neutrino the *spin vector* is in the same direction as its motion, while for the anti-neutrino the spin vector is opposite to the direction of motion.

Beta decay is a very common way for *fission product nuclides* to decay.

beta particle (symbol β—beta) An *elementary particle* emitted from a *nucleus* during *radioactive decay.* It has a single electrical charge and a mass equal to 1/1,837 (or 0.0005444) that of a *proton.* A negatively charged beta particle is identical to the *electron,* while a positively charged beta particle is called a *"positron."* Beta radiation may cause skin burns, and beta-emitters are harmful if they enter the body. Beta particles are easily stopped by a thin sheet of metal and, while more penetrating than *alpha particles,* are not as penetrating as *gamma rays.*

betatron A doughnut-shaped *accelerator* in which *electrons,* traveling in an *orbit* of constant radius, are accelerated by a changing magnetic field. Energies above 300 MeV have been attained.
See: accelerators

binary star A pair of stars orbiting about their common *center of mass* or *barycenter.* In a binary star system the star that is nearest the center of mass is called the "primary," while the smaller star of the system is called the "companion." Binary star systems can be divided as follows: visual binaries, eclipsing binaries, spectroscopic binaries and astrometric binaries. Visual binaries are those that can be resolved into two stars by an optical *telescope.* Eclipsing binaries occur when each star of the system alternately passes in front of the other, eclipsing it, and thereby causes their combined brightness to diminish. Spectroscopic binaries can be resolved by the *Doppler shift* of their *spectral* lines as the stars approach and then recede from the *Earth* while revolving about their common center

of mass. (Eclipsing binaries are typically also spectroscopic binaries.) In an astrometric binary one star cannot be visually observed, and its existence is inferred from the irregularities in the motion of the visible star of the system. Binary star systems are more common in the *Galaxy* than generally realized. Perhaps 50 percent of all stars are contained in binary systems. Typical mean separation distances of stars in such systems is on the order of 10 to 20 *astronomical units.*
See: star

binding energy The minimum *energy* required to dissociate into its component *neutrons* and *protons.* Neutron or proton binding energies are those required to remove a neutron or proton, respectively, from a nucleus. *Electron* binding energy is that required to remove an electron from an *atom* or a *molecule.*

biological shield A mass of absorbing material placed around a nuclear *reactor* or radioactive source to reduce *radiation* to a level that is safe for human beings.
See: absorber, shield, thermal shield

bionics The study of systems that function after the manner of, or in a manner characteristic of, or resembling, living systems.

biopak A container for housing a living organism in a habitable environment and for recording biological functions during space flight.

biopropellant A *rocket propellant* consisting of two unmixed or uncombined chemicals (*fuel* and *oxidizer*) fed to the *combustion chamber* separately.

biopropellant rocket A *rocket* using two separate *propellants* which are kept separate until mixing in the *combustion chamber.*

bipropellant valve A valve incorporating both *fuel* and *oxidizer* valving units driven by a common *actuator.*

biosensor A *sensor* used to provide information about a life process.

biosphere That transition zone between *Earth* and *atmosphere* within which most forms of *terrestrial* life are commonly found; the outer portion of the *geosphere* and inner or lower portion of the *atmosphere.*

biotechnology The application of engineering and technological principles to the life sciences.

biotelemetry The remote measurement and evaluation of life functions, as, e.g., in manned *spacecraft* and artificial *satellites*.

bird A colloquial *aerospace* term for a *rocket, satellite* or *spacecraft*.

bistable elements In *computer* terminology, a device which can remain indefinitely in either of two stable states.

blackout 1. In communications a fadeout, due to environmental factors, of radio and *telemetry* transmission between ground stations and aerospace vehicles traveling at high speeds in the *atmosphere*. It is caused by *attenuation* of the transmission signal as it passes through the *ionized boundary-layer* (plasma sheath) and *shockwave* regions generated by the high-velocity vehicle. 2. In *aerospace medicine* a physical condition in which vision is temporarily obscured by blackness, accompanied by a dullness of certain other senses. This is brought on by decreased blood pressure in the head and a consequent lack of oxygen. The condition may occur in pulling out of a high-speed dive in an airplane.

blade 1. One of a set of slat-like objects rigidly fixed to a rotatable shaft, each slat being carefully shaped as an *airfoil* such that (a) rotation of the shaft in a *fluid* creates a flow of fluid, or (b) fluid flow inpinging on the blades rotates the shaft. 2. A flat plate used to adjust flow in a *blade valve.*

blade valve 1. (rotary) The valve that controls flow by means of a flat plate that is rotated transversely in a slotted chamber between upstream and downstream lines. 2. (slide) The valve that controls flow by means of a flat plate that slides transversely in a slotted chamber between upstream and downstream lines.

blast-off A colloquial *aerospace* term for the lift-off of a *rocket* or *launch vehicle.*

blast wave A pulse of air, propagated from an explosion, in which the *pressure* increases sharply at the front of a moving air mass, accompanied by strong, transient winds.
See: shock wave

bleb A small *particle* of distinctive material.

bleed 1. The continuous flow of a *gas* through a *pilot valve.* 2. To remove or draw off *fluid* from a system.

blockage 1. The decrease in pump effective flow area due to the *blade* thickness and the *boundary layer* on the blades and end wall. 2. The restriction in a cavity or flow passage.

blockhouse (Also written "block house.") A reinforced-concrete structure, often built underground or partly underground, and sometimes dome-shaped, to provide protection against blast, heat, or *explosion* during *rocket* launchings or related activities; specifically, such a structure at a *launch site* that houses electronic control instruments used in launching a rocket.

blowdown system A closed propellant/pressurant system that decays in *ullage* pressure level as the *propellant* is consumed and ullage volume thereby is increased.

blueshift *Electromagnetic radiation* from a moving *celestial body* or object is called "blueshifted" when *spectral lines* in the visible part of the *spectrum* appear shifted toward the blue [higher *frequency*, shorter *wavelength*] part of the spectrum. This occurs when the object being observed is approaching the observer. Compare this term with redshift.
See: Doppler shift.

boattail The aft (rear) end of a *rocket* containing the propulsion system and its interface with vehicle tankage.

body burden The amount of radioactive material present in the body of a man or an animal.

boiling water reactor [abbr. BWR] A *nuclear reactor* in which (light or ordinary) water is used as both the *coolant* and *moderator*. In the BWR, the water is allowed to boil in the reactor *core* and the resulting steam is used directly to drive a *turbine* to generate electricity. (See Fig. 1) Therefore, it is a "direct cycle" system. The BWR typically operates at a *pressure* of 6895 kilopascals [1000 psi] and a temperature of 285°C (545°F).

boiloff The vaporization of a *cryogenic propellant* such as liquid oxygen or liquid *hydrogen,* as the *temperature* of the propellant

Fig. 1   Schematic of typical boiling water reactor.   (Drawing courtesy U.S. Department of Energy.)

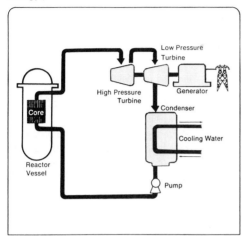

rises in the tank of a *rocket* being readied for launch.

Boltzmann constant [Symbol: k]   The ratio of the universal gas constant to Avogadro's number. It is equal to

$$1.381 \times 10^{-23} \text{ joules/degree Kelvin}$$

boost   1. Additional power, *pressure* or *force* supplied by a *booster,* as, for example, a hydraulic boost, or the extra propulsion given a flying vehicle during lift-off, climb or other part of its flight as with a *booster engine.*
2. Boost pressure.
3. To supercharge.
4. To *launch* or to push along during a portion of flight, as to boost a *ramjet* to flight speed by means of a *rocket,* or a rocket boosted to *altitude* with another rocket.

booster   Short for *"booster engine"* or *"booster rocket."*

booster engine   An engine, especially a *booster rocket,* that adds its *thrust* to the thrust of the sustainer engine.

booster pump   A pump in a fuel system, oil system or the like, used to provide additional or auxiliary *pressure* when needed or to provide an initial pressure differential before entering a main pump, as in pumping *hydrogen* near the boiling point.

booster rocket   1. A *rocket motor,* either solid or liquid, that assists the normal propulsive system or sustainer engine of a

rocket or aeronautical vehicle in some phase of its flight.
2. A rocket used to set a vehicle in motion before another engine takes over. In sense 2 the term launch vehicle is preferred.

booster separation motor [abbr. BSM]   The small, *solid-fueled* booster separation motors that "translate," or move, the *Space Shuttle Solid Rocket Boosters* away from the *Orbiter*'s still-thrusting *main engines* and the *External Tank.*
See: Solid Rocket Booster

boostglide vehicle   A vehicle (part *aircraft,* part *spacecraft*) designed to fly to the limits of the sensible *atmosphere,* then be boosted by *rockets* into the space above, returning to *Earth* by gliding under aerodynamic control.

bootstrap   Refers to a self-generating or self-sustaining process. An example would be the operation of a *liquid-propellant rocket engine* in which (during main stage operation) the *gas generator* is fed by the main propellants pumped by the *turbopump,* and the turbopump in turn is driven by hot gases from the gas generator system. Of course, the operation of such a system must be started by outside power or propellant sources. However, when its operation is no longer dependent on these external sources, the engine is said to be in "bootstrap" operation.
See: Space Shuttle Main Engine

borescope   An arrangement of mirrors or *fiber optics* that makes possible the inspection of otherwise inaccessible locations.

boss   A thickened protuberance in the wall of a *duct* or tank for the purpose of allowing attachment of components or the connection of other lines or instruments.

boundary layer   The layer of *fluid* in the immediate vicinity of a bounding surface; in fluid mechanics, the layer affected by *viscosity* of the fluid, referring ambiguously to the *laminar* boundary layer, *turbulent* boundary layer, planetary boundary layer, or surface boundary layer.

In *aerodynamics,* the boundary-layer thickness is measured from the surface to an arbitrarily chosen point, e.g., where the *velocity* is 99% of the stream velocity. Thus, in aerodynamics, the boundary layer can include only the laminar boundary layer or the laminar boundary layer plus all, or a portion of the turbulent boundary layer.

boundary-layer trip   Discontinuity or local turbulence in the *boundary layer* generated by a protrusion from the surface in contact with the boundary layer; tripping usually increases the severity of the thermal environment.

bow shock   A *shock wave* in front of a body, such as a *aerospace vehicle* or an *airfoil.*

bow wave   A *shock wave* in front of a body, such as an *airfoil,* or apparently attached to the forward tip of the body.

Boyle-Mariotte law   The empirical generalization that for an *ideal* or perfect *gas,* the product of pressure p and volume V is constant in an isothermal process:

$$- {_p}V = F(T)$$

where the function F of the temperature T cannot be specified without reference to other laws (e.g., Charles-Gay-Lussac law).

braking ellipses   A series of *ellipses,* decreasing in size due to aerodynamic drag, followed by a *spacecraft* in entering a *planetary atmosphere.*

breadboard   An assembly of preliminary circuits or parts used to prove the feasibility of a device, circuit, system or principle without regard to the final configuration or packaging of the parts.

breakaway   The action of a *boundary layer* separating from a surface.

breakaway disconnect   Separable connector that is disengaged by the separation force as the *launch vehicle* rises from the *launch pad* or a stage separates from a lower stage; also called a rise-off or staging disconnect.

breakoff phenomenon   The feeling that sometimes occurs during high-altitude flight or spaceflight of being totally separated and detached from the *Earth* and human society. Also called the breakaway phenomenon.

breeder reactor   A reactor that produces *fissionable fuel* as well as consumes it, especially one that creates more than it consumes. This new fissionable material is created by capture in *fertile materials* of neutrons from fission. The process by which

this occurs is known as breeding. Compare *converter reactor.*

breeding   *See:* breeder reactor

breeding ratio   The ratio of the number of *fissionable atoms* produced in a *breeder reactor* to the number of fissionable atoms consumed in the reactor. Breeding gain is the breeding ratio minus one. Compare *conversion ratio.*

bremsstrahlung   [A German word for "braking radiation."]

Electromagnetic radiation emitted (as *photons*) when a fast-moving charged particle (usually an *electron*) loses energy upon being accelerated and deflected by the electric field surrounding a positively charged atomic nucleus. *X-rays* produced in ordinary X-ray machines are bremsstrahlung.

bremsstrahlung effect   The emission of *electromagnetic radiation* as a consequence of the acceleration of charged elementary particles, such as electrons, under the influence of the attractive or repulsive force fields of atomic nuclei near which the charged particle moves. In cosmic-ray shower production, bremsstrahlung effects give rise to emission of *gamma rays* as electrons encounter atmospheric nuclei.

brennschluss   [A German word meaning combustion termination.] The cessation of burning in a *rocket,* resulting from consumption of the *propellants,* from deliberate shutoff, or from other cause; the time at which this cessation occurs.

See: burnout, cutoff

Brewster angle   The *angle of incidence* for which a wave polarized parallel to the plane of incidence is wholly transmitted.

bridge crane   A crane in which a beam or bridge carries the hoisting apparatus.

bridgewire   Resistance wire, attached to the leads of an electroexplosive device, whose function is to convert the electrical firing signal into thermal energy adequate to ignite the prime charge of an *igniter.*

Brinell hardness   Indentation hardness determined by pressing a hardened-steel sphere into the test material under a specified load for a specified time; the diameter of the impression produced is an index to the hardness.

brittle failure   Rupture of structural material that is not preceded by appreciable deformation of the material.

bubble chamber A device used for detection and study of *elementary particles* and *nuclear reactions.* Charged particles from an *accelerator* are introduced into a superheated liquid, each forming a trail of bubbles along its path. The trails are photographed, and by studying the photograph scientists can identify the particles and analyze the nuclear events in which they originate. Compare *cloud chamber, spark chamber.*

buffeting The beating of an aerodynamic structure or surfaces by unsteady flow, gusts, etc.; the irregular shaking or oscillation of a vehicle component owing to turbulent air or separated flow.

builtup shaft A shaft with a multiplicity of components such as collars, sleeves, and couplings.

bulkhead A dividing wall that provides access between internal sections and is sometimes designed to withstand *pressure* differences.

bungee A spring, elastic cord or other tension device used, for example, in an aircraft control system to balance an opposing force or in a landing gear to assist in retraction or to absorb shock.

burble A separation or breakdown of the *laminar flow* past a body; the eddying or turbulent flow resulting from this.

Burble occurs over an *airfoil* operating at an *angle of attack* greater than the angle of maximum lift, resulting in a loss of lift and an increase of drag.

burble point A point reached in an increasing *angle of attack* at which *burble* begins.

burn The firing of a *rocket engine.* For example, the "third burn" of the *Shuttle's orbital maneuvering subsystem* (OMS) engines would mean the third time during a *Shuttle flight* that the *OMS engines* had been fired.

burning surface The surface of a *solid-propellant* grain that is not restricted from burning at any given time during *propellant* combustion.

burnout angle The angle between the *local vertical* and the *velocity vector* at termination of *thrust* (i.e., at *burnout*).

burnout velocity The velocity of a *rocket* at

the termination of *thrust* (i.e., at *burnout*).

burn rate Literally, the rate at which a solid *propellant* burns, i.e., the rate of recession of a burning propellant surface, perpendicular to that surface, at a specified pressure and grain temperature; in grain design, the rate at which the web decreases in thickness during motor operation. Also called burning rate.

burn time For a solid *propellant*, the interval from attainment of a specified intitial fraction of maximum *thrust* or pressure level to web burnout.

burnup 1. (in nuclear energy) A measure of *reactor fuel* consumption. It can be expressed as (a) the percentage of fuel atoms that have undergone *fission,* or (b) the amount of energy produced per weight unit of fuel in the reactor (for example: megawatt-days per metric ton (MWd/t)). 2. (in aerospace) The vaporization and disintegration of a *satellite, spacecraft* or *rocket* or any of their components by aerodynamic heating upon entry into a planetary atmosphere.

burst disk Passive physical barrier in a *fluid* system that blocks the flow of fluid until ruptured by fluid *pressure.*

burst pressure *Fluid pressure* at which a pressurized component will rupture.

burst test *Pressure* test of a component to rupture to determine whether the component can withstand the calculated *burst pressure.*

butterfly valve Valve constructed to close off or throttle flow by rotation of a circular disk around a transverse axis within the flow passage.

C

C *coulomb*
C The symbol for degrees Celsius (formerly centigrade) in the (Celsius) *relative temperature scale.*
C symbol for the *velocity* of light.
—*curie*
calorie [symbol: cal] A unit of *thermal*

energy originally defined as the amount of heat required to raise the temperature of 1 gram of water through 1°C (the gram-calorie or small calorie).

Several calories are now in use: International Steam Table calorie = 4.1868 joules; mean calorie = 4.19002 joules; thermochemical calorie = 4.184 joules, 15°C calorie = 4.18580 joules; 20°C calorie = 4.1890 joules. The kilogram calorie or kilocalorie is 1000 times as large as a calorie.

camber Curvature of the mean line of an *airfoil*; the distance from the point of gratest curvature to the chord.

canard 1. Pertaining to an *aerodynamic vehicle* in which horizontal surfaces used for trim and control are forward of the main *lifting* surface.
2. The horizontal trim and control surfaces in such an arrangement.

captive firing Test firing of a propulsion system, in which the engine is operated at full or partial thrust while restrained in a test stand; the system is completely instrumented, and data to verify design and demonstrate performance are obtained.

captive test A holddown test of a propulsion subsystem, rocket engine or motor. Distinguished from *flight test*.

cargo The total complement of *payloads* (one or more) on any one *Space Shuttle* flight. Cargo includes everything contained in the *Orbiter cargo bay* plus other equipment, hardware and consumables located elsewhere in the Orbiter that are unique to the *user* and are not carried as part of the basic Orbiter payload support.

cargo bay The unpressurized mid-part of the *Orbiter* fuselage behind the cabin aft bulkhead where most payloads are carried. Its maximum usable *payload* envelope is 18.3 m (60 ft) long and 4.6 m (15 ft) in diameter. Hinged doors extend the full length of the bay. (See Fig. 1.)

cargo bay liner Soft protective material used to isolate sensitive *payloads* from the *Space Shuttle Orbiter*'s *cargo bay* structure.

carrier 1. In a semiconductor, a mobile conduction *electron* or *hole*.
2. In modulation of a signal, a *wave* suitable for being modulated as a sine wave, a recurring series of pulses, or a direct current.

3. In nuclear energy, a stable *isotope*, or a normal element, to which radioactive atoms of the same element can be added to obtain a quantity of radioactive mixture sufficient for handling, or to produce a radioactive mixture that will undergo the same chemical or biological reaction as the stable isotope. A substance in weighable amount which, when associated with a trace of another substance, will carry the trace through a chemical, physical or biological process.
See: radioactive tracer, tracer, isotopic.
4. In STS terminology, those flight elements not including the *Space Shuttle*—specifically, the *Spacelab, Space Tug* and the Inertial Upper Stage, etc.
See: payload carriers

carrier wave [Symbol: CW] A *wave* generated at a point in the transmitting system and modulated by the signal.

case 1. Structural envelope for the *solid propellant* in a *solid rocket motor*.
2. The outer portion of metallic materials.
See: Solid Rocket

case bonding Cementing of the *solid propellant* to the motor case through the insulation by use of a thin layer of adhesive (the liner).

case hardening Infiltration of a metallic surface with carbon so that the outer portion of the material (case) is made harder than the inner portion (core).

casing The part of the *pump* housing that surrounds the *impeller*.

Cassegrain telescope A reflecting telescope in which a small hyperboloidal mirror reflects the convergent beam from the paraboidal primary mirror through a hole in the primary mirror to an eyepiece in back of the primary mirror. Also called Cassegrainian telescope, Cassegrain.
See: Newtonian telescope; space telescope.

cathode The negative electrode in an electron (discharge) tube or electrolytic cell through which a primary stream of *electrons* enters a system. Compare with *anode*.
See: cold cathode, hot (thermionic) cathode, photocathode

cathode ray(s) A stream of *electrons* emitted by the negative electrode, or cathode, of a *gas-discharge tube* or by a hot filament in a *vacuum tube,* such as a television tube.

See: cathode ray tube

cathode ray tube [abbr. CRT] An electronic tube that permits the visual observation of electrical signals. The cathode ray tube consists essentially of an *electron gun* that produces a concentrated electron beam *(cathode rays),* a grid to control the beam intensity and therefore the display's brightness and a phosphorescent coating on the back of a viewing screen to convert the electron beam into *visible light.* The *excitation* of the phosphorescent coating produces light, the intensity of which is controlled by regulating the flow of *electrons.* Deflection of the electron beam is achieved either electromagnetically by currents in coils around the tube or electrostatically by *voltages* on internal deflection plates. In general electrostatic deflection is used when high frequency waves are to be displayed (that is, in most cathode ray oscilloscopes), whereas electromagnetic deflection is used when high velocity electron beams are needed to give a bright display, as for example, in radar screens and television screens.

Cauchy number A nondimensional number arising in the study of the elastic properties of a *fluid.* It may be written U^2p/E, where U is a characteristic velocity; p is the density; and E the modulus of elasticity of the fluid. It is the square of the *Mach number.*

cavitating venturi Convergent-divergent constriction in a line that produces *cavitation* at its throat; because of the cavitation effects, flow of the liquid in the line is a function only of pressure upstream of the constriction even though the downstream pressure varies.

cavitation The formation of bubbles, or cavities, in a fluid (liquid) that occurs whenever the *static pressure* at any point in the fluid flow becomes less than the fluid vapor pressure. In the flow of a liquid the lowest pressure reached is the vapor pressure. If the fluid velocity is further increased, the flow condition changes, *Bernoulli's theorem* breaks down and cavities are formed. The formation of these cavities, or vapor regions, alters the fluid flow path and there-

fore the performance of hydraulic devices and machinery, such as pumps. The collapse of these bubbles in down stream regions of high pressure creates local pressure forces that may result in pitting or deformation of any solid surface near the cavity at the time of collapse. Cavitation effects are most noticeable with high-speed hydraulic devices, such as a *rocket engine's turbopumps.*

celestial Of or pertaining to the heavens.

celestial body Any aggregation of matter in space constituting a unit for astronomical study, as the *Sun, Moon,* a *planet, comet, star, nebula,* etc. Also called heavenly body.

celestial coordinates Coordinates that define the position of a star in the sky. Right ascension locates a star in the east-west direction. Declination locates a star's position north or south of the celestial equator. Specifically, any set of coordinates used to define a point on the celestial sphere. The horizon, celestial equator, ecliptic and galactic systems of celestial coordinates are based on the *celestial horizon, celestial equator, ecliptic* and *galactic equator,* respectively, as the *primary great circle.*

celestial equator The great circle, 90° from the celestial poles, that is an extension of the *Earth's* equatorial plane as it cuts the *celestial sphere.*

celestial guidance The process of directing movements of an *aircraft* or *spacecraft,* especially in the selection of a *flight path,* by reference to *celestial bodies.* Also called automatic celestial navigation.

celestial horizon That *great circle* of the *celestial sphere* formed by the intersection of the celestial sphere and a plane through the center of the *Earth* and perpendicular to the *zenith*-nadir line.

celestial-inertial guidance The process of directing the movements of an aircraft or spacecraft, especially in the selection of a *flight path,* by an *inertial guidance system* which also receives inputs from observations of celestial bodies.

celestial latitude An angular distance north or south of the *ecliptic;* the arc of a circle of latitude between the ecliptic and a point on the *celestial sphere,* measured northward or southward from the ecliptic through 90°, and labeled N or S to indicate the direction of measurement.

celestial mechanics   The study of the theory of the motions of celestial bodies under the influence of gravitational fields.
See: gravitation

celestial meridian   A great circle of the *celestial sphere*, through the celestial poles and the *zenith*.

celestial navigation   The process of directing an *aircraft* or *spacecraft* from one point to another by reference to *celestial* bodies of known coordinates. Celestial navigation usually refers to the process as accomplished by a human operator. The same process accomplished automatically by a machine is usually termed *celestial guidance* or sometimes automatic celestial navigation.

celestial observation   In navigation, the measurement of the *altitude* of a celestial body, or the measurement of *azimuth*, or both altitude and azimuth. Also called sight. The expression may also be applied to the data obtained by such measurement.

celestial pole   Either of the two poles of intersection of the *celestial sphere* and the extended axis of the *Earth*, labeled N or S to indicate whether the north celestial pole or the south celestial pole.

celestial sphere   Concept proposed by ancient astronomers consisting of a hypothetical sphere of exremely large radius centered on *Earth* with the fixed stars lying on the outer boundaries.

central force   A force that for purposes of computation can . be considered to be concentrated at one central point with its intensity at any other point being a function of the distance from the central point. For example, *gravitation* is considered as a central force in celestial mechanics.

central force field   The spatial distribution of the influence of a *central force*.

centrifugal compressor   A compressor having one or more vaned rotary *impellers* which accelerate the incoming fluid radially outward into a *diffuser,* compressing by centrifugal force. Sometimes called a centrifugal-flow compressor. Compare *axial flow compressor.*

centrifuge   1. a device that swings material around in a circle. The centrifugal force on this material can be made to equal many g's. A high-speed centrifuge can separate cells in a liquid according to the density of each kind of cell.
2. A large motor-driven apparatus with a long arm at the end of which human and animal subjects or equipment can be revolved and rotated at various speeds to simulate (very closely) the (prolonged) accelerations encountered in high-performance *aircraft, rockets* and *spacecraft.*

centripetal acceleration   The acceleration on a particle moving in a curved path, directed toward the instantaneous center of curvature of the path, with magnitude v^2/R, where v is the speed of the particle and R the radius of curvature of the path. This acceleration is equal and opposite to the *centrifugal force* per unit mass.

2. A large motor-driven apparatus with a long arm at the end of which human and animal subjects or equipment can be revolved and rotated at various speeds to closely simulate the prolonged accelerations encountered in high-performance *aircraft, rockets* and *spacecraft.*

ceramic   An inorganic compound or mixture requiring heat treatment to fuse it into a homogeneous mass usually possessing high temperature strength but low ductility.

Cerenkov radiation   Light emitted when charged particles pass through a transparent material at a velocity greater than that of light in that material. It can be seen, for example, as a blue glow in the water around the *fuel elements* of pool *reactors*. P. A. Cerenkov was the Russian scientist who first explained the origin of this light.
See: radiation

cermet [ceramic + metal]   body consisting of ceramic particles bonded with a metal; used in *aircraft, rockets,* and *spacecraft* for high strength, high temperature applications. Also called ceremal (ceramic + alloy).

chain reaction   A reaction that stimulates its own repetition. In a *fission* chain reaction a

fissionable nucleus absorbs a *neutron* and fissions, releasing additional neutrons. These in turn can be absorbed by other fissionable nuclei, releasing still more neutrons. A fission chain reaction is self-sustaining when the number of neutrons released in a given time equals or exceeds the number of neutrons lost by escape from the system.
See: criticality

"Challenger" The name given to *Space Shuttle Orbiter* vehicle (OV) 99.

chamber coolant valve [abbr. CCV] A component of the *Space Shuttle Main Engine.*
See: Space Shuttle Main Engine

chamber filling interval Time period from complete ignition of the solid-propellant grain to achievement of equilibrium burning pressure.

chamber pressure [Symbol: P_c] The *pressure* of gases within the *combustion chamber* of a rocket engine.

chamber volume [Symbol: V_c] The volume of the rocket *combustion chamber,* including the convergent portion of the *nozzle* up to the throat.

channel construction Use of machined grooves in the wall of the *nozzle* or chamber to form *coolant* passages.

characteristic chamber length [Symbol L*] The length of a straight cylindrical tube having the same volume as the chamber of a *rocket engine* would have if it had no converging section.

characteristic exhaust velocity [Symbol: c*] Of a *rocket engine*, a descriptive parameter,

$$c* = V_e/C_F$$

where V_e is effective exhaust velocity and C_F is thrust coefficient. Also called characteristic velocity.

characteristic length Ratio of *combustion chamber* volume to *nozzle* throat area.

characteristic line Mathematical line inclined to the direction of flow, used to compute the flow field.

characteristic velocity Ratio of effective exhaust *velocity* to thrust coefficient; also called characteristic exhaust velocity.

characterization 1. Definition of physical or chemical properties of a material in relation to its application or use in a *propellant* formulation or *rocket* motor.
2. Definition of the total functional capability of a component system.

charged particle An *ion:* an *elementary charged particle* that carries a positive or negative electric charge.
See: plasma

charge factor A number derived from a formula for the appropriate STS *payload;* used to determine a shared-flight user's price or the price for a partial *Spacelab.*

charge spectrum The range and magnitude of electric charges with reference to *cosmic rays* at a specific *altitude.*

charring ablator An ablation material characterized by the formation of a carbonaceous layer at the heated surface which impedes thermal energy (*heat*) flow into the material by its insulating and reradiating characteristics.
See: ablation

chase pilot A pilot who flies an escort airplane advising a pilot who is making a check, training or research flight in another craft.
See: Approach and Landing Test (Shuttle)

chaser The space vehicle that maneuvers in order to effect a *rendezvous* with an *orbiting* object.

chatter Uncontrolled rapid seating and unseating of a flow-control device, usually at low-flow conditions.

check flight 1. A flight made to check or test the performance of an *aircraft, aerospace vehicle, rocket* or *spacecraft,* or a piece of equipment or component; or to obtain measurements or other data on performance; a test flight
2. A familiarization flight in an aircraft or aerospace vehicle, or flight in which a pilot or other aircrew member or members are tested or examined for proficiency.

checkout 1. A sequence of actions taken to test or examine a system or device as to its readiness for incorporation into a new phase of use, or for the performance of its intended function.
2. The sequence of steps taken to familiarize a person with the operation of an *airplane, aerospace* vehicle or other piece of equipment.

check valve   Flow-control device that allows flow in one direction only.

chemical dosimeter   A *detector* for indirect measurement of *radiation* by indicating the extent to which the radiation causes a definite change to take place. Compare *film badge, ionization chamber.*

chemical energy   Energy produced or absorbed in the process of a chemical reaction. In such a reaction, energy losses or gains usually involve only the outermost *electrons* of the atoms or *ions* of the system undergoing change; here a chemical bond of some type is established or broken without disrupting the original atomic or ionic identities of the consituents. Chemical changes, according to the nature of the materials entering into the change, may be induced by *thermal energy* (thermochemical), light (photochemical) and electric (electrochemical) energies.

chemical fuel   A *fuel* that depends upon an *oxidizer* for combustion or for development of *thrust*, such as liquid or solid rocket fuel or internal-combustion-engine fuel; distinguished from *nuclear fuel.*

chemical pressurization   The *pressurization* of *propellant* tanks in a *rocket* by means of high-pressure *gases* developed by the *combustion* of a *fuel* and *oxidizer* or by the decomposition of a substance.

chemical rocket   A *rocket* using *chemical fuel*, that is, fuel which requires an *oxidizer* for combustion, such as *liquid or solid rocket fuel.*

chemical shim   Chemicals, such as boric acid, which are placed in a *nuclear reactor coolant* to control the reactor by absorbing *neutrons*. Compare *burnable poison, shim rod.*

chemiluminescence   Any *luminescence* produced by chemical action.

chemisorption   The binding of a liquid or *gas* on the surface or in the interior of a solid by chemical bonds or forces.

chilldown   Cooling of all or part of a *cryogenic* engine system from *ambient temperature* to cryogenic temperature by circulating cryogenic fluid through the system prior to engine start.

choked flow   Flow in a *duct* or passage such that the flow upstream of a certain critical section cannot be increased by a reduction of downstream pressure.

choking Mach number   The *Mach number* at some reference point in a duct or passage (e.g., at the inlet) at which the flow in the passage becomes choked.
See: choked flow

chopper   A device used to interrupt the path of *radiation,* such as a beam of *light*, from a single source or to alternate it between two sources. Specifically, a rotating shutter for interrupting an otherwise continuous stream of *particles.* Neutron choppers can release short bursts of *neutrons* with known energies, used to measure nuclear *cross sections.*

chord   Straight line connecting the ends of an *arc;* usually, the line joining the leading and trailing edges of an *airfoil.*

chromosphere   The relatively thin region between the apparent solar surface (photosphere) and the base of the *corona.* It is the source of *solar prominences.*
See: Sun.

chugging   1. A form of *combustion instability* in a *rocket engine,* characterized by a pulsing operation at a fairly low frequency, sometimes defined as occurring between particular frequency limits.
2. The noise made in this kind of combustion.

circle of equal probability [abbr. CEP]   A measure of the accuracy with which a *rocket* or *missile* can be guided; a radius of the circle at a specific distance in which 50% of the reliable shots land. Also called circular error probable, circle of probable error.

circularize   To change an *elliptical orbit* into a circular one, usually by "*apogee* kicks."

circular velocity   At any specific distance from the primary, the orbital *velocity* required to maintain a constant-radius *orbit.*

circumferential seal   Seal, composed of a continuous ring or of one or more split or segmented rings, whose sealing surface is parallel to the centerline of the flow passage (also called radial seal).

circumlunar   Around the *Moon*, a term generally applied to *trajectories*.

cislunar   [Latin *cis*, on this side]. Of or pertaining to phenomena, projects or activity in the space between the Earth and *Moon*, or between the *Earth* and the Moon's *orbit*. Compare *translunar, circumlunar*.

cladding   1. In general, a coating placed on the surface of a material and usually bonded to the material. Also called clad.
2. In nuclear engineering, the outer jacket of *nuclear fuel elements*. It prevents corosion of the fuel and the release of *fission products* into the coolant. Aluminum or its alloys, stainless steel and zirconium alloys are common cladding materials.

clean room   A delimited space in which dust, *temperature* and humidity are controlled as necessary for the fabrication and/or assembly of critical components.

clevis   A fitting with a U-shaped end for attachment to the end of a pipe or rod.
See: External Tank, Solid Rocket Booster

clone   A genetic engineering term. Creature or plant descended from a single individual without recourse to the normal reproductive process.

closed-cycle reactor system   A nuclear *reactor* design in which the primary heat of fission is transferred outside the *reactor core* to do useful work by means of a coolant circulating in a completely closed system that includes a *heat exchanger*. Compare *direct-cycle reactor system, indirect-cycle reactor system, open-cycle reactor system*.

closed loop   Term applied to an electrical or mechanical system in which the *output* is compared with the *input* command signal, and any discrepancy between the two results in corrective action by the system elements.

closed-loop system   1. A system in which the output is used to control the input.
2. Sequential series of actions performed by a machine or system.

closed-loop telemetry   1. A *telemetry* system that is used as the indicating portion of a remote-control system.
2. A system used to check out test vehicle or telemetry performance without radiation of radio-frequency energy.

closed respiratory gas system   A completely self-contained system within a sealed cabin, capsule, spacecraft or space station that will provide adequate oxygen for breathing, maintain adequate cabin pressure, and absorb the exhaled carbon dioxide and water vapor.

closed system   1. In *thermodynamics*, a system so chosen that no transfer of mass takes place across its boundaries. Compare this term with *open system*.
2. In mathematics, a system of differential equations and supplementary conditions such that the values of all the unknowns (dependent variables) of the system are mathematically determined for all values of the independent variables (usually space and time) to which the system applies.
3. = closed ecological system.
4. A system which constitutes a *feedback* loop so that the inputs and controls depend on the resulting output.

closest approach   1. The point in time and space when two *planets* or other *celestial bodies* nearest to each other as they orbit about the sun or other primary.
2. The place or time of such an event.

closing rate   The speed at which two bodies approach each other, as, for example, two *spacecraft* "closing" for an orbital rendezvous or docking operation.

cloud chamber   A device in which the tracks of charged atomic particles, such as *cosmic rays* or *accelerator beams*, are displayed. It consists of a glass-walled chamber filled with a *supersaturated vapor*, such as wet air. When charged particles pass through the chamber, they trigger a process of *condensation*, and so produce a track of tiny liquid droplets, much like the vapor trail of a jet plane. This track permits scientists to study the particles' motions and interactions. Compare this term with *bubble chamber*.

cluster   Two or more *rocket motors* bound together so as to function as one propulsion unit.

coated optics   Optical elements (lenses, prisms, etc.) which have their surfaces covered with a thin transparent film to minimize reflection and loss of *light* in the system.

coaxial cable   A form of *waveguide* consisting of two concentric conductors insulated from each other.

coaxial injector   Type of injector in which one *propellant* surrounds the other at each injection point.

coherence   In radar, a relation between two *wave trains* such that, when they are brought into coincidence, they are capable of producing interpretable interference phenomena.

coherent   1. In *electromagnetic radiation,* being in phase, so that waves at various points in space act in unison, as, for example, in a *laser* producing coherent light.
2. Having a fixed relation between frequency and phase of input and output signal.

cohesive force   The force at the boundary of a liquid that pulls the liquid together.

cohesive fracture   Rupture or cracking of material within the body of the material, not at a bondline at the interface with other material.

coincidence circuit   An electronic circuit that produces a usable output pulse only when each of two or more input circuits receive pulses simultaneously or within an assignable time interval.

coincidence counting   A method for detecting or identifying radioactive materials and for calibrating their disintegration rates by counting two or more characteristic *radiation* events (such as gamma ray emissions) which occur together or in a specific time relationship to each other. This method is important in activation analysis, medical scanning, *cosmic ray* studies and low-level measurements.
See: counter, low-level counting.

coined groove   A narrow channel or depression stamped in a burst disk to provide localized thinning of material in a desire pattern.

co-investigator   A scientist working with the *principal investigator* on a *NASA* experiment.

cold cathode   A *cathode* whose operation does not depend on its *temperature* being above the *ambient* temperature.

cold flow   1. Term applied to a test of an engine or all or part of engine system, in which *fluid* is flowed through the test configuration without the engine being started; term also is applied to test of model fluid systems.
2. Permanent deformation of material caused by a compressive load that is less than the load necessary to yield the material; some time is required to obtain cold flow.

cold-flow test   A test of a *liquid rocket* without firing it to check or verify the efficiency of a propulsion subsystem, providing for the conditioning and flow of *propellants* (including tank pressurization, propellant loading and propellant feeding).

coldsoak   The exposure of a system or equipment to low *temperature* for a long period of time to ensure that its temperature is lowered to that of the surrounding atmosphere or operational environment.

cold working   Deforming metal plastically at a *temperature* lower than the recrystallization temperature.

collimate   1. To render parallel, as rays of *light.*
2. To adjust the line of sight of an optical instrument, such as a theodolite, in proper relation to other parts of the instrument.

collimation error   The angular error in magnitude and direction between two nominally parallel lines of sight; specifically, the angle by which the *line of sight* of an optical instrument or radar differs from what it should be.

collimator   A device for focusing or confining a *beam* of *particles* or *radiation,* such as *X rays.*

collision   A close approach of two or more *particles, photons,* atoms or nuclei, during which such quantities as energy, *momentum* and charge may be exchanged.
See: Compton effect, excited state, pair production, scattering

collision rate   The average number of collisions per second suffered by a *molecule* or other *particle* moving through a *gas.* Also called collision frequency.

colloidal system   An intimate mixture of two substances one of which, called the dispersed phase (or colloid) is uniformly distributed in a finely divided state through the second substance, called the dispersion medium (or dispersing medium). The dispersion medium may be a *gas,* a liquid or a solid, and the dispersed phase may also be any of these, with the exception that one does not speak of a colloidal system of one gas in another. Also called colloidal dispersion, colloidal suspension.

color temperature   1. An estimate of the temperature of an incandescent body, determined by observing the *wavelength* at which it is emitting with peak intensity (its color) and using that wavelength in Wien law.
2. The temperature to which a black body radiator must be raised in order that the light it emits may match a given light source in color. (Usually expressed in degrees Kelvin ($^\circ$K).)

color wheel   A device for describing visual color by number.

"Columbia"   The name given to *Space Shuttle Orbiter* vehicle (OV) 102, the first Orbiter to fly in space (12 April 1981).

column density   Density or *mass* of a column of given cross section and length. A column of water of 1 cm^2 section and 10 cm length has a column density of 10 gm/cm^2.

coma   1. The gaseous envelope that surrounds the *nucleus* of a *comet.*
2. In an optical system, a result of spherical aberration in which a point source of light, not on the axis, has a blurred, comet-shaped image.
See: comet

combined stresses   Stresses resulting from simultaneous action of all loads to which a structure is subject.

combustion chamber   Any chamber for the combustion of *fuel,* specifically that part of the *rocket* engine in which the combustion of *propellants* takes place at high pressure. The combustion chamber plus the diverging section of the *nozzle* comprise the rocket thrust chamber. Also called chamber, firing chamber.
See: Space Shuttle Main Engine

combustion devices   Devices located in those parts of a *liquid rocket engine* where controlled combustion, or burning, of the liquid *fuel* and *oxidizer* occurs. For the *Space Shuttle Main Engines,* there are five major components in this group: the *igni-tion system,* the *preburners,* the *main injector* the *main combustion chamber* and the *nozzle assembly.*
See: Space Shuttle Main Engine

combustion efficiency   The efficiency with which *fuel* is burned, expressed as the ratio of the actual energy released by the combustion to the potential *chemical energy* of the fuel.

combustion instability   Unsteadiness or abnormality in the *combustion* of fuel, as may occur, e.g., in a *rocket engine.*

combustion stabilization device   Contrivance in the combustion chamber that reduces or eliminates oscillatory combustion by reducing the coupling of the oscillations with the driving combustion processes or by increasing the damping inherent in the engine system.

combustion wave   A zone of burning propagated through a combustible medium.

comet   A dirty-ice "rock" orbiting in the Solar System which the Sun causes to vaporize, glow visibly and stream out a long luminous tail. Comets are generally regarded as samples of primordial material from which the planets were formed billions of years ago. The comet's nucleus is believed to consist of frozen gases and dust. As the comet approaches the Sun from the cold regions of deep space, the ices sublime and these vapors form an atmosphere or "coma" with a diameter that may reach 100,000 km (60,000 mi.).The nucleus has a diameter of only a few tens of kilometers.

command   A *signal* which initiates or triggers an action in the device which receives the signal. In computer operations also called "instruction".

command destruct   A command control system that destroys a flightborne test *rocket.* It is actuated on command of the range safety officer whenever the rocket performance indicates a safety hazard.

commander   The *Space Shuttle* crew member who has ultimate responsibility for the safety of the personnel aboard and has authority throughout the flight to deviate from the *flight plan,* procedures and personnel assignments as necessary to preserve crew safety or vehicle integrity. The commander is also responsible for the overall execution of the flight plan in compliance

with NASA policy, mission rules and *Mission Control Center* directives. During the *payload* operations phase of the Shuttle flight, the commander will, within the limitations prescribed for crew safety and vehicle integrity, be subject to the authority of the *mission specialist* in directing the allocation of *Space Transportation System* resources to accomplish the combined payload objectives, including consumables allocation, systems operation and flight plan modification. A commander plus a *pilot* or pilot-qualified mission specialist are always required to operate and manage the *Orbiter*.
See: pilot, mission specialist, Shuttle crew

command guidance   The guidance of a *spacecraft, aerospace vehicle,* or *rocket* by means of electronic *signals* sent to receiving devices in the vehicle.

communications   Sending and receiving messages by radio, television, teletype or telephone line, centered at the *Mission Control Center* during a space mission.
See: Oribiter communications and data systems.

communications satellite   A *satellite* designed to reflect or relay electromagnetic signals used for *communication.*

communications systems (Orbiter)
See: Orbiter avionics system; Orbiter communications and data systems

companion body   A nose cone, last-stage *rocket,* or other body that orbits along with an *Earth* satellite. Compare *afterbody.*

compliance (fluid)   Effective compressibility of a *fluid,* i.e., the change in fluid volume per unit pressure change.

component   An assembly; any combination of parts, subassemblies and assemblies; and assemblies mounted together. A component is normally capable of independent operation in a variety of situations.

composite materials   Structural materials of metals, *ceramics,* or plastics with built-in strengthening agents which may be in the form of filaments, foils, powders or flakes of a different compatible material.

composite propellant   solid *propellant* system comprising a discrete solid phase dispersed in a continuous solid phase.

compressibility   1. The property of a substance, such as air, by virtue of which its density increases with increases in pressure.

2. co-efficient of compressibility. In *aerodynamics,* this property of the air is manifested especially at high speeds (speeds approaching or higher than the speed of sound). Compressibility of the air about an aircraft may give rise to buffeting, aileron buzz, shifts in trim, and other phenomena not ordinarily encountered at low speeds; these are known generally as compressibility effects.

compressibility burble   A region of disturbed flow, produced by, and rearward of, a shock wave.
See: burble

compressible flow   In *aerodynamics,* flow at speeds sufficiently high that density changes in the *fluid* cannot be neglected.

compression chamber   A chamber of strong, sturdy construction in which the air pressure can be increased to two or more *atmospheres.*

compression set   Percent of deflection by which an *elastomer* fails to recover after a fixed time under specified squeeze and temperature.

compression system   Duct system wherein the fluid-column loads due to internal pressure are reacted by the support structure.

compressor blade   Either a rotor blade or a stator blade in an axial-flow compressor; sometimes used restrictively (and ambiguously) for a compressor rotor blade. Compare *impeller vane.*

Compton effect   Elastic scattering of *photons* (*X rays* or *gamma rays*) by *electrons.* In each such process, the electron gains energy and recoils, and the photon loses energy. This is one of three ways photons lose energy upon interacting with matter, and is the usual method with photons of intermediate energy and materials of low atomic number. It is named for A. H. Compton, American physicist, who discovered it in 1923.
See: collision, pair production, scattering

Compton electron   An orbital *electron* of an atom which has been ejected from its orbit as a result of an impact by a high-energy *quantum* of radiation (*X-ray* or *gamma ray*).

Compton wavelength [Symbol: λ_c]   Of a *particle,* the distance h/mc, where h is the Planck constant, m is the *mass* of the particle, and c is the velocity of light. The Compton wavelength of the *electron* (symbol λ_c) is 2.4261×10^{-10} centimeter; that of the *photon* (symbol $\lambda_{c.p}$) is 1.32140×10^{-13} centimeter.

concave surface A surface that curves inward; concave mirrors, for examples, are converging in action.

conceptual flight planning data package A data package produced by the *Space Transportation System* (STS) *flight operator* that contains the following information: a basic *trajectory,* a basic STS configuration and the flight constraints, the consumable allocations, the crew and vehicle in-flight *maintenance* requirements, the recommended *Shuttle crew* work/rest cycle, and an opportunity matrix used to establish the crew work/rest cycle.

condensation shock wave A sheet of discontinuity associated with a sudden *condensation* and fog formation in a field of *flow.* It occurs, e.g., on a wing, where a rapid drop in pressure causes the temperature to drop considerably below the dew point. Also called condensation shock.

condensation trail A visible trail of condensed vapor or ice particles left behind an *aircraft,* an *airfoil,* etc., in motion through the air. Also called a contrail or vapor trail.

There are three kinds of condensation trails: the aerodynamic type, caused by reduced pressure of the air in certain areas as it flows past the aircraft; the convection type, caused by the rising of air warmed by an engine; and the engine-exhaust, or exhaust-moisture, type, formed by the ejection of water vapor from an engine into a cold atmosphere.

conductance 1. In electricity, the ratio of the current flowing through an electric circuit to the difference of potential between the ends of the circuit, the reciprocal resistance.
See: conductivity.
2. In vacuum systems, the conditions divided by the measured difference in pressure p between two specified cross sections inside a pumping system:

$$G = Q/(p_1 - p_2)$$

conduction band The range of states in the energy spectrum of a solid in which *electrons* can move freely.

conductivity 1. The ability to transmit, as electricity, heat, sound, etc.
2. A unit measure of electrical conduction; the ease with which a substance conducts electricity, as represented by the current density per unit electrical-potential gradient in the direction of flow.

cone 1. A geometric configuration having a circular bottom and sides tapering off to an apex (as in a nose cone).
2. A type of light-sensitive cell in the retina. Cones are involved in color vision, high visual acuity and photopic vision.

configuration The technical and physical description required to fabricate, test, accept, operate, maintain, and logistically support *systems* or equipment.

configuration control Systematic evaluation, coordination, approval, or disapproval of all changes to the *baseline configuration.*

confluence The rate at which an adjacent flow is converging along an axis oriented normal to that flow. Compare *convergence.*

conic 1. A curve formed by the intersection of a plane and a right circular cone. Originally called *conic section.*

The conic sections are the *ellipse,* the parabola, and the hyperbola, curves that are used to describe the paths of bodies moving in space.

The circle is a special case of the ellipse, an ellipse with an eccentricity of zero.

The conic is the locus of all points the ratio of whose distances from the fixed point, called the *focus,* and a fixed line, called the directrix, is constant.
2. In reference to satellite orbital parameters, without consideration of the perturbing effects of the actual shape or distribution of mass of the primary. Thus conic *perigee* is the perigee the satellite would have if all the mass of the primary were concentrated at the primary's center.

connector Mechanical device for joining or fastening together two or more similar parts, e.g., lines or tubes in a fluid system or wires in an electrical system.

conservation of angular momentum The principle which states that absolute angular momentum is a property which cannot be created or destroyed but can only be transferred from one physical system to another through the agency of a net torque on the system. As a consequence, the absolute angular momentum of an isolated physical system remains constant.

The principle of conservation of angular momentum can be derived from the Newton second law of motion.

conservation of energy   The principle which states that the total *energy* of an isolated system remains constant if no interconversion of *mass* and energy takes place. This principle takes into account all forms of energy in the system; it therefore provides a constraint on the conversions from one form to another.

conservation of mass   The principle in *Newtonian mechanics* which states that mass cannot be created or destroyed but only transferred from one volume to another.
See: continuity equation

conservation of momentum   The principle that in the absence of forces absolute momentum is a property which cannot be created or destroyed.
See: Newton laws of motion.

console   An array of controls and indicators for the monitoring and control of a particular sequence of actions, as in the checkout of a rocket, a countdown action, or a launch procedure. A console is usually designed around desklike arrays. It permits the operator to monitor and control different activating instruments, data recording instruments or event sequences.

constant-level balloon   A balloon designed to float at a constant-pressure level. Also called constant-pressure balloon.

constellation   The easily identifiable configuration of the brightest visible stars in a moderately small region of the night sky.

contaminant tolerance   Amount (by weight) of a standard contaminant (added at the inlet of a filter under specified fluid flowrate, temperature and pressure) that causes the pressure loss in the filter to exceed a maximum allowable value.

contamination tolerance level   Value of contaminant *particle* size, or level of con-

tamination, in a fluid system at which the specified performance, reliability, or life expectancy of the components of the system is adversely affected.

continuity equation   In the *steady-flow* process, the mathematical statement of the principle of the *conservation of mass* by equation the flow at any section x, w_x, to the flow at any section y, w_y or $w_z = w_y$.

continuous spectrum   1. A spectrum in which *wavelengths, wave numbers* and *frequencies* are represented by the continuum of real numbers or a portion thereof, rather than by discrete sequence of numbers.
See: discrete spectrum.
2. For *electromagnetic radiation,* a spectrum that exhibits no detailed structure and represents a gradual variation of intensity with wavelength from one end to the other, as the spectrum from an incandescent solid. Also called continuum, continuum radiation.
3. For *particles,* a spectrum that exhibits a continuous variation of the momentum or energy.

continuous-wave radar   A general species of *radar* transmitting continuous waves, either modulated or unmodulated. The simplest form transmits a single frequency and detects only moving targets by the Doppler effect. This type of radar determines direction but usually not range. Also called CW radar. Compare *pulse radar.* Two advantages of CW radar are the narrow bandwidth and low power required. Range information may be obtained by some form of modulation, e.g., frequency modulation, pulse moderation.

continuous waves [Symbol: CW]   *Waves,* whose the successive oscillations are identical under steady-state conditions.

contraction ratio   Ratio of the area of the *combustion chamber* at its maximum diameter to the area of the *throat.*

contrast   1. In general, the degree of differential between different tones of an image. Where the degree is slight, the image is said to be flat. Where the difference is marked, it is said to be contrasty.
2. The difference in luminance between two portions of the visual field usually expressed as:

$$c = \frac{background - test\ field}{background} \times 100\%$$

Since this ratio can be negative for nearly black targets at a close range, and since the mathematical sign of the contrast has no psycho-physical significance, it is conventional to use only its absolute value.

contravane A vane that reverses or neutralizes the rotation of a *flow.* Also called a countervane.

control 1. A lever, switch, cable, knob, pushbutton or other device or apparatus by means of which direction, regulation or restraint is exercised over something.
2. In plural (a) A system or assembly of levers, gears, wheels, cables, boosters, valves, etc., used to control the *attitude,* direction, movement, power, and speed of an *aircraft, rocket, spacecraft,* etc. (b) Control surfaces or devices.
3. Sometimes capitalized. An activity or organization that directs or regulates an activity.
4. Specifically, to direct the movements of an aircraft, or *rocket* with particular references to changes in attitude and speed.

controllability The capability of an *aircraft, rocket* or *aerospace vehicle* to respond to control, especially in direction or *attitude.*

controlled thermonuclear reaction Controlled fusion, that is, *fusion* produced under research conditions, or for production of useful power.

controller In general, a device that converts an input signal from the controlled variable (temperature, pressure, level, or flowrate) to a valve *actuator* input (pneumatic, hydraulic, electrical, or mechanical) to vary the valve position to provide the required correction of the controlled variable.
See: Space Shuttle Main Engine.

control rocket A *vernier, retro-rocket,* or other such *rocket,* used to change the *attitude* of, to guide, or to make small changes in the speed of a rocket, *spacecraft* or the like.

control rod A rod, plate or tube containing a material that readily absorbs neutrons (hafnium, boron, etc.), used to control the power of a nuclear reactor. By absorbing neutrons, the control rod prevents them from causing further fission.

See: absorber, regulating rod, safety rod, shin rod

control vane A moveable *vane* used for control, especially a movable air vane or jet vane on a *rocket,* used to control flight *attitude.*

convection 1. In heat transfer, mass motions within a *fluid* resulting in transport and mixing of the properties of that fluid. The up-and-down drafts in a fluid or gas heated from below in one-g. Because the density of the heated fluid is lowered, the fluid rises. After cooling at the top, its density increases, and it sinks. Compare *conduction, radiation.*
2. In meteorology, atmospheric motions that are predominantly vertical. Compare *advection.*

convergence 1. The contraction of a *vector* field; also precise measure thereof.
Mathematically, convergence is negative divergence, and the latter term is used for both. Compare this term with *confluence.* (For mathematical treatment see *divergence.*)
2. The propery of a sequence or series of numbers or functions which ensures that it will approach a definite finite limit.
A series representation of a mathematical function exhibits convergence if the sum of the terms of the series approaches the value of the function more closely as more terms of the series are taken, the two agreeing in the limit of an infinite number of terms.
3. Decrease in area or volume.

convolution Longitudinal wave (corrugation) plastically formed in a thin-wall (usually metal) tube.

coolant A substance circulated through a *nuclear reactor* to remove or transfer heat. Common coolants are water, air, carbon dioxide, liquid sodium and sodium-potassium alloy (NaK).

coolant liner A part of the *main combustion chamber* of the *Space Shuttle Main Engine.*
See: Space Shuttle Main Engine

coolant manifolds A part of the *main combustion chamber* of the *Space Shuttle Main Engine.*
See: Space Shuttle Main Engine

coolant tube Relatively small-diameter thin-wall conduit attached to or forming the wall of a *regeneratively cooled combustion chamber* or *nozzle* and carrying *propellant* to cool the wall.

core   The central portion of a *nuclear reactor* containing the fuel elements and usually the moderator, but not the reflector.

core segment   The section of the pressurized *Spacelab module* that houses subsystem equipment and experiments.
See: Spacelab

coriolis effects   The physiological effects (nausea, vertigo, dizziness, etc.) felt by a person moving radially in a rotating system, as a rotating *space station*.

coriolis force   An inertial force on a moving body, or particles, produced by the movement of the masses involved, perpendicular to the axis of the primary rotating system.
See: inertial force, coriolis acceleration

coronagraph   An instrument used to study the solar corona. It uses occulting discs to form an artificial *eclipse* of the *Sun.*

corpuscular   Consisting of *particles,* specifically atomic particles.

correlation detection   A method of *detection* in which a *signal* is compared, point-to-point, with an internally generated reference.

The output of a correlation a detector is a measure of the degree of similarity between the input signal and the reference signal. The reference signal is constructed in such a way that it is at any time a prediction, or best guess, of what the input signal should be at that given time.

corridor
See: Reentry corridor

cosmic   Of or pertaining to the universe, especially that part of it outside the earth's atmosphere. Used by the USSR as equivalent to space, as in cosmic rocket, cosmic ship.

Cosmic Background Explorer [abbr. COBE]   A future astrophysics mission currently approved by NASA. It is designed to make all-sky maps of the diffuse infrared and microwave emissions of the Universe over the wavelength range from 1 mm to 13 mm. The radiation comes from the primeval big-bang, the earliest galaxies, interplanetary and intersteller dust and infrared galaxies. A spacecraft of the Explorer class may be used to put the COBE in an operational

orbit of 900 km, to maintain the pointing vector and 1-rpm spin, so that the sky coverage of the instruments is acceptable, and, to provide the instruments with power, telemetry and commands.

cosmic dust   Finely divided solid matter with particle sizes smaller than a *micrometeorite,* thus with diameters much smaller than a millimeter, moving in interplanetary *space.*

Cosmic dust in the *solar system* is thought to be concentrated in the plane of the ecliptic, thus causing the *zodiacal light.*

Cosmic Ray Observatory [abbr. CRO]   A possible NASA astrophysics mission designed to study the origin of the elements and to learn about the nucleosynthetic processes that produce the cosmic ray nuclei. The CRO investigations will be performed by a series of particle telescopes specifically designed to look at a given species of particles. A variety of techniques, such as transition radiation detectors, ionization loss counters, Cerenkov counters, superconducting magnets, ionization calorimeters, and trajectory devices will be used. The spacecraft will be inserted into a 400-km circular orbit at 56° inclination. It is designed to function for two years.

cosmology   The study of the origin, evolution and structure of the Universe.

cosmonaut   Term used by the Soviet Union to describe their *astronauts.*

Couette flow   The shearing flow of a *fluid* between two parallel surfaces that are in relative motion. It is two-dimensional steady flow without pressure gradient in the direction of flow and caused by the tangential movement of the bounding surfaces. The only practical type is the flow between concentric rotating cylinders (such as of the oil in a cylindrical bearing).

coulomb [Symbol: C]   The unit of quantity of electricity; the quantity of electricity transported in 1 second by a current of 1 *ampere.*

Coulomb damping   Dry-friction damping; friction *force* is constant in magnitude but always directed opposite to the *velocity.*

count   One pulse of current or voltage from a

The proposed Cosmic Ray Observatory.   (Drawing courtesy of NASA.)

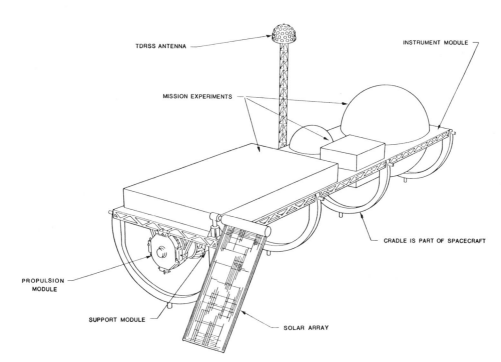

TDRSS ANTENNA

INSTRUMENT MODULE

MISSION EXPERIMENTS

CRADLE IS PART OF SPACECRAFT

PROPULSION MODULE

SUPPORT MODULE

SOLAR ARRAY

detector, indicating the passage of a *photon* or *particle* through the detector.

countdown   1. A step-by-step process that culminates in a climactic event, each step being performed in accordance with a schedule marked by a count in inverse numerical order; specifically, this process is used in leading up to the launch of a large or complicated *rocket* vehicle, or in leading up to a *captive test,* a readiness firing, a mock firing, or other firing test. 2. The act of counting inversely during the process.

counter   A general designation applied to *radiation detection* instruments or *survey meters* that detect and measure radiation in terms of individual ionizations, displaying them either as the accumulated total or as their rate of occurrence.

counterpermeation   Simultaneous migration of propellant vapor and *pressurant* (in opposite direction) across a permeable membrane.

count rate   The number of counts per second; a measurement of the intensity of the source.

coupled modes   Modes of vibration that are not independent but influence one another by energy transfer from one mode to the other.

coupling   Mechanical device that fastens together two parts of a *turbopump* shaft or connects the shaft to other components of the turbopump; also, a separable connector in a fluid system line or duct.

cracking   1. Thermal decomposition of heavy (complex) hydrocarbons into lighter and simpler hydrocarbons and other products. 2. Opening of a flow-control device to allow the flow of fluid.

cracking pressure   Effective differential pressure above which a flow-control device (e.g., a valve) will open and allow the flow of fluid.

craft   1. An *aircraft* or aircraft collectively. 2. Any vehicle or machine designed to fly through air or *space*.

crater   1. = lunar crater 2. The depression resulting from high speed solid *particle* impacts on a rigid material as, for example, a *meteoroid* impact on the skin of a *spacecraft.*

creep Permanent deformation of material caused by tensile or compressive load that is less than the load necessary to yield the material; some time is required to obtain creep.

creep strength Degree to which a given material resists creep; also the maximum load at which a given material will not exhibit a significant amount of creep.

crevice corrosion Corrosion that occurs in a narrow, relatively deep opening where two similar surfaces meet and trap a reactive fluid that acts as an *electrolyte;* corrosion occurs because of the concentration gradient of the reactive species established within the trapped fluid.

crew activity plan A timeline of activities to be performed in flight by the *Shuttle crew* to accomplish the objectives of a particular *Space Shuttle* flight.
See: crew activity planning, Shuttle crew

crew activity planning The analysis and development of activities to be performed in flight by the *Shuttle crew,* resulting in a timeline of these activities together with the necessary procedures and crew reference data to accomplish flight objectives.
See: crew activity plan, Shuttle crew

crew module [abbr. CM] A three-section pressurized working, living and stowage compartment in the forward portion of the *Orbiter.* It consists of the flight deck, the mid deck/equipment bay and an *airlock.*
See: Orbiter structure

critical Capable of sustaining a *chain reaction.*
See: criticality

critical assembly An assembly of sufficient fissionable material and *moderator* to sustain a *fission chain reaction* at a very low power level. This permits study of the behavior of the components of the assembly for various *fissionable* materials in different geometrical arrangements. Compare this term with *nuclear reactor.*

critical crack size Crack, or flaw size in a pressure vessel at or above which the crack, at a specified stress level, will grow and become unstable (i.e., will lend to *brittle failure*).

critical damping The minimum *damping* that will allow a displaced system to return to its initial position without *oscillation.*

critical flow capacity Point in the performance of a tank-pressurization *heat-exchanger* at which *pressurant* volumetric flowrate is at a maximum and an increase in pressurant produces a decrease in volumetric rate.

critical ignition pressure Pressure below which *propellants* cannot be ignited.

critical inspection and test method An inspection or test method which is used to verify a critical process.

criticality The state of a *nuclear reactor* when it is sustaining a *chain reaction.*
See: multiplication factor, prompt criticality, reactivity

critical mass The smallest mass of *fissionable* material that will support a self-sustaining *chain reaction* under stated conditions.

critical path That particular sequence of activities that has the greatest negative or least positive activity slack.

critical speed Shaft rotational speed at which a natural frequency of a *rotor/stator* system coincides with a possible forcing frequency.

critical temperature The *temperature* above which a substance cannot exist in a liquid state, regardless of the *pressure.*

critical velocity In rocketry, the *speed of sound* at the conditions prevailing at the *nozzle throat.* Also called throat velocity, critical throat velocity.

cross-pointer An instrument designed to provide information about a vehicle's position with respect to a *target* and display the information by means of intersecting pointers.

cross product
See: vapor product

cross section [Symbol: σ (sigma)] A measure of the probability that a nuclear reaction will occur. Usually measured in *barns,* it is the *apparent* (or effective) area presented by a target nucleus (or *particle*) to an on-coming particle or other nuclear radiation, such as a *photon* of gamma rays.

crosstalk Electrical disturbances in a communication channel as a result of *coupling* with other communication channels.

cryogenic Fluids or conditions at low

temperatures, usually at or below $=150°C$ (123K).

cryogenic fluid pump   A pump used for circulating *cryogenic propellants.* This term is **not** to be confused with *cryopump* or *cryogenic pump.*

cryogenics   The study and use of very low *temperatures.* This science may be concerned with practical engineering problems, such as producing the metric ton quantities of *liquid oxygen* (LO$_2$) and *liquid hydrogen* (LH$_2$) needed for *chemical rocket propellants,* or it may help a physicist investigate some basic properties of matter at very low temperatures. Cryogenic researchers work with temperatures down to within a millionth of a degree of *absolute zero.*

The large scale industrial technology of cryogenics is less than 50 years old. Currently, it is mainly concerned with the liquefaction of the "permanent" gases, such as hydrogen, helium or oxygen. When liquid all these substances are denser (i.e., more concentrated) and are consequently handled more easily than in the gaseous state. It took the demands of World War II and the space program to develop a global technology capable of producing LH$_2$ and LO$_2$ in metric ton quantities. The commercial production of liquid helium began in 1962.

Table 1 lists some important properties for the most common cryogenic liquids: oxygen, nitrogen, hydrogen and helium. The *critical point* shows the temperature above which it is impossible to liquefy the gas, no matter what *pressure* is applied.

Liquid oxygen is usually obtained by liquefying air and then separating it from the liquid nitrogen in a low-temperature fractionating column, taking advantage of the fact that the boiling points of oxygen and nitrogen are quite different. Next to fluorine oxygen is the strongest oxidizing agent known, and much of the liquid oxygen is used as an *oxidizer* in chemical rockets.

Liquid nitrogen has become increasingly plentiful as a by-product of the liquid oxygen manufacturing process. It is chemically inert and is used to provide a safe, nonreactive cryogenic environment for both research and industrial purposes.

Liquid hydrogen is a chemically reactive substance. For example it will react explosively with oxygen (even that amount present in the air) to form water in one of the most energetic chemical reactions known. For this reason hydrogen and oxygen are frequently used in chemical rockets, like the *Space Shuttle Main Engines.* An entirely different property of hydrogen would be used in a *nuclear rocket.* The effectiveness of any rocket can be measured in terms of a quantity called the *specific impulse* (I_{sp}), which is the ratio of the *thrust* to the rate of propellant consumption. When a hot propellant gas expands in a rocket *nozzle,*

$$I_{sp} = \text{(a constant)} \times \sqrt{T/MW}$$

where T   is the temperature (of the propellant gas)
MW is the molecular weight (of the propellant gas)

Hydrogen has the lowest molecular weight (MW = 2) and will therefore produce the greatest specific impulse when heated in a *nuclear rocket reactor.* (Of course material limitations imposed by the removal of heat from the *nuclear reactor's core* will limit the maximum operating temperature of such a

Table 1 Properties of some common cryogenic liquids.

Substance	Critical Temperature (K)	Boiling Point (K)	Melting Point (K)	Density at Boiling Point (gm/cc)	Heat of Vaporization (J/l)
oxygen (O$_2$)	154.8	90.2	54.4	1.14	243,000
nitrogen (N$_2$)	126.1	77.4	63.2	0.80	161,300
hydrogen (H$_2$) (99.7% para)	32.9	20.3	13.8	0.071	31,600
helium-4 (^{4}He)	5.2	4.2	—	0.125	2,720
helium-3 (^{3}He)	3.32	3.19	—	0.0586	480

Table 2 Comparison of (Theoretical) rocket engine performance data.

Source of Power	Specific Impulse (I^{sp}) (seconds)
chemical rockets	
liquid oxygen + kerosene	300
liquid oxygen + liquid hydrogen	390
liquid fluorine + liquid hydrogen	410
nuclear rockets (solid core)	
liquid hydrogen + nuclear reactor at 1,500 K	720
liquid hydrogen + nuclear reactor at 3,000 K	1,020

system.) (See Table 2.)

Molecular hydrogen exists in two different forms, depending on whether the *nuclear spins* of the two *atoms* are in the same direction (**ortho** hydrogen) or in opposite directions (**para** hydrogen). There is a difference in energy between the two forms. Therefore, they act as if they were different chemical species, and we can think of the processes by which changes from ortho to para hydrogen occur as we would a chemical reaction. For example, just as in a chemical reaction, the *equilibrium* ratio of the number of ortho to para hydrogen molecules in a mixture depends on and varies with temperature. At room temperature the equilibrium concentration is 75 percent ortho hydrogen, whereas at 20 K the equilibrium concentration has changed to 99.8 percent para hydrogen. Liquefaction of room-temperature hydrogen gas leads to an unstable mixture that slowly converts to para hydrogen, giving off enough heat in the process to boil away 70 percent of the liquid product. This situation used to make the storage of liquid hydrogen very difficult. However, today most liquid hydrogen is passed over a catalyst to bring about ortho-to-para conversion during liquefaction. The resulting almost-pure liquid para hydrogen can be stored without large evaporation losses.

Liquid helium has the lowest boiling point of the common cryogenic liquids. It is chemically inert and is an extremely valuable substance for much laboratory and space research. Helium will remain liquid down to the lowest temperatures unless a pressure of 25 *atmospheres* is applied to it to make it become solid. Liquid helium exhibits many unusual "superfluid" properties.

cryogenic materials   Metals and alloys that are usable in systems or structures operating at very low *temperatures* and usually possess improved physical properties at such temperatures.

cryogenic propellant   A rocket *fuel, oxidizer* or propulsion fluid that is liquid only at very low temperatures. *Liquid hydrogen* and *liquid oxygen* are examples of cryogenic propellants.

cryogenic pump   A pump that uses *cryopumping* to create a vacuum.

cryogenic seal   Seal that must function effectively at *temperatures* below =150° (123K).

cyrogenic temperature   In general a temperature range below the boiling point of nitrogen—77.4 K (−195 C). More particularly temperatures within a few degrees of *absolute zero*.

cryopump   An exposed surface refrigerated to *cryogenic temperature* for the purpose of pumping gases in a vacuum chamber by condensing the gases and maintaining the condensate at a temperature such that the *equilibrium vapor pressure* is equal to or less than the desired ultimate *pressure* in the chamber.
See: cryopumping

cryopumping   The process of removing gases from an enclosure by condensing the gases on surfaces at very low (*cryogenic*) temperatures.
See: cryopump, cryogenic temperature

curvilinear coordinates   Any linear coordinates which are not *Cartesian coordinates*. Examples of frequently used curvilinear coordinates are *polar coordinates* and *cylin-*

drical coordinates

cutoff or cut-off   1. An act or instance of shutting something off; specifically, in rocketry, an act or instance of shutting off the *propellant* flow in a *rocket,* or of stopping the *combustion* of the *propellant.* Compare this term with *burnout.*
2. Something that shuts off, or is used to shut off.
See: fuel shutoff.
3. Limiting or bounding as, for example, in cutoff frequency.

cybernetics   The study of methods of *control* and communication which are common to living organisms and machines.

cyborg   A term from science fiction literature meaning an artificially produced human being.

cycle   1. The complete sequence of values of a periodic quantity that occur during a period.
2. One complete wave, a *frequency* of 1 wave per second.
3. Any repetitive series of operations or events.

cycle life   The number of times a unit may be operated (e.g., opened and closed) and still perform within acceptable limits.

cyclic vibration   Vibration mode, induced by rough combustion in a *rocket engine,* that periodically produces severe g loads at one predominant frequency.

cylindrical barrel section   An assembly of the *liquid oxygen* tank of the *Space Shuttle External Tank.*
See: External Tank

cylindrical coordinates   A system of *curvilinear coordinates* in which the position of a point in space is determined by (a) its perpendicular distance from a given line, (b) its distance from a selected reference plane perpendicular to this line, and (c) its angular distance from a selected reference line when projected onto this plane. The coordinates thus form the elements of a cylinder, and, in the usual notation, are written r, θ, and z where r is the radial distance from the cylinder's axis z and θ is the angular position from a reference line in a cylindrical cross section normal to z. Also called cylindrical polar coordinates, circular coordinates.

The relations between the cylindrical coordinates and the rectangular *Cartesian coordinates* (x,y,z,) are x = r cos θ, y = r sin θ, z = z.

cylindrical grain   Solid-propellant grain in which the internal cross section is a circle.

cyclonic   Having a sense of *rotation* about the *local vertical* the same as that of the *Earth*'s rotation: that is, as viewed from above, counterclockwise in the Northern Hemisphere, clockwise in the Southern Hemisphere, undefined at the equator; the opposite of *anticyclonic.*

cyclotron   A *particle accelerator* in which charged particles receive repeated synchronized accelerations by electrical fields as the particles spiral outward from their source. The particles are kept in the spiral by a powerful magnetic field. Compare this term with *synchrocyclotron.*

cyclotron radiation   The *electromagnetic radiation* emitted by charged *particles* as they orbit in a magnetic field. The radiation arises from the centripetal acceleration of the particle as it moves in a circular *orbit.*

cynthion   Combining form "pertaining to the *Moon*". For example, "apocynthion corresponds to *apogee.*"

D

d   *dyne.*

Dalton law   The empirical generalization that, for many *perfect gases* (ideal gas), a mixture of these gases would have a pressure equal to the sum of the partial pressures that each of the gases would have as sole component with the same volume and temperature, provided there is no chemical interaction.

dam   Flat plate inserted perpendicularly into a *fluid* membrane in order to partially or fully separate two streams approaching from opposite directions.

damped   The progressive decrease, or decay, of a free *oscillation* due to the expenditure of *energy* by friction.

damped wave   Any wave whose *amplitude*

decreases with time, or whose total energy decreases by the transfer of energy to other portions of the wave spectrum.

damping   The suppression of *oscillations* or disturbances; the dissipation of *energy* with time.

damping factor   The ratio of the *amplitude* of any one of a series of damped *oscillations* to that of the following one at the same phase.

dart configuration   A configuration of an *aerodynamic vehicle* in which the control surfaces are at the tail of the vehicle. Contrast this term with canard.

data link   Any communications channel or circuit used to transmit data from a *sensor* or a *computer*, a readout device, or a storage device.

data point   A unit of fundamental information obtained through the processing of raw data.

data systems (Orbiter)
See: Orbiter avionics system; Orbiter communications and data systems.

datum line   Any line which can serve as a reference or base for the measurement of other quantities.

datum plane   A plane from which angular or linear measurements are reckoned. Also called reference plane.

datum point   Any point which can serve as a reference or base for the measurement of other quantities.

daughter   A *nuclide* formed by the *radioactive decay* of another nuclide, which in this context is called the *parent*.

dead reckoning [abbr. DR]   In navigation, determination of *position* by advancing a previous known position for *courses* and distances.

dead spot   In a control system, a region about the neutral control position where small movements of the *actuator* do not produce any response in the system.

dead time   In a *radiation* counter, the time interval, after the start of a count, during which the *counter* is insensitive to further *ionizing events*.

debond   Localized failure of adhesive at the interface of two components cemented together.

De Broglie wavelength   For a particle of *mass* m and *velocity* v, the De Broglie wave-

length, $\lambda = h/mv$, where h is the Planck constant.

Debye length   A theoretical length which describes the maximum separation at which a given *electron* will be influenced by the electric field of a given positive *ion*. Sometimes referred to as the Debye shielding distance or *plasma* length. It is well known that charged *particles* interact through their own electric fields. In addition, Debye has shown that the attractive *force* between an electron and ion which would otherwise exist for very large separations is indeed cut off for a critical separation due to the presence of other positive and negative charges in between. This critical separation or Debye length decreases for increased plasma density.

decay constant   1. Attenuation constant.
2. (symbol λ) A constant relating the instant rate of *radioactive decay* of a *radioactive* species to the number of atoms N present at a given time t. Thus,

$$-(dN/dt) = \lambda N$$

If zero is the number of atoms present at time then

$$N = No e^{-\lambda t}$$

decayed object   An object once, but no longer, in *orbit*.

decay heat   The heat produced by the decay of *radioactive nuclides*.
See: afterheat; decay, radioactive.

decay product   A *nuclide* resulting from the *radioactive* disintegration of a *radionuclide*, being formed either directly or as the result of successive transformations in a *radioactive series*. A decay product may be either radioactive or *stable*. Also called daughter, daughter element.

decay, radioactive   The spontaneous transformation of one nuclide into a different nuclide or into a different energy state of the same nuclide. The process results in a decrease, with time, of the number of the original *radioactive atoms* in a sample. It involves the emission from the nucleus of *alpha particles, beta particles* (or electrons, or *gamma rays;* or the nuclear capture or ejection of orbital electrons; or fission). Also called *radioactive disintegration*.
See: half-life, nuclear reaction, radioactive series.

decay time   1. In *computer* operations, the time required for a *pulse* to fall to one-tenth of its peak value.
2. In charge-storage tubes, the time interval during which the magnitude of the stored charge decreases to a stated fraction of its initial value. The fraction is usually 1/e where e is the base of natural logarithms.
3. Approximately the lifetime of an orbiting object in a nonstable *orbit*. Decay time is usually applied only to objects with short orbit lifetimes caused by *atmospheric drag*.
4. In nuclear energy, the time required for a radioactive substance to decay to stable state.
See: radioactive decay

declination   1. (sybmol σ) Angular distance north or south of the *celestial equator;* the *arc* of an *hour circle* between the celestial equator and a point on the *celestial sphere,* measured northward or southward from tne celestial equator through 90°, and labeled N or S to indicate the direction of measurement.
2. Magnetic declination.

decoder   A device for translating electrical signals so that the signals can perform predetermined functions.

decommutator   Equipment for separation, demodulation, or demultiplexing commutated signals.

decompression   The reduction of atmospheric *pressure;* particularly, various techniques for preventing *caisson disease.*

decompression sickness   An *aerospace medicine* term for a disorder experienced by deep sea divers and aviators caused by reduced atmospheric pressure and evolved gas bubbles in the body, marked by pain in the extremities, pain in the chest (chokes), occasionally leading to severe central nervous symptoms and neurocirculatory collapse.

decrement   A decrease in the value of a variable.
See: increment

dedicated flight   A *Space Shuttle* flight assigned to a single *user*. In this type of mission, the user pays all costs of the launch and associated services plus options.

dedicated Spacelab   An extension module devoted to a single discipline, such as life

sciences, that may fly more than once a year for several years and may be assigned to a *payload* development center.

deep space   An area in space which is not in the vicinity of *Earth.*

Deep Space Network [abbr. DSN]   The communication and tracking system for all automated scientific spacecraft circumnavigating the solar system. DSN takes over communications from other near Earth facilities shortly after the spacecraft has been injected onto the proper trajectory toward its target planet. The network consists of three multi-station complexes, each equipped with a 64 meter-diameter (210 feet) antenna station and two 26-meter (85 feet) antenna stations. In addition, each station is equipped with transmitting, receiving, data handling and interstation communication equipment. The nerve center at the Jet Propulsion Laboratory controls and monitors the DSN performance.
   The DSN has provided tracking and data acquisition support to NASA's moon exploration projects, its missions to Mars, Venus and Mercury and interplanetary and Jupiter exploration flights. It is currently monitoring the Voyager flights to the outer planets. Should the Voyagers continue to send back data through the late 1980s, the DSN will be receiving data across 4½ to 6 billion kilometers. In addition to tracking and acquiring data for flight projects, the network has been used for numerous radio experiments: natural radio sources, planetary radar studies of planet surfaces, etc. In the 1980s it will be used in the search for extra-terrestrial intelligence.

deep space probes   Spacecraft designed for exploring *deep space* to the vicinity of the *Moon* and beyond. Deep space probes with specific missions may be referred to as "lunar probe," "Mars probe," "solar probe," etc.

deep-space vacuum   A term applied to *pressure* less than 10^{-11}mm Hg ($1.333 \times 10^{-13} \text{N/cm}^2$).

Defense Meteorological Satellite Program [DMSP]   Accurate weather forecasts have always been a basic need of military commanders. While there are other meteorological satellites in use by the civilian community, the Defense Meteorological

Fig. 1   Defense Meteorological Satellite Program Block 5D-2 Satellite   (Illustration courtesy of the U.S. Air Force.)

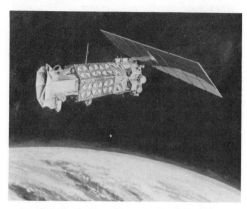

Satellite Program satellites are designed to meet unique military requirements for worldwide weather information.

Through the DMSP satellites, military weather forecasters can detect and observe developing patterns of weather and can track existing weather systems over remote areas, including oceans. The data can help identify severe weather such as thunderstorms, locate and determine the intensity of hurricanes and typhoons, and gather imagery used to form three-dimensional cloud analyses which are the basis for computer simulation of various weather conditions.

Current DMSP satellites (Block 5D-1) are being replaced by the Block 5D-2 (See Fig. 1) satellites. The last of the 5D-1 satellites was launched in 1980, and the first 5D-2 satellite in 1981. The 5D-2 satellite structure looks very similar to 5D-1; however, many changes have been incorporated to support larger and more numerous remote sensor systems, and to increase the operational life span of the satellites. Both 5D-1 and 5D-2 satellites provide near constant resolution of the visible and *infrared* data collected.

The satellites (there are normally two in orbit at any one time) *orbit* at an altitude of approximately 833 km (450 nm) in a *near-polar, sun-synchronous* orbit. They take about 101 minutes to complete their orbits, and each scans a 2963 km (1,600 nm) wide area. Thus, each satellite can cover the entire earth in about 12 hours.

The functions of a *launch vehicle upper stage* have been integrated into the orbital satellite, providing for ascent phase guidance for the launch vehicle from lift-off through orbit insertion, as well as electrical power, *telemetry, attitude control*, and propulsion for the second and third stages. Each satellite is divided into four major sections: (1) a precision mounting platform for sensors and other equipment requiring precise alignment; (2) an equipment support module which encloses the bulk of the electronics; (3) a reaction control equipment support structure containing the spent third stage rocket motor and supporting the ascent phase reaction control equipment; and (4) a *solar cell* array.

The Operational Linescan System is the primary sensor on board the satellite, providing visual and infrared imagery. The flexible design of the satellite also allows the addition of mission sensors. The major mission sensors are of two basic types; temperature/moisture sounders and auroral detection.

The infrared temperature/moisture sounder measures *infrared radiation* emitted from different heights within the *atmosphere*, allowing forecasters to plot curves of *temperature* and *water vapor* at various altitudes. The microwave temperature

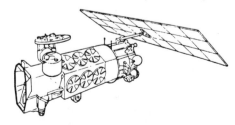

Fig. 2   The special microwave imagery sensor (SSMI), swings out from the DMSP Block 5D-2 satellite when in use.   (Drawing courtesy of U.S. Air Force.)

sounder is also used on Block 5D-1 and 5D-2 satellites, and measures *microwave radiation* emitted from different heights within the atmosphere, even through cloud cover. This instrument allows forecasters to plot curves of temperature versus altitude even over cloudy regions of the globe. The microwave imager, a passive microwave sensor placed on Block 5D-2 satellites, will provide precipitation, soil moisture, sea state and ice cover information.

A special microwave imagery sensor will be flown on Block 5D-2 satellites in the mid 1980s. (See Fig. 2) Called the SSMI (special sensor, microwave imagery), the sensor is a passive microwave radiometer that detects thermal energy emitted by the earth-atmosphere system in the microwave portion of the *electro-magnetic spectrum*. Meteorologists at the Air Force Global Weather Central and the Navy's Fleet Numerical Oceanography Center will use SSMI sensor data to measure ocean surface wind speed, ice coverage and age, areas and intensity of precipitation, cloud water content and land surface moisture.

Defense Satellite Communications System [DSCS] The Defense Satellite Communications System (DSCS) is a worldwide satellite communications program sponsored by the Department of Defense. In Phase I, 26 small (0.92m (3 ft) diameter, weighing approximately 45.4 kg (100 lb$_m$) each) *communications satellites* were launched between June 1966 and June 1968. In this Initial Defense Communications Satellite Program (IDCSP), each satellite relayed voice, imagery, computerized *digital* data, and teletype transmissions. Even though these early satellites were designed to last 18 months, one of the 26 is still operating ten years after launch.

The IDCSP has since been replaced by DSCS Phase II. The DSCS II satellites carry many times the communications load of IDCSP with substantial increase in transmission strength and doubled lifetime expectancy. The DSCS II operational system will consist of four active satellites and two spares orbiting the *Earth*. Phase II satellites contain propulsion systems for orbit repositioning to support contingency operations. The two *dish-shaped antennas*

on DSCS II are steerable by ground command. The antennas can concentrate their electronic beams on small areas of the Earth's surface for intensified coverage to link small, portable ground stations into the communication system.

In Phase III, now under development, new longer-lasting more powerful DSCS satellites will be operational in the 1980s. Phase III satellites will be designed to last twice as long as DSCS II with six active communications transmitter channels instead of four. *Antenna* design for DSCS III allows the user to switch between fixed, earth-coverage antennas and multiple beam antennas. The latter will provide an earth coverage *beam* as well as electrically steerable area and narrow-coverage beams. In addition, a steerable transmit *dish antenna* will provide a spot beam with increased radiated power for users with small receivers. In this way the communications beams are tailored to suit the needs of different size user terminals almost anywhere on the surface of the *Earth*.

Ten years after launch in Phase I, one IDSCP satellite is still operating; a total of 16 DSCS II satellites has been procured. The first two DSCS II spacecraft, launched November 2, 1971, no longer operate. Of the second two, launched December 13, 1973, one is still operating over the western Pacific Ocean. The third pair did not achieve orbit because of a failure in the Titan III guidance system on May 20, 1975. The first two of the replacement set were launched on May 12, 1977, and are providing operational communications. Two more satellites, F9 and F10, were launched on March 25, 1978, but did not achieve orbit because of a failure in the Titan III second stage. On December 1978 DSCS II satellites 11 and 12 were successfully launched and placed in orbit providing, for the first time, global user coverage. In 1977 four additional DSCS II satellites, modified to double EIRP, were procured for launch in 1979 and early 1980's. Spacecraft 13 and 14 were placed in space during November 1979. The first DSCS III was launched along with a modified DSCS II in 1981.

See: U.S. Air Force Role in Space

deflagration Burning process in which large quantities of *gas* and *energy* are released rapidly. In a deflagration, the reaction front advances at less than *sonic velocity* and gaseous products move away from unreacted material; a deflagration may, but need not, be an explosion.

deflector A plate, *baffle,* or the like that diverts something in its movement or flow, as: (a) a plate that projects into the airstream on the underside of an airfoil to divert the airflow, as into a slot—sometimes distinguished from a *spoiler*; (b) a conelike device placed or fastened beneath a *rocket* launched from the vertical position, to deflect the exhaust *gases* to the sides; (c) any of several different devices used on jet engines to reverse or divert the exhaust gases; (d) a baffle or the like to deflect and mingle *fluids* prior to combustion, as in certain *jet engines.*

degas To remove *gas* from a material, usually by heating under high *vacuum.* Compare get.

degassing The deliberate removal of *gas* from a material, usually by application of heat under high *vacuum.*

degradation Gradual deterioration in performance.

degree of freedom 1. A mode of motion either angular or linear, with respect to a *coordinate system,* independent of any other mode.

A body in motion has six possible degrees of freedom, three linear and three angular. 2. Specifically, of a gyro the number of *orthogonal axes* about which the spin *axis* is free to rotate. 3. In an unconstrained dynamic or other system, the number of independent variables required to specify completely the state of the system at a given moment. If the system has constraints, i.e., kinematic or geometric relations between the variables, each such relation reduces by one the number of degrees of freedom of the system. In a continuous medium with given boundary conditions, the number of degrees of freedom is the number of normal modes of *oscillation.* 4. Of a mechanical system, the minimum number of independent generalized coordinates required to define completely the posi-

tions of all parts of the system at any instant of time.

In general, the number of degrees of freedom equals the number of independent generalized displacements that are possible.

de Laval nozzle [After Dr. Carl Gustaf Patrik de Laval (1845-1913, Swedish engineer.] A converging-diverging *nozzle* used in certain *rockets.* Also called Laval nozzle.

delay The time (or equivalent distance) displacement of some characteristic of a *wave* relative to the same characteristic of a reference wave; that is, the difference in phase between the two waves. In one-way radio propagation, for instance, the phase of the delay of the reflected wave over the direct wave is a measure of the extra distance traveled by the reflected wave in reaching the same receiver.

delayed neutrons *Neutrons* emitted by excited nuclei in a radioactive process, so called because they are emitted an appreciable time after the *fission.* Compare this term with *prompt neutrons.*

delayer A substance mixed in with *solid rocket propellants* to decrease the rate of combustion.

Delta Class Payloads weighing approximately 900 kg to 1,100 kg (2,000 lb$_m$ to 2,500 lb$_m$).

delta nitrogen tanks Optional flight kit used to provide additional nitrogen for the *Orbiter* living space atmosphere.

Delta-V velocity change.

delta waste tanks Optional flight kit used when the number of crewmembers or the number of days on orbit exceeds the *baseline.* The tanks collect wastewater generated by the crew.

delta wing A triangularly shaped wing of an *aircraft.*

deluge collection pond A facility at a *launch site* into which water used to cool the *flame deflector* is flushed as the *rocket* begins its ascent.

demodulation The process of recovering the modulation *wave* from a modulated carrier.

demodulator An electronic device which operates on an input of a modulated carrier to recover the modulating *wave* as an output.

densitometer An instrument for the measurement of *optical density* (photographic transmission, photographic reflection, visual transmission, etc.) of a material, generally of a photographic image.

density   1. The ratio of the mass of a quantity of matter to its volume.

2. (photographic) The blackening of an exposed, developed photographic film negative, measured (with a *densitometer*) as the fraction of a light beam absorbed by the darkened film. The positive print from a negative reverses to black and white.

density function   The number of particles per unit of volume.

de-orbit burn   A *retrograde rocket engine* firing by which vehicle *velocity* is reduced to less than that required to remain in *orbit*.

depleted uranium   In nuclear technology *uranium* (U) that has a smaller percentage of the isotope uranium-235(^{235}U) than the 0.7% found in *natural uranium*. It is obtained from the *spent (used) nuclear fuel* elements or as by-product tails, or residues, of uranium *isotope separation*.

depletion layer   In a *semiconductor* a region in which the mobile carrier charge density is insufficient to neutralize the net fixed charge density of *donors* and *acceptors*.

deployable radiators   An assembly of the *Orbiter* mechanical subsystems associated with the *payload bay* doors.
See: Orbiter structure

deployment   The process of removing a *payload* from the *cargo bay* and releasing it to a position free of the *Orbiter*.

descending node   That point at which *planet,* planetoid, or *comet* crosses to the south side of the *ecliptic;* that point at which a *satellite* crosses to the south side of the equatorial plane of its primary. Also called southbound node. The opposite is ascending node or northbound node.

destruct   The deliberate action of destroying a *rocket* vehicle after it has been launched, but before it has completed its course. Destructs are executed when the rocket gets off its plotted course or functions in a way so as to become a hazard.

destruct line   On a *rocket* test range, a boundary line on each side of the downrange course beyond which a rocket cannot fly without being destroyed under destruct procedures, or a line beyond which the *impact point* cannot pass.
See: command destruct.

detached shock wave   A *shock wave* not in contact with the body which originates.
See bow wave. Also called detached shock.

detector   Material or a device that is sensitive to *radiation* and can produce a response signal suitable for measurement or analysis. A radiation detection instrument.

detonation   A rapid chemical reaction which propagates at a *supersonic velocity*.

detonation wave   A *shock wave* in a combustible mixture, which originates as a *combustion wave*.

deuterium [Symbol: ^{2}H or D]   An *isotope* of hydrogen whose nucleus contains one neutron and one proton and is therefore about twice as heavy as the nucleus of normal hydrogen, which is only a single proton. Deuterium is often referred to as *heavy hydrogen*; it occurs in nature as 1 atom to 6500 atoms of normal *hydrogen*. It is nonradioactive.

device, nuclear   A *nuclear explosive* used for peaceful purposes, tests or experiments. The term is used to distinguish these explosives from nuclear weapons, which are packaged units ready for transportation or use by military forces. Compare this term with nuclear weapons.

Dewar (flask)   A double-walled container with the interspace completely evacuated of gas to prevent the container's contents from gaining or losing thermal energy (i.e., heat) through (gaseous) *conduction* or *convection*. This device, named for the Scottish physicist and chemist Sir James Dewar (1842–1923) is frequently used for storing liquefied gases *(cryogenic fluids)*. To further isolate the contents of the container from the external thermal environment, the inner walls are silvered to reduce *radiation heat transfer*.

dewetting   Phenomenon in *solid propellants,* in which the binder *(fuel)* breaks free from the embedded *oxidizer* and metal particles.

diamonds   The patterns of *shock waves* often visible in a *rocket* exhaust which resembles a series of diamond shapes placed end to end.

diaphragm   1. A thin membrane that can be used as a seal to prevent *fluid* leakage or as an actuator to transform an applied pressure to linear force.

2. A positive expulsion device used to expel *porpellant* from a tank in zero-g conditions.

dielectric   A substance that contains few or no free charges which can support electrostatic stresses.

In an electromagnetic field, the centers of the nonpolar molecules of a dielectric are displaced, and the polar molecules become oriented close to the field. The net effect is the appearance of charges at the boundaries the dielectric. The frictional work done in orientation absorbs energy from the field which appears as heat. When the field is removed the orientation is lost by thermal agitation and so the energy is not regained. If free-charge carriers are present they too can absorb energy.

A good dielectric is one in which the absorption is a minimum. A *vacuum* is the only perfect dielectric. The quality of an imperfect dielectric is its dielectric strength; and the accumulation of charges within an imperfect dielectric is termed dielectric absorption.

differential pressure   The pressure difference between two systems or volumes.

diffracted wave   A *wave* whose front has been changed in direction by an obstacle or other nonhomogeneity in a medium, other than by *reflection* or *refraction.*

diffraction   The process by which the direction of *radiation* is changed so that it spreads into the geometric shadow region of an opaque or refractive object that lies in a radiation field.

Diffraction is an optical edge effect.

At present, reference to *Huygen principle* is a common means of explaining diffraction. Analysis of the interference between individual Huygen wavelets which originate in the vicinity of the edge of an irradiated body reveals that detectable amounts of radiant *energy* must invade the nominal shadow zone of the object and there, by interference, set up characteristic energy distributions known as diffraction patterns. The amount of diffractive bending experience by a ray is a function of *wavelength;* thus dispersion occurs, although dispersion

is in the opposite sense to that produced by *refraction.*

diffraction-limited   limited in resolution by the *wavelength* of light.

diffuse background   The *microwave radiation* that pervades all of *space* and is believed to be a remnant of the initial *big-bang* which marked the birth of the universe.

diffuse radiation   Radiant energy propagating in many different directions through a given small volume of space; to be contrasted with *parallel radiation.*

diffuse reflection   Any reflection process in which the reflected *radiation* is sent out in many directions usually bearing no simple relationship to the *angle of incidence;* the opposite of *specular reflection.*

diffuse reflector   Any surface which reflects incident rays in a multiplicity of directions, either because of irregularities in the surface or because the material is optically unhomogeneous, as a paint, although optically smooth; the opposite of a *specular reflector.*

diffusion   1. In general, the movement of *atoms* or *molecules* of one *fluid* into another fluid.

2. In an atmosphere, or in any gaseous system, the exchange of fluid parcels between regions, in apparently random motions of a scale too small to be treated by the *equations of motion.*

3. In materials, the movement of *atoms* of one material into the *crystal* lattice of an adjoining material, e.g., penetration of the atoms in a *ceramic* coating into the lattice of the protected metal.

4. In *ion engines,* the migration of neutral atoms through a porous structure incident to *ionization* at the emitting surface.

diffusion bonding   Method of joining two metals, where *temperature* and *pressure* create intermolecular bonds.

diffusion velocity   1. The relative mean molecular *velocity* of a selected *gas* undergoing *diffusion* in a gaseous atmosphere, commonly taken as a nitrogen (N_2) atmosphere.

The diffusion velocity is a molecular phenomenon and depends upon the gaseous concentration as well as upon the pressure and temperature gradients present.
2. The velocity or speed with which a turbulent diffusion process proceeds, as evidenced by the motion of individual *eddies.*

diffusivity   A measure of the rate of *diffusion* of a substance, expressed as the diffusivity coefficient K. When K is constant, the diffusion equation is

$$\frac{\delta q}{\delta t} = K \nabla^2 q$$

where q is the substance diffused; ∇^2 is the Laplacian operator; and t is time. The diffusivity has dimensions of a length times a velocity; it varies with the property diffused, and for any given property it may be considered a constant or a function of temperature, space, etc., depending on the context. In the case of molecular diffusion the length dimension is the mean free path of the molecules. By analogy, in eddy diffusion. length becomes the mixing length. The coefficient is then called the eddy diffusivity, and is in general several orders of magnitude larger than the molecular diffusivity. Also called coefficient of diffusion.

diluent   *Fluid* (often excess fuel) added to the exhaust *gas* to cool the gas below the *temperature* resulting from chemically balanced *combustion;* also, any substance added to a material to attenuate one or more properties of the material.

diplexer   A device permitting an *antenna* system to be used simultaneously or separately by two transmitters.

dip   *See:* altitude

dipole   1. A system composed of two, separated, equal electric or magnetic charges of opposite sign.
2. Dipole antenna

dipole antenna   A straight *radiator,* usually fed in the center, and producing a maximum of radiation in the place normal to its *axis.* The length specified is the overall length. Common usage in *microwave* antennas considers a dipole to be a metal radiating structure which supports a line current distribution similar to that of a thin straight wire, a half *wavelength* long, so energized that the current has two nodes, one at each of the far ends.

directional antenna   an *antenna* that radiates or receives radio signals more efficiently in some directions than in others.
   A group of antennas arranged for this purpose is called an *antenna array.*

discharge coefficient   Ration of the actual flowrate to the ideal flowrate, calculated on the basis of one-dimensional *inviscid* flow.

disconnect   Short for a quick-disconnect—a separable connector characterized by two separable halves, an interface seal and, usually, a latch-release locking mechanism; it can be separated without the use of tools in a very short time.

"Discovery"   The name given to *Space Shuttle Orbiter* vehicle (OV) 103.

discrete spectrum   A *spectrum* in which the component *wavelengths* (and wave numbers and frequencies) constitute a *discrete* sequence of values (finite or infinite in number) rather than a *continuum* of values. *See:* continuous spectrum.

discriminator   An electronic circuit which selects pulses according to their *pulse height* or *voltage.* It is used to delete extraneous radiation or background radiation, or as the basis for energy spectrum analysis.

dish   A colloquial term for a parabolic radio or radar *antenna,* whose shape is roughly that of a soup bowl.

dish antenna   A parabolic radio or radar antenna, shaped roughly like a soup bowl—therefore, the use of the word "dish."

disintegration, radioactive   Equivalent to *radioactive decay.*

dispersion   1. In *rocketry,* deviation from a prescribed *flight path;* specifically, circular dispersion.

2. A measure of the scatter of *data points* around a mean value or around a regression curve.

Usually expressed as a *standard-deviation* estimate, or as a standard error of estimate. Note that the scatter is not centered around the true value unless systematic errors are zero.

3. The process in which *radiation* is separated into its component *wavelengths.*

Dispersion results when an optical process, such as diffraction, refraction or scattering, varies according to *wavelength.*

4. (in spectroscopy) A measure of the resolving power of a *spectroscope* or spectrograph.

5. As applied to materials, a scattering of very fine particles (e.g., *ceramics*) within the body of a metallic material usually resulting in overall strengthening of the composite material.

displacement   A *vector* quantity that specifies the change of position of a body or particle usually measured from the *mean* position or position of rest. Displacement can be represented by a rotation vector or translation vector, or both.

displacement thickness   Distance by which the outer *streamlines* in fluid flow are shifted (displaced) as a result of the formation of the boundary layer.

dissociation   Separation of a compound into chemically simpler components.

divergence   1. The expansion or spreading out of a *vector* field.

2. A static instability of a *lifting surface* or of a body on a vehicle wherein the aerodynamic loads tending to deform the surface or body are greater than the elastic restoring forces.

divergence efficiency   Ratio of *thrust* calculated for the actual *nozzle* contour (potential flow) to the thrust of an ideal-flow nozzle.

See: ideal nozzle

Dobson spectrophotometer   A photoelectric *spectrophotometer* that is used in the determination of the ozone content of the *atmosphere.* The instrument compares the solar energy at two *wavelengths* in

Fig. 1   Artist's conception of an Orbiter vehicle docking with a space-based power conversion system being assembled in low Earth orbit.   (Illustration courtesy of NASA.)

the *absorption band* of ozone by permitting the two radiations to fall alternately upon a *photocell.* The stronger *radiation* is then attenuated by an optical wedge until the photoelectric system of the photometer indicates equality of incident radiation. The ratio of radiation intensity is obtained by this process and the *ozone* content of the atmosphere is computed from the ratio.

docking   In general the act of joining two or more *spacecraft* or orbiting objects. More specifically the process of sealing two manned spacecraft together in *orbit* with latches and sealing rings so that two hatches can be opened between them without losing cabin atmosphere to permit crew members to move from one spacecraft to the other.

docking module [abbr. DM]   The optional *Space Shuttle flight kit* the *Orbiter* vehicle allows to positively intercept, engage with and release other orbiting elements/vehicles equipped with similar docking mechanisms. (*See* Fig. 1.)

Previously, a special component added to the *Apollo spacecraft* so that it could be joined with the *Soyuz spacecraft.*

See: Apollo-Soyuz Project

dogleg   A directional turn made in the launch *trajectory* to produce a more favorable *orbit inclination.*

dollar   A unit of *reactivity.* One dollar is the maximum amount of reactivity in a reactor due to *delayed neutrons* alone.

dose   *See:* absorbed dose, biological dose, maximum permissible dose, threshold dose.

dose rate   The radiation dose delivered per unit time and measured, for instance, in *rems* per hour.

dosimeter   A device that measures and indicates the amount of *radiation* absorbed.

double-base propellant   A *solid rocket propellant* using two unstable compounds, such as nitrocellulose and nitroglycerin.

The unstable compounds used in a double-base propellant do not require a separate *oxidizer*.

double-entry compressor   A *centrifugal compressor* that takes in air or fluid on both sides of the *impeller*, with *vanes* on each side to accelerate the *fluid* into the *diffuser*. The double-entry compressor is not a *multistage compressor*.

doublet   *Injector* orifice pattern consisting of one or more pairs of orifices that produce converging streams.

doubling time   The time required for a *breeder reactor* to produce as much *fissionable material* as the amount usually contained in its *core* plus the amount tied up in its *fuel cycle* (fabrication, reprocessing, etc.) It is estimated as 10 to 20 years in typical reactors.

See: breeder reactor

downcomer   1. Axial feed passage from the rear of the *injector*.
2. Vertical feedline that conveys *fluid* from a higher to a lower location on a vehicle.
3. Coolant tube in which *coolant* flows in the same direction as exhaust *gas*.

downweight   Landing weight; specifically, STS *payloads* and all *items* required by specific payloads.

drag [Symbol: D]   1. A retarding *force* acting upon a body in motion through a *fluid*, parallel to the direction of motion of the body. It is a component of the total forces acting on the body. *See: aerodynamic force.*
2. Atmospheric resistance to the orbital motion of a *spacecraft*. The effect of drag is to lower the *orbit*. For *Earth* orbits above 200 kilometers, the altitude decreases very slowly; below 150 kilomters, the orbit, "decays" rapidly.

drag coefficient [Symbol: CD]   A coefficient representing the *drag* on a given *airfoil* or other aerodynamic body, or a coefficient representing a particular element of drag.

drag parachute   1. Drogue parachute.
2. Any of various types of parachutes attached to high-performance *aircraft* that can be deployed, usually during landings, to decrease speed and also, under certain flight conditions, to control and stabilize the aircraft.

drag pump   Pump whose rotor consists of a disk with many short radial blades. The flow enters radially and is carried within the *blade* passages around the disk and is discharged radially through a port.

drift   1. The lateral divergence from the prescribed *flight path* of an *aircraft*, a *rocket*, or *aerospace vehicle*, due primarily to the effect of a crosswind.
2. A slow movement in one direction of an instrument pointer or other marker.
3. A slow change in *frequency* of a radio transmitter.
4. The angular deviation of the spin axis of a *gyro* from a fixed reference in space.
5. In *semiconductors*, the movement of *carriers* in an electric field.

drift rate   The amount of *drift* per unit time.

Drift rate has many specific meanings in different fields. The type of drift rate should always be specified.

drift mobility   In a *semiconductor*, the average drift *velocity* of *carriers* per unit *electric field*. In general, the mobilities of *holes* and *electrons* are different.

drift velocity   The average *velocity* of a charged particle in a *plasma* in response to an applied *electric field*.

The motion of an individual *particle* is quite erratic as it bounces off other particles and has its direction continually changed. On the average, however, a particle will slowly work its way in the direction of the applied electric force. This velocity is usually much smaller than the random velocity of the particle between collisions.

drogue parachute   A small parachute used specifically to pull a larger parachute out of stowage; a small parachute used to slow down a descending *space capsule, spacecraft* or airplane.

See: Solid Rocket Booster

drogue recovery   A type of recovery system used for *space capsules* after initial atmospheric entry, involving the deployment of one or more small parachutes to diminish speed, to reduce aerodynamic heating and to stabilize the vehicle so that larger recovery parachutes can be safely deployed at lower altitudes without too great an opening shock.

See also: Solid Rocket Boosters

drone   A remotely controlled *aircraft*, missile or light vehicle.

dry criticality   Nuclear *reactor* criticality achieved without *coolant*. Compare *wet criticality*.

See: criticality

dry cycle   Operation of a valve or similar component without propellant or test fluid in the flow passages.

dry emplacement   A *launch emplacement* that has no provision for water *coolant* during launch. Compare *wet emplacement*.

dry-film lubricant   Material that reduces rolling or sliding friction between mating surfaces by coating one or both of the surfaces with a slippery film, often permanently bonded to the surface; also called solid lubricant.

dry weight   The weight of a *rocket* vehicle without its fuel.

This term, appropriate especially for *liquid rockets*, is sometimes considered to include the *payload*. Compare *take-off weight*.

dual-cycle reactor system   A reactor turbine system in which part of the steam fed to the *turbine* is generated directly in the *reactor* and part in a separate heat exchanger. A combination of *direct-cycle* and *indirect-cycle reactor systems*.

dual modulation   The process of modulating a single *carrier wave* or *subcarrier* by two different types of *modulation* (e.g., amplitude- and frequency-modulation), each conveying separate information.

dual-purpose reactor   A nuclear *reactor* designed to achieve two purposes, for example to produce both electricity and new fissionable material.

duct   Specifically, a tube or passage that confines and conducts a *fluid*, as a passage for the flow of air to the *compressor* of a *gas-turbine engine*, a pipe leading air to a supercharger, etc.

ducted fan
1. A fan enclosed in a *duct*.
2. *Ducted-fan engine*.

ducted-fan engine   An *aircraft* engine incorporating a *fan* or propeller enclosed in a *duct*; especially a *jet engine* in which an enclosed fan or propeller is used to ingest ambient air to augment the gases of combustion in the jetstream.

The air may be taken in at the front section of the engine and passed around the combustion engine section, or it may be taken in aft of the combustion chamber. In the former case the ducted fan may be considered a type of bypass engine.

ductile failure   Rupture of structural material after plastic deformation; also, unacceptable dimensional change without fracture.

ducting   The trapping of an *electromagnetic wave*, in a *waveguide* action, or between a layer of *atmosphere*, or between a layer of the atmosphere and the *Earth*'s surface.

duct propulsion   A means of propelling a vehicle by ducting a surrounding *fluid* through an engine, adding momentum by mechanical or thermal means, and ejecting the fluid to obtain a reactive force. Compare *rocket propulsion*.

dummy antenna   A device which has the necessary impedance characteristics of an *antenna* and the necessary power-handling capabilities, but which does not radiate or receive radio waves. Also called artificial antenna.

dump cooling   Method of reducing heat transfer by flowing the *turbine* exhaust gas down the *nozzle* coolant passages and discharging the *gas* at the exit, expansion at the exit being used to increase performance.

duplexer   A device that permits a single *antenna* system to be used for both transmitting and receiving.

Duplexer should not be confused with diplexer, a device permitting an antenna to be used simultaneously or separately by two transmitters.

dust   (In meteor terminology) finely divided solid matter, with particle sizes in general smaller than *micrometers,* as meteoric dust, meteoritic dust.

dynamic imbalance   Distribution of *rotor* mass such that the principal *inertia axis* of the rotor is rotationally misaligned with the bearing axis. Moments are generated when the rotor rotates about the bearing axis. Dynamic imbalance, also referred to as moment imbalance, requires measurement and correction in two or more planes perpendicular to the rotor axis.

dynamic load A load imposed by dynamic action, as distinguished from a static load. Specifically, with respect to *rockets, spacecraft* or aircraft, a load due to an acceleration of the vehicle imposed by maneuvering, landing, thrusting (firing rockets), etc.

dynamic model A scale model of a system, such as a *rocket, spacecraft* or airplane, that has its linear dimensions, mass and *moments of inertia* reproduced in proportion to the original.

dynamic pressure [symbol q] 1. The *pressure* exerted by a fluid, such as air, by virtue of its motion. It is equal to one-half the fluid density (ρ) times the fluid veolicty squared (V^2), that is, $(\frac{1}{2}) \rho V^2$.
2. The pressure exerted on a body by virtue of its motion through a fluid, as for example, the pressure exerted on a *rocket* or *Shuttle vehicle* as it moves through the *atmosphere*.

dynamic seal A mechanical device used to minimize leakage of *fluid* from the flowstream region of a fluid-system component when there is relative motion of the sealing interfaces.

dynamic similarity The relationship that exists between a model and its prototype when, by virtue of the similarity between their geometric dimensions and mass distributions or elastic characteristics, the motion of the model is in some respect (such as linear velocity, acceleration, vibration or flutter) similar to the motion of the prototype; also, the similarity between the fluid flows about a scale model and its prototype when the flows have the same *Reynolds number*. In *aerodynamics* dynamic similarity is used to determine the effects of fluid flow on a system, such as a *rocket* or an airplane, by observing the effect of similar flow upon a scale model.
See: dynamic model

dynamic viscosity Viscous action in a flowing fluid can be viewed as a shear stress (τ) between fluid layers. This shear stress is assumed to be proportional to the normal velocity gradient $\varphi v/\varphi z$. Where u is the fluid velocity, which is a function of the distance (z) from some surface or boundary, then the proportionality constant (μ) is

called the dynamic viscosity. See Fig. 1. *Momentum* exchange is considered to be the physical mechanism of viscosity; essentially *molecules* transport momentum from regions of high-bulk (macroscopic) *velocity* (u) to regions of low velocity (especially near the stationary surface). This transport phenomenon has the tendency to pull the slower-moving fluid layers along, and the resultant "drag" or shear stress is the quantity τ, which is simply the resistive force per unit area.

dyne [Symbol: d] That unbalanced *force* that acting for 1 second on a body of 1 gram mass produces a *velocity* change of 1 centimeter per second.

dysbarism An *aerospace medicine* term describing a condition of the human body resulting from the existence of a *pressure* differential between the total ambient pressure of dissolved and free *gases* within the body tissues, fluids and cavities.

Characteristic symptoms, other than *hypoxia*, caused by decreased barometric pressure are *bends* and abdominal gas pains at altitudes above 620 m (25,000 feet) to 9145m (30,000 feet). Increased barometric pressure as in descent from high altitude, is characterized by painful distension of the ear drums.

E

Earth The third planet in the Solar System and the only one known to contain life. It has a radius of 6378 km and one natural satellite. Earth's atmosphere is composed primarily of nitrogen and oxygen. The planet's interior is divided into three layers: the crust (to a depth of 32 km.) consisting largely of sedimentary rock on a base of igneous rock; the mantle (to a depth of 2900 km.) composed of silicate rock; and, a core with a large nickel component. Satellites have shown the planet is surrounded by a radiation zone called the Van Allen Radiation Belts. Earth's magnetic field is tear shaped as result of the solar wind, with the narrow end pointing away from the Sun.

Earth Albedo Neutrons *Neutrons* produced by *cosmic rays* interacting with the *atmosphere* that are reflected away from the *Earth.*

eccentricity [Symbol: e]   A measure of the ovalness of an *orbit.*
1. Of any *conic,* the ratio of the length of the radius *vector* through a point on the conic to the distance of the point from the directrix.
2. Of an *ellipse,* the ration of the distance between the center and focus of an ellipse to its semimajor axis.

The eccentricity e of an ellipse can be computed by the formula:

$$e = \sqrt{1 - b^2/a^2}$$

where a is semimajor axis and b is semiminor axis.
3. Of an ellipse, the distance between the center and the focus.

eccentric orbit   An *orbit* that deviates from a circle, thus forming an *ellipse.*

eclipse   1. The reduction in visibility or disappearance of a nonluminous body by passing into the shadow cast by another nonluminous body.
2. The apparent cutting off, wholly or partially, of the light from a luminous body by a dark body coming between it and the observer.

The first type of eclipse is exemplified by a lunar eclipse, the *Moon* passing through the shadow cast by the *Earth*; or by the passage of a *satellite* into the shadow cast by its *planet*; but when the satellite actually passes directly behind its planet, it may properly be termed an *occultation.*

The second type of eclipse is exemplified by a solar eclipse, caused by the Moon passing between the *Sun* and the Earth. If the relative positions and distances are such that at a point on the Earth the Sun is completely obscured, the eclipse is total; if the distances are such that, when in line with the Sun, the Moon is surrounded by a ring of light, the eclipse is annular; and when the

Moon passes to one side of a straight line, it is a partial eclipse.

ecliptic   The apparent annual path of the *Sun* among the stars; the intersection of the plane of the *Earth's orbit* with the *celestial sphere.*

The ecliptic is a great circle of the *celestial sphere* incline at an angle of about 23° 27′ to the *celestial equator.*

ecliptic plane   The plane of the *orbit* of the *Earth* in its annual motion around the *Sun.*

ecosphere   1. Biosphere
2. A volume of space surrounding the *Sun,* extending from the *orbit* of *Venus* past the orbit of *Mars,* in which some exobiologists believe conditions are favorable for the development and maintenance of life.

eddy   In a *fluid,* any circulation drawing its *energy* from flow of much larger scale and brought about by *pressure* irregularities.

eddy velocity   The difference between the mean *velocity* of fluid flow and the instantaneous velocity at a point. For example,

$$u_e = u - u$$

where u_e is the eddy velocity; u is instantaneous velocity; and u is mean velocity. Also called fluctuation velocity.

eddy viscosity   The turbulent transfer of *momentum* by eddies giving rise to an internal *fluid* friction, in a manner analogous to the action of molecular *viscosity* in laminar flow, but taking place on a much larger scale.

E-D nozzle   Short term for expansion-deflection *nozzle,* which has an annular throat that discharges exhaust *gas* with a radial outward component.

effective area   Actual area acted on that results in a *force* in any device; for example, the area of a bellows joint at the main diameter of the convolutions: internal *pressure* exerted over this area creates axial or end thrust (pressure separating force) tending to elongate the bellows.

effective heat ablation   Figure of merit for a given material subjected to steady-state heating conditions and undergoing steady-state ablation; the quality represents the heat dissipated per unit mass of ablated material under specified conditions.
See: ablation.

effector Any device used to maneuver a *rocket* in flight, such as an aerodynamic surface, a *gimbaled* motor or a jet.

efficiency Ratio of *energy output* to energy *input*.

ejection capsule 1. In an *aircraft* or manned *spacecraft*, a detachable compartment serving as a cockpit or cabin, which may be ejected as a unit and parachuted to the ground.
2. A *satellite, probe* or unmanned spacecraft, a box-like unit, usually containing recording instruments or records of observed data, which may be ejected and returned to *Earth* by a parachute or other deceleration device.
Except for its orbital flight tests, the *Space Shuttle* will not carry ejection seats.

elastic collision A collision between two *particles* in which no change occurs in the *internal energy* of the particles, or in the sum of their *kinetic energy*. Commonly referred to as a billiard-ball collision.

elasticizer An elastic substance or *fuel* used in *solid rocket propellant* to prevent cracking of the *propellant grain* and to bind it to the *combustion-chamber* case.

elastic limit Maximum stress that can be applied to a body without producing permanent deformation.

elastomer Polymeric material that at room *temperature* can be stretched to approximately twice its original length and on release return quickly to its original length.

electrical decay The time period from initial electrical signal to first motion of the activated part.

electric discharge The flow of electricity through a *gas*, resulting in the emission of *radiation* that is characteristic of the gas and of the intensity of the current. Also called discharge, gaseous electric discharge, gaseous discharge.

electric field A region in which a charged *particle* would experience an electrical force; the geometric array of the imaginary electric lines of force that exist in relation to points of opposite charge. An electric field is a *vector* field in which magnitude of the vector is the electric-field strength and the vector is parallel to the lines of force.

electric-field strength The electrical force exerted on a unit positive charge at a given point in space. Electric-field strength is expressed, in the practical system of *electrical* units, in terms of volts/centimeter, or volts/meter. It is a *vector* quantity, being the magnitude of the electric-field vector. Also called electric field, electric intensity, electric field intensity, electric potential gradient, field strength.
The electric-field strength of the *atmosphere* is commonly referred to as the atmospheric electric field.

electric potential In *electrostatics*, the work done in moving unit positive charge from infinity to the point specified. Sometimes shortened to potential.

electrode 1. A terminal at which electricity passes from one medium into another. The positive electrode is called *anode*; the negative electrode is called *cathode*.
2. In a *semiconductor* device, an element that performs one or more of the functions of emitting or collecting electrons or holes, or of controlling their movements by an electric field.
3. In electron tubes, a conducting *element* that performs one or more of the functions of emitting, collecting or controlling, by an electromagnetic field, the movements of *electrons* or *ions*.
See: anode; cathode.

electrodynamics The science dealing with the *forces* and *energy* transformations of electric currents, and the *magnetic fields* associated with them.

electroexplosive device [EED] Contrivance in which electrically insulated terminals are in contact with, or adjacent to, a composition that reacts chemically when the required electrical energy level is discharged through the terminals.

electroluminescence Emission of *light* caused by an application of electric fields to solids or gases. In gas electroluminescence, light is emitted when the *kinetic energy* of *electron* or *ions* accelerated in an electric field is transferred to the *atoms* or *molecules* of the gas in which the discharge takes place.

electromagnetic Having both magnetic and electric properties.

electromagnetism 1. *Magnetism* produced by an electric current.
2. The science dealing with the physical

relations between electricity and magnetism.

electromotive force [abbr. emf]   The sum (algebraic) of the potential differences acting in an electrical circuit. The emf is measured by the energy liberated when a unit electric charge passes completely around the circuit in the same direction as the resultant emf.

electron [Symbol: e]   An elementary particle with a unit negative electrical charge and a mass 1/1837 that of the *proton*. Electrons surround the positively charged *nucleus* and determine the chemical properties of the *atom*. Positive electrons, or *positrons*, also exist. Compare this term with *antimatter*. *See:* pair production.

electron avalanche   The process in which a relatively small number of free *electrons* in a *gas* that is subjected to a strong electric field accelerate, *ionize* gas *atoms* by *collision*, and thus form new free electrons to undergo the same process in cumulative fashion.

electron-beam welding   The process in which a controlled stream of high-velocity *electrons* produces the heat for fusion (note: this is *not* thermonuclear fusion) by striking the workpiece in a vacuum.

electron capture [Symbol: EC]   A mode of *radioactive decay* of a *nuclide* in which an orbital *electron* is captured by and merges with the *nucleus*, thus forming a new nuclide with the *mass number* unchanged but the *atomic number* decreased by one. *See:* K-capture

electron gun   A device that produces an *electron* beam. This is generally a narrow beam of high-velocity electrons, the intensity of which is controlled by the electrodes in the gun. The beam created by the electron gun is an essential component of many electronic instruments, such as *cathode ray tubes*.

electron tube   A device in which conduction by *electrons* takes place through a *vacuum* or gaseous medium within a gastight envelope.

electron volt   The amount of *kinetic energy* gained by an *electron* when it is accelerated through an electric potential of 1 volt. It is equivalent to 1.603×10^{-12} erg. It is a unit of energy, or work, not voltage.

electrophoresis   The separation of cells in a

liquid by an electric field or a *voltage*.

elementary particles   The simplest particles of *matter* and *radiation*. Most are short-lived and do not exist under normal conditions (exceptions are *electrons, neutrons, protons,* and *neutrinos*). Originally this term was applied to any particle that could not be subdivided, or to constituents of atoms; now it is applied to nucleons (protons and neutrons), electrons, *mesons muons, strange particles*, and the *antiparticles* of each of these, and to *photons*, but not to *alpha particles* or *deuterons*. Also called fundamental particles.

elevon   A horizontal aerodynamic control surface that combines the functions of an elevator and an aileron. It provides flight control during the *Orbiter*'s atmospheric flight. *See:* Orbiter structure

ellipse   A smooth, oval curve accurately fitted by the *orbit* of a *satellite* around a much larger mass. Specifically, a plane curve constituting the locus of all points the sum of whose distances from two fixed points called focuses or foci is constant; an elongated circle.

The orbits of *planets*, satellites, planetoids, and *comets* are ellipses, the primary being at one focus. *See:* conic section

ellipsoid   A surface whose plane sections (cross sections) are all *ellipses* or circles, or the solid enclosed by such a surface.

ellipticity [symbol: e]   The amount by which a *spheroid* differs from a sphere or an *ellipse* differs from a circle, calculated by dividing the difference in the length of the axes by the length of the major *axis*. Also called compression.

emission   With respect to *electromagnetic radiation*, the process by which a body emits electromagnetic radiation as a consequence of its temperature only. With respect to *electric propulsion* and energy conversion, the sending out of charged *particles* from a surface causing the generation of these particles; e.g., emission of *ions* from an ionizing surface in *ion engines*. *See:* electric propulsion.

emission line   A small band of *wavelengths* emitted by a low-density *gas* when it glows. The pattern of several emission lines is characteristic of the gas and is the same as the *absorption lines* absorbed by that gas from the light passing through it.

emission spectrum   The array of *wavelengths* and relative intensities of *electromagnetic radiation* emitted by a given radiator. Each radiating substance has a unique, characteristic emission spectrum, just as every medium of transmission has its individual *absorption spectrum*.

emissivity [Symbol: e or ϵ]   The ratio of the emittance of a given surface at a specified *wavelength* and emitting *temperature* to the emittance of an ideal *black body* at the same *wavelength* and temperature. The greatest value that an emissivity may have is unity, the least value zero. It is a corollary of *Kirchhoff's law* that the emissivity of any surface at a specified temperature and wavelength is equal to the *absorptivity* of that surface at the same temperature and wavelength. The spectral emissivity is for a definite wavelength, while the total emissivity is for all wavelengths.

emittance [Symbol: ϵ]   1. The radiant *flux* per unit area emitted by a body.
2. The ratio of the emitted radiant flux per unit area of a sample to that of a *black body* radiator at the same *temperature* and under the same conditions.

Spectral emittance refers to emittance measured at a specified *wavelength*.

emulsion   In photography, a light-sensitive coating on a film, plate, or paper. Some emulsions are specially manufactured to detect tracks of *ions*.

encounter   a close *flyby* or *rendezvous* of a *spacecraft* with a *target body*.

endurance limit   Maximum stress at which a material presumably can endure an infinite number of stress reversals (cycles).

end wall   Surface of the housing and *rotor* hub between adjacent blades on a *pump* rotor.

"energetic" propellants   1. Liquid *propellants* that contribute *energy* to the exhaust gas through exothermal decomposition prior to oxidation reactions.
2. Solid propellant with added ingredients that raise energy output above the norm for that class of propellant.

energy   The capacity for doing work. *Nuclear energy* is released by nuclear reactions.

Kinetic energy is the energy of motion of a mass and is $\frac{1}{2}mv^2$, where m is the mass and v is the velocity. Energy is also radiated in the form of *photons* moving at *velocity* c. Each photon has energy $E = hf = hc/\lambda$, where h is the Planck constant, f is the frequency, c is the velocity and λ is the *wavelength* of light or *X-rays*. The energy arriving per second from a distant light source on a unit area is the *intensity I*. Photon energy is measured in *electronvolts*. X-rays have photon energies from about 100 electronvolts (soft) to 50 kilo-electronvolts (hard). High-energy *astrophysics* is the study of high-energy photons.

energy conversion efficiency   In *rocketry*, the efficiency with which a *nozzle* converts the *energy* of the working substance into *kinetic energy*, expressed as the ratio of the kinetic energy of the jet leaving the nozzle to the kinetic energy of a hypothetical ideal jet leaving an *ideal nozzle* using the same working substance at the same initial state and under the same conditions of *velocity* and expansion.

energy (flux) density   The amount of (radiant) energy per unit volume.

energy level   Any one of different values of *energy* which a *particle, atom* or *molecule* may adopt under conditions where the possible values are restricted by *quantum theory* conditions. During transitions from one energy level to another, quanta of radiant energy are emitted or absorbed, their frequency depending on the difference between the energy levels.

energy management   In *rocketry* the monitoring of the expenditure of *fuel* for flight control and navigation.

energy release system   Portion of a solid rocket igniter that provides the heat necessary to ignite the *propellant* and raise it to a selfsustaining combustion level.

engine   A machine or thermodynamic device that converts *energy*, especially *thermal energy*, into useful *work*.
See: rocket engine

engine bell   The conical *thrust chamber* enclosure, or *nozzle*, of a *rocket engine*. It is located behind the *combustion chamber*.

engine mount   A structure used for attaching an *engine* to a vehicle.
See: Space Shuttle Main Engines, Orbiter

structure

engine spray The part of a *launch pad deluge system* that cools a *rocket*'s engine (or engines) during launch.

enhancement A term found in *robotics* describing increased functional utility of human beings through mechanical/electronic means.

enriched material Material in which the percentage of a given *isotope* present in a material has been artificially increased, so that it is higher than the percentage of that isotope naturally found in the material. For example, enriched *uranium* contains more of the fissionable isotope uranium-235 than the naturally occurring percentage (0.7%). *See:* isotopic enrichment

"Enterprise" The name given to *Space Shuttle Orbiter* vehicle (OV) 101.

See: Space Shuttle, Space Shuttle names

entropy [Symbol: S, s] 1. A measure of the extent to which the *energy* of a system is unavailable. A mathematically defined thermodynamic function of state, the increase in which gives a measure of the *energy* of a system which has ceased to be available for work during a certain process:

$$ds = \frac{dq}{T} \text{ rev}$$

where s is specific entropy, T is absolute *temperature,* and q is heat per unit of mass, for reversible processes.
2. In communication theory, average information content.
3. The statistical *thermodynamics* definition of entropy, based on the work of Gibbs, is

$$S = -k \, \Sigma \text{ pi ln pi}$$

where k is the Boltzmann constant, pi is the probability of the ith *quantum* state.
See: 2nd and 3rd Law of Thermodynamics

entry Region of the *thrust chamber* where the contour of the chamber converges to the *nozzle* throat.

envelope External boundary defining the limits on the dimensions of the component, subsystem, or system.

environmental chamber A chamber in which *humidity, temperature, pressure,* fluid contents, noise and movement may be

controlled so as to simulate different *environments.*

epithermal neutron An intermediate neutron.

epithermal reactor An intermediate reactor.

equal-percentage characteristic The relation between valve flow and the valving element stroke in which a percentage change in opening at any stroke increment will cause an equal percentage change in flow.

equation of state An equation relating *temperature, pressure* and *volume* of a system in thermodynamic equilibrium.

A large number of such equations have been devised to apply equally to gaseous and liquid phases throughout a wide range of temperatures and pressures. Of these, the simplest are the ideal or *perfect gas law.*

equations of motion A set of equations that give information regarding the motion of a body or of a point in space as a function of time when initial position and initial velocity are known.
See: Newton laws of motion

equator The *primary great circle* of a sphere or spheroid, such as the *Earth*, perpendicular to the *polar axis*, or a line resembling or approximating such a circle.

equatorial orbit A *satellite* or *spacecraft orbit* that is in the plane of the *Earth*'s equator.

equatorial satellite A *satellite* whose *orbit* plane coincides, or almost coincides, with the *Earth*'s equatorial plane.

equilibrium composition Chemical composition that the exhaust *gas* would attain if given sufficient time for reactants to achieve chemical balance.

equivalent airspeed Indicated, or measured, airspeed corrected for position error (*angle of incidence*) and air compressibility.

ERBS Earth Radiation Budget Satellite.

erg The unit of *energy* or *work* in the *centimeter-gram-second* unit system. An erg is the work performed by a *force* of one *dyne* acting through a distance of one centimeter. It is described by the following formula:

$$1 \text{ erg} = 10^{-7} \text{ joule}$$

erosive burning The increased burning of *solid propellant* that results from combustion products moving at high *velocity* over the burning surface.

Eulerian coordinates Any system of coordinates in which the properties of a fluid are assigned to points in space at each given time, without attempted to identify individual fluid packets or parcels from one time to the next. Compare this term with *Lagrangian coordinates.*

See: equations of motion

Eulerian equations Any of the fundamental equations of *hydrodynamics* expressed in *Eulerian coordinates.*

European Space Agency [abbr. ESA] An international body, composed of 11 member nations, (Belgium, Denmark, France, Germany, Ireland, Italy, Netherlands, Spain, Sweden, Switzerland and the United Kingdom) designed to give Europe the independent capability of exploring space and using it for a wide range of peaceful purposes: communications, industrialization, etc. The agency, through its various subdivisions, designs and tests spacecraft

(which are built by European industry), monitors the missions and disseminates collected information to the various member states.

Since 1968 the ESA has successfully launched 15 scientific missions. Satellites in the ESRO, HEOS, COS-B, GEOS, ISEE and IUE classes have studied the polar ionosphere, auroral phenomena, solar particles, gamma rays, cosmic rays, solar x-rays, the magnetosphere and solar winds. Satellites in the OTS class have been launched for communications and those in the METEOSAT class for meterologic study.

Spacelab, a manned reusable space laboratory, has been developed by the European Space Agency for flight on board the United States' *Space Shuttle.* [See Fig. 1] It is the most important joint ESA/NASA program and one that promises to truly open up the use of space to all mankind. [See: Spacelab]

The *Ariane* launcher is intended to give

Fig. 1   An artist's concept of the shuttle/Spacelab configuration. Spacelab, developed by ESA, will permit scientists to conduct their research on orbit.   (Illustration courtesy of NASA.)

Fig. 2   GEOS-2—Europe's first geostationary scientific satellite.   (Drawing courtesy of the European Space Agency.)

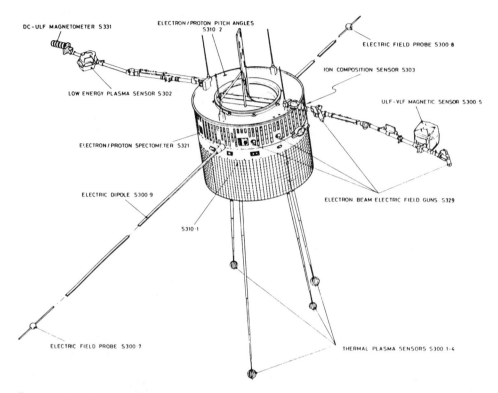

DC-ULF MAGNETOMETER S331

ELECTRON/PROTON PITCH ANGLES
S310 2

ELECTRIC FIELD PROBE S300 8

ION COMPOSITION SENSOR S303

LOW ENERGY PLASMA SENSOR S302

ULF-VLF MAGNETIC SENSOR S300 5

ELECTRON/PROTON SPECTOMETER S321

ELECTRIC DIPOLE S300 9

ELECTRON BEAM ELECTRIC FIELD GUNS S329

S310 1

ELECTRIC FIELD PROBE S300 7

THERMAL PLASMA SENSORS S300 1-4

Europe a launch capability for its own applications and scientific *satellites*. Ariane is designed to place satellites of up to 970 kg (2139 lb$_m$) in *geosynchronous orbit*.

eutectic   A mix of two materials that has a lower melting point than either material alone. When the molten mix solidifies, one material freezes in a regular pattern throughout the other.

evaporative freezing   Freezing that can occur when a *liquid* leaks into a hard vacuum and expands to pressures below the *triple point* of the liquid.

excess reactivity   More reactivity than that needed to achieve *criticality*. Excess reactivity is built into a *nuclear reactor* (by using extra fuel) in order to compensate for fuel *burnup* and the accumulation of *fission-product poisons* during operation.
See: reactivity

excitation   1. An external *force,* or other input, applied to a system that causes the system to respond in some way. Also called *stimulus.*

2. The increase in the *internal energy* of an atomic or molecular system caused by a *collision* with another particle of greater *gravity.*

excited atom   An *atom* with one or more of its bound *electrons* in an increased *energy* level.

excited state   The state of a *molecule, atom, electron* or *nucleus* when it possesses more than its normal *energy*. Excess nuclear energy is often released as a *gamma ray*. Excess molecular energy may appear as *fluorescence* or heat. Compare this term with *ground state.*

exhaust plume   Hot gas ejected from the *thrust chamber* of a *rocket* engine; the plume expands as the vehicle ascends, this exposing the engine and vehicle to greater radiative area.

exhaust plume blowback   A condition in which *ambient pressure* drives the *nozzle* exhaust gases forward, over the motor and its attachments.

exhaust stream   The stream of gaseous,

atomic, or radiant *particles* that emit from the *nozzle* of a *rocket* or other reaction engine.

exhaust velocity   The *velocity* of gaseous or other *particles (exhaust stream)* that exhaust through the *nozzle* or a *reaction engine,* relative to the nozzle.

exit   The aft end of the divergent portion of the *nozzle,* the plane at which the exhaust gases leave the nozzle; also called exit plane.

exit pressure   Pressure of the exhaust gas at the *nozzle exit.*

exobiology   The field of biology that involves the study of *extraterrestrial* environments for living organisms, the recognition of evidence for the possible existence of life in these environments and the study of any nonterrestrial life forms that may be found. The challenges of exobiology are being approached from several different directions. First, materials from other *celestial bodies* can be directly sampled—as was accomplished during the U.S. *Apollo Program* and its manned lunar expeditions (1969–1972) and the U.S. *Viking mission* to *Mars* (1976). *Lunar rocks and soil samples* have not revealed any traces of life, and the results involving the robotic sampling of Martian soils are still unclear. In particular Viking has given us some chemical information about the Martian soil, but we still do not know enough about its nature to predict what reactions will occur when water and nutrients are added to the soil in Viking's biological laboratory. Even if the Martian soil is completely lifeless, it is possible that some reactions with the added water and nutrients are just imitating biological activity. Because of these uncertainties scientists are being cautious in their current interpretations of the Viking lander's biological experiments data.

A second approach involves using *terrestrial* laboratories to conduct experiments that attempt either to simulate the primeval conditions that led to the formation of life on *Earth* (and then extrapolate such results to other planetary environments) or to study (terrestrial) living matter under conditions that simulate other planetary atmospheres.

A third approach is an attempt to communicate with (or at least detect signals from) other intelligent life forms within our *Galaxy* (i.e., the *Milky Way Galaxy*). This effort is called "SETI" or the *"search for extraterrestrial intelligence."* At present, the principal aim of SETI activities throughout the world is to listen for evidence of intelligent generation of extraterrestrial *microwave* signals. Current interpretations of stellar formation processes leads many scientists to believe *planets* are normal and frequent companions of most *stars.* Considering that the Milky Way Galaxy contains some 100 billion to 200 billion stars, present theories on the origin and evolution of life indicate that it is probably not unique to Earth but may in fact be widespread throughout the Galaxy. Scientists also speculate that life elsewhere might have evolved to the point of exhibiting intelligence, curiosity and the ability to build the tools required for interstellar transmission and reception of (intelligent) signals.

To date, however, none of the above approaches has yielded any distinctly positive evidence that life (simple or intelligent) exists beyond our own terrestrial *ecosphere.* Nonetheless, in light of the tremendous impact that such a discovery would have on man and his concept of himself in the Universe, exobiology is a truly fascinating field of study—well worth continued support and participation.

exothermic (reaction)   A chemical or physical reaction in which thermal energy (heat) is released.

exotic fuel   Any *fuel* considered to be unusual, as for example, a boron-base fuel.

expandable space structure   A structure which can be packaged in a small volume for launch and then erected to its full size and shape outside the *Earth's atmosphere.*

expansion geometry   Contour of the *nozzle* from throat to exit plane.

expansion wave   A simple *wave* or progressive disturbance in the *isentropic flow* of a compressible fluid, such that the *pressure* and *density* of a fluid particle decrease on crossing the wave in the direction of its motion. Also called rarefaction wave.

experiment racks   Removable and reusable assemblies in the *Spacelab module* that provide structural mounting and connec-

tions to supporting subsystems (power, *thermal control*, data management, etc.) and *experiment* equipment.
See: Spacelab

experiment segment   The section of the pressurized *Spacelab module* that houses experiments and sensors.
See: Spacelab

exploding bridge-wire [abbr. EBW]   Initiator consisting of a small wire (1 to 4 mils) running between two terminals and exploded by application of high *voltage.*

Explorer I   America's first successful satellite, launched on January 31, 1958 by a Jupiter-C rocket. It measured three phenomena—cosmic ray and radiation levels (data that led to the discovery of the Earth's radiation belts), the temperature in the vehicle, and the frequency of collisions with micrometeorites. There was no provision for data storage, and therefore the satellite transmitted its information continually.

Explorer I—the first U.S. satellite discovered the first of two radiation belts around the Earth. (Illustration courtesy of NASA.)

explosion turbine   A *turbine* rotated by gases from an intermittent combustion process taking place in a constant-volume chamber.

explosive-actuated devices   All major *Space Shuttle Solid Rocket Booster* (SRB) functions (except steering), from *launch pad* release through Solid Rocket Booster/ *External Tank* separation to the recovery of SRB segments, depend on the use of explosive-actuated devices (i.e., on electrically initiated *pyrotechnics*).
See: Solid Rocket Booster

explosive bolt   A bolt incorporating an explosive which can be detonated on command, thus destroying the bolt. Explosive bolts are used, for example, in separating a *satellite* from a *launch vehicle.*

explosive decompression   A very rapid reduction of air pressure inside an *aircraft*

or *spacecraft* cabin, coming to a new static condition of balance with the external pressure.

explosive valve   Valve having a small explosive charge that when detonated provides high-pressure *gas* to change valve position (also known as a squib valve).

expulsion efficiency   The index of the ability of an expulsion device to expel the liquid from a tank: the ratio of expelled volume to loaded volume.

external expansion   Expansion of the exhaust gases from the *nozzle* throat directly without a controlled-expansion wall.

External Tank (ET)   A tank that contains the *propellants* for the three *Space Shuttle Main Engines* and forms the structural backbone of the *Shuttle flight system* in the launch configuration. At *lift-off* the External Tank absorbs the total 28,580-kilonewton (6,425,000-lb$_f$) *thrust* loads of the three main engines and the two *Solid Rocket Boosters* (SRBs). When these Solid Rocket Boosters separate at an altitude of approximately 44 km (27 miles), the *Orbiter,* with the main engines still burning, carries the External Tank piggyback to near *orbital velocity,* approximately 113 km (70 miles) above the Earth. There, 8.5 minutes into the *mission,* the now nearly empty tank separates and falls in a preplanned *trajectory* into the Indian Ocean. The External Tank is the only major expendable element of the *Space Shuttle.* (See Fig. 1.)

The three main components of the External Tank are an oxygen tank (located in the forward position), a hydrogen tank located in the aft position) and a collar-like intertank, which connects the two propellant tanks, houses instrumentation and processing equipment and provides the attachment structure for the forward end of the Solid Rocket Boosters.

The hydrogen tank is 2.5 times larger than the oxygen tank but weighs only one-third as much when filled to capacity. The reason for the difference in weight is that the *liquid oxygen* is 16 times heavier than the *liquid hydrogen.*

The skin of the External Tank is covered with a *thermal protection system,* which is a nominal 2.54-cm (1-in.)

thick coating of spray-on polyisocyanurate form. The purpose of the thermal protection system is to maintain the propellants at an acceptable temperature, to protect the skin surface from *aerodynamic heat* and to minimize ice formation.

The External Tank includes a propellant feed system to convey the propellants to the Orbiter main engines, a pressurization and vent system to regulate the tank pressure, an environmental conditioning system to regulate the *temperature* and render the *atmosphere* in the intertank area inert, and an electrical system to distribute power and instrumentation signals and provide lightning protection. Most of the fluid control components (except for vent valves) are located in the Orbiter to minimize throwaway costs (i.e., the costs for materials that must be expended, or discarded, during a *flight*).

The structure of the External Tank is designed to accommodate complex load effects and pressures from the propellants as well as those from the two Solid Rocket Boosters and the Orbiter. Primarily constructed of aluminum alloys, the tank contains 917.6 m (3,010.5 linear feet) of weld. The basic structure is made of 2024, 2219 and 7075 aluminum alloys, and the thickness ranges from 0.175 cm to 5.23 cm (0.069 in. to 22.06 in.).

The liquid oxygen tank (see Fig. 3) contains 541,482 liters (143,060 gallons) of *oxidizer* at 90° K (−183°C or −297°F). It is 16.3 m (53.5 ft) long and 8.4 m (27.5 ft) in diameter. The empty weight is 5,695 kg (12,555 lb$_m$); loaded with propellant it weighs 622,188 kg (1,371,697 lb$_m$). The liquid oxygen tank is an assembly of preformed fusion-welded aluminum alloy segments that are machined or chemically milled. It is composed of gores, panels, machined fittings and ring chords. Because the oxygen tank is the forwardmost component of the External Tank and also of the Space Shuttle vehicle, its nose section curves to an *ogive* (or pointed arch shape) to reduce aerodynamic drag. A short cylindrical section joins the ogive-shaped section to the aft *ellipsoidal* dome section. A ring frame at the juncture of the dome and cylindrical section contains an integral

flange for joining the liquid oxygen tank to the intertank.

The major assemblies that comprise the liquid oxygen tank are the nose cap and cover plate, the ogive nose section, the cylindrical barrel section, the slosh *baffles* and the aft dome.

The conical nose cap that forms the tip of the liquid oxygen tank is removable and serves as an aerodynamic *fairing* (i.e., a protective covering that reduces drag) for the propulsion and electrical system components. The cap contains a cast aluminum lighting rod that provides protection for the Shuttle launch vehicle. The cover plate serves as a removable pressure *bulkhead* and provides a mounting location for propulsion system components.

The ogive nose section is fusion-welded and consists of a forward ring, 8 forward gores and 12 aft gores. It connects to the cylindrical barrel section, which is fabricated from four chemically milled panels.

Slosh baffles are installed horizontally in the liquid oxygen tank to prevent the sloshing of oxidizer. The baffles, which minimize liquid residuals and provide *damping* of fluid motion, consist of eight rings tied together with longitudinal stringers and tension straps. Slosh baffles are required only in the liquid oxygen tank, because liquid

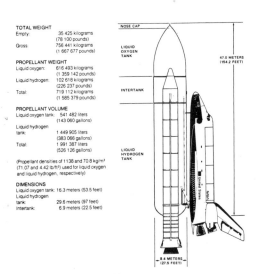

External Tank

TOTAL WEIGHT
Empty: 35 425 kilograms (78 100 pounds)
Gross: 756 441 kilograms (1 667 677 pounds)

PROPELLANT WEIGHT
Liquid oxygen: 616 493 kilograms (1 359 142 pounds)
Liquid hydrogen: 102 618 kilograms (226 237 pounds)
Total: 719 112 kilograms (1 585 379 pounds)

PROPELLANT VOLUME
Liquid oxygen tank: 541 482 liters (143 060 gallons)
Liquid hydrogen tank: 1 449 905 liters (383 066 gallons)
Total: 1 991 387 liters (526 126 gallons)

(Propellant densities of 1138 and 70.8 kg/m³ (71.07 and 4.42 lb/ft³) used for liquid oxygen and liquid hydrogen, respectively)

DIMENSIONS
Liquid oxygen tank: 16.3 meters (53.5 feet)
Liquid hydrogen tank: 29.6 meters (97 feet)
Intertank: 6.9 meters (22.5 feet)

Fig. 1   Space Shuttle External Tank data summary.   (Drawing courtesy of NASA.)

oxygen, which is 12 percent heavier than water—1,137 kg/m^3 (71 lb$_m$/ft^3) compared to 1,025 kg/m^3 (64 lb$_m$/ft^3)—could slosh and throw the vehicle out of control. The density of liquid hydrogen is low enough that baffles are not required.

Anti-vortex baffles are installed in both propellant tanks to prevent gas from entering the engines. The baffles minimize the rotaing action as the propellents flow out the bottom of the tanks. Without them the propellants would create a vortex similar to a whirlpool in a bathtub drain. The slosh baffles form a circular cage assembly; the anti-vortex baffles look more like fan blades.

The dome section of the liquid oxygen tank consists of a ring frame, 12 identical gore segments and a dome end cap 355.6 cm (140 in.) in diameter. The end cap contains a propellant feed outlet, an electrical connector and a 91.4-cm (36-in.) manhole for access to the tank.

The liquid hydrogen tank (see Fig. 4) is the largest component of the External Tank. Its primary functions are to hold 1,449,905 liters (383,066 gallons) of liquid hydrogen at a temperature of 20° K (−253°C or −423°F) and to provide a mounting platform for the Orbiter and the Solid Rocket Boosters. The aluminum alloy structure is 29.9 m (97 ft) long and 8.4 m (27.5 ft) in diameter and is composed of a series of barrel sections, ellipsoidal domes and ring frames. The weight when empty is 14,402 kg (31,750 lb$_m$); when loaded, it weighs 1,107,020 kg (257,987 lb$_m$).

The liquid hydrogen tank is a fusion-welded assembly of four barrel sections, five main ring frames and two domes. Thirteen intermediate ring frames stabilize the barrel skins, and two *longerons* are installed in the aft barrel section to receive Orbiter thrust loads. The integrally stiffened skin of the tank is designed to be non-buckling at limit load. Five major ring frames are used to join the dome and barrel sections, to receive and distribute loads and to provide connections to the other structural elements.

The intertank is not a tank in itself but serves as a mechanical connection between the liquid oxygen and liquid hydrogen tanks. The primary functions of the intertank are to provide structural continuity to the propellant tanks, to serve as a protective compartment to house instruments, and to receive and distribute loads from the Solid Rocket Boosters.

The intertank is a 6.9-m (22.5-ft) long cylinder consisting of two machined thrust panels and six stringer stiffened panels. Unlike the propellant tanks the intertank is constructed machanically without welds. On the *launch pad* the intertank links a portion of the ET instrumentation to the ground through an umbilical panel.

The use of the intertank also makes it possible for the External Tank to have separate propellant tank bulkheads (domes), avoiding the design complexity and added operational constraints associated with having a common bulkhead. The lower dome of the liquid oxygen tank extends downward into the intertank, and the upper dome of the liquid hydrogen tank extends upward into the intertank. The intertank includes a door 117 cm (46 in.) wide by 132 cm (52 in.) high for ground personnel access to the inside of the intertank, the forward manhole opening of the liquid hydrogen tank and the aft manhole opening of the liquid oxygen tank.

In addition to the propellants and the various subsystems required to feed propellant to the Orbiter, the External Tank also contains a tumble system that assures that the tank will break up upon re-entry and fall within the designated ocean impact area after separation from the Orbiter. Because the tank is expendable most fluid controls and valves for the main propulsion system operation are located in the reusable Orbiter. Thus, the attachment hardware is an integral part of the structural system.

Before oxidizer loading, the liquid oxygen and liquid hydrogen tanks are purged with gaseous helium to dry the tanks and remove residual air. Propellant is supplied to the tanks from ground storage facilities through the same feedlines that deliver fuel to the Orbiter during launch. The loading operation is controlled from the ground by the use of ET-mounted propellant-level sensors. Ten sensors are mounted in each tank to indicate the level of propellant as the tanks fill. Should a problem occur during or after loading, both tanks can be drained either simultaneously or sequentially with the vent-relief valves closed and the tanks pressurized. The propellant feed system is divided into four primary subsystems: liquid oxygen feed; liquid hydrogen feed; pressurization, vent relief and tumbling; and environmental conditioning.

At launch liquid oxygen is fed to the Orbiter engines at a rate of 72,340 kg/min (159,480 lb_m/min) or 63,588 liters/min (16,800 gal/min). The feedline is a 43.2 cm (17 in.) in diameter insulated pipe made of aluminum and corrosion-resistant steel. It consists of eight sections of flexible and straight lines and elbows running from the aft dome of the liquid oxygen tank, through the intertank, along the outside of the liquid hydrogen tank and to the base of the Orbiter through an umbilical disconnect plate. A major component of the subsytem is an anti-geyser line that runs alongside the liquid oxygen feedline and provides a circulation path to reduce accumulation of gaseous oxygen in the feedline. The buildup of a large gaseous oxygen bubble could push the liquid out into the tank, emptying the line. The subsequent refilling of the line from the tank could cause pressure in the line exceeding its design capability. This phenomenon is commonly known as a "water hammer effect." Another adjacent line provides helium, which is injected into the anti-geyser line to maintain liquid oxygen circulation.

Liquid hydrogen is fed to the Orbiter engines at 12,084 kg/min (26,640 lb_m/min) or 171,396 liters/min (45,283 gal/min) through a 43.2 cm (17 in.) in diameter feedline made of aluminum and non-corrosive steel. One section of the line is insulated and runs from the aft dome to the base of the Orbiter through the umbilical disconnect plate. The other section is located inside the liquid hydrogen tank. This uninsulated internal feedline section consists of a bellows segment and a siphon segment; the siphon is incorporated inside the liquid hydrogen tank to maximize propellant use. Also incorporated into the subsystem is a liquid hydrogen recirculation line to prevent formation of liquid air.

During propellant loading some of the liquid hydrogen and oxygen converts into gas. A dual-purpose valve for each propellant tank vents the gas and prevents excessive pressure buildup. Pressurized helium is used to pneumatically open the valves before propellant loading, approximately two hours before launch. During the terminal sequence for launch, the valves are closed so the tank can be pressurized. After closing at launch the valves act as safety-relief mechanisms to protect against tank over-pressurization. The maximum operating pressure of the liquid oxygen tanks is 152 kN/m^2 (22 psi), and the maximum operating pressure of the liquid hydrogen tanks is 234 kN/m^2 (34 psi). During standby operations the liquid oxygen and liquid hydrogen tanks are pressurized with gaseous helium to maintain a nominal positive pressure before loading and launch to avoid possible structural damage that could result from thermal and atmospheric pressure changes.

Approximately three minutes before launch the tanks are pressurized until liftoff with helium piped from a ground facility. Following engine ignition at a *T-time* of about T − 4 seconds, the *ullage* pressure is supplemented using propellant gases vaporized in the *engine heat exchangers* and routed to the two ET propellant tanks. The tank pressure is maintained based on data inputs from ullage pressure sensors in each tank to control valves in the Orbiter. A combination of ullage and propellant pressure provides the necessary net positive suction pressure to start the engines. The net positive suction pressure is the pressure needed at the main pump inlets to cause the pumps to work properly. The pumps, in turn, supply high-pressure liquid oxygen and liquid hydrogen to the *thrust chamber*. Acceleration pressure is added for operation. Fuel is forced to the engines primarily by tank pressures and, to a lesser degree, by *gravity*. Tank pressurization pushes fuel out of the tanks much like squeezing air out of a balloon.

The External Tank is *jettisoned* 10 to 15 seconds after Orbiter main engine cutoff. At separation the ET tumble system is activated. The tumble system prevents *aerody-*

namic skip during re-entry, ensuring that tank debris will fall within the preplanned disposal location in the Indian Ocean. The tumble system is initiated by the firing of a *pyrotechnic* valve upon receipt of the separation signal. This valve, located in the nose cap, releases the pressurized gas from the liquid oxygen tank, causing the ET to spin at a minimal rate of 10 degrees per second.

The environmental conditioning system is required to purge the inside of the intertank and sample gas composition within the tank during propellant loading. During loading and until launch the intertank is purged with dry gaseous nitrogen from the ground facility to render the compartment inert and to avert any buildup of hazardous gases. A ground gas detection system, using a *mass spectrometer,* samples the environment during propellant loading to detect any hazardous oxygen and/or hydrogen gas concentrations. Should such a concentration be detected, emergency actions would be initiated.

The electrical system provides propellant level and pressure sensing, instrumentation functions, electrical power distribution, tumbling initiation and lightning protection. Basically the system incorporates the five categories of instrumentation and sensors and associated cabling. All instrumentation data are recorded in the Orbiter for transmittal to *ground stations.* All ET electrical power is provided by the Orbiter.

The instrumentation is comprised of 38 sensors, which control ullage temperature, ullage pressure, liquid level, liquid hydrogen depletion and vent valve position. The tumbling system is activated before separation by signals from the Orbiter to a pyrotechnic valve inside the liquid oxyen tank. This system includes a relay to prevent inadvertent firing.

The cabling subsystem consists of cables, wiring, connectors, disconnect panels and protected wire splices between the External Tank and the Orbiter. The cabling is designed to protect wiring before and during launch. Cable trays on the outside of the tank protect the cabling and are part of the ET structure. Five cables are routed across the External Tank from the Orbiter to the Solid Rocket Boosters to handle control

signals and to monitor the condition of the boosters.

The tip of the nose cap forms a lightning rod to protect the tank during launch. Approximately 50.8 cm (20 in.) long, this rod is electrically bonded to the fairing over the gaseous oxygen line and then to the gaseous oxygen line itself. In addition conductive paint strips provide an electrical path from the rod to the gaseous oxygen line. The current is then carried to the vehicle skin, across to the Orbiter and out through the engine exhaust. The lightning protection, which is primarily required for the liquid oxygen tank, is provided by the *launch site* until lift-off. Thereafter, the aforementioned lightning rod protects the External Tank from the direct and indirect effects of lightning.

The outside of the External Tank is covered with a multilayered thermal protective coating to enable it to withstand the extreme temperature variations experienced during prelaunch, launch and early flight. The materials used are an outer spray-on polyurethane foam that covers the entire tank and an *ablating* material that provides additional protection for the portions of the tank subject to very high temperatures. Although the outer surface is covered with a thermal coating approximately 2.5 cm (1 in.) thick, the exact type of material, the thickness and the application vary at different locations on the tank.

Polyisocyanurate foam insulation is applied over the oxygen tank, the intertank and the hydrogen tank. This insulation primarily reduces the boiloff rate of the propellants; it also eliminates ice formation on the outside of the tanks due to the extremely cold propellants within.

A high-temperature ablating material (which ablates rather than chars) is applied to the ET nose cone, the aft dome of the liquid hydrogen tank, portions of the liquid hydrogen barrel and areas where projections are subject to high aerodynamic heating during flight.

During the ascent phase of a Shuttle flight, the thermal protection system maintains the primary structure and subsystem components within the design temperature limits and minimizes unusable liquid hydrogen resulting from thermal stratification. The system also aids the fragmentation process as the External Tank re-enters the atmosphere after separation from the Orbiter. It affects the structural and gas temperatures, which ensure the necessary debris size, re-entry trajectory and desired impact area.

As the largest and most central element of the Space Shuttle, the External Tank provides interconnections, or interfaces, with the Orbiter, the Solid Rocket Boosters and the ground supports. These interfaces are made through fluid and electrical umbilical links and large structural accommodations.

Four attachment points on each side of the External Tank link it with the two Solid Rocket Boosters. Each booster is connected by one forward attachment located on the side of the intertank and three aft stabilization connection points attached to the ET aft major ring frame. Adjustment provisions are located on the sides of each SRB interface to align the ET and the SRB centerlines on the same geometric plane. Pullaway electrical interfaces are located on the aft stabilization struts. The boosters contain the attachment hardware, consisting of bolts and pins, to receive the interfaces and accomplish separation. The External Tank side of these interfaces remains passive at separation.

Three structural interfaces, two aft and one forward, link the External Tank and the Orbiter. The two aft interfaces are tripods located at the ET aft major ring frame and the liquid hydrogen tank longerons. The forward interface is supported from the liquid hydrogen tank forward ring frame. Numerous pinned and spherical joints allow for multidirectional motion and minimize bending induced by thermal and structural loads.

All ET/Orbiter structural interface attachment hardware is provided by the Orbiter. Orbiter systems control the separation of the External Tank from the Orbiter vehicle following main engine cutoff. The ET/Orbiter fluid and electrical interfaces are located at two aft umbilical assemblies adjacent to the two ET aft structural interfaces. These umbilical assemblies consist of clustered disconnects

that mate with the ET fluid lines and electrical cables. Both are used to provide redundancy (i.e., backup) for ET/Orbiter and Orbiter/SRB electrical interfaces.

The External Tank and the ground fluid and pneumatic systems are linked through an umbilical connector from the intertank. This interface, called the ground umbilical carrier plate, controls vent valve actuation, helium injection into the liquid oxygen anti-geyser line, atmosphere monitoring and conditioning of the intertank cavity. It also conveys gaseous hydrogen boiloff.

The ground umbilical carrier plate is disconnected at Solid Rocket Booster ignition by explosion of a *pyrotechnic separation bolt*. A lanyard system, which is provided to back up the pyrotechnic system, pulls the bolt out on the flight side as the vehicle lifts off. The umbilical plate is made of cast aluminum and weighs approximately 60 kg (130 lb$_m$). It interfaces with the External Tank through a peripheral *Teflon* seal and rests against the intertank outer skin when mated.

External Tank (ET) thermal protection system (TPS)   The outside of the *Space Shuttle External Tank* is covered with a multilayered thermal protective coating to enable the tank to withstand the extreme *temperature* variations experienced during prelaunch, launch and early flight. The materials used are an outer spray-on polyurethane foam that covers the entire tank and an *ablating* material that provides additional protection for those portions of the ET subject to very high temperatures. This system helps maintain the *cryogenic propellants* at an acceptable temperature, protect the skin surface from *aerodynamic* heating and minimize ice formation.

extraterrestrial   From outside the *Earth*.

extraterrestrial life   Life forms evolved and existing outside the terrestrial *biosphere*. *See:* exobiology

extraterrestrial radiation   In general *solar radiation* received just outside the Earth's *atmosphere*.

extravehicular mobility unit [EMU]   The extravehicular mobility unit (EMU) comprises the space suit, the life-support subsystem, the displays and controls module, the manned maneuvering unit, and several other crew items designed for *extravehicular activity*, together with emergency life support and rescue equipment.

The extravehicular mobility concept for the *Space Shuttle* is more comprehensive than for any other space program, including the highly mobile lunar exploratory missions. Crewmembers must be prepared to exit the *spacecraft* for inspection of the *Orbiter* or *payload;* for photography; for possible manual override of Orbiter or payload systems; for installation, removal, and transfer of film cassettes on payload sensors; for operation of outside equipment; for cleaning optical surfaces; and for repair, replacement, or calibration of modular equipment either on the Orbiter or in the payload bay.

For later missions, *shuttle crewmembers* will reposition themselves and other objects from the Orbiter to locations hundreds of meters distant from the spacecraft. In the event of an emergency resulting in a disabled Orbiter, crewmembers will use the personnel rescue system to safely transport shirt-sleeved colleagues from the disabled craft to the rescue craft. The extravehicular mobility unit has been designed to accommodate a wide variety of interchangeable systems that interconnect easily and securely and that do not require two-man operation for either normal or emergency use.

In contrast to the *Apollo* lunar suit, the Space Shuttle space-suit assembly costs less and is more flexible. The Shuttle suit is not customized for the wearer; instead, it is fitted from differently sized component parts. Basic to the extravehicular suit is the liquid cooling and ventilation garment (LCVG), a mesh one-piece suit made of *spandex* and zippered for front entry. The cooling garment serves to remove metabolic heat produced by the crewmember and is connected to the primary life-support subsystem, where the heat is actually removed. It also serves to ventilate the limbs. The liquid cooling water and the return air travel through a harness to the inside front of the hard upper torso, where they connect to the life-support subsystem.

Under the cooling garment, the crewmember wears a urine collection device, which receives and stores up to 950 milliliters of urine for transfer to the *Orbiter waste management system.* [*See:* Orbiter crew accommodations] The inner clothing ensemble is completed with the addition of the in-suit drink bag, which contains 0.6 liter (21 ounces) of potable water to be used by the crewmember. The communications carrier assembly, called a "Snoopy Cap," fits over the head and chin of the crewmember and snaps in place with a chinguard. It contains headphones and a microphone for two-way communications and caution-and-warning tones.

The space suit consists of the hard upper torso, gloves, the lower torso, the helmet, and the extravehicular visor assembly that fits over the helmet. The lower torso is available in various sizes and consists of a waist ring attached to the leg/boots. The hard upper torso is made in five sizes with a matching hard waist ring. Connecting gloves are available in 15 sizes. The helmet is available in only one size. The suit is easier to don and doff than the Apollo suit because of the hard waist ring. Each component connects by hard snap-ring retainers; there are no zippers. The hard upper torso has enabled the use of bearings

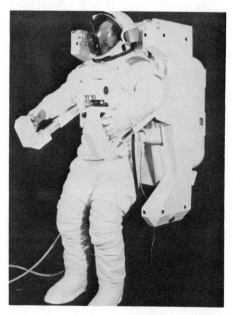

Fig. 1   The Shuttle's Manned Maneuvering Unit.   (Illustration courtesy of NASA.)

in the shoulder and arm joints, greatly facilitating the crewmember's freedom of movement in those axes. Bending, leaning, and twisting motions of the torso can all be done with relative ease.

To ensure not only mobility but also freedom of balance, the center of gravity of the suit assembly is within 10 centimeters (4 inches) vertically and 7.6 centimeters (3 inches) horizontally of that of a nude standing crewmember. The extravehicular visor assembly snaps onto the outside of the helmet and provides protection from micrometeoroids and from ultraviolet and infrared radiation from the Sun. An electrical harness connects the communications carrier assembly and the biomedical instrumentation subsystem to the hard upper torso where internal connections are routed to the extravehicular communicator. The cable routes signals from the ECG sensors attached to the crewman through the bioinstrumentation system to the extravehicular communicator as well as routing *caution-and-warning signals* and *communications* from the communicator to the crew headset.

The extravehicular communicator is a separate subassembly that attaches to the upper portion of the life-support system at the back of the hard upper torso. The controls are located on the displays and controls module mounted at the front of the upper torso. The extravehicular communicator provides radio communication between the suited crewmember and the Orbiter.

The function of the portable life-support subsystem (PLSS) is to provide a constantly refreshed atmosphere for crewman breathing and suit pressurization and to remove metabolically produced heat from the crewman through the liquid-cooling and ventilation garment. The system also provides communications for the crewman, *lightemitting diode (LED) displays,* and a caution-and-warning system for alerting the crewman of any system failure of abnormal condition with the life-support system.

The portable life-support subsystem is attached, in modular form, to the back of the hard upper torso. It includes the portable oxygen bottles; water tanks; a fan/separator/pump motor assembly; a sublimator; a contaminant control cartridge; various regulators, *valves,* and sensors; and a communications, bioinstrumentation, and microprocessor module. The secondary oxygen pack attaches to the bottom of the portable life-support subsystem.

The displays and controls module (DCM) is an integrated assembly that attaches directly to the front of the hard upper torso. The module contains a series of mechanical and electrical controls, a microprocessor, and an *alphanumeric LED* display easily seen by a crewman wearing the space suit. The function of the displays and control module is to provide the crewman control capability for the PLSS and the secondary oxygen pack and information in the form of a visible and audible status of the PLSS, the suit, and when attached, the manned maneuvering unit.

The manned maneuvering unit [See Fig. 1] is a one-man propulsive backpack that snaps onto the back of the spacesuit's portable life-support system. It allows a suited crewmember to reach many otherwise inaccessible areas outside the Orbiter. In addition, it can be used to support *payloads* by enabling a crewmember to perform inspections, servicing, adjusting, or repairing onorbit. In the event of a disabled Orbiter, an MMU-fitted crewmember would assist in the rescue of crewmen trapped in the disabled Orbiter.

The MMU contains both attach points and power outlets for equipment such as lights, cameras, and power tools; thus, the MMU serves as a portable space work station.

extreme ultraviolet [abbr. EUV] *Wavelengths* between 100 and 1000 *angstroms* (10 and 100 nanometers).

Extreme Ultraviolet Explorer [abbr. EUVE] A mission planned by NASA to obtain accurate data on the spectral energy distribution of all detectable extreme ultraviolet (EUV) sources near the Sun. These data are used to check and refine theories for the late stages of stellar evolution and to analyze the effects of EUV radiation on the interstellar medium. In its operational orbit of 550 km at 28.5° inclination, the spacecraft will spin about the Sun line, while the

The Extreme Ultraviolet Explorer   (Drawing courtesy of NASA.)

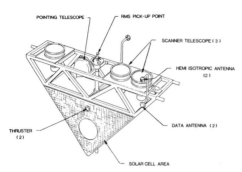

instruments take data in the Earth's shadow. Three 60-cm.-diameter grazing incidence telescopes, with different filters, are mounted perpendicular to the Sun line. One multifilter telescope is canted slightly from the antisolar direction. As the spacecraft rotates, the grazing incidence telescopes examine a great circle of the celestial sphere and the multifilter examines a small region of the celestial equator. As the Earth moves around the Sun, the area examined shifts 1°/day until the entire celestial sphere is examined.

eyeballs in, eyeballs out   Terminology used by test pilots to describe the *acceleration* experienced by the person being accelerated. Thus the acceleration experienced by an *astronaut* at a *liftoff* is "eyeballs in" (positive *g* in terms of vehicle acceleration), and the acceleration experienced when *retrorockets* fire is "eyeballs out" (negative *g* in terms of vehicle acceleration).

F

F   symbol for degree(s) Fahrenheit.

Fabry-Perot interferometer   A form of filter that operates by destructive interference on all *wavelengths* except that which it is designed to pass.
See: interferometer

fail-operational   The ability to sustain a failure and retain full operational capability for safe mission continuation.

fail-safe   The ability to sustain a failure and retain the capability to successfully terminate the mission. For *GSE,* the ability to sustain a failure without causing loss of vehicle systems or loss of personnel capability.

fail safe   Term for philosophy in the design of propulsion system hardware that seeks to avoid the compounding of failures; fail-safe design provisions ensure that the component (e.g., a valve element) will move to a predetermined "SAFE" position if electrical, pneumatic or hydraulic power is lost.

fail-safe device   A device used to minimize risk in case of a malfunction or *abort* of a system. It usually involves an automatic mechanism that makes the system cease operation whenever a failure in a critical component or subsystem is detected. One of the most common examples of a fail-safe device is the electric fuse.

fairing   A structural component of a *rocket* or *aerospace vehicle* designed to reduce drag or air resistance by smoothing out non-streamlined objects or sections.

fallaway section   A section of a *launch vehicle* or *rocket* that is cast off and separates from the vehicle during *flight,* especially a section that falls back to *Earth.*
See: External Tank, Solid Rocket Boosters

fanning beam   A radiant beam of *energy,* such as a *radar beam,* that sweeps back and forth over a limited arc.

farad (symbol F)   The SI unit of electrical *capacitance.* It is defined as the capacitance of a *capacitor* whose plates have a potential difference of one volt when charged by a quantity of electricity equal to one *coulomb.* Since the farad is too large a unit for typical applications, submultiples—such as the *microfarad* (10^{-6}F), the *nanofarad* (10^{-9}F) and the *picofarad* (10^{-12}F)—are frequently encountered.

fast breeder reactor   A *nuclear reactor* that operates with *fast neutrons* and produces more *fissionable material* than it consumes.
See: breeder reactor, fast neutron, fast reactor

fast neutron   A *neutron* with energy greater than approximately 100,000 *electron volts.* Compare this term with "intermediate neutron," "*prompt neutron*" and "*thermal* neutron."

fast reactor A *nuclear reactor* in which the *fission chain reaction* is sustained primarily by *fast neutrons* rather than by *thermal or intermediate neutrons.* Fast reactors contain little or no *moderator* to slow down the *neutrons* from the speeds at which they are ejected from fissioning *nuclei.* Compare this term with *"intermediate reactor"* and *"thermal reactor."*

fatigue 1. A weakening or deterioration of metal or other material occurring under load, especially under repeated cyclic or continued loading. Self-explanatory compounds of this term include: "fatigue crack," "fatigue failure," "fatigue load," "fatigue resistance" and "fatigue test."
2. The state of the human organism after exposure to any type of physical or psychological *stress* (as for example, pilot fatigue).

fatigue life Number of cycles of stress, under stated test condition, that can be sustained by a material prior to failure.

fatigue limit
See: fatigue strength

fatigue strength The maximum *stress* that can be sustained for a specified number of cycles without failure, the stress being completely reversed within each cycle unless otherwise stated. Also called "fatigue limit."

feed 1. The location or point at which a signal enters a device, as for example, antenna feed.
2. The act of providing a signal.

feedback 1. The return of a portion of the output of a device to the input. Positive feedback adds to the input; negative feedback subtracts from the input.
2. Information concerning results or progress returned to an originating source.
3. In a aeronautics the transmittal to the cockpit controls of *forces* initiated by *aerodynamic* action on control surfaces.

feed materials Refined *uranium* or *thorium* metal or their pure compounds in a form suitable for use in *nuclear reactor fuel* elements or as feed for uranium enrichment processes.

ferry flight An (in the atmosphere) flight of a *Space Shuttle Orbiter* mated on top of the Boeing 747 *Shuttle carrier aircraft.*
See: Approach and Landing Test(s)

fertile material A material. not itself *fis-*sionable by *thermal neutrons,* that can be converted into a *fissile material* by irradiation in a *nuclear reactor.* There are two basic fertile materials: uranium-238 ($^{238}_{92}$U) and thorium-232 ($^{232}_{90}$Th). When these fertile materials capture *neutrons,* they are partially converted into fissile plutonium-239 ($^{239}_{94}$Pu) and uranium-233 ($^{233}_{92}$U), respectively. Compare this term with "fissile materials."

fiber optics A bundle of glass fibers that transports light from one point to another very much as a copper wire transmits electricity. This bundle of fibers can be bent, wound or turned to permit an otherwise inaccessible object to be viewed. Fiber optic tubes were developed for viewing inaccessible space vehicle hardware and events, such as fuel tank domes, *rocket stage* separation and engine burn.

filler Substance added to another material (usually an *elastomer* or plastic) to improve material properties, alter one or more specific properties (e.g., change hardness) or extend or dilute the material.

film cooling The cooling of a body or surface, such as the inner surface of a *rocket combustion chamber,* by maintaining a thin *fluid* layer over the affected area. Compare this term with *"ablation."*

filter 1. A thin slab of selective material placed in front of a detector that lets through only a selected color, group of *wavelengths* or group of *photon* energies.
2. A device that removes suspended particulate matter in a liquid or gas. The removal is achieved by forcing the *fluid* (including suspended solid materials) through a material that retains the solid matter while permitting the passage of the fluid.
3. In electrical applications a network that has been designed to freely transmit currents or signals with frequencies that occur within one or more bands of frequencies (called "pass bands") and to attenuate all other current or signal frequencies.

fin 1. In aeronautics a fixed or adjustable *airfoil* attached longitudinally to an *aircraft, rocket* or similar body to provide a stabilizing effect.
2. In heat transfer a projecting flat plate or structure that facilitates thermal *energy* transfer, such as a cooling fin.

fire   1. In *aerospace* technology a term meaning to ignite a *rocket engine*.
2. Also meaning in aerospace technology to *launch* a rocket *vehicle*.

fissile material   While sometimes used as a synonym for *fissionable material,* this term has also acquired a more restricted meaning, namely, any material fissionable (i.e., capable of undergoing *fission*) by *neutrons* of all energies, including (and especially) *thermal* (slow) neutrons as well as *fast* neutrons. Uranium-235 ($^{235}_{92}$U), uranium-233

fission   The splitting of a heavy *nucleus* into two approximately equal parts (that are nuclei of lighter elements), accompanied by the release of a relatively large amount of *energy* and generally one or more *neutrons.* Fission can occur spontaneously but usually is caused by nuclear absorption of neutrons, *gamma rays* or other particles. Compare this term with *"fusion."*
See: chain reaction, nuclear reaction

fissionable material   A term commonly used as a synonym for *fissile material.* The meaning has also been extended to include material that can be fissioned by *fast neutrons* only, as for example, uranium-238 ($^{238}_{92}$U). The expression is frequently used in *nuclear reactor* technology to mean *fuel.* Compare this term with *"fertile material"* and *"fissile material."*

fission fragments   The two nuclei that are formed by the *fission* of an atomic *nucleus.* Also referred to as primary fission products. They are of medium *atomic weight* and are *radioactive.*
See: fission products

fission-product poisoning   The absorption or capture of *neutrons* by *fission products* in a *nuclear reactor,* decreasing its *reactivity.*
See: poison

fission products   The *nuclei (fission fragments)* formed by the fission of heavy elements, plus the *nuclides* formed by the fission fragments' *radioactive decay.*

fission yield   The amount of *energy* released by *fission* in a *thermonuclear* (fusion) explosion as distinct from that released by *fusion.*
Also, the amount (percentage) of a given *nuclide* produced by fission. Compare this term with *"yield."*

fixed-area exhaust nozzle   On a *jet engine* an exhaust *nozzle* exit that remains constant in area. Compare this term with *"variable-area exhaust nozzle."*

flame bucket   A deep, cavelife construction built beneath a *launch pad.* It is open at the top to receive the hot gases from the *rocket* positioned above it and is also open on one or three sides below, with a thick metal fourth side bent toward the open sides so as to the deflect the exhausting gases.
See: flame deflector

flame deflector   1. In a vertical *launch,* any of the obstructions of various designs that intercept the hot exhaust gases of the *rocket engine(s),* thereby deflecting them away from a structure or the ground. The flame deflector may be a relatively small device fixed to the top surface of the *launch pad,* or it may be a heavily constructed piece of metal mounted as a side and bottom of a *flame bucket.* In the latter case the deflector may be perforated with numerous holes connected with a source of water, bending at an angle of approximately 45° into the line of exhaust stream. During *thrust* buildup and the beginning of the launch, a deluge of water pours from the holes in such a deflector to keep it from melting.
2. In a captive *(hot) test* of a rocket engine, an elbow in the exhaust conduit or flame bucket that deflects the flame into the open.

flame spreading interval   Time period from first ignition of the *solid propellant grain* to the ignition of the entire grain surface.

flashback   A reversal of flame in a system, counter to the usual flow of the combustible mixture.

flash burn A skin burn caused by a flash of *thermal radiation*. It can be distinguished from a flame burn by the fact that it occurs on unshielded parts of the body that are in direct line-of-sight with the thermal radiation source.

See: ionizing radiation, thermal burn

flash evaporator A form of flash boiler in which water is evaporated as it is pumped through by a feed pump.

See: Orbiter environmental control and life support system (ECLSS).

flashpoint The temperature at which a substance (e.g., *fuel* or oil) will give off a vapor that will flash or burn momentarily when ignited.

Fleet Satellite Communications (FLTSATCOM) System The Fleet Satellite Communications (FLTSATCOM) system is designed to provide an operational near-global satellite communications network to support high priority communications requirements of the United States Navy and Air Force. The Naval Electronics Systems Command manages the overall FLTSATCOM program, while the *Air Force Space Division* (AFSD) directs acquisition of the program's space segment. Satellites in the FLTSATCOM system allow communications between naval aircraft, ships, submarines, ground stations, Strategic Air Command, and the presiden-

Fig. 1   Orbiting the Earth from its geosynchronous outpost in space a Fleet Satellite Communications System spacecraft relays voice, teletype, and computer data between military commanders and ships at sea. (artist concept)   (Illustration courtesy of the U.S. Air Force.)

tial command networks. Each of the five FLTSATCOM satellites procured by Space Division will be placed in *geostationary orbits* 35,807 km (22,250 miles) above the earth's *equator*. (See Fig. 1)

Each satellite has 23 communication channels in the ultra-high (UHF) and super-high frequency (SHF) bands. Ten of the channels are used exclusively by the Navy for communications among its land, sea, and air forces around the world, while 12 other channels aboard each *spacecraft* are used by the Air Force as part of its *AFSATCOM system* for command and control of nuclear capable forces. A single 500 kHz channel on the satellite is allotted to the National Command Authorities.

The satellites are three-axis stabilized in geosynchronous orbits 35,807 km (22,250 miles) above the Earth's equator. Body-fixed antennas point constantly at the Earth, while the satellites' solar arrays track the sun line. Ground-commanded station keeping by the *Air Force Satellite Control Facility* maintains the spacecraft within 1 degree of their operational longitude. Hydrazine jets allow for changing the satellites' operating locations in space.

flexible hose Pliant conduit consisting of a flexible inner core of convoluted metal or plastic tubing and an outer braided wire sleeve that is attached to fixed ends to prevent buckling and separation when the core is *pressurized*.

flexible joint (flexible section) Nonrigid connector such as metal *bellows, flexible hose,* or *ball-boint* assembly that joins two *duct* sections and permits relative motion between the ducts in one or more planes; includes both the flexible member and the restraint linkage.

flight control room [FCR] A facility used by the *flight control team* for direct support of a *Space Shuttle* flight from prelaunch countdown through landing rollout. In function this facility corresponds to the mission operations control room of previous U.S. manned space programs.

flight control team [FCT] Personnel in the *Mission Control Center* on duty to provide support for the duration of each *Space Shuttle* flight. The team is composed of the dedicated flight control room teams and the multipurpose support room personnel.

flight crew   Any personnel on the *Space Shuttle* engaged in flying the *Orbiter* and/or managing resources on board (e.g., *commander, pilot, mission specialist*).
See: Shuttle crew

flight data file [FDF]   The onboard file of crew activity plans, procedures, reference material and test data available to the crew for carrying out the flight. There will normally be a *Space Transportation System* flight data file for Shuttle crew activities and a *payload* flight data file for payload crew activities.

flight deck (FD)   A portion of the crew station module of the *Space Shuttle Orbiter*.
See: Orbiter structure

flight-dependent training   Preparation of a *mission* or *payload specialist* (or payload specialists, depending on the mission goals) for a specific flight. Part of the training involves integrated simulations with the rest of the *Shuttle flight crew* and ground terms.
See: mission specialist, payload specialist

flight design   The analysis of trajectory, consumables, attitude and pointing, and navigation necessary to support the planning of a *Space Shuttle flight*.

flight kit   Optional *hardware* (including consumables) that provides additional, special or extended services to *Shuttle payloads*. Kits are packaged in such a way that can be easily installed and removed.

flight path   The path made or followed in the air or in space by an *aircraft, launch vehicle, rocket, spacecraft*, etc.; the continuous series of positions occupied by a flying body.

flight readiness firing [FRF]   1. A test operation of a rocket consisting of the complete firing of the *liquid propellant* engines while the vehicle is restrained at the *launch pad*. This test is conducted to verify the readiness of the rocket for a flight test or operational mission.
2. In a *Space Shuttle* flight readiness firing, the Shuttle *flight vehicle* is stacked on the launch pad and a countdown demonstration test performed. This test is designed to duplicate to the fullest extent possible an actual launch countdown. Propellant load-

ing occurs in a normal launch sequence, which then culminates in a 20-second flight readiness firing. Engine shutdown after 20 seconds of sustained firing completes this *milestone.*

flight simulation   A training session in which the *Shuttle flight crew* and/or ground operations support personnel practice a portion of a flight.

flight test   1. A test to see how an *aircraft, spacecraft, launch vehicle* or *aerospace vehicle* performs in *flight.*
2. A test of a component part of a flying vehicle, or of an object carried in such a vehicle, to determine its suitability or reliability in terms of its intended function. This is accomplished by making the component or object endure actual flight conditions.

flight test vehicle   A test *vehicle* for conducting *flight tests*, involving either the vehicle's own capabilities or the capabilities of equipment needing flight test that is carried aboard the vehicle.

flip-flop   A device having two stable states and two input terminals (or types of input signals), each of which corresponds with one of the two states. The device remains in either state until caused to change to the other state upon application of an appropriate signal. This causes the circuit to "flip" into one state, remaining there until another signal on the other input terminal causes the device to "flop" into the other state. Flip-flops are used extensively in computer systems and *binary counters.*

flow   A stream or movement of *air* or other *fluid;* the rate of fluid movement in the open or in a duct, pipe or passage.

flowfield   Aerodynamic and thermodynamic states of the *gas* flow in a chamber or *nozzle.*

flow limiter   Device to control flowrate at or below a specified value over a range of *pressures,* in particular restricting flow during pressure surges.

flow separation (separated flow)   Detachment of the flow from the wall of the flow passage.

flow-to-close valve   *Valve* in which the flow direction and forces acting on the valving element provide a closing force.

flow-to-open valve   *Valve* in which the flow direction and forces acting on the valving element provide on opening force.

fluid   A substance that, when in static *equilibrium,* cannot sustain a *shear* stress. A "perfect fluid" is one that has zero *viscosity*— that is, it offers no resistance to shape change. Actual fluids only approximate this behavior. The term "fluid" is used to refer to both liquids and gases.

fluid-cooled   Term applied to a *thrust chamber* or *nozzle* whose walls are cooled by *fluid* supplied from an external source, as in *regenerative cooling, transpiration cooling,* or *film cooling.*

fluid-film bearing   Type of *bearing* wherein separation of the bearing and journal depends on the shearing of a lubricating film in the clearance between parts; *viscous forces* within the *fluid* support the bearing load.

fluid interface   Common boundary of two or more surfaces exposed to *fluid* (e.g., mating flanges of a duct) or the interface between a fluid and containing device (e.g., tube wall).

fluidized bed reactor   A *nuclear reactor* design in which the *fuel* ranges in size from small particles to pellets. Although the fuel particles are solid, their entire mass behaves like a *fluid,* because a stream of liquid or gas *coolant* keeps them in motion.
See: nuclear reactor

fluorescence   Many substances can absorb *energy* (as for example, from *X-rays, ultraviolet* light or *radioactive* particles) and immediately emit this energy as an *electromagnetic photon,* often of *visible light.* This emission is called "fluorescence." The emitting substances are said to be fluorescent. Compare this term with "*luminescence*" and "*scintillation.*"
See: excited state

flux   1. In general the rate of transport or *flow* of some quantity, frequently used in reference to the flow of some form of *energy.*
2. *Neutron* flux is a measure of the intensity of neutron radiation. It is the number of neutrons passing through one square centimeter of a given target in one second and is expressed as nv, where n equals the number of neutrons per cubic centimeter and v equals the neutron velocity in centimeters per second.

See: integrated neutron flux, intensity, neutron density

flux density   The *flux* (i.e., rate of *flow*) of any quantity, usually a form of *energy,* through a unit area of specified surface. Please note that this term, unlike *radiant density,* is not an indication of volumetric density. Compare this term with "*luminous density.*"

flyby   An interplanetary *mision* in which the *spacecraft* passes close to the target *planet* but does not impact it or go into *orbit* around it.

flying test bed   A *rocket, aircraft* or other flying *vehicle* used to carry objects or devices being flight tested.

folding optics   A mirror system used to reduce the physical length of a *telescope* or optical instrument while maintaining its optical length.

footprint   An area within which a *spacecraft* is intended to land.

forced-separation disconnect   Separable connector that is disengaged by explosive, *hydraulic,* or *pneumatic pressure.*

forced vortex flow   Flow in which the *fluid* tangential *velocity* is forced to vary in a manner other than inversely with radius of the flow passage.

forcing function   Vibration or alternating stress that imposes an oscillation on a system.

forward assembly   A component of the *Solid Rocket Booster* that consists of the nose cap, the frustum, the ordnance ring and the forward skirt.
See: Solid Rocket Booster

forward fuselage   The portion of the *Space Shuttle Orbiter structure* that contains the "cockpit," living quarters and experiment operator's station. This area houses the pressurized *crew module* and provides support for the nose section, the nose gear and the nose gear wheel wells and doors.
See: Orbiter structure

forward skirt   The component of the *forward assembly* of the *Space Shuttle Solid Rocket Booster* that houses flight *avionics,* rate gyro assemblies, range safety system panels and systems tunnel components.
See: Solid Rocket Booster

Fourier analysis   The representation of

physical or mathematical data by the use of the *Fourier series* or *Fourier integral*.

Fourier integral   The representation of a function f(x) for all values of x in terms of infinite integrals in the form

$$f(x) = \frac{1}{2\pi} \int_{-\infty}^{\infty} \int_{-\infty}^{\infty} f(t) \cos [u (t - x)] \, dt \, du$$

See: Fourier transform, Fourier series

Fourier series   The representation of a function f(x) in an interval (−L, L) by a series consisting of sines and cosines with a common period 2L, in the form,

$$f(x) = A_o + \, , \, \Sigma \, A_n \cos \frac{n\pi x}{L} + B_n \sin \frac{n\pi x}{L}$$

where the Fourier coefficients are defined as

$$A_o = \frac{1}{2L} \int_{-L}^{L} f(x) \, dx$$

$$A_n = \frac{1}{L} \int_{-L}^{L} f(x) \cos \frac{n\pi x}{L} \, dx$$

and

$$B_n = \frac{1}{L} \int_{-L}^{L} f(x) \sin \frac{n\pi x}{L} \, dx$$

When f(x) is an even function, only the cosine terms appear; when f(x) is odd, only the sine terms appear.

The conditions on f(x) guaranteeing convergence of the series are quite general, and the series may serve as a root-mean-square approximation even when it does not converge.

If the function is defined on an infinite interval and is not periodic, it is represented by the Fourier integral. By either representation, the function is decomposed into periodic components whose frequencies constitute the spectrum of the function. The Fourier series employs a discrete spectrum of wavelengths $2L/n(n = 1, 2,...)$; the Fourier integral requires a continuous spectrum.

Fourier transform   An analytical transformation of a function f(x) obtained (if it exists) by multiplying the function by e^{-iux} and integrating over all x,

$$F(u) = \int_{-\infty}^{\infty} e^{-iux} f(x) \, dx \, (-\infty < u < \infty)$$

where u is the new variable of the transform

F(u) and $i^2 = -1$. If the Fourier transform of a function is known, the function itself may be recovered by use of the inversion formula:

$$f(x) = \frac{1}{2}\pi \int_{-\infty}^{\infty} e^{iux} F(u) \, du \, (-\infty < x < \infty)$$

The Fourier transform has the same uses as the Fourier series: for example, the integrand F(u) exp (iux) is a solution of a given linear differential equation, so that the integral sum of these solutions is the most general solution of the equation.

When the variable u is complex, the Fourier transform is equivalent to the *Laplace transform*.

See: Fourier integral

four-way valve   A valve having four controlled working passages such that there are two simultaneous flow paths through the valve; commonly used to control double-acting *actuators*.

fracture toughness   Capability of a material to resist *brittle failure*.

free flight   Unconstrained or unassisted flight, as: (a) the flight of a *rocket* after consumption of its *propellant* or after motor shutoff; (b) the flight of an unguided projectile.

free flying system   Any satellite or *payload* that is detached from the *Orbiter* during operational phases and is capable of independent operation. Also called 'free flyer.'

free stream   1. Length of the jet from the orifice exit to the point of impingement on another jet or surface.

2. The central flow region in a flow passage where flow is unimpeded by any constraints.

free-stream capture area   The cross-sectional area of a column of air swallowed by a *ramjet* engine.

free-vortex flow   Flow in which the *fluid* axial *velocity* is constant from the hub to tip while the fluid tangential velocity varies inversely with the radius of the flow passage.

friability   The tendency of a crystalline structure to crumble.

front-end avionics   *Avionics* receiver-stage circuitry.

Froude number   A *nondimensional number*

defined as the ratio of inertia force to gravity force.

$$\frac{\text{Froude}}{\text{Number}} = \frac{\text{inertia force}}{\text{gravity force}} = \frac{V^2}{gl}$$

where V is the velocity, g is the acceleration of gravity, l is the characteristic length.

frozen composition Exhaust-gas chemical composition that does not change during expansion in the *nozzle.*

frustration threshold The point at which an individual feels or shows frustration over inability to achieve an objective.

frustum The component of the *forward assembly* of the *Space Shuttle Solid Rocket Booster* that houses the three main parachutes of the recovery system, the *altitude* switch and frustum location aids and the floatation devices.
See: Solid Rocket Booster

fuel 1. Any substance used to liberate *thermal energy,* either by chemical or *nuclear* reaction, as used, e.g., in a *heat engine.*

With a liquid-propellant rocket engine, fuel is ordinarily distinguished from *oxidizer* where these are separate.

2. In nuclear technology the term "fuel" applies to *fissionable* material used or usable to produce energy in a *nuclear reactor.* Also applied to a mixture, such as *natural uranium,* in which only part of the atoms are readily fissionable, if the mixture can be made to sustain a *chain reaction.*
See: fissionable material

fuel binder Continuous phase that contributes the principal structural condition to *solid propellant* but does not contain any oxidizing element, either in solution or chemically bonded.

fuel bleed valve (FBV) A component of the *Space Shuttle Main Engine.*
See: Space Shuttle Main Engine

fuel cell A device that converts chemical *energy* directly into electrical energy by reacting continuously supplied chemicals. The reaction is between a fuel, such as hydrogen, and an oxidant, such as oxygen (air). It is noncombustible reaction promoted by the use of a *catalyst.* (See Fig. 1.) The reaction is the exact opposite of the electrolysis of water. The fuel cell is made up of several individual cells stacked together. The number of cells determines

the voltage, just as the number of battery cells determines a battery's voltage. The concept of using an electrochemical process to generate and store electrical energy in flashlight batteries is familiar to most people. In a flashlight battery the chemical energy is stored inside the cell, whereas in a fuel cell the chemical energy is stored outside the cell. Fuel cells will produce electricity as long as fuel and oxidant are supplied.

The idea of deriving electrical energy directly from oxidizable fuels has attracted attention for many years because this method enables electricity to be produced without *heat engine* or mechanical devices. It is possible to achieve very high conversion efficiencies using such fuel cells. Some key advantages of fuel cells in electric energy production are: (1) low noise and low pollution emissions, (2) reliability and long life due to the absence of moving parts and (3) high energy conversion efficiency.
See: Orbiter power generation system

fuel consumption The using of *fuel* by an *engine* or power plant; the rate of this consumption, measured e.g., in kg/sec or kg/min.

fuel cycle The series of steps involved in supplying *fuel* for *nuclear power reactors.* It includes mining, milling, the original fabrication of fuel elements, their use in a *reactor,* chemical processing to recover the *fissionable* material remaining in the *spent fuel,* reenrichment of the fuel material, and refabrication into new fuel elements.

fuel element A rod, tube, plate, or other mechanical shape or form into which nuclear *fuel* is fabricated for use in a *nuclear reactor.* (Not to be confused with chemical element.)

fuel reprocessing The processing of *nuclear reactor* fuel to recover the unused *fissionable material.*

fuel shutoff The action of shutting off the flow of liquid *fuel* into a *combustion chamber* or of stopping the combustion of a solid fuel; the event or time marking this action. Compare this term with *cutoff.*

fully ionized plasma The state of a *plasma* where all the neutral *particles* have lost at least one *electron.*

function time   In an *electroexplosive device,* the time period between application of initiating energy and some later function such as bridgewire break, ignition of output charge, or start of pressure rise.

fuse   Igniting device consisting of a detonating or *deflagrating* train for propagation of ignition energy.

fusion   is the formation of a heavier *nucleus* by combining two lighter ones. The *Sun*'s *energy* and the energy of other *stars* is produced by *thermonuclear* fusion. Fusion reactions that are brought about by means of high temperatures are called "thermonuclear reactions".

fusion weapon   An *atomic weapon* using the energy of nuclear *fusion,* such as a *hydrogen bomb.*

G

g   The symbol representing the *acceleration* due to *gravity.* The acceleration of gravity at the Earth's surface (sea level) is approximately 9.8 m/sec^2 (32.2 ft/sec^2), that is, "one g." When a *spacecraft* or vehicle is accelerated during launch or decelerated during atmospheric re-entry, everything inside it experiences a *force* that may be as high as "several g's." For example, the *Space Shuttle Main Engines* can be throttled over a thrust range of 65 percent to 109 percent of the nominal thrust level, thereby limiting the Shuttle *flight* vehicle's acceleration level to a maximum of 3 g's during ascent.

gage pressure [Symbol: pg]   In engineering terminology, a term used to indicate the difference between *atmospheric pressure* and *absolute pressure.*

gain   1. A general term used to denote an increase in *signal power* in transmission from one point to another. Gain is usually expressed in *decibels* and is widely used to denote *transducer gain.*
2. An increase or amplification. In radar work there are two general uses of this term: (a) Antenna gain, or gain factor, is the ratio of the power transmitted along the beam axis to that of an *isotropic radiator* transmitting the same total power. (b) Receiver gain, or video gain, is the amplification given a signal by the receiver.

galactic   1. Pertaining to the *Milky Way* Galaxy.
2. Pertaining to the *galactic system* of *coordinates,* as galactic latitude.

galactic equator   The great circle, 90° from the galactic poles, on the *celestial sphere* that represents the path of the *Galaxy.*

galactic longitude   *Longitude* of a *star* or other object in the sky measured in degrees (like longitude on *Earth.*) The "equator" is the plane of the *Milky Way.* The starting point (O°) like the *Greenwich* meridian is the direction toward the center of the galaxy as seen from Earth.

An astronomical coordinate system using *latitude* measured north and south from the galactic equator and *longitude* meausred in the sense of increasing *right ascension* 0° to 360°.

galactic system of coordinates   An astronomical coordinate system using *latitude* measured north and south from the galactic equator and *longitude* measured in the sense of increasing *right ascension* from 0° to 360°.
the plane of the *Milky Way.* The starting point (0°) like the *Greenwich* meridian is the direction toward the center of the galaxy as seen from Earth.

An astronomical coordinate system using *latitude* measured north and south from the galactic equator and *longitude* meausred in the sense of increasing *right ascension* 0° to 360°.

galling   Progressive surface damage of mating sliding surfaces under high loads, the result being increased friction and possible seizure.

galvanic corrosion   Surface damage due to generation of an electric current resulting from the exposure of electrically connected dissimilar metals to an electrolyte; the metal that is lower on the *EMF* scale is rapidly attacked.

Gamma Ray Observatory (GRO) A future *astrophysics* mission currently being planned by *NASA*. It will be launched by the *Space Shuttle* and have a 400 km (216 n.mi) Earth-orbit at 28.5° *inclination* as its operational location. The objective of the GRO is the study of the most energetic *photons* originating in our *galaxy* and beyond. These photons provide the most direct means of studying the largest transfers of energy occurring in astrophysical processes. The 11,000kg [24,250 lb$_m$] GRO will carry five large instruments into a 400km *circular* orbit for an observation period of two years. The GRO will be launched by a Shuttle and supported by the *Tracking and Data Relay Satellite System*. The scientific instruments will include: a gamma ray spectroscopy experiment, *scintillation* spectrometer, imaging Compton telescope, gamma ray *telescope,* burst and transient source experiment.

gamma rays [Symbol: γ] High-energy, short-wavelength *electromagnetic radiation*. Gamma radiation frequently accompanies *alpha* and *beta* emissions and always

Fig. 1 A modified Saturn V rocket, topped by the Skylab space station, lifts off on 14 May 1973 from complex 39A at the Kennedy Space Center. The massive gantry structure is visible to the right of the launch vehicle. (Photograph courtesy of NASA.)

accompanies *fission*. Gamma rays are very penetrating and are best stopped or shielded against by dense materials, such as lead or depleted uranium. Gamma rays are essentially similar to *X-rays*, but are usually more energetic, and are nuclear in origin.
See: decay radioactive, excited state, photon

gantry A frame that spans over something, such as an elevated platform that runs astride a work area, supported by wheels on each side. Specifically, the term is short for *"gantry crane"* or *"gantry scaffold."* (See fig. 1.)

gantry crane A large crane mounted on a platform that usually runs back and forth on parallel tracks astride a work area. Often shortened to *"gantry."*

gantry scaffold A massive scaffolding mounted on a bridge or platform supported by a pair of towers or trestles that normally run back and forth on parallel tracks. It is used to assemble and service a large *rocket* as the rocket rests on its *launch pad*. The term is often shortened to *"gantry."*

gas The state of matter in which the *molecules* are practically unrestricted by intermolecular forces so that the molecules are free to occupy any space within an enclosure.
See: ideal gas; perfect gas

gas cap The *gas* immediately in front of a *meteoroid* or *reentry body* as it travels through the *atmosphere*; the leading portion of a *meteor*. This gas is compressed and adiabatically heated to incandescence.

gas centrifuge process A method of isotopic separation in which heavy gaseous *atoms* or *molecules* are separated from light ones by *centrifugal force*.
See: isotope separation

gas chromatograph An instrument used to identify *gases* and volatile molecular species through their residence times in *adsorption columns* of various lenghts, diameters and materials.

gas constant [Symbol: R] The constant factor in the *equation of state* for *perfect (ideal) gases*. The universal gas constant is

$$R_o = 8.314 \text{ joules/K-mol}$$

The gas constant for a particular gas, speific gas constant,

$$R = R./\,mw$$

where mw is the molecular weight of gas. *See:* Boltzmann constant.

gas-cooled reactor   A *nuclear reactor* in which a *gas* is the coolant.

gas distributor   Passive device that determines the flow pattern of the gas entering an *ullage* space.

gas generator

Assemblage of parts similar to a small *rocket engine* in which *propellant* is burned to provide hot exhaust gases to (1) drive the *turbine* in the *turbopump* assembly of a rocket vehicle, or (2) pressurize liquid propellants, or (3) provide thrust by exhausting through a nozzle.

gasket   Deformable element used to prevent leakage between two relatively static surfaces in a fluid system.

gas-metal-arc (GMA) welding   Inert-gas welding process using as a heat source an electric arc between a bare consumable filler wire and the workpiece.

gas permeativity   Capability of a gas to penetrate or diffuse through another substance.

gas scrubbing   The contacting of a gaseous mixture with a liquid for the purpose of removing gaseous contaminants or entrained liquids or solids.

gas solubility   Capability of a given *gas* to dissolve in a given *fluid* under specified conditions.

gas-tungsten-arc (GTA) welding   Inert-gas welding process wherein heat is produced by an electric arc between a nonconsumable *electrode* and the work; a filler metal is optional.

gas turbine   1. *turbine* rotated by expanding gases, as in a *turbojet engine* or in a turbo-supercharger.

2. A *gas-turbine engine.*

gas-turbine engine   An engine incorporating as its chief element a *turbine* rotated by expanding gases. In its most usual form, it consists essentially of a rotary air *compressor* with an air intake, one or more *combustion* chambers, and an exhaust outlet.

gate   1. To control passage of a signal as in the circuits of a *computer*.

2. A *circuit* having an output and inputs so designed that the output is energized only when a definite set of input conditions are met. In computers, called AND-gate.

gating   The process of selecting those portions of a *wave* that exist during one or more selected time intervals or which have magnitudes between selected limits.

Geiger-Muller counter (Geiger-Muller tube)   A *radiation* detection and measuring instrument. It consists of a gas-filled (Geiger-Muller) tube containing *electrodes*, between which there is an electrical *voltage* but no current flowing. When *ionizing radiation* passes through the tube, a short, intense pulse of current passes from the negative electrode to the positive electrode and is measured or counted. The number of pulses per second measures the intensity of radiation. It is also known as Geiger counter; it was named for Hans Geiger and W. Muller who invented it in the 1920s. *See:* counter

gel   A jellylike substance that offers little resistance to liquid *diffusion* but prevents fluid currents and small solid *particles* (*crystals*) from moving.

Gemini Project   Project Gemini (1964-1966) was the beginning of sophisticated manned space flight. It expanded and refined the scientific and technological endeavors of Mercury, adding a second crew member and a maneuverable spacecraft. New objectives included rendezvous and docking techniques with orbiting spacecraft and extravehicular walks in space. In all, 10 two-man launches occurred, successfully placing 20 astronauts in orbit and returning them safely to Earth.

generalized coordinates   Any set of coordinates specifying the state of the system under consideration. Usually employed in problems involving a finite number of *degrees of freedom*, the generalized coordinate are chosen so as to take advantage of the constraints of the system in reducing the total number of coordinates. Also called Lagrangian coordinates.

generalized transmission function   In atmospheric-radiation theory, a set of values, variable with *wavelength*, each one of which represents an average *transmission coefficient* for a small wavelength interval and for a specified optical path through the absorbing *gas* in question.

general perturbations In *orbital* determinations, a method of calculating perturbative effects by expanding and integrating in series.
See: perturbation

generation In any technological development, as of a *missile, jet engine,* or *aerospace vehicle,* a stage or period that is marked by features or performances not marked, or existent, in a previous period of development or production, as in the first generation of *rockets* using liquid *propellants.*

generation time The mean time for the *neutrons* produced by one *fission* to produce fissions again in a *chain reaction.*

genetic effects of radiation *Ionizing radiation* effects that can be transferred from parent to offspring. Any radiation-caused changes in the genetic material of sex cells. Compare this term with radiomutation, somatic effects of radiation.

GEO Geosynchronous Earth orbit.

geocentric Relative to the *Earth* as a center; measured from the center of the Earth.

geocentric diameter The diameter of a *celestial body* measured in seconds of *arc* as viewed from the *Earth*'s center.

geocentric latitude Of a position in the Earth's surface; the angle between a line to the center of the *Earth* and the plane of the equator.

Because the Earth is approximately an oblate spheroid, rather than a true sphere, this differs from geographic latitude, the maximum difference being 11.6 minutes of arc at latitude 45°.

geocentric parallax The difference in the apparent direction or position of a *celestial body* measured in seconds of *arc*, as observed from the center of the earth and a point on its surface.
See: parallax

geodesic line The shortest line on a mathematically derived surface, between two points on the surface.

A geodesic line on the spheroidal earth is called a geodetic line. Also called geodesic.

geodesy The science which deals mathematically with the size and shape of the *Earth*, and the earth's external *gravity* field, and with surveys of such precision that overall size and shape of the Earth must be taken

into consideration.

geodetic Pertaining to *geodesy*, the science that deals with the size and shape of the *Earth*.

geodetic coordinates Quantities that define the position of a point on the *spheroid* of reference with respect to the planes of the geodetic equator and of a reference meridian.

geodetic datum A datum consisting of five quantities, the *latitude* and *longitude* and elevation above the reference *spheroid* of an initial point, a line from this point, and two constants that define the reference spheroid. *Azimuth* or orientation of the line, given the longitude, is determined by astronomic observations. Alternatively, the datum may be considered as three rectangular coordinates fixing the origin of a *coordinate system* whose orientation is determined by the *fixed stars*, and the reference spheroid is an arbitrary coordinate surface of an orbiting *ellipsoidal coordinate system*.

geodetic latitude Angular distance between the plane of the *equator* and a normal to the *spheroid*. It is the *astronomical* latitude corrected for the meridional component of the deflection of the vertical.

This is the latitude used for charts. Also called geographic latitude, topographical latitude.

geodetic longitude The angle between the plane of the reference *meridian* and the plane through the polar axis and the normal to the *spheroid*. It is the *astronomical longitude* corrected for the prime vertical component of the deflection of the vertical divided by the cosine of the *latitude*.

This is the longitude used for charts. Also called geographic longitude.

geodetic position A position of a point on the surface of the *Earth* expressed in terms of *geodetic latitude* and *geodetic longitude*.

A geodetic position implies an adopted *geodetic datum*, which must be stated for a complete record of the position.

geoid The figure of the *Earth* as defined by the *geopotential surface* that most nearly coincides with mean sea level over the entire surface of the earth.

Because of variations in the direction of gravity, to which it is everywhere perpendicular, the geoid is not quite an *ellipsoid of revolution*, the sea-level surface being higher under mountainous areas.

geomagnetism   1. The *magnetic* phenomena, collectively considered, exhibited by the *Earth* and its *atmosphere* and by extension the magnetic phenomena in *interplanetary space*.

2. The study of the *magnetic field* of the Earth.

geophysics   The physics of the *Earth* and its environment, i.e., earth, air, and (by extension), space.

Classically, geophysics is concerned with the nature of physical occurrences at and below the surface of the Earth including, therefore, geology, oceanography, *geodesy*, seismology, hydrology, etc. The trend is to extend the scope of geophysics to include *meteorology, geomagnetism, astrophysics* and other sciences concerned with the physical nature of the *universe*.

geopotential   The potential energy of a unit mass relative to sea level, numerically equal to the *work* that would be done in lifting the unit mass from sea level to the height at which the mass is located; commonly expressed in terms of *geopotential height*.

The geopotential Φ at height Z is given mathematically by the expression,

$$\Phi = {}_0\!\int^Z g\, dZ$$

where g is the acceleration of gravity. Compare gravitational potential.

geoprobe   A rocket vehicle designed to explore space near the *Earth* at a distance of more than 6437 km (4000 mi) from the Earth's surface. Rocket vehicles operating lower than 6437 km are termed "*sounding rockets*."

GEOS   Geodynamic Experimental Ocean Satellite.

geosphere   The solid and liquid portions of the *Earth*; the *lithosphere* plus the *hydrosphere*.

Above the geosphere lies the *atmosphere* and at the interface between these two regions is found almost all of the *biosphere*, or zone of life.

geostrophic wind   That horizontal wind velocity for which the *coriolis acceleration* exactly balances the horizontal pressure force.

geosynchronous Earth orbit (GEO)   An *orbit* 35,900 km (19384 nm) above the *Earth*'s equator—or 42,400 km (22,894 nm) from the Earth's center—in which a *satellite* or *spacecraft* revolves about the Earth at the same rate at which the Earth rotates on its axis. Consequently, when observed, from the Earth, the satellite appears to be stationary over a point on the Earth's surface. Also called "geosynchronous orbit" and "synchronous orbit."

get   To remove gas from a vacuum system by *sorption*.

Getaway Special (GAS)   The Office of Space Transportation Operations at NASA Headquarters is responsible for establishing and conducting a program to fly small experiments, which take advantage of any Shuttle space or weight availability opportunities as they arise.

The *Space Transportation System* has been designed to provide frequent, routine access to space. The getaway special program takes advantage of the fact that on many *Shuttle flights* the primary payload will not require the full Shuttle volume/weight-into-orbit capability. As a result, small self-contained payloads can be accommodated on a volume/weight availability basis and will gnerally be flown on a first come, first served basis—using, of course, the most appropriate Shuttle flight to satisfy payload requirements.

The GAS program is intended to stimulate and encourage the use of space by researchers, educational institutions, and private individuals. This program also increases the overall awareness of Shuttle Era space technology by *non-aerospace* industries and academic institutions and will stimulate a degree of enthusiasm for space activities. The small self-contained payload program can also serve as a logical spring-board for more sophisticated space experiments and demonstrations. Properly utilized by universities, aerospace and non-aerospace industries, and future space entrepreneurs, the GAS program will help demonstrate a wide variety of activities which take advantage of space and the STS—thereby supporting, in an initial way at least, the "humanization of space."

getter   1. A material which is included in a vacuum system or device for removing gas by *sorption*.

2. To remove gas by sorption

geysering   The accumulation of gas in a line and the subsequent expulsion of liquid from the line by a gas bubble.

See: External Tank

gibbous   A term used to describe a phase of a moon or planet when more than half, but not all, of the illuminated disk can be viewed by the observer.

Gibbs function [Symbol: g]   A mathematically defined *thermodynamic* function of state, that is constant during a reversible *isobaric-isothermal* process. In symbols the specific Gibbs function g is

$$g = h - Ts$$

where h is specific enthalpy; T is Kelvin temperature; and s is specific entropy. By use of the first law of thermodynamics for reversible processes,

$$dg = s \, dT + dp$$

Also called Gibbs free energy, thermodynamic potential. Compare *Helmholtz function.*

giga [Symbol: G]   A prefix meaning multiplied by 10^9.

gigahertz 10^9 *hertz.*

gimbal   1. A device, on which an *engine* or other object may be mounted, that has two mutually perpendicular and intersecting *axes of rotation* and thus gives free angular movement in two directions.

2. In a *gyro,* a support that provides the *spin axis* with a *degree of freedom.*

3. To move a *reaction engine* about on a gimbal so as to obtain *pitching* and *yawing correction moments.*

4. To mount something on a gimbal.

gimbal bearing assembly   A component of the *Space Shuttle Main Engine* that provides the mechanical interface with the *Orbiter* vehicle for transmitting *thrust* loads and permits angulation of the actual *thrust vector* (force) about each of two vector control axes.

See: Space Shuttle Main Engine

gimbaled motor   A *rocket engine* mounted on a *gimbal.*

See: Space Shuttle Main Engines

GIOTTO Mission   Mission planned by the European Space Agency to encounter and study Halley's Comet. Scheduled for launch in early 1985 and encounter in March 1986, it is designed to study the composition of the comet's coa and dust particles; analyze the physical processes and chemical reactions that occur in the cometary atmosphere and ionosphere, measure the total gas production rate and the dust flux and size/mass distribution: investigate the

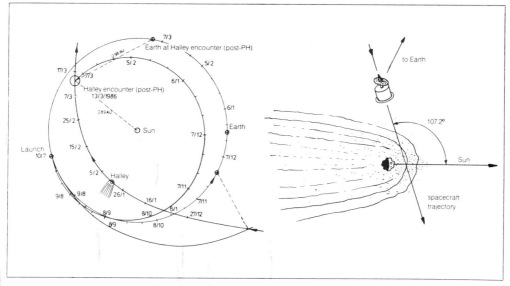

Reference flight profile of the ESA GIOTTO mission.   (Drawing courtesy of the European Space Agency.)

The GIOTTO spacecraft.   (Drawing courtesy of the European Space Agency.)

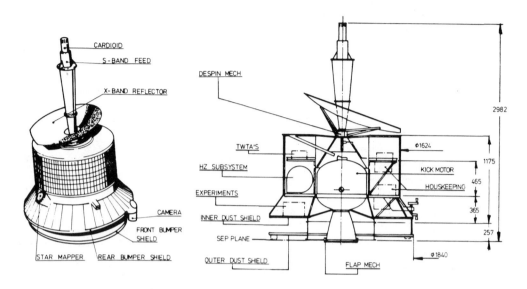

macroscopic system of plasma flow resulting from the interaction between the cometary plasma and the solar wind; and, provide images of the comet nucleus. (See Fig. 1)

gland Cavity in which an O-ring is installed; includes the groove and the surface of the mated part that together confine the O-ring.

glide 1. A controlled descent by a heavier-than-air *aeronautical* or *aerospace vehicle* under little or no engine *thrust* in which forward motion is maintained by *gravity* and vertical descent is controlled by *lift* forces.
2. A descending flight path of a glide, sense 1, as a shallow glide.
3. To descend in a glide, sense 1.

glide path 1. The *flight path* of an *aeronautical* or *aerospace vehicle* in a *glide,* seen from the side.
2. The path used by an aircraft or spacecraft in approach procedure and that is generated by an instrument-landing facility.

glider A fixed-wing aircraft specially designed to *glide,* or to glide and soar. This kind of aircraft ordinarily has no power-plant.

glide ratio The ratio of the horizontal distance traveled to the vertical distance descended in a *glide.* Also called *gliding ratio.*

glove box A sealed box in which workers, using gloves attached to and passing through openings in the box, can handle toxic or *radioactive* materials safely from the outside.

gradient The space rate of decrease of a function. Of a function in three space dimensions, the *vector* normal to surfaces of constant value of the function and directed toward decreasing values, with magnitude equal to the rate of decrease of the function in this direction. The gradient of a function f is denoted by $-\nabla f$, and is itself a function of both space and time. The ascendent is the negative of the gradient.

grain Integral piece of a molded or extruded *solid propellant* that comprises both *fuel* and *oxidizer* in a solid rocket motor and is shaped to produce, when burned, a specified performance-vs-time relation.
See: solid propellant rocket

grain anomaly Nonhomogeneity in cured propellant grain (e.g., a void, or a fuel-rich pocket).

grain conditioned temperature Uniform temperature of the propellant grain just before ignition.

Grashof number [N_{GRA}] A *nondimensional number* that is found in the study of fluid convection in the vicinity of a hot body. It is defined as

$$N_{GRA} = \frac{\text{(inertia force)} \times \text{(buoyant force)}}{\text{(viscous force)}^2}$$

$$= \frac{\rho^2 g \, l^3 \beta \, \Delta T}{\mu^2}$$

where ρ is the (mass) density, g is the gravitational acceleration, l is the characteristic length, β is the temperature coefficient of volume expansion, ΔT is the temperature difference, μ is the absolute viscosity.

gravimeter An instrument for measuring g by the extension of a spring.

gravitational potential 1. The potential associated with the force of *gravitation* arising from the attraction between mass points, e.g., the earth's center and a particle in space.

The gravitational potential, associated with the force of gravitation should not be confused with the *geo-potential* associated with the force of gravity. The latter is equal to the former plus the centrifugal force due to the earth's rotation. The potentials of the three forces are related in the same manner. 2. At any point, the work needed to remove an object from that point to infinity.

gravity well An analogy in which the *gravitational* field is considered as a deep pit out of which a *space vehicle* has to climb to escape from a *planetary* body.

gray body A hypothetical body that absorbs some constant fraction, between zero and one, of all *electromagnetic radiation* incident upon it, which fraction is the *absorptivity* and is independent of *wavelength*. As such, a gray body represents a surface of absorptive characteristics intermediate between those of a *white body* and a *black body*.

grazing incidence Light arriving at a reflecting surface from an extreme oblique angle.

great circle The intersection of a *sphere* and a plane through its center. Also called orthodrome.

The intersection of a sphere and a plane which does not pass through its center is called a *small circle.*

ground 1. The earth's surface, especially the *Earth's* land surface. Used in combination to form adjectives, as in *ground-to-air, ground-to-ground.*
See: surface
2. The domain of nonflight operations that normally take place on the earth's surface or

in a vehicle or on a platform that rests upon the surface, as in *ground support.*
3. = electrical ground.

ground-effect machine A machine that hovers or moves just above the ground by creating a cushion of supporting air between it and the ground surface and by varying the *thrust* vector and magnitude to regulate direction and rate of motion.

ground elapsed time (GET) The time since launch.

ground environment 1. The environment that surrounds and affects a system or piece of *aerospace* equipment while it operates on the ground.
2. The system or part of a system, as of a guidance system, that functions on the ground; the aggregate of equipment, conditions, facilities and personnel that go to make up a system, or part of a system, functioning on the ground.

ground-handling equipment Equipment on the ground used to move, lift or transport a *space vehicle,* a *rocket* or component parts. Such equipment includes the *gantry,* the transporter and the forklift.
See: ground support equipment

ground state The state of a *nucleus, atom* or *molecule* at its lowest (normal) energy level. Compare *excited state.*

ground support equipment (GSE) Any non-flight (i.e., ground-based) equipment used for launch, checkout or in-flight support of a space vehicle or project. More specifically, GSE consists of non-flight equipment, implements and devices that are required for the handling, servicing, inspection, testing, maintenance, alignment, adjustment, checkout, repair and overhaul of an operational *end item,* a subsystem or a component thereof. In general GSE is not considered to include land or buildings but may include equipment required to support another item of GSE.

groundtrack The path followed by a *spacecraft* over the *Earth's* surface.

ground umbilical carrier plate (GUCP) The *Space Shuttle External Tank* and the ground fluid and pneumatic systems are linked through an umbilical connector from the intertank. This interface, called the ground umbilical carrier plate, controls vent valve actuation, helium injection into the

liquid oxygen anti-geyser line, atmosphere monitoring and conditioning of the inter-tank cavity. It also carries gaseous hydrogen boiloff.
See: External Tank

group velocity   The velocity of a *wave* disturbance as a whole, i.e., of an entire group of component *simple harmonic waves.* The group velocity U is related to the phase speed c of the individual *harmonic* waves of length l by the frequency equation

$$U = c - l(dc/dl)$$

The phase speed c is thus equal to the group velocity only in the case of nondispersive waves, i.e., when dc/dl = O.

g-suit or G-suit   A suit that exerts *pressure* on the abdomen and lower parts of the body to prevent or retard the collection of blood below the chest under positive acceleration.
See: space suit

g-tolerance   A tolerance in a person or other animal, or in a piece of equipment, to an *acceleration* of a particular value.

guidance   The process of directing the movements of an aeronautical vehicle or space vehicle, with particular reference to the selection of a flight path.

In preset guidance a predetermined path is set into the guidance mechanism and not altered, in inertial guidance accelerations are measured and integrated within the craft, in command guidance the craft responds to information received from an outside source. Beam-rider guidance utilizes a beam; terrestrial-reference guidance, some influence of the *Earth;* celestial guidance, the celestial bodies and particularly the stars; and homing guidance, information from the destination. In active homing guidance the information is in response to transmissions from the craft, in semiactive homing guidance the transmissions are from a source than the aircraft, and in passive homing guidance natural radiations from the destinations are utilized. Midcourse guidance extends from the end of the launching phase to an arbitrary point enroute and terminal guidance extends from this point to the destination.
*See:*Orbiter guidance, navigation and control systems

from this point to the destination.
See: Orbiter guidance, navigation and control systems

guided missile   1. Broadly, any *missile* that is subject to, or capable of, some degree of guidance or direction after having been launched, fired or otherwise set in motion. 2. Specifically, an unmanned, self-propelled flying vehicle (such as a pilotless aircraft or rocket) carrying a destructive payload and capable of being directed or of directing itself after launching or take-off, responding either to external direction or to direction originating from devices within the missile itself.
3. Loosely, by extension, any steerable projectile.

gyro   A device that utilizes the angular momentum of a spinning mass (rotor) to sense angular motion of its base about one or two axes orthogonal (mutually perpendicular) to the spin axis. Also called "gyroscope."

gyroscope
See: gyro

gyroscopic inertia   The property of a spinning mass (rotor) of resisting any *force* that tends to change its *axis of rotation.*
See: gyro

gyroscopic moment   *Moment* induced on rotating components by the angular displacement of the rotating axis, as in a *gyroscope.*

H

habitable payload   A *payload* with a pressurized compartment suitable for supporting a crewperson in a *shirtsleeve environment.*

half-angle   Angle between the *nozzle* center line and a line parallel to the inner surface of the nozzle exit cone; also called divergence angle.

half-life   The time in which half the atoms of

a particular radioactive substance disintegrate to another nuclear form. Measured half-lives vary from millionths of a second to billions of years. The half life $t_{1/2}$ is given by

$$t_{1/2} = (\ln 2)/\lambda$$

where λ is the decay constant.
See: decay, radioactive

half-life, effective   The time for a *radionuclide* contained in a biological system, such as a man or an animal, to reduce its activity by half as a combined result of radioactive decay and biological elimination. Compare *biological half-life.*
See: half-life

half-thickness   The thickness of any given *absorber* that will reduce the intensity of a bean of *radiation* to one-half its initial value.

half-value layer   The thickness of any particular material necessary to reduce the *dose rate* of an *X-ray beam* to one-half its original value.

hammerhead crane   A heavy-duty crane with a horizontal counterbalanced jib, or protecting arm.

hangfire   A faulty condition in the ignition system of a *rocket engine.*

hard landing   An impact landing of a spacecraft on the surface of a *planet* or natural *satellite* destroying all equipment except possibly a very rugged package.

hard line   Line or *duct* that incorporates no flexible joints but is provided with deflection capability by the use of loops and elbows and low-modulus or thin-wall material.

hardness   Of *X-rays* and other *radiation* of high energy, a measure of penetrating power. For example, radiation which will penetrate a 10-centimeter thickness of lead is considered "hard radiation."
See: hard radiation

hard radiation   Radiation of high penetrating power; that is, *radiation* of high *frequency* and short *wavelength.*

A 10-centimeter thickness of lead is usually used as the criterion upon which the relative penetrating power of various types of radiation is based. Hard radiation will penetrate such a shield; soft radiation will not.

hard sealing surface   Surface fabricated of material (metal, ceramic, or cermet) that does not permanently yield or deform except with wear (flexible metal disks are a special type of sealing surface).

hard vacuum   A term applied to *pressure* less than 10^{-8} mm Hg (1.333×10^{-10} N/cm^2).

harmonic   1. An integral multiple or submultiple of a given *frequency;* a *sinusoidal* component of a periodic wave.
2. A signal having a frequency which is a harmonic (sense 1) of the fundamental frequency.

harmonic analysis   A statistical method for determining the *amplitude* and *period* of certain *harmonic* or *wave* components in a set of data with the aid of *Fourier series.*

harmonic analyzer   A device that resolves a *periodic* curve into its *harmonic* constituents.

harmonic distortion   Nonlinear *distortion* characterized by the appearance in the output of multiples of the *fundamental* when the input wave is *sinusoidal.*

harmonic function   Any solution of the *Luplace equation.*

harmonic motion   The projection of circular motion on a diameter of the circle of such motion. Simple harmonic motion is produced if the circular motion is of constant speed. The combination of two or more simple harmonic motions results in compound harmonic motion.

hatch   A door in the pressure hull of a *spacecraft.* The hatch is sealed tightly to prevent the cabin atmosphere from escaping to the outside vacuum.
See: Orbiter structure

hat section   Extruded, machined or formed metal *stringers* for stiffening *spacecraft* skin.

head or headrise   Increase in *fluid pressure* supplied by a pump; the difference between pressure at the pump inlet and pressure at pump discharge, fluid pressure being expressed as equivalent height (in feet) of a fluid column.

head coefficient   Measure of *headrise* related to *impeller* discharge tip speed.

heading   The horizontal direction in which a craft is pointed, expressed as angular distance from a reference direction, usually 0° at the reference direction clockwise through 360°.

Heading is often designated as true, magnetic, compass, or grid as the reference direction is true, magnetic, compass, or grid north, respectively.

heat   *Energy* transferred by a *thermal* process. Heat can be measured in terms of the dynamical units of energy, as the erg, joule, etc., or in terms of the amount of energy required to produce a definite thermal change in some substance, as, for example, the energy required per degree to raise the temperature of a unit mass of water at some temperature (calorie, Btu).

heat balance   1. The equilibrium that exists on the average between the *radiation* received by a *planet* and its *atmosphere* from the *Sun* and that emitted by the planet and atmosphere.

2. The equilibrium that is known to exist when all sources of heat gain and loss for given region or body are accounted for. In general this balance includes advective, evaporative (etc.) terms as well as a radiation term.

heat engine   A system that receives *energy* in the form of heat and that, in the perfomance of an energy transformation, does *work*.

See: thermodynamic efficiency,
Carnot engine

heater blanket   An electrical heater in sheet form wrapped around all or a portion of a *cryogenic* component (e.g., an *acuator*) to prevent the *temperature* within the component from falling below a stated operating minimum.

heat exchanger   1. In general a device for transferring *thermal energy* (heat) from one fluid to another without intermixing the fluids.

2. In the *Space Shuttle Main Engines,* a component that converts *liquid oxygen* to *gaseous oxygen* for vehicle oxygen tank and *pogo*-system accumulator pressurization.

See: regenerator, radiator, Space Shuttle Main Engines

heat of ablation   Total of the incident heat that an ablative material dissipates per unit mass ablated.

See: ablation

heat shield   1. Any device that protects something from heat.

2. Specifically, the protective structure necessary to protect a *reentry* body or *aerospace vehicle* from *aerodynamic heating*.

See: heat sink, Orbital thermal protection system

heat sink   1. In *thermodynamic* theory, a means by which *heat* is stored, or is dissipated or transferred from the system under consideration.

2. A place toward which the heat moves in a system.

3. A material capable of absorbing heat; a device utilizing such a material and used as a thermal protection device on a *spacecraft* reentry vehicle or *aerospace vehicle*.

4. In nuclear propulsion, any thermodynamic device, such as a *radiator* or condenser, that is designed to absorb the excess heat energy of the *working fluid*. Also called heat dump.

heat-sink chamber   *Combustion chamber* in which the heat capacity of the chamber wall limits wall temperature (effective for short-duration firing).

heat soak   The increase in temperature in rocket-engine components after firing has ceased, the result of heat transfer through contiguous parts when no active cooling exists.

Heat soakback   Thermal energy on surface or skin flowing back into the vehicle structure, usually by means of conduction heat transfer.

heat-transfer coefficient   Analytically or empirically determined *parameter* that expresses the rate of *heat transfer* per unit area per unit temperature difference between two substances.

heat treatment   Heating and cooling a solid metal or alloy in such a way as to obtain desired conditions or properties.

Heating for the sole purpose of hotworking is excluded from the meaning of this definition.

heavy hydrogen   *Deuterium.*

heavy-water-moderated reactor   A reactor that uses *heavy water* as its *moderator*. Heavy water is an excellent moderator and thus permits the use of inexpensive natural (unenriched) uranium as a fuel. Compare light water reactor.

heliocentric   Relative to the *Sun* as a center, as a heliocentric orbit.

heliocentric parallax   The difference in the apparent positions of a *celestial body* outside the solar system, as observed from the *Earth* and *Sun*. Also called stellar parallax. *See: parallax*

Helmhltz function [Symbol: a]   A mathematically defined *thermodynamic* function of state, the decrease in which during a reversible *isothermal process* is equal to the *work* done by the system. The Helmholtz function is

$$a = u - Ts$$

where u is specific internal energy; T is Kelvin temperature; and s is specific entropy. By use of the first law of thermodynamics for reversible processes,

$$da = -s \, dT - dw$$

where dw is the work done per unit mass by the system. Also called Helmholtz free energy, work function. Compare *Gibbs function.*

hermetic seal   A seal evidencing no detectable leakage or permeation of *gas* or moisture.

hertz [Symbol: Hz]   The SI unit of *frequency,* one oscillation (cycle) per second; 1 kilohertz = 1000 cycles/sec, 1 megahertz = 10^6 cycles/sec; 1 gigahertz = 10^9 cycles/sec.

Hertz stress   Maximum compressive stress due to the *pressure* between contacting load-carrying elastic bodies, at least one of which is a curved body.

heterogeneous decomposition   Separation of a substance into simpler components that differ in phase.

heterogeneous reactor   A *nuclear reactor* in which the fuel is separate from the *moderator* and is arranged in discrete bodies, such as *fuel elements.* Most reactors are heterogeneous. Compare *homogeneous reactor.*

high-cycle fatigue   Life-cycle capability determined by the elastic strain range; generally greater than 10^4 cycles.

high-pass filter   A *wave filter* having a single

transmission *band* extending from some critical or cutoff *frequency,* not zero, up to infinite frequency.

hold   1. During a *countdown:* To halt the sequence of events until an impediment has been removed so that the countdown can be resumed, as in "T minus 40 and holding."

hole   A mobile vacancy in the electronic valence structure of a *semiconductor* that acts like an *electron* with a positive charge.

homing guidance   Guidance in which a craft or *missile* is directed toward a destination by means of information received from the destination.

It is active homing guidance if the information received is in response to transmissions from the craft, semiactive homing guidance if in response to transmissions from a source other than the craft, and passive homing guidance if natural radiations from the destination are utilized.

homogeneous reactor   A *nuclear reactor* in which the *fuel* is mixed with or dissolved in the *moderator* or *coolant.* Compare *heterogeneous reactor.*

homokinetic plane   In universal joints (or gimbal joints), the plane that is normal to the plane containing the shafts and bisecting the angle between them.

hot cathode   A *cathode* that functions primarily by the process of *thermionic emission.* Also called thermionic cathode.

hot cell   A heavily shielded enclosure in which *radioactive* materials can be handled by persons using remote manipulators and viewing the materials through shielded windows or periscopes. *See: shield*

hot-core injector   *Injector* that produces a central hot-gas combustion region surrounded by a "cold" fuel sheath.

hot fire   Term applied to a test of an engine system in which the engine is started (ignited) and operated while performance of the system and its components is observed and measured; period of operation need not be full operational duration. *See: hot test*

hot firing *See: hot test*

hot-gas manifold (HGM)   The structural backbone of the *Space Shuttle Main Engine,* it interconnects and supports the *preburners, high-pressure fuel turbopumps, main combustion chamber* and *main injector.*

See: Space Shuttle Main Engine

hot-gas valve   Valve that controls the flow of hot gases, opening at low power levels but restricting the flow at mainstage; it operates at temperatures in excess of 200°F (366K) and as high as 1000°F (811K) or higher.

hot laboratory   A laboratory designed for the safe handling of *radioactive* materials, and usually containing one or more *hot cells.*

hot streaking   Stratification of burning gases in a *combustion chamber* into longitudinal zones of high-temperature gases that do not break up and mix with cooler gases. The term derives from the localized heat marks visible on the chamber wall after firing has ended.

Fig. 1   The hot test firing of "Columbia's" Space Shuttle Main Engines on 20 February 1981 as a rehearsal for its maiden flight.   (Photograph courtesy of NASA.)

hot test   A propulsion system test conducted by actually firing the *propellants* (usually for a short period of time). Also called "hot firing." In figure 1 flames shoot out from the nozzles of the *Space Shuttle "Columbia's"* three main liquid engines during the successful 20-second static firing on 20 February 1981. This hot test capped a formal rehearsal for the Columbia's maiden flight of 12 April 1981. During this test, the three

main engines reached 100 percent power. Holddown bolts secured the Shuttle vehicle to the *mobile launcher platform.* Compare this term with *cold-flow test.*

housing   The physical structure that forms the containing envelope for an assembly.

humidity   1. The amount of *water vapor* in the *air.*

2. Specifically, relative humidity.

See: absolute humidity, dew point

Huygens principle   A very general principle applying to all forms of *wave motion,* stating that every point on the instantaneous position of an advancing *phase front* (wave front) may be regarded as a source of secondary spherical wavelets. The position of the phase front a moment later is then determined as the envelope of all the secondary wavelets (ad infinitum).

This principle, stated by the Dutch physicist Christian Huygens (1629-95), is extremely useful in understanding effects due to refraction, reflection, diffraction and scattering of all types of radiation, including sonic radiation as well as *electromagnetic* radiation and applying even to ocean-wave propagation.

Huygens wavelets   The assemblage of secondary waves asserted by Huygens to be set up at each instant at all points on the advancing surface of a wave, or *phase front.*

Many phenomena of wave optics can be neatly explained on this assumption (*Huygens principle*) of the continual creation of new wavelets and the subsequent destructive or constructive interference between the wavelets to set up the next-imagined state of the advancing wave front.

hydraulic   1. Operated, moved, or effected by *liquid* used to transmit energy.

2. A system or device using liquid as the operating field.

hydraulic dashpot   A device used to reduce the velocity of the *actuator* as it approaches a fixed stop, so that impact energy levels are reduced.

hydrogen bomb   A *nuclear weapon* that derives its energy largely from *fusion.*

See: thermonuclear reaction

Hydrogen—the universal fuel of tomorrow   Hydrogen is used as the fuel for the Space Shuttle Main Engines. It also has the exciting potential of becoming the universal terrestrial fuel of tomorrow. Aerospace technology involving the safe handling, storage and application of hydrogen directly supports this future development.

Plentiful, clean, high in *energy* content, adaptable to power generation and to industrial, residential, and transportation users—this could be a description of the perfect fuel. In fact it is a description of hydrogen, the lightest and one of the most abundant chemical elements, found in water and in all the earth's organic matter. Pure hydrogen is a clean fuel: Its only combustion product when burned with oxygen is water. Even burned in air it yields almost no pollutants.

Hydrogen's energy content per cubic meter is less than one-third that of natural gas. But its energy content per kilogram, almost three times that of gasoline, is the highest of any fuel known.

When cooled to a liquid [*See:* cryogenics] so that it takes up less than 1/700 as much space, hydrogen thus becomes a natural for space propulsion, which requires high-energy, low-weight fuel. The *Space Shuttle* and the rockets propelling the *Apollo* missions to the *Moon* burn liquid hydrogen. Hydrogen's high energy content could also make it a desirable fuel for more ordinary transportation, as well as for home and industrial use, helping to shift us away from dependence on scarcer fossil fuels.

Counterbalancing hydrogen's desirable properties is this major drawback: It is extremely rare in its elemental form. Though abundant, hydrogen is almost invariably locked (bonded) into chemical compounds. Two common examples are water, which covers 70 percent of the earth's surface, and all organic matter. Releasing the hydrogen stored in these materials requires expending significant amounts of energy.

Since energy must be invested in hydrogen before energy can be gotten out—that is, before hydrogen is useful as a fuel—hydrogen is considered a means of storing energy. The heat or electrical energy required to separate it from the *elements* to which it is bonded is in effect stored in hydrogen until that fuel is burned.

Hydrogen, discovered in 1766, served as a buoyant gas in *balloons* and as an agent for extracting metals from raw materials. Late in the 19th century, people began to burn "town gas" or "manufactured gas," a half-hydrogen, half-carbon-monoxide fuel made from coal. Networks for distributing town gas are still in use in several countries, including Brazil and Germany.

Currently, several million tons of hydrogen are produced annually in the United States, primarily for use in petroleum refining and in making ammonia and methyl-alcohol, two major industrial chemicals. Most hydrogen is produced by reacting natural gas or light oil with steam at a high temperature. Small amounts of very pure hydrogen are produced by electrolysis: An electric current passing through water (H_2O) splits it into its two components, hydrogen and oxygen. This more expensive material is used in special applications like food processing that require a higher purity gas than can be inexpensively produced from natural gas or light oil.

A much bigger role is now envisioned for hydrogen as a storage medium. Energy storage has come to be recognized as vital to using energy resources wisely. Storage can make generating electricity both more efficient and more economical. It can also help to make the best use of variable solar energy. And hydrogen could be used for both of these storage applications.

The demand for electricity produced at power plants is variable—higher during the day than at night, during weekdays (when businesses are in operation) than on weekends, and, in many parts of the country, during summer (when cooling needs are up) than the rest of the year. Peak demands which threaten to exceed a power plant's generating capacity alternate with demands so low that much of that generating capacity is idle. Being able to store the excess energy that could be produced during low demand times to supplement the electrical output during periods of peak demand—a process called load-leveling—enables power plants to operate more efficiently and economically: Much more of the power they provide can be produced by their large-capacity nuclear and coal-fired equipment. Inefficient older generating equipment and petroleum- or natural-gas-fired turbines do not have to be brought on line to keep up with peak demands. And the expense of building additional generating capacity to cover those peaks might be avoided.

Electricity itself is difficult to store economically, but it can be converted into a more easily storable form, like hydrogen. A utility could produce hydrogen with excess electricity during off-peak times, store it, and reconvert it to electricity, probably in *fuel cells,* for use during peak demand times.

Practical use of *solar* energy also requires a way of storing energy for backup use when there is no sunlight. Hydrogen produced with electricity during high sunlight hours could be burned later to produce supplemental heat for solar heating systems or to drive electrical generating systems.

As an energy carrier and storage medium, hydrogen has several potential advantages over electricity and devices that convert and store electrical energy. Hydrogen can be transported long distances by pipelines instead of by expensive overhead transmission lines that require wide right-of-ways. Hydrogen can probably be stored as cheaply underground as natural gas is, providing longer term storage (from week to week and season to season) than storage devices like *flywheels* and *superconducting magnets* can. And it is relatively simple, compared to "going electric," to modify present automobiles, home furnaces, power plants, and industrial plants so that they can burn hydrogen instead of fossil fuels. Hydrogen production in effect converts energy from sources as varied as coal, the sun, and nuclear materials to a uniform, widely useful fuel.

Taking advantage of hydrogen's versatility, however, will require developing less expensive, more efficient production, storage, and distribution methods, for at present hydrogen is much more expensive than the fossil fuels it might replace. If new technology is developed and if fossil fuel prices continue to increase—and both are likely— then hydrogen may become a practical and affordable solution to part of the global energy problem.

At this time, cheaper and more efficient

hydrogen production methods are the greatest need. Since natural gas is the scarcest of the fossil fuels, its use in making hydrogen will probably give way to using abundant coal. Hydrogen is necessarily produced as an intermediate step in plants making synthetic gas or oil from coal. However, it will probably be just as easy to make those synthetic fuels as to make hydrogen and more convenient to transport and burn them in existing facilities. Making hydrogen from fossil fuels makes sense primarily for those applications requiring hydrogen's light-weight, clean-burning qualities.

To minimize the depletion of fossil fuel reserves used as both the source of hydrogen and the source of the power needed to produce it, research facilities are now working on making hydrogen from water by electrolytic and thermochemical methods that can use solar and nuclear energy.

Since water is a poor conductor of electricity, making an electrolytic cell to separate hydrogen from water requires adding an electrolyte, a substance that does conduct electricity. Essentially, an electrolytic cell is made of two electrodes suspended in an electrolyte like potassium hydroxide, a corrosive chemical. The electrodes are made of expensive, corrosion-resistant metals such as platinum or palladium. When water is pumped into the cell and electric current is passed through it, hydrogen gas collects at one electrode and oxygen at the other. An asbestos cloth separator between the two electrodes permits current to pass but prevents the two gasses from mixing.

Since generating electricity is a rather inefficient process (60-65 percent of the fuel energy consumed is lost) [*See:* thermodynamics], it is important that systems using high-quality electrical energy be very efficient users of that power. Two approaches are currently being studied to improving present electrolytic cells, which convert only about 65 percent of the electrical energy they use into hydrogen. The first approach is to improve the performance of the potasium hydroxide cell by increasing the *temperature* at which electrolysis occurs and to reduce the cost by using less expensive materials, especially for the electrodes. Since asbestos fails at about 80°C (176°F), other separator materials are also being tested as possible substitutes. A porous mat made of thin fibers of a synthetic polymer or an oxide material is one possibility. It might allow operating temperatures as high as 150°C (302°F), thus increasing conversion efficiency to 90 percent.

The second approach is to use a thin sheet of an acidic fluorocarbon plastic as the electrolyte. Tiny particles of a platinum alloy pressed onto both sides of the solid electrolyte serve as electrodes. The raw material, water, is the only liquid used. Researchers have used this technology to convert electricity to hydrogen with 90 percent efficiency in a small laboratory module.

Electricity needed to electrolyze water or hydrogen production can come from today's fossil or nuclear power plants, or it could be produced with solar energy. At one kind of solar power plant, a field of solar collectors would concentrate the sun's rays on a boiler atop a tall "power tower," creating the high temperature steam needed to generate electricity. Another type of solar power system would generate electricity directly from arrays of *photovoltaic cells.*

The *sun's* energy could also be used to produce hydrogen using a process that would shortcut the electricity-producing step. Preliminary studies of *photolysis,* the process by which light decomposes materials, suggest that *electrodes* made of semiconducting materials can absorb sunlight and split water at the electrode surface. A solar-powered "electrolyzer" based on this process would need little outside electricity.

Production of hydrogen using thermochemical cycles is a newer concept than electrolytic production and hence has not been as well studied. A thermochemical cycle is a series of linked chemical reactions that produces hydrogen and oxygen from water and heat; all other chemical compounds produced are completely recycled. Though thermochemical cycles are unlikely to be developed for large-scale use in this century, they hold special promise. Because they could use the heat from solar and nuclear plants [*See:* nuclear reactors] directly, they are potentially more efficient and cheaper

than electrolysis, which requires the energy-wasting step of converting heat into electricity. Thermochemical cycles could produce hydrogen directly from heat energy with 40-75 percent efficiency, compared to electrolysis with an overall heat→electricity→hydrogen conversion efficiency of 35-55 percent.

Many thermochemical cycles are theoretically possible, but the most efficient ones require very high temperatures, 800°C (1472°F) and above. For efficient, inexpensive hydrogen production, researchers are looking for cycles that involve only a few, rapid reactions; that use inexpensive, easily available, and noncorrosive chemicals; that require few extra processes (such as separation of a solid from a liquid); and that lose little energy in linking the reactions into a complete cycle.

Cycles involving sulfur or halides are especially promising. Researchers are presently focusing on two sulfur-based cycles. One cycle involves reacting sulfur dioxide, water, and iodine to produce sulfuric acid and hydrogen iodide, which then decompose to hydrogen and oxygen, plus sulfur dioxide and iodine. The other cycle, involving sulfurous acid, is a "hybrid" production method that couples a high-temperature thermochemical step with an electrolytic step. The electrolysis replaces a difficult thermochemical step and as a result reduces materials handling, the number of reactions, and the number of extra separation processes. Although the hybrid process requires some electricity in addition to heat, the quantities of electricity are much smaller than those necessary for electrolysis alone.

The 800°C-plus temperatures required to make thermochemical processes efficient can be supplied by *nuclear reactors* or solar concentrators. The leading reactor design for production of hydrogen is the *high-temperature gas-cooled reactor.* An experimental gas-cooled reactor has reached temperatures up to 1100°C (1922°F). Similar reactors have been used to produce electric power although not at full commercial scale. The concept of using high-temperature nuclear reactors to provide the heat for thermochemical cycles is still in its infancy and is being developed mainly in Italy, the United States, Germany, and

Japan.

Concentrated solar energy might also be coupled with thermochemical cycles to split water. The necessary high temperatures have already been achieved in demonstrations with concentrating mirrors which focus the sun's rays on a reaction vessel.

Whether hydrogen is produced by electrolytic or thermochemical methods, whether the source of energy is a coal-fired or nuclear plant, a solar power tower or a field of solar electric cells, the resulting fuel is an energy carrier much more convenient and versatile than the coal, uranium, or solar energy from which it derives.

Hydrogen's low energy content per cubic meter as a gas and the very low temperature needed to liquify it ($-267°$C, or $-423°$F, at normal pressure) cause unique storage problems. A number of solutions are in various stages of development.

NASA has developed procedures for storing liquid hydrogen under normal pressure in refrigerated metal tanks, a method also used in industrial plants, and has built a 1-million-gallon (3.785-million liter) tank, the world's largest refrigerated tank, at the *Kennedy Space Center*. However, converting hydrogen gas to a liquid and keeping it refrigerated is expensive, consuming the equivalent of 25-30 percent of hydrogen's energy value.

Even larger amounts of hydrogen could perhaps be stored as gas under high pressure in natural cavities such as depleted oil and gas fields, mines, and caverns. This high-pressure storage would require less energy than refrigerating hydrogen liquid. Scientists are presently comparing the underground storage and properties of hydrogen with those of natural gas and will later conduct field studies to assess factors like storage losses and contamination and to survey promising hydrogen storage sites.

Storing large amounts of hydrogen gas at very high pressure in metal tanks is another possibility, but it appears to be much more expensive to build tanks to withstand high pressure than to use existing natural storage sites.

A new and promising means of storing hydrogen is in hydride compounds. Certain metals and their alloys, like magnesium-nickel, magnesium-copper, and iron-titanium, can absorb hydrogen gas to form chemical compounds referred to as hydrides. When heated, these compounds release the hydrogen again. This storage method avoids having to contain large volumes of hydrogen as a gas or to maintain the special pressures and temperatures needed to store hydrogen as a compressed gas or as a liquid. Though they can store twice the hydrogen in the same space required by liquid hydrogen tanks, hydrides store relatively little energy per kilogram. Research underway to find lighter hydrides or hydrides with even larger storage capacities could lead to convenient use of hydrogen in vehicles.

Small amounts of liquid or gaseous hydrogen are now routinely transported on land or water in special storage vessels. The large amounts envisioned in the future can be transported only by pipeline or ocean tanker. Systems of pipelines have transported gas mixtures containing hydrogen for many years. However, improved equipment will be necessary to handle pure hydrogen gas at high pressures because the pipe materials commonly used become brittle under such conditions. Stainless steel, while it does not become brittle, is prohibitively expensive for long-distance pipelines. If pure hydrogen gas is to be transported in pipelines, inexpensive, embrittlement-resistant materials, or new techniques for designing pipelines with existing materials, must be developed. Adding certain impurities, or "inhibitors," to hydrogen could also eliminate embrittlement.

Visionary planners sees hydrogen coming into wider use perhaps as early as the 1980s, the timing depending largely on economics. The first new application may be as a raw material rather than as a fuel. Substituting hydrogen, produced from water or coal, in the manufacture of various chemicals now made from natural gas would stretch out dwindling natural gas supplies. Adding hydrogen to natural gas used for fuel can also stretch supplies. There appear to be no major technical problems in using mixtures containing 10-20 percent hydrogen, although at the present price of hydrogen, consumer gas bills would increase.

Using hydrogen for load-leveling in coal-fired, nuclear, or solar power plants might become a reality in this century. Power plants could use their excess electricity to produce hydrogen electrolytically during off-peak times, then reconvert the hydrogen, in fuel cells or gas-turbine-driven generators, to electricity for use during peak times. Though this electricity→hydrogen→electricity conversion is only 40-45 percent efficient with today's equipment, efficiencies of 60-65 percent would be possible with new electrolyzers and fuel cells being developed. Supplementing power plant generating capacity with such peak power systems based on hydrogen could save scarce fossil fuels.

In addition to its regular use for *rocket propulsion,* hydrogen has also been used experimentally as a fuel for airplanes, naval vessels, and motor vehicles. For use as an airplane fuel, hydrogen's light weight is an obvious plus, although the large storage space needed (four times the space jet fuel requires) would require modifications in airplane design. However, the reduced weight of the fueled aircraft would reduce fuel requirements by about one-third.

A conventional automobile engine can operate on hydrogen with relatively simple changes in the carburetor. Though they require twice the space, liquid hydrogen tanks may have an advantage over hydride storage for cars because they weigh only $1/10$ as much. However, since hydrogen is extremely flammable, techniques to prevent static sparks in hydrogen storage and handling areas need to be developed. Since hydrides release their hydrogen slowly when heated, there is little risk of explosion from damage to a hydride storage tank. Hydrides may thus become a safer and more convenient alternative for transportation. Hydride storage has already been used successfully in a city bus operating between Provo and Orem, Utah.

Technology aside, hydrogen may be slow to take over as a substitute for gasoline because the vast existing network for distribution of gasoline will have to be modified to handle hydrogen. Furthermore, several million automobile drivers and service station attendants will have to be educated in the safe use of hydrogen. Hydrogen is fundamentally no more dangerous than gasoline or natural gas, but it has different properties that require different precautions. For example, since hydrogen burns with an invisible flame, a flame colorant may be required. With the exception of nitrous oxide (NO_x), exhaust emission from hydrogen-fueled vehicles should be environmentally acceptable (no lead, sulfur, smoke, or odor, very little carbon monoxide, and few hydrocarbons).

Widespread use of hydrogen as an all-purpose fuel would be the last stage in the evolution of hydrogen energy systems. The heat produced by solar and nuclear plants could provide the energy to generate hydrogen. Instead of using that hydrogen at the power plant to generate electricity, it could be piped to urban centers, where it would fuel vehicles and provide heat for homes and for industrial processes. Electricity, which now accounts for about one-quarter of U.S energy production, would have a different role than its does now. Most of the electricity needed for specialized uses, like running motors and lights, could be generated in dispersed, hydrogen-fueled substations, eliminating the need for massive cross-country transmission lines.

If such a "hydrogen economy" becomes a reality, it will likely be well into the 21st century. Considerable new technology must be developed, and other factors can stall introduction of that technology, including concern about safety and the environment, and the large investments that would have to be made in new facilities if hydrogen is to become a common fuel.

Thus it will probably be a long time before the world runs completely on hydrogen, if it ever does. In the meantime, the hydrogen stockpiled in overwhelming quantities in every body of water could help to replace petroleum and natural gas by powering cars, providing heat for homes and industry, and serving as a chemical feedstock.

Hydrogen—rocket propulsive key to man's conquest of the *Solar System*—is also his key to terrestrial energy abundance.

Using hydrogen for load-leveling in coal-fired, nuclear, or solar power plants might become a reality in this century. Power plants could use their excess electricity to produce hydrogen electrolytically during off-peak times, then reconvert the hydrogen, in fuel cells or gas-turbine-driven generators, to electricity for use during peak times. Though this electricity→hydrogen→electricity conversion is only 40-45 percent efficient with today's equipment, efficiencies of 60-65 percent would be possible with new electrolyzers and fuel cells being developed. Supplementing power plant generating capacity with such peak power systems based on hydrogen could save scarce fossil fuels.

hydrogen embrittlement   Decrease in a metal's tensile strength, notched tensile strength, fatigue strength, resistance to crack growth, and especially ductility as a result of absorption by the metal of newly formed gaseous hydrogen.

hydroponics   The technique of growing plants without soil, in nutrient solutions.

hydrosphere   That part of the *Earth* that consists of the oceans, seas, lakes and rivers.

hydrostatic bearing   Fluid-film bearing wherein the *pressure* required to maintain separation of the surfaces is externally supplied.

hydrostatic equation   In numerical equations, the form assumed by the vertical component of the vector equation of motion when all *coriolis,* earth-curvature, frictional, and vertical-acceleration terms are considered negligible compared with those involving the vertical *pressure* force and the *force* of *gravity.* Thus,

$$dP = -\rho g dZ$$

where P is the atmospheric pressure; ρ the density; g is the acceleration of gravity; and Z is the geometric height.

For cyclonic-scale motions the error committed in applying the hydrostatic equation to the *Earth's* atmosphere is less than 0.01 percent. Strong vertical accelerations in thunderstorms and mountain waves may be 1 percent of gravity or more in extreme situations.

hydrostatic equilibrium   1. The state of a *fluid* whose surfaces of constant *pressure* and constant mass (or density) coincide and are horizontal throughout. Complete balance exists between the force of *gravity* and the *pressure* force.
See: hydrostatic equation
2. Of a rotating body, a state in which the body maintains, or returns to, the figure generated by this rotation in spite of small disturbances.

hydrostatic pressure   *Fluid pressure* due to *gravitational force.*
See: hydrostatic equation

hydrostatic seal   Seal that incorporates features that maintain an interfacial film thickness by means of *pressure* provided either by an external source or by the pressure differential across the seal.

hyperbaric   Pertaining to breathing *atmosphere pressures* above sea level normal.

hyperbarism   An aerospace medicine term describing disturbances in the body resulting from an excess of the *ambient pressure* over that within the body fluids, tissues and cavities.

hyperbola   An open curve with two branches, all points of which have a constant difference in distance from two fixed points called focuses.

hyperbolic   Of or pertaining to a *hyperbola.*

hyperbolic velicity   A velocity sufficient to allow escape from the *solar system.*

Comets unless captured by the sun have *hyperbolic* velocities and their *trajectories* are hyperbolas.

hypergolic   Self-igniting.

hypergolic fuel (hypergol)   A *rocket fuel* that ignites spontaneously when brought into contact with an oxidizing agent (*oxidizer*).

hypergolic ignition   Ignition that involves no external energy source, but results entirely from the spontaneous reaction of two materials when they are brought into contact; materials may be two liquids or a liquid impinging on a solid.

hypergolic propellants   Rocket *propellants* (bipropellants) that ignite spontaneously when mixed together.

hyperon   One of a class of short-lived *elementary particles* with a mass greater than that of a *proton* and less than that of a *deuteron.* All hyerons are unstable and yield a *nucleon* as a decay product.

hypersonic 1. Pertaining to *hypersonic flow.*
2. Pertaining to speeds of Mach 5 or greater.

hypersonic flow In *aerodynamics,* flow of a *fluid* over a body at speeds much greater than the speed of sound and in which the *shock waves* start at a finite distance from the surface of the body. Compare *supersonic flow.*

hypersonic glider An unpowered vehicle specifically an atmospheric entry vehicle, designed to fly at *hypersonic* speeds.

hypersonics That branch of *aerodynamics* that deals with *hypersonic flow.*

hypervelocity Extremely high velocity.
Applied by physicists to speeds approaching the speed of light, but generally implies speeds of the order of *spacecraft* or *satellite* speed and greater, e.g. 5-10 km/sec or greater.

hysteresis 1. Any of several effects resembling a kind of internal friction, accompanied by the generation of *heat* within the substance affected.
Magnetic hysteresis occurs when a ferromagnetic substance is subjected to a varying magnetic intensity; electric hysteresis occurs when a dialectric is subjected to a varying electric intensity. Elastic hysteresis is the internal friction in an elastic solid subjected to varying stress.
2. The delay of an indicator in registering a change in a parameter being measured.

I

ice frost A thickness of ice that gathers on the outside of a *rocket* or *aerospace vehicle* over surfaces supercooled as by *cryogenic propellants* (E.G., liquid oxygen) inside the vehicle.
The ice frost is quickly shaken loose and falls to the ground once the rocket begins its ascent.
See: external tank

ideal exhaust velocity The exhaust *velocity* of an *ideal rocket.*

ideal fluid 1. Perfect fluid.
2. *Inviscid* fluid

ideal gas The *pressure* (p), volume (V) and *temperature* (T) behavior of many gases at low pressures and moderate temperatures is approximated quite well by the ideal (perfect) gas *equation of state,* which is

$$p \ V \ = N \ R_u T$$

where N is the number of *moles* of gas

R_u is the *universal gas constant*

(8.315 kJ/kg-mole K),
(1.986 Btu/lb$_m$-mole R)

This equation is based on the experimental work originally conducted by Boyle (*Boyle's Law*), Charles (*Charles's Law*) and Gay-Lussac (the *Gay-Lussac Law*). The ideal gas equation and its many equivalent forms is of particular interest and has wide application in *thermodynamics,* as for example, in describing the performance of an *ideal rocket.*

ideal nozzle The nozzle of an *ideal rocket,* or a nozzle designed according to the *ideal gas laws.*

ideal rocket A theoretical *rocket* postulated for parameters that are corrected in practice.
An ideal rocket assumes a homogeneous and invariant propellant, observance of the *ideal* (perfect) *gas laws,* no friction, no heat transfer across the rocket wall, an axially directed *velocity* of all exhaust gases, a uniform gas velocity across every section normal to the *nozzle axis,* and chemical equilibrium established in the *combustion chamber* and maintained in the nozzle.

ideal velocity The *velocity* acquired by an *ideal rocket* in field free *space,* under the influence of no external *forces* except the *thrust* force.

idiot pin A locating pin that matches two parts in correct orientation so that mating components cannot be mated incorrectly.

igloo A pressurized container for *Spacelab pallet* subsystems when no pressurized *module* is used.
See: Spacelab

igniter A device used to begin combustion such as a spark plug in the combustion chamber of a *jet engine,* or a *squib* used to ignite the *fuel* in a *rocket*.

ignition   The attainment of self-sustaining combustion of *propellants* in a *rocket engine* or motor.

ignition delay   In *solid rocket motors*, the time period from the moment of arrival of the thermal energy from the *igniter* at the *propellant grain* surface until an *exothermic* gas-phase reaction is self-sustaining (i.e., the propellant is burning); in *liquid rocket engines,* the time from initial contact of *fuel* and *oxidizer* until a measurable *pressure* is generated.

ignition lag time

The time period from initiation of the *igniter* until the first ignition of the *solid propellant.*

ignition temperature   The surface *temperature* of a *solid-propellant grain* at the moment combustion begins; the temperature depends on the extent of *pyrolysis,* on heating rate, and on pressure, especially at low pressures and high flux levels.

illuminance   The total *luminous flux* received on a unit area of a given real or imaginary surface. Illuminance is analogous to *irradiance,* but is to be distinguished from the latter in that illuminance refers only to light and contains the luminous efficiency weighting factor necessitated by the nonlinear wavelength-response of the human eye. Compare this term with *luminous intensity.*

images from space   In the last seventeen years, *spacecraft* imaging systems have made most previous visual planetary data obsolete. Taking advantage of close flybys, orbits, and landings, spacecraft have provided scientists with exceptionally clear and close views of *planets* and *moons* as far away from the Earth as *Saturn.* Pictures *telemetered* back from space by *radio* have enabled planetary scientists to discover a Moon-like surface on *Mercury* and circulation patterns in the atmosphere of *Venus.* Pictures of *Mars* have shown craters, giant canyons, and volcanoes on the planet's surface. Details of *Jupiter's* atmospheric circulation have been revealed, active volcanoes on the Jovian moon Io, previously unknown moons, and a ring circling the planet were discovered. New moons orbiting Saturn were found and the rings

once thought to be four in number, were resolved into more than 1000 concentric ring features.

These new discoveries about the planets and their moons as well as thousands of other discoveries like them were made possible by the development of a spacecraft technology for picture transmission. The views we see of the Martian surface or of Saturn, through this technology, are facsimile images and not true photographs. A scanning system on board the spacecraft converts the reflected light from a planet or moon that enters the spacecraft's optical system, into numerical data. The data is transmitted to *Earth* via radio waves where *computers* assemble the received information into pictures.

Most planetary spacecraft produce images of the planets and moons by using slow-scan television cameras. These cameras take a much longer time to form and transmit images than do commercial television systems. Although taking longer to generate, the images are of a much higher quality and contain more than twice the amount of information visible on a home television screen.

Since the successful flyby of Mars by the *Mariner 4* spacecraft in 1964, major improvements in spacecraft imaging systems have been accomplished. Transmitting at a data *bit* rate of under 10 bits per second (a bit is a one or a zero), a week was required for Mariner 4 to transmit enough information to produce 21 pictures of Mars: the same amount of data in just one Voyager picture of Jupiter transmitted in 1979 in only 48 seconds.

Using the *Voyager* spacecraft imaging system as an example, the process of producing finished pictures of planets and moons is accomplished in five general steps:

On Spacecraft	1. Image Scanning 2. Data Storage and Transmission
On Earth	3. Data Reception 4. Data Storage 5. Image Reconstruction

The Voyager spacecraft carries a dual television camera system on a science instrument platform that can be tilted in any direction for precise aiming. On command, a subject can be viewed with either wide-angle or narrow-angle telephoto lenses. Reflected light from the subject enters the lenses and falls on the surface of a selenium-sulfur vidicon television tube, 11 millimeters square. Unlike most standard television cameras, a shutter controls the amount of light reaching the tube. Exposure periods can vary from 0.005 second for very bright subjects to 15 seconds or longer when searches are being made for very faint objects such as previously unknown moons. The televison tube temporarily retains the image until it can be scanned or measured for brightness levels. During the scanning process, the vidicon tube surface is divided into 800 lines each consisting of 800 points. Individual points are called pixels, a contraction of the term "picture element." The total number of pixels into which the image is divided is 800^2 or 640,000.

As each pixel is scanned for brightness, it is assigned a number from 0 to 255. The measured range from white to black is 256 or 2^8. To express the assigned brightness in computer terms, each number is converted into binary language.

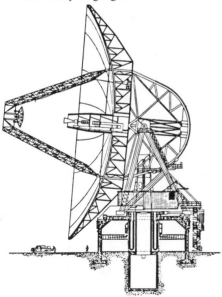

Fig. 1   Dish-shaped radio antenna (64-meters in diameter) used to receive radio signals from planetary probes and spacecraft.   (Drawing courtesy of NASA.)

Following the scanning of pixels and the conversion of light levels into binary form, the bit information is either stored on tape for later transmission or transmitted directly to Earth in *real time*. At the distance of Jupiter, data is transmitted at a rate of more than 100,000 bits per second. For each image, 5,120,000 bits (640,000 pixels × 8 binary bits) must be sent. Data storage capability is used when the spacecraft passes out of sight behind a planet or moon and radio communications are temporarily eclipsed. With the Voyager spacecraft, data equivalent to 100 images can be stored for later transmission.

On Earth, the Voyager radio signals are received by one of three large radio antennas. Each antenna is dish-shaped and has a diameter of 64 meters. (See Fig. 1). A motor drive system precisely aims the antenna toward Voyager and compensates for Earth rotation. To maintain constant contact with the spacecraft, the antennas are located approximately 120° apart at Goldstone, California, Madrid, Spain, and Canberra, Australia. [*See:* Deep Space Network] As one antenna loses contact, due to Earth rotation, a second antenna rotates into view and takes over the job of receiving spacecraft data. While not in contact with Voyager, the remaining two antennas are directed toward other tasks.

Upon receiving the data from the spacecraft, computers at the *NASA Jet Propulsion Laboratory* in Pasadena, California simultaneously store the data for future use and reassemble it into images. In forming the images, the computer converts the binary bit sequences for each pixel into small squares of light. The brightness of the squares is determined by the numerical value of the pixel. The squares are displayed on a television screen and organized into a grid 800 by 800 pixels in size. The resulting image formed by all the lighted squares on the television screen is a black and white facsimile picture of the object being studied.

To obtain color images, considerably more information is needed from the spacecraft than for black and white images. A wheel with a variety of colored filters is rotated in front of the television tube on the spacecraft during the acquisition of the image. In rapid succession, three separate images of the same subject are taken through blue, green, and orange filters. For each image, the television tube is scanned and the resulting binary bits are transmitted to Earth. By the time the scanning of the third image is complete, 15,360,000 binary bits (3 × 640,000 × 8) are traveling towards Earth via radio waves.

Each of the filters affects the amount of light reaching the television tube. The orange filter, for example, is transparent to orange light but blue light appears much darker than normal. On Earth, the three filtered images are given colors and blended together to form a "true" color image.

A final step may be added to the production of images from space. To the unaided eye, some images appear nondescript. Shading differences in planetary surfaces or cloud tops may be too subtle to be detected on just visual examination. A selective enhancement of portions of the image may be necessary to bring out details. Pixels of a particular numerical value can be given unusual color to make them stand out. Two almost identical shades of yellow, for example, can be colored red and blue, thereby exaggerating their differences. The process of image enhancement is similar to adjusting the color and brightness controls on a color television set.

Images of the planets and moons of our Solar System have been the most valuable source of spacecraft-generated information available to planetary scientists. Data gathered at close range, and from above the filtering effects of the Earth's atmosphere, produce views that are far better in quality and detail than pictures taken through *telescopes* held fast to the Earth. The unprecedented quantity of data received from these systems has greatly aided planetary scientists in reworking current theories on the nature and origins of our *Solar System.*

impact   1. A single *collision* of one mass in motion with a second mass which may be either in motion or at rest.
2. Specifically, the action or event of an object, such as a *rocket,* striking the surface of a planet or natural satellite, or striking another object; the time of this event, as in from launch to impact.

3. To strike a surface or an object.
4. Of a rocket or fallaway section: To collide with a surface or object, as in the rocket impacted 20 minutes after launch.

impact acceleration   The *acceleration* generated by very sudden starts or stops of a vehicle.

impact area   The area in which a *rocket* strikes the *Earth's* surface.

Used specifically in reference to the "impact area" of a rocket range.

impact melt   1. Part of the *spacecraft* or target material that is melted by the impact of a *hypervelocity* projectile.
2. Rock melt produced on or within a planetary body by the hypervelocity impact of a *meteorite, asteroid* or *comet.*

impact pressure   1. *Pressure* of a moving *fluid* brought to rest that is in excess of the pressure the fluid has when it does not flow, i.e., *total pressure* less *static pressure.*

Impact pressure is equal to dynamic pressure in *incompressible* flow, but in *compressible* flow impact pressure includes the pressure change owing to the compressibility effect.
2. A measured quantity obtained by placing an open-ended tube, known as an impact tube or *pitot tube,* in a gas stream and noting the pressure in the tube on a suitable *manometer.*

Since the pressure is exerted at a *stagnation point,* the impact pressure is sometimes referred to as the stagnation pressure or total pressure.

impeller   A device that imparts motion to a fluid; specifically in a *centrifugal compressor,* a rotary disk, faced on one or both sides with radial vanes, that accelerates the incoming fluid outward into a *diffuser.*

impingement rate   The rate per unit area per unit time (e.g., per square meter per second) that *molecules* strike a plane surface in a *gas* at rest. Also called rate of incidence.

impinging-stream injector   In a *liquid-propellant rocket engine,* a device that injects the *fuel* and *oxidizer* into the *combustion chamber* in such a manner that the streams of fluid intersect one another.

implosion   The rapid inward collapsing of the walls of an evacuated system or device as the result of failure of the walls to sustain the ambient pressure.

implosion weapon   A nuclear weapon in which a quantity of *fissionable material,* less than a *critical mass* at ordinary *pressure,* has its volume suddenly reduced by compression (a step accomplished by using chemical explosives) so that it becomes *supercritical,* producing a nuclear explosion.
See: supercritical mass

impulse   1. The product of a *force* and the time during which the force is applied; more specifically, the impulse (I) is defined as

$$I \equiv \int_1^2 F dt$$

where the force F is time dependent and equal to zero before time t_1 and after time $t_2.$
2. [Symbol: I_t] *total impulse.* Compare this term with *specific impulse.*

impulse stage   Stage in a *pump* or *turbine* in which there is no change in *static pressure* across the rotor.

impulse turbine   A type of *turbine* having *rotor* blades shaped so that the wheel is turned from the impact the *fluid* against the blades, with no *pressure* drop occurring across the blades.

incandescence   Emission of *light* due to high *temperature* of the emitting material. Any other emission of light is called *luminescence.*

incidence   1. Partial coincidence, as a circle and a tangent line.
2. The impingement of a ray on a surface.

inclination (symbol i)   1. The maximum angle between the plane of an *orbit* and a reference plane, usually the *equatorial plane.* Standard *Space Shuttle* inclinations are 28.5° and 57°.
2. The angle between the orbital plane of a satellite, comet or planet and the plane of the *ecliptic.* Inclination is one of the *orbital elements* and varies between 0° and 180°.
See: orbit

incompressible fluid   A fluid for which the density is constant.

index of refraction [symbol: n]   1. A measure of the amount of *refraction* (a property of a *dielectric* substance). It is the ratio of the *wavelength* or phase *velocity* of an *electromagnetic wave* in a vacuum to that in the substance. Also called refractive index,

absolute index of refraction.
It can be a function of wavelength, *temperature* and *pressure.* If the substance in nonabsorbing and nonmagnetic at any wavelength, then n^2 is equal to the dielectric constant at that wavelength.
The complex index of refraction is obtained when the attenuation of the wave per radian, called the absorptive index K, is paired with the index of refraction. It is written

$$n^* = n(1 - iK)$$

When the wave passes from one medium n_1 to another n_2, the angle of incidence $\emptyset$ and the angle of refraction $\emptyset'$, both measured with respect to the normal to the interface are related by

$$\sin \emptyset / \sin \emptyset' = n_1^*/n_2^* = \text{constant}$$

which becomes, for a nonabsorbing medium, the ratios of the (noncomplex) indices of refraction. In the particular case that medium 2 is a *vacuum,* this ratio is the index of refraction of medium 1. This is known as Snell law.
2. A measure of the amount of refraction experienced by a ray as it passes through a refractive interface, i.e., a surface separating two media of different densities. It is the ratio of the absolute indices of refraction of the two media (see sense 1 above). Also called refractive index, relative index of refraction.

indirect-cycle reactor system   A nuclear reactor system in which a *heat exchanger* transfers heat from the reactor *coolant* to a second fluid which then drives a *turbine.* Compare this term with *closed-cycle reactor system, direct-cycle reactor system.*

induced magnetism   *Magnetism* acquired by a piece of magnetic material while it is in a magnetic field.
See: permanent magnetism

induced radioactivity   *Radioactivity* that is created when substances are bombarded with *neutrons,* as from a nuclear explosion or in a reactor, or with charged particles produced by *accelerators.*
See: activation

inducer   An auxiliary *pump* with a spiral *impeller,* mounted at the inlet of a main pump, whose function is to raise the *fluid pressure* at the inlet by an amount sufficient to preclude *cavitation* in the main pump.

inelastic collision   A *collision* between two *particles* in which changes occur both in the *internal energy* of one or both of the particles and in the sums, before and after collision, of their *kinetic energies.*

inertance   The impeding effect of *fluid inertia* on the transmission of *oscillations* in a fluid-filled conduit.

inert atmosphere   A gaseous medium that because of its lack of chemical reaction is used to enclose tests or equipment.

inert gas   Any one of six gases, helium, neon, argon, kryton, xenon and radon, all of whose shells of planetary *electrons* contain stable numbers of electrons so that the *atoms* are almost completely chemically inactive.

   All these gases are found in the *Earth's atmosphere* but, with the exception of argon, are found only in very small amounts. Also called rare gas, noble gas

inertial coordinate system   A system in which the *momentum* of a particle is conserved in the absence of external *forces.* Thus, only in an inertial system can *Newton laws of motion* be appropriately applied.

   When relative coordinate systems are used, moving with respect to the inertial system, apparent forces arise in Newton laws, such as the *coriolis force.*

inertial force   A force in a given coordinate system arising from the *inertia* of a parcel moving with respect to another coordinate system. The inertial force is proportional and directionally opposite to the accelerating force.

   For example, the *coriolis acceleration* on a parcel moving with respect to a coordinate system fixed in space becomes an inertial force, the *coriolis force,* in a coordinate system rotating with the Earth. Also called inertia force.

inertial guidance   Guidance by means of the measurement and integration of *acceleration* from within the *aircraft, spacecraft* or *aerospace vehicle.*

inertial orbit   The type of *orbit* described by all *celestial bodies,* in conformance with *Kepler laws* of *celestial motion.*

   This applies to all *satellites* and *spacecraft*

providing they are not under any type of propulsive power.

inertial navigation   *Dead reckoning* performed automatically by a device which gives a continuous indication of position by integration of *accelerations* since leaving a starting point.

inertial upper stage   A solid fuel system used to boost payloads from the Space Shuttle's low earth orbit to higher orbits, or into interplanetary trajectories. During the launch phase of the Shuttle flights, the passive upper stage will be in the Orbiter's payload bay. After Shuttle is in its low orbit, the upper stage—satellite combination is placed in space. The inertial upper stage then functions as a launch vehicle.

infinity [Symbol: ∞]   1. A point, line or region, beyond measurable limits.
   A source of light is regarded as at infinity if it is at such a great distance that rays from it can be considered parallel.
See: parallax.
2. Any quantity larger than the largest quantity that can be stored in a register of a specific computer.

inflection   Reversal of direction of curvature.
   A point at which reversal takes place is called point of inflection or inflection point.

in-flight start   An engine ignition sequence after take-off and during flight.
   This term includes starts both within and above the sensible *atmosphere.*

infrasonic frequency   A frequency below the *audiofrequency* range.
   The word infrasonic may be used as a modifier to indicate a device or system intended to operate at an infrasonic frequency.

inhibitor   Anything that inhibits; specifically, a substance bonded, taped or dip dried onto a *solid propellant* to restrict the burning surface and to give direction to the burning process.

inhibitor gate   In *telemetry,* a device that when triggered, prevents information pulses from passing.

initial mass   The *mass* of a *rocket vehicle* at launch.

initial-value problem   A dynamical problem whose solution determines the state of a system at all times subsequent to a given time at which the state of the system is specified by given initial conditions. Also called transient problem.

The initial-value problem is contrasted with the steady-state problem, in which the state of the system remains unchanged in time.

See: boundary-value problem

initiation   The process of starting combustion, explosion or detonation of materials by such means as impact, friction, electrostatic discharge, shock, fragment impact, flame, or heat.

initiator or initiation system   The part of the solid *rocket igniter* that converts a mechanical, electrical, or chemical input stimulus to an *energy* output that ignites the energy release system.

injection   1. The introduction of *fuel,* fuel and air, and *oxidizer,* water, or other substance into an engine induction system or *combustion chamber.*
2. The time following launching when nongravitational forces *(thrust, lift,* and *drag)* become negligible in their effect on the *trajectory* of a *rocket* or spacecraft.
3. The process of putting a *spacecraft* up to *escape velocity* or an *artificial satellite* into orbit.

injection cooling   The method of reducing *heat transfer* to a body by *mass-transfer cooling* accomplished by injecting a *fluid* into the local flow field through openings in the surface of the body.

injector   A device that propels *fuel* or *propellant* into a *combustion chamber* under *pressure* other than atmospheric.

See: impinging-stream injector

inlet   An entrance or orifice for the admission of *fluid.*

Frequently used in compounds, such as inlet air, inlet air temperature, inlet casing, inlet duct, inlet guide vane, inlet port, inlet valve, etc.

in phase   The condition of two or more *cyclic* motions that are at the same part of their cycles at the same instant. Also called in step.

Two or more cyclic motions that are not at the same part of their cycles at the same instant are said to be out of phase or out of step.

input   1. The path through which *information* is applied to any device.
2. The means for supplying information to a machine.
3. Information or *energy* entering into a system. Compare this term with output.
4. The quantity to be measured, or otherwise operated upon, that is received by an instrument. Also called input signal.

insertion   The process of putting an *artificial satellite, aerospace vehicle* or *spacecraft* into *orbit.*

instantaneous readout   Transmission of data simultaneously with the computation of data to be transmitted.

See: real time

instrument   To provide a *satellite, rocket, aircraft, aerospace vehicle,* etc., or component with instrumentation.

instrumentation   1. The installation and use of electronic, gyroscopic, and other instruments for the purpose of detecting, measuring, recording, telemetering, processing or analyzing different values or quantities as encountered in the flight of a *rocket* or *spacecraft.*
2. The assemblage of such instruments in a rocket, spacecraft or the like.
3. A special field of engineering concerned with the design, composition and arrangement of such instruments.

instrument module (IM)   A combination of hardware consisting of the *mission* instruments, mounting structure and all supporting and interfacing equipment that combine with the *Multimission Modular Spacecraft* at the *transition adaptor.*

intact abort   An *abort* of the *Shuttle* mission in which crew, *payload* and *Orbiter* vehicle are returned to a landing site.

integral tank   A *fuel* or *oxidizer* tank built within the normal contours of an *aircraft* or *rocket* vehicle and using the skin of the vehicle as a wall of the tank.

integrated neutron flux   *Flux* multiplied by time, usually expressed as nvt, when n = the number of *neutrons* per cubic centimeter, v = their *velocity* in centimeters per second, and t = time in seconds.

See: flux

integrated simulation   A *Space Shuttle* training session which includes the crew, flight operations support personnel at the

Mission Control Center and the Payload Operations Control Center personnel in combined operations simulating a portion of a flight.

integrating accelerometer   A *transducer* designed to measure, and capable of measuring, *velocity* by means of a time integration of *acceleration*.

interdigitate   To aline two or more sets of fins or projections on a *rocket* so that each fin or projection of one set lies in a plane between the planes established by fins or projections of the other set or sets.

interface   1. In general a common boundary, whether material or non-material, between two parts of a system.

2. In the *Space Transportation System* a mechanical, electrical or operational common boundary between two elements of a system.

3. In fluid mechanics a surface separating two fluids across which there is a dissimilarity of some fluid property, such as density or velocity, or of some derivative of these properties. The equations of motion do not apply at the interface but are replaced by the *boundary conditions*.

interference fit   The retention of a component in a mounting solely by virtue of the friction between the *interfaces;* the degree of friction (and hence retention force) is governed by the relative dimensions of the mating parts; also called press fit.

interferometer   An apparatus used to produce and measure *interference* from two or more *coherent* wave trains from the same source.

Interferometers are used to measure *wavelengths,* to measure angular width of sources, to determine the angular position of sources (as in *satellite* tracking), and for many other purposes.

interior ballistics   That branch of *ballistics* that deals with the propulsion of projectiles, i.e., the motion and behavior of projectiles in a gun barrel, the *temperatures* and *pressures* developed inside a gun barrel or *rocket,* etc. In dealing with *missiles* rockets and *launch vehicles,* interior ballistics treats reactions inside the vehicle occurring during light, as for example the shift of *center of mass* as a function of propellant consumption. Sometimes called *internal ballistics.*

intermediate (epithermal) neutron   A *neutron* having energy greater than that of a *thermal neutron* but less than that of a *fast neutron.* The range is generally considered to be between about 0.5 and 100,000 *electron volts.* Compare this term with *fast neutron* and *thermal neutron.*

intermediate (epithermal) reactor   A *nuclear reactor* in which the *chain reaction* is sustained mainly by intermediate *neutrons.* Compare this term with *fast reactor, thermal reactor.*

internal energy   A mathematically defined *thermodynamic function* (of state), interpretable through *statistical mechanics* as a measure of the molecular activity a microscopic energy modes of the system. It appears in the *first law of thermodynamics* (for a closed system) as:

$$du = dq - dw$$

where du is the increment of specific internal energy, dq is the increment of heat, and dw the increment of work done by the system per unit mass. The differential du is a perfect differential.

interstage section   A section of a *missile* or *rocket* that lies between stages.

interstellar   Between or among the *stars.*

interstellar gas and dust   Material in the space between *stars.* Low-density hydrogen and other gases are detected from their *absorption* and *emission* of specific *wavelengths* of *light* and *radio waves.* Fine dust *particles* scatter light like smog does.

interstellar matter   Gas and dust in the *galaxy* between the *stars.*

intertank   A component of the *Space Shuttle External Tank* that is not a tank in itself but serves as a mechanical connection between the *liquid oxygen* and *liquid hydrogen* tanks. The primary functions of the intertank are to provide structural continuity to the propellant tanks, to serve as a protective compartment to house instruments and to receive and distribute *thrust* loads from the *Solid Rocket Boosters.*

See: External Tank

intravehicular activity (IVA)   Shuttle crew activities taking place inside the *Orbiter,* within a *payload* module carried in the *cargo bay* or within the cargo bay when the doors are closed. The term "intravehicular activity" refers to where the activity is performed rather than to the local atmospheric *pressure* under which it is performed.

intrinsic power   The energy radiated per second from an *X-ray* (or light) source. The brightness that we see is the intrinsic power divided by the square of the source distance, minus the absorption by interstellar smog.

inverse-square law   A relation between physical quantities of the form: x proportional to $1/y^2$ where y is usually a distance; and x terms are of two kinds, *forces* and *fluxes.*

inverter   1. A device for changing direct current to alternating current.

2. In computers, a device or circuit which inverts the *polarity* of a *pulse.*

inviscid   Not viscous, not clinging or sticky; frictionless, as in inviscid flow.

inviscid fluid   Ideal (perfect) fluid.

ion density   (In atmospheric electricity), the number of *ions* per unit volume of a given sample of air; more particularly, the number of ions of a given type (positive small ion, negative small ion, positive large ion, etc.) per unit volume of air. Also called ion concentration.

ion exchange   A chemical process involving the reversible interchange of various *ions* between a solution and a solid material, usually a plastic or a resin. It is used to separate and purify chemicals, such as *fission products, rare earths, etc., in solutions.*

ionization   The process of adding one or more *electrons* to, or removing one or more electrons from, *atoms* or *molecules,* thereby creating *ions.* High temperatures, electrical discharges, or nuclear radiations can cause ionization.

ionization by collision   The removal of an orbital *electron* from an *atom* or *molecule* by an impacting *particle* (often, by the *absorption* of a *photon*). The atom or molecule is then left with an excess positive charge, i.e., it is positively ionized.

ionization chamber   An instrument that detects and measures *ionizing radiation* by measuring the electrical current that flows when radiation ionizes *gas* in a chamber, making the gas a conductor of the electricity.

ionization potential   The *energy* required to *ionize* an *atom* or *molecule.* The energy is usually given in terms of *electron volts.*
See: work function

ionizer   A filament, grid or porous body in an *ion engine* or other device which strips an *electron* from the outer shell of a neutral *atom* to form a positively charged *ion.*
See: electric propulsion

ionizer efficiency   The ratio of the number of *ions* emitted from an *ionizer* to the number of neutral *atoms* entering the ionizer.

ionizing event   Any occurrence in which an *ion* or group of ions is produced; for example, by passage of a charged *particle* through *matter.*

ionizing radiation   Any *radiation* displacing *electrons* from *atoms* or *molecules,* thereby producing *ions.* Examples: *alpha,* beta, gamma radiation, short-wave ultraviolet light. Ionizing radiation may produce severe skin or tissue damage.
See: radiation burn, radiation illness

ion mobility   In gaseous *electric conduction,* the average *velocity* with which a given *ion* drifts through a specified *gas* under the influence of an *electric field* of unit strength. Mobilities are commonly expressed in units of centimeters per second per volt per centimeter. Also called ionic mobility.

ionospheric storm   Disturbances of the *ionosphere,* resulting in anamalous variations in its characteristics and effects on radio communication.

ion pair   A closely associated positive *ion* and negative ion (usually an *electron*) having charges of the same magnitude and formed from a neutral *atom* or *molecule* by *radiation.*

irradiation   Exposure to radiation, as in a nuclear reactor.

isentropic   Of equal or constant *entropy* with respect to either space or time.

isobar   1. One of two or more *nuclides* having about the same *atomic mass* but different *atomic numbers,* hence different chemical properties. Example: $^{14}_{6}C$, $^{14}_{7}N$, and $^{14}_{8}O$ are isobars. Compare this term with *isotope.*

2. A line of equal or constant *pressure,* specifically such a line in a weather map.

isobaric   Of equal or constant *pressure*, with respect to either space or time.

isochoric   Of equal or constant volume, usually supplied to a *thermodynamic* process during which the volume of the system remains unchanged.

isomer   1. One of two or more *nuclides* with the same numbers of *neutrons* and *protons* in their nuclei, but with different energies; a nuclide in the *excited state* and a similar nuclide in the *ground state* are isomers.
2. One of two or more *molecules* having the same atomic composition and molecular weight, but differing in geometric configuration.

isomeric transition   A radioactive transition from one nuclear isomer to another of lower energy.

isothermal process   Any *thermodynamic* change of state of a system that takes place at constant temperature.

isotope   One of two or more atoms with the same *atomic number* (the same chemical element) but with different *atomic weights*. An equivalent statement is that the nuclei of isotopes have the same number of *protons* but different numbers of *neutrons*. Thus $^{12}_{6}C$, $^{13}_{6}C$, and $^{14}_{6}C$ are isotopes of the element carbon, the subscripts denoting their common atomic numbers, the superscripts denoting the differing mass numbers, or approximate atomic weights. Isotopes usually have very nearly the same chemical properties, but somewhat different physical properties. Compare this term with *isobar, isotone, nuclide*.
See: radioisotope

isotope separation   The process of separating *isotopes* from one another, or changing their relative abundances, as by gaseous diffusion, electromagnetic separation or by using *lasers*. All systems are based on the mass differences on the isotopes. Isotope separation is a step in the *isotopic enrichment* process.
See: mass spectrometer

isotopic enrichment   A process by which the relative abundances of the *isotopes* of a given *element* are altered, thus producing a form of the element which has been enriched in one particular isotope. For example, enriching natural uranium in the uranium-235 isotope.

See: enriched material, gaseous diffusion

isotropic   The same in all directions. In general, pertaining to a state in which a quantity or spatial derivatives thereof are independent of direction. Uniform alloys are isotropic. *Eutectics* are nonisotropic because parallel fibers all extend in one direction.

isotropic radiation   *Diffuse radiation* which has exactly the same intensity in all directions.

isotropic radiator   An *energy* source that radiates uniformly in all directions.

isotropy   The condition in a material in which properties are the same in all directions.

J

jacket   1. A covering or casing of some kind.
2. Specifically, a shell around the *combustion chamber* of a *liquid-fuel rocket,* through which the propellant is circulated in *regenerative cooling.*
3. A coating of one material over another to prevent oxidation, micro-meteroid penetration, etc.

Jacobian   The determinant formed by the n^2 *partial derivatives* of n functions of n variables, when the derivatives of each function occupy one row of the determinant. For the case of two functions f(x,y) and g(x,y), the Jacobian J (f,g) is

$$J(f,g) = \begin{vmatrix} \dfrac{\delta f}{\delta x} & \dfrac{\delta f}{\delta y} \\ \dfrac{\delta g}{\delta x} & \dfrac{\delta g}{\delta y} \end{vmatrix} = \frac{\delta f}{\delta x}\frac{\delta g}{\delta y} - \frac{\delta f}{\delta y}\frac{\delta g}{\delta x}$$

Sometimes written

$$J\frac{f,g}{x,y} \quad \text{or} \quad \frac{\delta(f,g)}{\delta(x,y)}$$

jamming   Intentional transmission or reradiation of *radio signals* in such a way as to interfere with reception of desired signals by the intended receiver.

jerk   A vector that specifies the time rate of change of the *acceleration;* the third derivation of displacement with respect to time.

jerkmeter   An instrument for measuring the magnitude of the time rate of change of *acceleration.*

jet   1. A strong well-defined stream of *fluid* either issuing from an orifice or moving in a contracted duct, such as the jet combustion gases issuing from a *reaction engine,* or the jet in the test section of a wind tunnel.
2. A tube, *nozzle,* or the like through which fluid passes, or from which it issues, in a jet, such as a jet in a carburetor.
3. A *jet engine.*

jet engine   1. Broadly, any *engine* that ejects a *jet* or stream of *gas* or fluid, obtaining all or most of its *thrust* by reaction to the ejection.
See: reaction engine.
2. Specifically, an *aircraft* engine that derives all or most of its thrust by reaction to its ejection of combustion products (or heated air) in a jet and that obtains oxygen from the *atmosphere* for the combustion of its fuel, distinguished in this sense from a *rocket engine.* A jet engine of this kind may have a *compressor,* commonly turbine-driven, to take in and compress air *(turbojet),* or it may be compressorless, taking in and compressing air by other means *(pulsejet, ramjet).*

jet nozzle   A *nozzle,* usually specially shaped, for producing a *jet,* such as the exhaust nozzle on a jet or *rocket* engine.
See: rocket nozzle

jet propulsion   1. The propulsion of a *rocket* or other craft by means of a *reaction engine.*
2. = Duct propulsion.
Duct propulsion and rocket propulsion are the two forms of jet propulsion.

jetstream   A *jet* issuing from an orifice into a medium with much lower velocity, such as the stream of combustion products ejected from a *reaction engine.*
In the meteorological sense jet stream is two words, see the following definition.

jet stream   A strong band of wind or winds in the upper *troposphere* or in the *stratosphere,* moving in a general direction from west to east and often reaching velocities of hundreds of kilometers (miles) an hour.

jet thrust   The *thrust* of a *fluid,* especially as distinguished from the thrust of a propeller.
The thrust of a rocket engine is calculated in the same manner as gross thrust of a *jet engine.*

jettison   To discard. When the fuel in a *booster rocket* is used up, the no-useless booster is disconnected from the *spacecraft* and allowed to fall back to *Earth.*
See: Solid Rocket Boosters.

jet vane   A vane, wither fixed or movable, used in a *jetstream,* especially in the jetstream of a *rocket,* for the purposes of stability or control under conditions where external *aerodynamic* controls are ineffective.

jitter   1. Instability of the signal or trace of a *cathode-ray tube.*
2. Small rapid variations in a *waveform* due to deliberate or accidental electrical or mechanical disturbances or to changes in the supply *voltages,* in the characteristic of components, etc.

Joule-Thomson effect   The decrease in *temperature* which takes place when a *gas* expands through a throttling device as a *nozzle.* Also called Joule-Kelvin effect.
The rate of change of temperature T with pressure p in the Joule-Thomson effect is called the Joule-Thomson coefficient (symbol μ):

$$\mu = \left[\frac{dT}{dp}\right]_h$$

where h denotes constant *enthalpy.*

Jovian   Of or pertaining to the planet Jupiter; associated with Jupiter; or similar to Jupiter, as in *Jovian planet.*

Jovian planet   Any one of the giant planets: *Jupiter, Saturn, Uranus, or Neptune.* Usually in plural Jovian planets.

JP-4   A liquid fuel for *jet* and *rocket engines,* the chief ingredient of which is kerosene.

junction   In a *semiconductor* device, a region of transition between semiconducting regions of different electrical properties.

Jupiter   Largest planet in the Solar System, with more than twice the mass of all other planets combined and a diameter of 142,000 km. It is about 1000 times as large as the Earth. Yet its density is less than a quarter of Earth's, indicating that it is made up largely of hydrogen and helium. The planet is a vast ocean, mostly liquid hydrogen, topped by an approx. 1000 km (600 mi.) deep atmosphere, made up mainly of gaseous hydrogen. The temperature at Jupiter's core is theorized to be about 30,000 C (54,000 F). The planet is surrounded by alternatively light and dark bands called zones and belts respectively. Pioneer 10 and 11 spacecraft pictures and temperature measurements suggest that the bands are actually convection cells distorted by Jupiter's high speed rotation (35,000 km—22,000 mi—per hour). The planet has at least 15 satellites: 1979J1 (discovered by Voyager II), Amalthea, Io, Europa, Ganymede, Callisto, Leda, Himalia, Elara, Lysithea, Ananke, Carme, Pasiphae and Sinope.

Jupiter-C   A modified version of the Redstone Ballistic Missile and a direct descendant of the V-2 rocket developed in Germany during World War II. The three stage rocket carried the first successful American artificial Earth satellite, Explorer I, into orbit in January 1958.

K

kaon   An *elementary particle* (contraction of K-meson). A heavy *meson* with a mass about 970 times that of an electron.

K-capture   The capture by an atomic nucleus of a orbital *electron* from the first (innermost) orbit or *shell,* or K-shell, surrounding the nucleus.
See: atom, capture, electron capture

Keplerian   Pertaining to motion in conformance with *Kepler laws,* as Keplerian *trajectory,* Keplerian *ellipse.*

Kepler's laws   The three empirical laws governing the motion of *planets* in their *orbits,* discovered by Johannes Kepler (1571-1630). These are: (a) the orbits of the planets are *ellipses,* with the *Sun* at a common focus; (b) as a planet moves in its orbit, the line joining the planet and Sun sweeps over equal areas in equal intervals of time (also called law of equal areas); (c) the squares of the *sidereal* periods of revolution of any two planets are proportional to the cubes of their mean distances from the Sun.

kinematics   The brance of mechanics dealing with the description of the motion of bodies or *fluids* without reference to the *forces* producing the motion.

kinematic viscosity   A coefficient defined as the ratio of the *dynamic viscosity* of a *fluid* to its density. The units of kinematic viscosity are meters squared per second, m^2 sec (ft^2 sec). It is used to modify *perfect fluid* equations of motion to include phenomena characteristic of a real fluid.

kinetic energy [Symbol: KE]   The *energy* which a body possesses as a result of its motion, defined as one-half the product of its *mass* m and the square of its speed v, $\frac{1}{2}mv^2$. The kinetic energy per unit volume of a fluid parcel is thus $\frac{1}{2}p\,v^2$, where p is the density and v the speed of the parcel.

kinetic performance   That portion of the *nozzle* performance that depends on the *equilibrium state* of the chemically reacting system during *gas* expansion.

Kirchhoff law   The *radiation* law which states that at a given temperature the ratio of the *emissivity* to the *absorptivity* for a given *wavelength* is the same for all bodies and is equal to the emissivity of an ideal *black body* at that *temperature* and *wavelength.*

Loosely put, this important law asserts that good absorbers of a given wavelength are also good emitters of that wavelength. It is essential to note that the Kirchhoff law relates absorption and emission at the same wavelength and at the same temperature. Also called the Kirchhoff radiation law.

klystron An *electron tube* for converting direct-current energy into radio frequency (rf) energy by alternately speeding up and slowing down electrons.

knot A nautical mile per hour, 1,1508 statute mile per hour. [1.852 km/hr]

Knudsen flow The flow of *gases* through ducts and tubes under conditions intermediate between *laminar viscous flow* and *molecular flow*. Also called transition flow.

Knudsen number A number used to describe the flow of a low density *gas,* equal to the ratio λ / l where λ is the mean free path of the gas molecule and l is a characteristic length, such as boundary layer thickness, or apparatus dimension.

The Knudsen number is used most commonly to define the extent to which the gas behaves like a collection of independent particles (free-molecule regime, Knudsen number much larger than unity) or like a viscous fluid (continuous regime, Knudsen number much smaller than unity). Intermediate regimes are termed transition region, and slip-flow region. See: rarefied gas dynamics.

Ku-band A *frequency* band extending from approximately 15.35 to 17.25 *gigahertz*.
See also: Orbiter avionics system;
Orbiter guidance, navigation and control system;
Orbiter displays and controls;
Orbiter communications and data systems.

Kuiper Airborne Observatory A Lockheed C141 Starlifter jet transport used by NASA for airborne infrared astronomy. The aircraft is modified to accept tele-

NASA's Kuiper Airborne Observatory—a Lockheed C-141 jet transport equipped with a 91.5 cm. telescope.   (Photograph courtesy of NASA.)

scopes, including, most recently, a 91.5 centimeter (36-inch) aperture telescope. The observatory is being used to gather information about the Universe, information which is providing scientists with a better understanding of the birth and evolution of the planets, the stars, and six stellar systems, including entire galaxies. The facility is used primarily for observations in the infrared spectrum, much of which is not possible from ground-based telescopes because of the attenuating action of the Earth's atmosphere.

L

lag 1. The delay between change of conditions and the indication of the change on an instrument.
2. Delay in human reaction.
3. The amount one *cyclic* motion is behind another, expressed in degrees. The opposite is lead.

Lagrangian coordinates A system of *coordinates* by which *fluid parcels* are identified for all time by assigning them coordinates that do not vary with time. Examples of such coordinates are: (a) the values of any properties of the fluid conserved in the motion; (b) more generally, the positions in space of the parcels at some arbitrarily selected moment. Subsequent positions in space of the parcels are then the dependent variables, functions of time and of the Lagrangian coordinates. Also called material coordinates. Compare *Eulerian coordinates*.
See: Lagrangian equations

Lagrangian correlation The correlation between the properties of a *flow* following a single parcel of fluid through its space and time variations. Compare *Eulerian correlation*.

Lagrangian equations Any of the fundamental equations of *Hydrodymanics* expressed in *Lagrangian coordinates*.

Lagrangian point One of the five solutions by Lagrange to the three-body problem in which three bodies will move as a stable configuration. In three of the solutions the bodies are in line; in the other two the bodies are at the vortices of equilateral triangles.

Lagrange predicted in 1772 that if the three bodies form an equilateral triangle revolving about one of the bodies, the system would be stable. This prediction was fulfilled in 1908 with the discovery of the asteroid Achilles approximately 60° ahead of Jupiter, and Jupiter's orbit. Since then other asteroids have been discovered 60° ahead and 60° behind Jupiter.

Lambert law   A law of physics which states that *radiant intensity* (flux per unit solid angle) emitted in any direction from a unit radiating surface varies as the cosine of the angle between the normal to the surface and the direction of the radiation. The *radiance* (or luminance) or a radiating surface is, therefore, independent of direction. Also called Lambert cosine law.

Lambert law is not obeyed exactly by most real surfaces, but an ideal *black body* emits according to this law. This law is also satisfied (by definition) by the distribution of radiation from a perfectly diffuse radiator and by the radiation reflected by a perfectly diffuse reflector. In accordance with Lambert law, an incandescent spherical black body when viewed from a distance appears to be simply a uniformly illuminated disk. This law does not take into account any effects that may alter the radiation after it leaves the source.

laminar boundary layer   In *fluid flow*, layer next to a fixed boundary. The fluid velocity is zero at the boundary but the molecular *viscous* stress is large because the velocity gradient normal to the wall is large. *See:* turbulent boundary layer

The equations describing the flow in the laminar boundary layer are the Navier-Stokes equations containing only the inertia and molecular viscous terms.

laminar flow   In *fluid flow*, a smooth flow in which no crossflow of fluid particles occur between adjacent *stream lines*; therefore, a flow conceived as made up of layers— commonly distinguished from *turbulent flow*.

Landau damping   The *damping* of a space

charge *wave* by *electrons* which move at the *phase velocity* of the wave and gain energy transferred from the wave.

lander   A spacecraft or mission which lands on another *celestial body;* e.g. Surveyor on the Moon, Viking on Mars.

landing gear   The apparatus comprising those components of an aircraft, *spacecraft* or *aerospace vehicle* that support and provide mobility for the craft on land, water or other surface. The landing gear consists of wheels, floats, skis or other devices, together with all associated struts, bracing, shock absorbers, etc.
See: Orbiter structure

Langmuir probe   A small metallic *conductor* or pair of conductors inserted within a *plasma* in order to sample the plasma current.

In some cases, the plasma density, electron temperature, and plasma potential can be inferred from a measurement with a Langmuir probe.

lanthanide series   The series of elements beginning with lanthanum, Element No. 57, and continuing through lutetium, Element No. 71, which together occupy one position in the *Periodic Table* of the elements. These are the *"rare earths"*, which all have chemical properties similar to lanthanum. They also are called the "lanthanides". Compare *actinide series.*

Laplace equation   The elliptic partial differential equation

$$\nabla^2 \varphi = 0$$

where c is a scalar function of position, and V^2 is the Laplacian operator. In rectangular Cartesian coordinates x, y, z, this equation may be written

$$\frac{\delta^2 \varphi}{\delta x^2} + \frac{\delta^2 \varphi}{\delta y^2} + \frac{\delta^2 \varphi}{\delta z^2} = 0$$

The Laplace equation is satisfied, for example, by the velocity potential in an irrotational flow, by gravitational potential in free space, by electrostatic potential in the steady flow of electric currents in solid conductors, and by the steady-state temperature distribution in solids.

A solution of the Laplace equation is called a *harmonic function.* Compare *Poisson equation.*

Laplace transform   An integral transform of a function obtained by multiplying the given function $\int(t)$ by e^{-pt}, where p is a new variable, and integrating with respect to t from $t = 0$ to $t - ^\infty$. Also called Laplace transformation.

Thus, the Laplace transform of $\int(t)$ is

$$L\,(\textstyle\int) = \int_0^\infty e^{-pt} f(t)\;dt$$

and may be denoted by the symbol $^L\,(\int)$. The Laplace transform is especially useful in solving initial-value problems associated with inhomogeneous linear differential equations with constant coefficients and is also quite valuable in final value analysis.
See: Fourier transform

Laplacian operator   The mathematical operator $\nabla^2 = \nabla \bullet \nabla$ (or sometimes written Δ) where ∇ is the del-operator. In rectangular Cartesian coordinates, the Laplacian operator may be expanded in the form

$$\nabla^2 = \frac{\delta^2}{\delta x^2} + \frac{\delta^2}{\delta y^2} + \frac{\delta^2}{\delta z^2}$$

Also called Laplace operator.
See: Laplace equation

Laplacian speed of sound   The *phase speed* of a *sound wave* in a *compressed fluid* if the expansions and compressions are assumed to be *adiabatic*. this speed a is given by the formula

$$a^2 = (c_p/c_v)\;RT$$

where c_p and c_v are the specific heats at constant pressure and volume, respectively; R is the gas content; and T is the Kelvin temperature. The value of this speed under standard consitions in dry air is 331 meters per second.

lapse rate   The decrease of an atmospheric *variable* with height, the variable being temperature, unless otherwise specified.

Large Area Modular Array of Reflectors [abbr. LAMAR]   A possible *NASA astrophysics* mission designed to conduct an *all-sky survey* of discrete *X-ray* sources. It is intended that LAMAR will be utilized as a national observatory. The large (3750 kg.) instrument consists of a modular array of coaligned *grazing incidence* telescopes with an effective area of approximately 3 x 10^4 cm^2 at 1.5 keV.

If developed, the LAMAR observatory will be launched on the Space Shuttle and placed in a 400-km, 28.5° inclination orbit (lower orbit/inclination is highly desirable). All instruments on LAMAR will have their principal axes pointing in the same direction and must be coaligned to within 1 arc-min. The LAMAR can be pointed (1 arc-min accuracy) to any area of interest in the sky and rotated slowly to obtain scientific data.

Larmor orbit   The circular motion of a charged particle in a uniform magnetic field.

Whereas the motion of the particle is unimpeded along the magnetic field, motion perpendicular to the field is always accompanied by a force perpendicular to the direction of motion and the field. The *electron* or *ion* will orbit in a plane perpendicular to the magnetic field. By adding any arbitrary velocity along the magnetic field, the total path looks like a helix. The size of the Larmor orbit or helix is proportional to the particle velocity divided by the magnetic field. In a 1-gauss field, a 1-volt electron has an orbit of about 3 centimeters, and a 1-volt *proton* an orbit of about 1 meter.

latch   1. A mechanical or magnetic device that maintains a flow-controlled component in a desired position without the constant application of power.
2. A device that fastens one part to another but is subject to release so the parts can be separated.

latent heat   The unit quantity of *heat* required for *isothermal* change in phase of a unit mass of matter.

Latent heat is termed heat of *fusion*, heat of *sublimation*, heat of *vaporization*, depending on the change of state involved.

latent heat of vaporization   The quantity of (thermal) energy required to achieve *evaporation* or vaporization of a unit mass of liquid. For a *simple compressible substance*, the energy transfer as heat required to evaporate a unit mass at constant *pressure* is given by the expression:

$$h_{fg} = h_g - h_f$$

where h_{fg} is the latent heat of vaporization (also called the *enthalpy* of evaporation)

h_g is the enthalpy at *saturated vapor* conditions

h_f is the enthalpy at *saturated liquid* conditions

lateral   1. Of or pertaining to the side; directed or moving toward the side.
2. Of or pertaining to the *lateral axis*; directed, moving, or located along, or parallel to, the lateral axis.

latitude   Angular distance from a *primary great circle* or plane.

Terrestrial latitude is angular distance from the equator, measured northward or southward through 90° and labeled N or S to indicate the direction of measurement; astronomical latitude is angular distance between the direction of *gravity* and the plane of the equator; *geodetic* (or topographical) latitude is angular distance between the plane of the equator and a normal to the spheroid; *geocentric* latitude is the angle between a line to the center of the *Earth* and the plane of the equator. Geodetic and sometimes astronomical latitude are also called geographic latitude. Geodetic latitude is used for charts. Assumed latitude is the latitude at which an observer is assumed to be located for an observation of computation. Fictitious latitude is angular distance from a fictitious equator. Grid latitude is angular distance from a grid equator. Transverse or inverse latitude is angular distance from a transverse equator. Oblique latitude is angular distance from an oblique equator. Difference of latitude is the shorter arc of any meridian between the parallels of two places, expressed in angular measure. Magnetic latitude, magnetic inclination, or magnetic dip is angular distance between the horizontal and the direction of a line of force of the earth's magnetic field at any point. Geomagnetic latitude is angular distance from geomagnetic equator. A parallel of latitude is a circle (or approximation of a circle) of the earth, parallel to the equator, and connecting points of equal latitude; or a circle of the celestial sphere, parallel to the ecliptic. Celestial latitude is angular distance north or south of the ecliptic. Galactic latitude is angular distance north or south of the galactic equator.

lattice   1. An orderly array of pattern of

nuclear *fuel elements* and moderator in a *reactor* or critical assembly.
2. Also, the arrangement of *atoms* in a crystal.
See: geometry

launch (n)   1. The action taken in launching a *rocket* or *aerospace vehicle* from the surface.
2. The resultant of this action, i.e., the transition from static repose to dynamic flight by the rocket.
3. The action of sending forth a rocket, probe or other object from a moving vehicle, such as an *aircraft* or *spacecraft*.
See: lift-off

launch (v)   1. To send off a *rocket* vehicle under its own rocket power, as in the case of guided aircraft rockets, artillery rockets, and *space vehicles*.
2. To send off a *missile* or *aircraft* by means of a catapult, as in the case of the *V-1*, or by means of *inertial force*, as in the release of a bomb from a flying aircraft.
3. To give a space probe an added boost for flight into space just before separation from its launch vehicle.
See: lift-off

launch azimuth   The initial compass heading of a powered vehicle at *launch*.

launch complex   In general the site, facilities and equipment used to launch a *rocket* or *aerospace vehicle*.
See: launch site
The complex differs according to the type of rocket or particular rocket, or according to whether land launched or ship launched. The term is sometimes considered to include the launch crew.

launch configuration   The combination of *boosters, spacecraft,* and launch escape system (if appropriate) that must be lifted off the ground at launch.
See: Space Shuttle Flight System

launch crew   A group of engineers and technicians that prepare and launch a *rocket* or *aerospace vehicle*.

launcher   1. Specifically, a structure or device often incorporating a tube, a group of tubes or a set of tracks, from which self-propelled *missiles* are sent forth and by means of which the missiles usually are aimed or imparted initial guidance — distinguished in this specific sense from a catapult.

2. Broadly, a structure, machine or device, including the catapult, by means of which *airplanes, rockets*, or the like are directed, hurled or sent forth.

launching angle   The angle between a horizontal plane and the longitudinal axis of a *rocket, aerospace vehicle*, etc. being launched.

launching rail   A rail that gives initial support and guidance to a *rocket* launched in a non-vertical position.

launch pad (LP)   1. In general the load-bearing base or platform from which a rocket is launched. Frequently called simply the "pad."

2. In the *Space Transportation System* the area where the stacked *Shuttle flight vehicle* undergoes final prelaunch checkout and *countdown* and from which it is launched.

launch ring   The metal ring on the *launch pad* on which a *missile* stands before launch.

launch site   1. A defined area from which an *aerospace* or *rocket* vehicle is launched, either operationally or for test purposes.

2. More broadly, a launching base.

launch stand   A facility or station at which a *rocket* vehicle is launched normally incorporating a *launch pad* with *launcher*. Compare *test stand*.

launch window   An interval of time during which an *aerospace vehicle* or *rocket* can be launched to accomplish a particular purpose.

L-band   A *frequency* band extending from approximately 0.39 to 1.55 *gigahertz*.
See also: Orbiter guidance, navigation and control system.

L/D ratio   *Lift-drag ratio.*

leakage   In nuclear engineering, the escape of *neutrons* from a *reactor core*. Leakage lowers a reactor's *reactivity*.

leakage rate   The quantity of *fluid* escaping past a seal in a given length of time.

"leak-before-burst" condition   Condition requiring that cracks penetrating a pressure-vessel wall do so without reaching the (critical) size for unstable propagation (i.e., the size that leads to *brittle fracture*).

left-handed polarized wave   An ellipitically polarized transverse *electromagnetic wave* in which the rotation of the *electric field* vector is counterclockwise for an observer looking in the direction of propagation. Also called counterclockwise polarized wave.

length [Symbol: l, *l*]   The dimension of an *aircraft, rocket* or aerospace vehicle from nose to tail; the measure of this dimension. Compare this term with *span*.
See: Orbiter structure

lens   A set of pieces of glass or quartz accurately shaped to focus *light* from a distant object to form an image of that object.

lepton   One of a class of light *elementary particles* (having small mass). Specifically, an *electron*, a *positron*, a *neutrino*, an *antineutrino*, a *muon* or an antimuon. Compare this term with *baryon* and *meson*.

lethal dose   A dose of *ionizing radiation* sufficient to cause death. Median lethal dose (MLD or LD-50) is the dose required to kill within a specified period of time (usually 30 days) half of the individuals in a large group of organisms similarly exposed. The LD-50/30 for man is about 400-450 *roentgens*.

lever link   Mechanical linkage between the *actuator* and the valving element of *rotary valve* that consists of a lever or crank on the rotary member and a link with *clevis* connections from the lever to the actuator shaft.

Lewis number [Symbol: Le]   A *dimensionless number* (or parameter) describing the ratio of mass diffusivity to thermal diffusivity

$$Le \equiv \frac{\text{mass diffusivity}}{\text{thermal diffusivity}} = \frac{D\rho C_p}{k}$$

where D is mass diffusivity, p is the mass density, C_p is the specific heat at constant pressure, and k is thermal conductivity.

lift [Symbol L]   1. That component of the total *aerodynamic force* acting on a body perpendicular to the undisturbed *airflow* relative to the body.

2. To lift off, or take off in a verticle ascent. Said of a rocket vehicle.
See: lift-off

lift coefficient [Symbol: C_L]   A *coefficient* representing the *lift* of a given *airfoil* or other body.

The lift coefficient is obtained by dividing the lift by the free-stream dynamic pressure and by the representative areas under consideration.

lift-drag ratio The ratio of *lift* to *drag* obtained by dividing the lift by the drag or the lift coefficient by the drag coefficient. Also called L/D ratio.

lift-off The action of a *rocket* or *aerospace vehicle* as it separates from its *launch pad* in a vertical ascent. Compare *take-off.*

Please note that lift-off is applicable only to vertical ascent; take-off is applicable to ascent at any angle. A lift-off is action performed by a rocket; a launch is action performed upon a rocket or upon a satellite or spaceship carried by a rocket.

light Visible *radiation* [about 0.4 to 0.7 micrometers (microns)] in *wavelength* considered in terms of its *luminous efficiency,* i.e., evaluated in proportion to its ability to stimulate the sense of sight.

light-emitting diode (LED) A semiconductor device that radiates in the visible region (of the electromagnetic spectrum). It is used in alphanumeric displays.

light flash A momentary flash seen with the astronaut's eyes closed when a high-speed *ion* passes through the eye.

lightning protection The tip of the *Space Shuttle External Tank* nose cap forms a lightning rod to protect the tank during launch. Lightning protection, which is primarily required for the *liquid oxygen* tank, is provided by the launch site until *lift-off.* Thereafter, the lightning rod protects the External Tank from the direct and indirect effects of lightning.
See: External Tank

light water Ordinary water (H_2O), as distinguished from heavy water (D_2O).

line (or duct) Enclosed passageway (usually circular in cross section with relatively thin walls) that conveys fluid under pressure.

linear acceleration [Symbol: a] The rate of change of linear *veolocity.*
See: acceleration

linear accelerator A long straight tube (or series of tubes) in which charged *particles* (ordinarily *electrons* or *protons*) gain in *energy* by the action of oscillating *electromagnetic fields.*
See: accelerator

linear characteristic Straight-line relation, as for example, the relation between valve flow and valving element stroke at a con-

stant value of *pressure* drop.

line of apsides The line connecting the two points of an *orbit* that are nearest and farthest from the center of attraction, as the *perigee* and *apogee* of the *Moon* or the *perihelion* and *aphelion* of a *planet*; the major *axis* of any *elliptical* orbit and extending indefinitely in both directions.

line of flight The line in air or space along which an *aircraft, spacecraft* or *aerospace vehicle* flies or travels.

line of force A line indicating the direction that a *force* acts, as in a magnetic field.

line of nodes The straight line connecting the two points of intersection of the *orbit* of a *planet,* planetoid, or *comet* and the *ecliptic,* or the line of intersection of the planes of the orbits of a *satellite* and its *primary.*

line of position In navigation, a line representing all possible locations of a craft at a given instant. In *space* this concept can be extended to sphere of position, plane of position, etc.

line of sight [abbr. LOS] 1. The straight line between the eye of an observer and the observed object or point. Also called optical path.
2. Any straight line between one point and another, or extending out from a particular point.
3. In radio, a direct propagation path that does not go below the radio horizon.

liner 1. Thin layer of *adhesive* specifically used to bond *solid propellant* to the motor case or to the insulation.
2. *Ablative material* used to line the inner wall of the *combustion chamber* or nozzle to reduce heat transfer to the wall.
See: solid propellant rocket

line spectra The spontaneous emission of *electromagnetic radiation* from the bound *electrons* as they jump from high to low *energy* levels in an *atom.*

This *radiation* is essentially at a single *frequency* determined by the jump in energy. Each different jump in energy level, therefore, has its own frequency and the net radiation is referred to as the line spectra. Since these line spectra are characteristic of the atom, they can be used for identification purposes.

line spectrum A *spectrum* that contains a finite number of components within a specified *frequency* range.

line width   The finite width, expressed either in *wavelength* units or *frequency* units, of a *spectral line* (e.g., an absorption line).

It is customary to employ, as a convenient measure of this quantity, the *half-width*, which is the width of the spectral line measured between the two points at which its intensity is just half the peak intensity of the line center.

lip seal   Pressure-actuated seal that prevents leakage in dynamic and static applications by employing a scraping or wiping action on the mating surface.

liquid   A substance in a state in which the individual particles move freely with relation to each other and take the shape of the container, but do not expand to fill the container. Compare this term with *fluid*.

liquid fuel   A *rocket fuel* that is *liquid* under the conditions in which it is utilized in the rocket.
See: liquid propellant

liquid hydrogen (LH$_2$) tank   The largest component of the *Space Shuttle External Tank*. Its primary functions are to hold, 1,449, 905 liters (383,066 gallons) of *liquid hydrogen* at a *temperature* of 20 K (-253 C or -423 F) and to provide a mounting platform for the *Orbiter* and *Solid Rocket Boosters*.
See: External Tank

liquid length   The distance along a chamber wall over which film coolant remains in a liquid state.

liquid oxygen (LO$_2$) tank   An assembly of the *Space Shuttle External Tank* that contains 541,482 liters (143,060 gallons) of *oxidizer* at 90 K (-183 C or -297 F).
See: External Tank, liquid oxygen

lithium hydroxide (LiOH)   A chemical compound used for removing carbon dioxide from a closed atmosphere.
See: Orbiter environmental control and life support system

"little Green men" (LGM)   A colloquial expression (originating in the science fiction literature) for *extraterrestrial* visitors, presumably intelligent.
See also: unidentified flying object

live testing   The testing of a *rocket* engine, *aerospace vehicle* or *missile* by actually launching it. Compare this term with *static testing*.

load factor   Ratio of vehicle *thrust* to its overall mass.

lobe   An element of a *beam* of focused electromagnetic (e.g., radio) *energy*. Lobes define surfaces of equal power density at varying distances and directions from the radiating *antenna*.

Their configuration is governed by two factors: (a) the geometrical properties of the antenna reflector and feed system; and (b) the mutual interference between the direct and reflected rays for an antenna situated above a reflecting surface. In addition to the major lobes of an antenna system, there exist side lobes (or minor lobes) that result from the unavoidable finite size of the reflector. They exist at appreciable angles from the *axis* of the beam, and, while objectionable, they normally contain much less energy than that in the major lobe.
See: radiation pattern

local fallout
See: fallout

local vertical   At a particular point, the direction that the *force* of *gravity* acts.

lockwire   1. Flexible slender rod or thread of material (usually metal) that is passed through matching holes in two (rotating) parts so as to fasten them together securely. 2. To fasten securely in place by means of a wire.

locomotion   Act of moving from place to place.

logarithm   The power to which a fixed number, called *the base* usually 10 or e (2.7182818), must be raised to produce the value to which the logarithm corresponds.

An antilogarithm or inverse logarithm is the value corresponding to a given logarithm. A cologarithm is the logarithm of the reciprocal of a number.

logarithmic   Pertaining to *logarithms*; in a proportion corresponding to the logarithms of numbers, as a logarithmic scale.

Long Duration Exposure Facility (LDEF)
The Long Duration Exposure Facility (LDEF) is being developed by *NASA* to accommodate many science, technology and application experiments which require a *free-flying* long term exposure in space. The LDEF approach to resource efficient space operations has been engineered and

Fig. 2 The Long Duration Exposure Facility in the Orbiter's payload bay (artist concept). (Illustration courtesy of NASA.)

standardized with the *STS* operations. It is the intention of the LDEF Project Office to take care of "getting into space" via the *Shuttle,* while individual LDEF investigators devote their creative energies to specific experiments.

The LDEF, though not considered an STS element, makes use of the Shuttle and supports a class of experiments which require free-flying exposure in space. It is a reusable, unmanned, gravity gradient stabilized, free-flying structure. (See: Fig. 1). The LDEF is placed in *Earth orbit* by the Shuttle and remains there for an extended period of time.

The deployment sequence starts with the *Orbiter* maneuvering into a nose-down, *gravity gradient* alignment orientation. The Orbiter's *remote manipulator system* (RMS) then engages the LDEF's grapple fitting, four trunnion retention fittings unlatch, and the RMS removes the LDEF from the Orbiter's *payload bay.* [See: Fig. 2] Before release, the longitudinal axis of the LDEF is aligned with the *local* Earth *vertical,* other required orientations established and the angular velocities brought within specified limits. Once LDEF is released, the Shuttle moves away in a manner that minimizes any contamination effects from the *reaction control system's* exhaust plumes. This entire deployment sequence is completed in one orbital period.

After release, gravity gradient *torques* and *viscous damping* from a *magnetic* damper null the initial angular motions. Within 8 days the *steady state* pointing remains within 2° of local Earth vertical and *oscillations* about the longitudinal axis

are kept within 5°. The maximum *acceleration* level at release is 1×10^{-3} g and the maximum steady state acceleration is 1×10^{-6} g. The LDEF's *orbital period* ranges from 92.8 to 95.6 minutes.

The LDEF remains on orbit for 6 to 9 months, exposing its experiment trays to the space environment. During this period, the orbital *altitude* will decay about 37 km (20 n mi). When the planned exposure period has been completed, another Shuttle flight recovers the LDEF and returns it to Earth. After landing, LDEF is removed from the payload bay. Experiment trays are then examined, removed and returned to their sponsors for postflight analysis. All LDEF flights will originate from the *Kennedy Space Center.* In a typical delivery

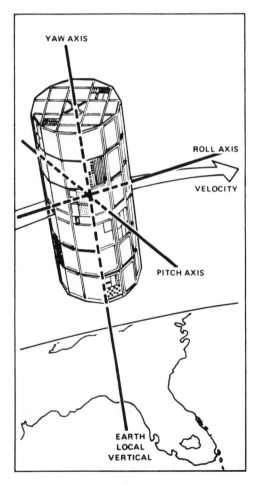

Fig. 1 Orientation of the LDEF in free flight. (Drawing courtesy of NASA.)

Fig. 3   Deployment sequence for the Long Duration Exposure Facility.   (Drawing courtesy of NASA.)

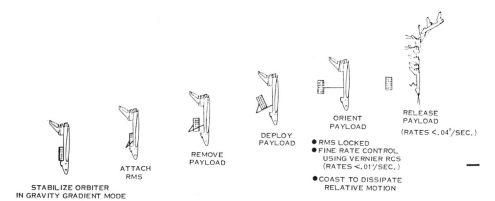

STABILIZE ORBITER
IN GRAVITY GRADIENT MODE

ATTACH
RMS

REMOVE
PAYLOAD

DEPLOY
PAYLOAD

ORIENT
PAYLOAD
● RMS LOCKED
● FINE RATE CONTROL
   USING VERNIER RCS
   (RATES <.01°/SEC.)
● COAST TO DISSIPATE
   RELATIVE MOTION

RELEASE
PAYLOAD
(RATES <.04°/SEC.)

flight the LDEF is placed in a circular *orbit* of approximately 556 km (300 n mi) altitude with an *inclination* between 28.5° and 57°.

The LDEF includes: the primary structure, experiment trays, a viscous magnetic motion damper, LDEF retention and ground handling fittings, and a grapple fixture (for the RMS). The LDEF structure consists of a 12-sided regular polygon framework, 9.14 m (30 ft) in length and 4.27 m (14 ft) across the major diagonal. This frame work is composed of ring frames and longerons made from aluminum extrusions. Experiment trays are mounted into the bays formed by the ring frames and *longerons*. The LDEF can accomodate 72 experiment trays around its circumferential surface, 6 trays at one end, and eight trays at the other. The LDEF structure weighs approximately 3629 kg (8,000 lbm) and is designed to support trays and experiments weighing 5,450 kg (12,000 lbm). [See: Fig. 3]

longeron   The main longitudinal member of a fuselage or *nacelle*.
See: Orbiter structure

longitude   1. Angular distance, along a *primary great circle*, from the adopted reference point; the angle between a reference plane through the polar axis and a second plane through that *axis*.

Terrestrial longitude is the arc of a parallel, or the angle at the pole, between the prime meridian and the meridian of a point on the *Earth*, measured eastward or westward from the prime meridian through 180°, and labeled E or W to indicate the direction of measurement. Astronomical

longitude is the angle between the plane of the reference meridian and the plane of the *celestial meridian; geodetic* longitude is the angle between the plane of the reference meridian and the plane through the polar axis and the normal to the spheroid. Geodetic and sometimes astronomical longitude are also called geographic longitude. Geodetic longitude is used for charts. Assumed longitude is the longitude at which an observer is assumed to be located for an observation or computation. Difference of longitude is the smaller angle at the pole or the shorter *arc* of a parallel between the meridians of two places, expressed in angular measure. Fictitious longitude is the arc of the fictitious equator between the prime fictitious meridian and any given fictitious meridian. Grid longitude is angualr distance between a prime grid meridian and any given grid meridian. Oblique longitude is angular distance between a prime oblique meridian and any given oblique meridian. Transverse or inverse longitude is angular distance between a prime transverse meridian and any given transverse meridian. Celestial longitude is angular distance east of the *vernal equinox*, along the *ecliptic*. Galactic longitude is angular distance east of *sidereal hour angle* 80°, along the galactic equator. 2. Of a *planet* in *solar system*, the sum of two angles; the *celestial longitude* of the *ascending node* of the planetary orbit, and the angle measured eastward from the ascending node along the *orbit* to the position of the planet.

longitudinal axis   The fore-and-aft line through the *center of gravity* of a craft.

longitudinal slot   Slot inside a propellant grain parallel to the axis of the grain.
See: solid propellant rocket

look angles   The *elevation* and *azimuth* at which a particular *satellite* is predicted to be found at specified time. Look angles are used for satellite tracking and data acquisition to minimize the amount of searching needed to acquire the satellite in the telescope field of view or the antenna beam.

Lorentz contraction   The hypothesis that all measuring rods (lengths) contract in the direction of motion as

$$\beta = \sqrt{1 - (v^2/c^2)}$$

where v is the speed of motion, and c is the speed of light.
See: relativity

Lorentz force   The *force* affecting a charged *particle* due to the motion of the particle in a *magnetic field*. The Lorentz force is

$$F_L = q(v \times B)$$

where q is the charge on the moving object; v is the *velocity* of the object; and B is the magnetic induction *vector*.

low-cycle fatigue   Life-cycle capability determined by the plastic range; generally less than 10^4 cycles.

low-level analysis (low level counting)   A procedure to measure the *radioactive* content of materials with very low levels of activity, using sensitive detecting instruments and with good shielding to eliminate the effects of *background radiation* and *cosmic rays*.
See: coincidence counting, counter

low-pass filter   A *wave filter* having a single transmission band extending from zero *frequency* up to some critical or bounding frequency, not infinite.

lox-hydrogen engine   an engine using liquid *hydrogen* as *fuel* and liquid oxygen as *oxidizer*.
See: Space Shuttle Main Engine

L-Sat Program   A program devised by the European Space Agency to develop commercially viable multi-purpose satellite systems for future telecommunication. (See Fig. 1) This may eventually include direct-to-home television broadcasting. British Aerospace is the prime contractor. Launch is scheduled for 1985.

lumen [Symbol: lm]   The *SI unit* of *luminous flux*. It is defined as the luminous flux emitted by a uniform point source, of intensity one *candela* (cd), in a cone of *solid angle* one steradian.

$$1 \text{lm} \equiv \frac{1 \text{cd}}{(4\pi)}$$

luminance   In *photometry*, a measure of the intrinsic *luminous intensity* emitted by a source in a given direction; the illuminance produced by *light* from the source upon a unit surface area oriented normal to the line of sight at any distance from the source, divided by the solid angle subtended by the source at the receiving surface. Also called brightness but luminance is preferred.
See: Lambert law.

It is assumed that the medium between source and receiver is perfectly transparent; therefore, luminance is independent of *extinction* between source and receiver. The source may or may not be self-luminous.

Luminance is a measure only of light; the comparable term for *electromagnetic radiation* in general is *radiance*.

luminescence   Emission of *light* produced by the action of biological or chemical processes or by *radiation*, or any other cause except high *temperature* (which produces *incandescence*). Electroluminescence is luminescence from electrical discharges—such as sparks or *arcs*. *Excitation* in these cases results mostly from *electron* or *ion collision* by which the *kinetic energy* of electrons or ions, accelerated in an *electric field*, is given up to the atoms or molecules of the gas present and causes light emission. Chemiluminescence results when energy, set free in a chemical reaction, is converted to light energy. The light from many chemical reactions and from many flames is of this type. Photoluminescence, or fluorescence, results from excitation by absorption light. The term phosphorescence is usually applied to luminescence which continues after excitation by one of the above methods has ceased.

luminous   1. In general, pertaining to the emission of *visible radiation*.

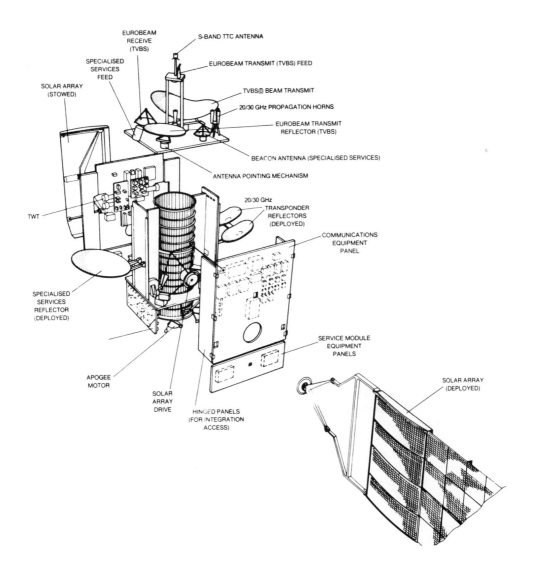

Fig. 1   The L-SAT, en exploded view, highlighting the modularity of its design.   (Drawing courtesy of the European Space Agency.)

2. In *photometry*, a modifier used to denote that a given physical quantity, such as luminous emittence, is weighted according to the manner in which the response of the human eye varies with the *wavelength* of the light.

luminous density   The instantaneous amount of *luminous energy* contained in a unit volume of the propagating medium; to be distinguished from *radiant density* in that it is weighted in accordance with the characteristics of the human eye in its non-uniform response to different *wavelengths* or *light*. Compare this term with *flux density, illuminance.*

luminous efficiency   For a given *wavelength* of *visible radiation*, the ratio of the *flux* that is effectively sensed by the human eye to the flux that is intrinsic in the *radiation*. It may be represented as a dimensionless ratio, e.g., *lumens* per watt.

   Thus, luminous efficiency is a weighting factor that is applied to radiation quantities so that they are related physiologically to the response of the human eye, which varies as a function of wavelength. All quantities that are weighted in this manner should be modified by the term luminous (e.g., luminous emittence, luminous flux, etc.)

luminous emittance   The emittance of *visible radiation* weighted to take into account the different response of the human eye to different *wavelengths* of *light*.

   In *photometry*, luminous emittance is always used as a property of a self-luminous source, and therefore should be distinguished from luminance.
See: luminous efficiency

luminous energy   The *energy* of *visible radiation*, weighted in accordance with the *wavelength* dependence of the response of the human eye. Also called light energy.
See: luminous efficiency

luminous flux   *Luminous energy* per unit time; the *flux* of *visible radiation*, so weighted as to account for the manner in which the response of the human eye varies with the *wavelength* of radiation.

   The basic unit for luminous flux is the *lumen.*
See: luminous efficiency

luminous intensity   Luminous energy per unit time per unit solid angle; the intensity (flux per unit *solid angle*) of *visible radiation* weighted to take into account the variable response of the human eye as a function of the *wavelength* of *light*; usually expressed in *candela* (cd).
See: luminous efficiency

lumped mass   Concept in engineering analysis wherein a mass is treated as if it were concentrated at a point.

lunar   Of or pertaining to the *Moon.*

Lunar Module [abbr. LM]   Vehicle used for manned Moon landing. The module lifts off from Earth enclosed in a compartment of the Saturn 5 launch vehicle, below the command-service module that houses the astronauts. The command module pulls the LM from its storage area once the craft are on their way to the Moon. Once in lunar orbit, the crew undocks the module for descent on the Moon. The silver and black ascent stage, containing the astronauts' pressurized compartment and the clusters of rockets that control the craft, fits on top of the shiny gold descent stage that actually touches down on the Moon. The descent stage contains a main, centrally located rocket engine. This segment of the craft remains on the Moon as the crew lifts off in the ascent stage to rejoin the command module. After the crew transfers to the command module, the ascent stage is also left behind.

lunar orbit   *Orbit* of a *spacecraft* around the *Moon.*

Lunar Orbiter Probes   NASA project (between August 1966 and August 1967) designed to take and transmit both wide-angle and closeup images of the Moon in preparation for a manned landing. The Orbiters were used to select the best landing sites for the Apollo mission and also showed that the Moon's gravitational field permitted stable orbits. The probes were sent into orbit around the Moon to gather information and then purposely crashed at the end of the mission to prevent interference with future projects. There were five missions in the program. All were successful.

lunar parallax   The horizontal *parallax* or the *geocentric* parallax of the *Moon.*

lunar probe   A probe for exploring and

reporting on conditions on or about the *Moon.*

lunar soil and rock samples (for display)   The landings of the *Apollo astronauts* on the *Moon* (1969–72) was a magnificient technological triumph for the United States and for all mankind. These missions ushered in a new era in our study of the *solar system* and the *universe* around us. For the first time in human history, we could examine another world at close range. Astronauts stood on the Moon, photographed the small details of its mysterious surface and set up instruments to probe its interior. From *orbit* around the Moon, other sensitive instruments in the *Apollo spacecraft* measured the chemical composition, *gravity* and magnetism of the Moon.

The actual return of lunar rock and soil samples to Earth was a major scientific achievement of the Apollo program. Only by studying "moon rocks" with all the resources of laboratories here on Earth could scientists determine the exact nature of the Moon, unravel its long and complicated history and learn how the Moon had recorded the history of the solar system for 4.5 billion years.

The six successful Apollo landings returned more than 2,000 different samples of the Moon—342 kg (842 lb$_m$) in all! From these samples we have learned that the Moon is not a monotonous and uniform world but rather a complex *planet* with its own unique history.

M

Mach angle   The angle between a *Mach line* and the direction of movement of undisturbed flow.
See: Mach wave

Mach cone   1. The cone-shaped *shock wave* that theoretically emanates from an infinitesimally small particle moving at *supersonic* speed through a fluid. It is the *locus* of the *Mach lines.*
2. The cone-shaped shock wave generated

by a sharp-pointed body, as at the nose of a high-speed aircraft.
See: Mach wave

Mach line   A line representing a *Mach wave;* a Mach wave.

Machmeter   An instrument that measures and indicates speed relative to the *speed of sound,* that is, a device that indicates *Mach number.*

Mach number (symbol M)   A number expressing the ratio of the speed of a body or of a point on a body with respect to the surrounding air or other fluid, or the speed of a flow, to the *speed of sound* in the medium. It is named after the Austrian scientist Ernst Mach (1838–1916).

If the Mach number is less than one, the flow is called **subsonic,** and local disturbances can propagate ahead of the flow. If the Mach number is greater than one, the flow is called **supersonic,** and disturbances cannot propagate ahead of the flow, with the result that *shock waves* form.
See: Cauchy number

Mach wave   1. A *shock wave* theoretically occurring along a common line of intersection of all the pressure disturbances emanating from an infinitesimally small particle moving at *supersonic* speed through a fluid. Such a wave is considered to exert no changes in the condition of the fluid through which it passes. The concept of the Mach wave is used to define and study certain disturbances in a supersonic flow field.
2. A very weak shock wave appearing, for example, at the nose of a very sharp body where the fluid undergoes no substantial change in direction.

magnetic field   A region of space in which any *magnetic dipole* experiences a *magnetic force* or *torque.* It is the field of force that surrounds a magnetic pole or a current flowing through an electrical conductor in which there is a *magnetic flux.*

magnetic field strength [Symbol: H]   The strength of the magnetic *force* on a unit magnetic pole in a region of space affected by other magnets or electric currents. The integral of the magnetic field strength along a closed path is equal to the *magnetomotive force,* a quantity measured in *amperes* or ampere-turns. The magnetic field strength, also called magnetic intensity, is measured in amperes per meter.

magnetic flux [Symbol: Φ]   A measure of the total size of a *magnetic field*. It is defined as the scalar product of the *magnetic flux density* [B] and the area. The *weber* is the *SI-unit* of magnetic flux.

magnetic flux density [Symbol: B]   The *magnetic flux* that passes through a unit area of a *magnetic field* in a direction at right angles to the magnetic force. The vector product of the magnetic flux density [$\vec{B}$] and the current [$\vec{I}$] in a conductor gives the *force* per unit length of conductor. The *tesla* is the *SI unit* of magnetic flux density, being defined as one *weber* of *magnetic flux* per meter squared. Also called magnetic induction.

magnetomotive force   The closed path (circular) integral of the *magnetic field strength*.

$$F_m = \oint \vec{H}\, dx$$

Also called magnetic potential, it is measured in *amperes* or ampere-turns.

main combustion chamber (MCC)   A component of the *Space Shuttle Main Engine* whose primary function is to receive the mixed *propellants* from the *main injector*, accelerate the hot combusted gases to *sonic velocity* through the *throat* and expand these gases supersonically through the *nozzle*.
See: Space Shuttle Main Engine

main injector   A component of the *Space Shuttle Main Engine* that performs the vital function of mixing the *liquid oxygen* and *liquid hydrogen* together as thoroughly and uniformly as possible to produce efficient combustion.
See: Space Shuttle Main Engine

main propellant valves (MPVs)   The main propellant valves of the *Space Shuttle Main Engine* consist of the *main oxidizer valve,* the *main fuel valve, the oxidizer preburner oxidizer valve,* the fuel *preburner oxidizer valve* and the *chamber coolant valve.*
See: Space Shuttle Main Engine

Main Stage   1. In a *multistage rocket,* the stage that develops the greatest amount of *thrust,* with or without *booster engines.*
2. In a single-stage rocket vehicle powered by one or more engines, the period when full thrust (at or above 90%) is attained.
3. A *sustainer engine,* considered as a stage after booster engines have fallen away, as in "the main stage of the *Atlas.*"
See: Space Shuttle Main Engine

mainstage   Attainment of 90 percent or more of the steady-state rated *thrust* level of a *rocket engine.*

Maintenance   The actions taken to maintain an item in a specified condition, including systematic inspection, detection of flaws and servicing to prevent failure. Also the actions taken to restore an item to a specified operational condition, including fault identification, item replacement, repair and verification of serviceability.

main valve   The valve, usually located just upstream of the *thrust chamber* injector, that controls the flow of the propellant to the injector.

mandrel   The tool that forms the geometry of the central cavity in the casting of a *solid-propellant grain.*

maneuver pad   Data or information on *spacecraft attitude, thrust* values, event times, etc., that is transmitted in advance of a maneuver.

manifold   Fluid-flow enclosure that distributes the flow in a desired manner from an inlet or inlets to an outlet or outlets.

manipulators   Mechanical devices used for handling of *radioactive materials.* Frequently they are remotely operated from behind a protective shield.
See: hot cell, remote manipulator system, teleoperators.

manometer   An instrument for measuring pressure of *fluids* both above and below *atmospheric pressure.*

marmon clamp   A ring-shaped clamp, consisting of three equal length segments held together by *explosive bolts,* used to couple the main subsections of a *rocket* vehicle.

Mars   The fourth planet in the Solar System with a diameter of 6794 km. Mars is rugged with huge volcanoes and deep chasms. It is pockmarked by craters of thousands of meteorites that have crashed in the planet. Most of our information on the planet comes from the Mariner 9 and Viking probes. Cameras on Mariner discovered what appeared to be dried riverbeds, suggesting the onetime presence of water, and encourging the theory that there may be life on Mars. But the Viking probe left the question unresolved. All life-detecting instruments gave ambiguous results. The atmosphere is 100 times thinner than Earth's, but fierce windstorms rage in the summer. Mars's polar caps, composed mostly of carbon dioxide, recede and advance with the Martian seasons.

mascon   Mass concentration; a region of high *density* below the *Moon's* surface.

maser   An *amplifier* utilizing the principle of microwave amplification by stimulated *emission* of *radiation.* Emission of *energy* stored in a molecular or atomic system by a *microwave* power supply is stimulated by the *input signal.*

mass [Symbol: m]   The quantity of matter in a body. Often used as a synonym for weight, which, strictly speaking, is the force exerted by a body under the influence of *gravity.*
See: atomic mass unit, atomic weight.

mass defect   The difference between the *atomic mass* and the mass *number* of a *nuclide.*

mass-energy equation (mass-energy equivalence) (mass energy relation)   The statement developed by Albert Einstein that "the mass of a body is a measure of its energy content," as an extension of his 1905 Special Theory of Relativity. The statement was subsequently verified experimentally by measurements of *mass* and *energy* in *nuclear reactions.* The equation, usually given as: $E = mc^2$, shows that when the energy of a body changes by an amount, E, (no matter what form the energy takes) the mass, m, of the body will change by an amount equal to E/c^2. (The factor c^2, the square of the speed of light in a vacuum, may be regarded as the conversion factor relating units of mass and energy,) This equation predicted the possibility of releasing enormous amounts of energy (in the atomic bomb) by the conversion of mass to energy. It is also called the Einstein equation.

mass flow rate [Symbol: ṁ]   The mass of *fluid* flowing through or past a reference point per unit time. Typical units are kg/sec [lb_m/sec].

mass flux   Mass flowrate through a given area expressed as the ratio of mass flowrate to flow area.

mass number (symbol A)   The number of *nucleons* (i.e., the number of *protons* and *neutrons*) in an atomic *nucleus.* It is the nearest whole number to an atom's *atomic weight.* For example, the mass number of the isotope uranium-235 is 235.
See: atomic number, nuclide designation

mass ratio   The ratio of the *mass* of the *propellant* charge of a *rocket* to the total mass of the rocket when charged with the propellant.

mass spectrometer   An instrument used for the precise measurement of the *atomic masses* or *isotopes.* It separates *nuclei* that have different charge-to-mass ratios by passing a stream of charged particles through electrical and magnetic fields. The mass spectrometer can therefore analyze a gas or *vapor* according to the atomic mass of its components.

mass transfer cooling   This can be categorized as either (1) *transpiration cooling,* (2) *film cooling* or (3) *ablation cooling.* Aerospace designers have long recognized mass transfer to the *boundary layer* as a technique for significantly altering the *aerodynamic heating* phenomena associated with a body in *hypersonic flight* through a planetary atmosphere.

mass-velocity ratio   A quantity m_v/m_r expressing the relativistic variation of *mass* with *velocity.*
$$m_v = \frac{m_o}{[1 - (v^2/c^2)]}$$
where m_v is moving mass, m_r is rest mass, v is velocity, and c is the velocity of *light.*

This ratio becomes important only at speeds approaching the speed of light.

mating   The act of fitting together two major components of a system as mating of a *launch vehicle* and a *spacecraft.*

matter   The substance of which a physical object is composed. All materials in the *universe* have the same inner nature, that is, they are composed of *atoms,* arranged in different (and often complex) ways; the specific atoms and the specific arrangements identify the various materials.

maximum credible accident   The most serious *nuclear reactor* accident that can reasonably be imagined from any adverse combination of equipment malfunction, operating errors, and other foreseeable causes. The term is used to analyze the safety characteristics of a reactor. For example, *space nuclear reactors* are designed to be safe even if a maximum credible accident should occur.

matrix   In general, an array of elements composed of rows and columns; items arranged in a pattern. Specifically, an array of numbers or mathematical symbols that can be manipulated according to certain rules. Consider, for example, the linear transformation (system of linear equations)

$$y_1 = a_{11}\ x_1 + a_{12}\ x_2$$
$$y_2 = a_{21}\ x_1 + a_{22}\ x_2$$

where x_1, x_2 and y_1, y_2 are variable quantities, while the coefficients a_{11}, a_{12}, a_{21} and a_{22} are given numbers. By arranging the coefficients in the way they occur in the transformation and enclosing them in parentheses, an array of the form

$$\begin{matrix} a_{11} & a_{12} \\ a_{21} & a_{22} \end{matrix}$$

is obtained. This array is an example of a matrix. Therefore, a rectangular array of numbers of the form

$$\begin{matrix} a_{11} & a_{12} & \dots & a_{1n} \\ a_{21} & a_{22} & \dots & a_{2n} \\ \cdot & \cdot & \cdot & \cdot \\ \cdot & \cdot & \cdot & \cdot \\ \cdot & \cdot & \cdot & \cdot \\ a_{m1} & a_{m2} & \dots & a_{mn} \end{matrix}$$

is called a "matrix." The numbers a_{11}, $a_{12}\dots$, a_{mn} are called the "elements" of the matrix. The horizontal lines are called "rows" or "row vectors"; while the vertical lines are called "columns" or "column vectors." A matrix with "m" rows and "n" columns is called an (m x n) matrix [expressed "m" by "n" matrix].

matrix mechanics   A form of *quantum mechanics.*

max-Q   An aerospace term describing the condition of maximum dynamic *pressure*— the point in the flight of a *launch vehicle* when it experiences the most severe *aerodynamic forces.*

maximum evaporation rate   The maximum rate at which molecules could emerge from a surface, deduced from measurements of saturated *vapor pressure* at the same temperature.

Maxwellian distribution   The *velocity* distribution, as computed in the *kinetic theory of gases,* of the *molecules* of a *gas* in thermal equilibrium.

In nuclear engineering this distribution is often assumed to hold for *neutrons* in thermal equilibrium with the *moderator* (thermal neutrons).

mean anomaly [Symbol: M]   The angle from *periapsis* through which an orbiting body would move in a specified period of time if it moved at its mean angular rate.

mean free path [Symbol: λ]   The average distance traveled by a particle, atom, or molecule between *collisions* or interactions.

mean life   The average time during which an *atom,* an excited *nucleus,* a *radionuclide* or a *particle* exists in a particular form.

median lethal dose   The amount of radiation required to kill, within a specified period, 50% of the individuals of a group of animals or organisms.
See: lethal dose

Mercury   Innermost planet in the Solar System, with a diameter of 488 km. The Mariner 10 spacecraft revealed a desolate, heavily cratered surface, in many ways resembling the Moon. As Mercury's interior cooled and shrank, it compressed the crust creating huge scarps (cliffs) that crisscross the planet. They are as high as 2 km (1.2 mi.) and as long as 1500 km (900 mi.). The planet, like Earth, appears to have a crust of silicate rock. Mariner discovered a trace of an atmosphere, a trillionth of the density of the Earth's. It is composed chiefly of argon, neon and helium. Mercury's magnetic field is only one hundredth that of Earth. Temperatures range from 510° C (950° F) on the sunlit side to −210°C (−350° F) on the dark side.

Mercury Project   America's pioneering project to put man in orbit. The series of six suborbital and orbital flights was designed to demonstrate that man could withstand the high acceleration of a rocket launching, a prolonged period of weightlessness and then a period of high deceleration during reentry. The program lasted from May 1961 to May 1963. It included two one-man sub-orbital flights followed by four orbital manned missions. Two space boosters were combined with the Mercury capsule: the first, a modified Redstone, was used for the suborbital portion; the second, the Mercury-Atlas combination, was used for the orbital missions.

meson   One of a class of medium-mass, short lived *elementary particles* with a mass between that of the *electron* and that of the *proton.* Examples: pi-mesons (pions) and K-mesons (kaons). Compare this term with *baryon* and *lepton.*

metagalaxy   The entire system of *galaxies* including the *Milky Way;* that is, the entire contents of the *universe* together with the region of *space* it occupies.

metallurgy   The study and practice of producing, fabricating, heat treating and alloying metals.

metastable atom   An *atom* with the *electron* excited to an *energy level* where simple *radiation* is forbidden and thus the atom is momentarily stable.
See: forbidden line

meterorological rocket   A *rocket* designed for routine upper *air* observation (as opposed to research) especially that portion inaccessible to *balloons,* i.e., above 30480m [100,000 feet]. Also called rocketsonde.

method of attributes   In *reliability* testing, measurement of quality by noting the presence or absence of some characteristic (attribute) in each of the units in the group under consideration and counting how many do or do not possess it.

metric data (downlink)   Information regarding the condition of telemetered systems.

metrology   The science of dimensional measurement; sometimes includes the science of weighing.

micromanometer   A *manometer* capable of measuring very small *pressure* changes or differences.

micrometeorite penetration   Penetration of the thin outer shell (skin) of space vehicles by small particles travelling in space at high velocities.

micrometer   One of a class of instruments for making precise linear measurements in which the displacements measured corres-pond to the travel of a screw of accurately known pitch.

micrometeroids   Particles of *meteoritic* dust in *space* ranging in size from 1 to 200 or more micrometers (microns) in size.

microwave   Of or pertaining to radiation in that region of the *electromagnetic* (radio) *spectrum* between approximately 1,000 and 300,000 *megahertz* (30 cm to 1 mm *wavelength*).

midcourse guidance   *Guidance* of a *rocket* from the end of the launching phase to some arbitrary point or at some arbitrary time when *terminal guidance* begins. Also called incourse guidance.
See: guidance

mid deck (MD)   A portion of the crew station module of the *Space Shuttle Orbiter.* It contains provisions and stowage facilities for four crew sleeping stations. Stowage of the *lithium hydroxide canisters* and other gear, the *waste management system,* the *personal hygiene station* and the work/dining table are also located in the mid deck.
See: Orbiter structure

mid fuselage   A structure that interfaces with the forward fuselage, the aft fuselage and the wings and, in addition to forming the *payload bay* of the *Orbiter,* supports the payload bay doors, hinges and tie-down fittings; the forward wing glove; and various Orbiter system components.
See: Orbiter structure

Mie scattering   Any scattering produced by spherical *particles* without special regard to comparative size of radiation *wavelength* and particle diameter.
See: Mie theory.

Mie theory   A complete mathematical-physical theory of the scattering of *electromagnetic radiation* by spherical particles, developed by G. Mie in 1908. In contrast to *Rayleigh scattering,* the Mie theory embraces all possible ratios of diameter to *wavelength.*

The Mie theory is very important in meteorological optics, where diameter-to-wavelength ratios of the order of unity and larger are characteristic of many problems regarding haze and cloud scattering. Scattering of radar energy by raindrops constitutes another significant application of the Mie theory.

milestone An important event or decision point in a program or plan. The term originates from the use of stone markers set up on roadsides to indicate the distance in miles to a given point. Milestone charts are used extensively in *aerospace* programs and planning activities.

minimum ionizing speed The speed with which a free electron must move through a given *gas* to be able to *ionize* gas atoms or *molecules* by *collision*. In *air* at standard conditions, this speed is about 100 km/sec.

Milky Way An immense accumulation of millions of stars. It is the galaxy to which our Sun belongs.

mirage A *refraction* phenomenon in the *atmosphere* wherein an image of some object is made to appear displaced from its true position.

Simple mirages may be any one of three types, the inferior mirage, the superior mirage, or the lateral mirage, depending respectively, on whether the spurious image appears below, above, or one side of the true position of the object. Of the three, the inferior mirage is the most common, being usually discernible over any heated street in daytime during summer. The abnormal refraction responsible for mirages is invariably associated with abnormal temperature distributions that yield abnormal spatial variations in the refractive index.

missile Any object thrown, dropped, fired, launched, or otherwise projected with the purpose of striking a target. Short for ballistic missile, guided missile.

Missile should not be used loosely as a synonym for *rocket, spacecraft,* or *launch vehicle.*

missilry The art or science of designing, developing, building, launching, directing, and sometimes guiding a rocket missile; any phase or aspect of this art or science. Also called missilery.

mission The performance of a set of investigations or operations in space to achieve program goals. In contrast to a *flight,* which is a single *Space Shuttle* round trip, a mission may require only part of a flight or several flights to complete its goals.

Mission Control Center [abbr. MCC] The NASA facility responsible for providing total support for all phases of a flight—prelaunch, ascent, reentry and landing. It provides systems monitoring and contingency support for all STS elements, maintains communications with the crew and onboard systems, performs flight data collection and coordinates flight operations. The payload on an STS mission is the responsibility of the Payload Operations Control Center. MCC is located at the Johnson Space Center in Houston, Tex.

mission specialist The *Space Shuttle* crew member responsible for coordinating overall *payload/Space Transportation System* (STS) interaction and, during the payload operation phase of a flight, directing the allocation of STS and crew resources to accomplish the combined payload objectives. The mission specialist is responsible to the *user* or users when carrying out scientific assignments and operates in compliance with mission rules and *Payload Operations Control Center* directives. When so designated by the STS user or users, the mission specialist also has authority to resolve conflicts regarding payloads and to approve deviations from the flight plan such as may arise from payload equipment failures or other factors.

The mission specialist is therefore a "career" astronaut proficient in payload/experiment operations, requirements, objectives and supporting equipment. He or she is also knowledgeable about the *Orbiter* and attached payload support systems and is the prime crew member for *extravehicular activity.* Finally, the mission specialist is responsible for coordinating overall Orbiter operations in the areas of crew activity planning, consumables usage and other activities affecting payload operations. Under certain circumstances the mission specialist may perform special payload handling or maintenance. He or she may also assist in managing payload operation, at the discretion of the user, and may, in certain cases, serve as a *payload specialist.* In fact, the mission specialist has prime responsibility for experiments to which no payload specialist is assigned and assists the payload specialist when appropriate.

During launch and recovery the mission specialist is responsible for monitoring and controlling the payload to assure payload

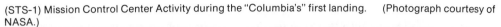

(STS-1) Mission Control Center Activity during the "Columbia's" first landing.   (Photograph courtesy of NASA.)

integrity and vehicle safety. He or she also assists the *commander* and *pilot* during these phases as required. The mission specialist occupies the *mission station,* which is located on the Orbiter's *flight deck,* just behind and to the right of the pilot's seat. This station contains the controls to manage the Orbiter's interfaces with payloads and their equipment that are critical to the Orbiter's safety. The mission station is equipped to monitor, command, control and communicate with payloads attached to the Orbiter or flying nearby. A *caution and warning* display alerts the mission specialist to malfunctions in payload systems or components. Orbiter functions that are not immediately critical to the flight can also be managed by the mission specialist from this station.
See: Shuttle crew, commander, payload specialist, pilot

mission station (MS)   A position located aft of the pilot's station on the right side of the *Orbiter*'s *flight deck* that has displays and controls for Orbiter-to-*payload* inter-

faces and payload subsystems and an auxiliary caution-and-warning display that alerts the crew to critical malfunctions in the payload systems. Payload support operations are performed from this location, usually by the *mission specialist.*
See: Mission specialist, Orbiter, Orbiter structure

mixed-flow compressor   A rotary *compressor* through which the acceleration of *fluid* is partly radial and partly axial.

mixed payloads   *Shuttle cargo* which contains more than one type of *payload.*

mixing length   A mean length of travel characteristic of a particular motion in a *fluid* over which an *eddy* maintains its identity; analogous to the *mean free path* of a *molecule.*

mixture ratio   *Mass flow rate* of *oxidizer* divided by *mass flow rate* of *fuel.*

MKS system   A system of units based on the meter, the kilogram, and the second.
See: SI units

mobile launch platform (MLP)   The structure on which the elements of the

Space Shuttle are stacked in the *Vehicle Assembly Building* and then moved to the *launch pad* at the *Kennedy Space Center*. *See:* Launch pad, Kennedy Space Center

mobility aid Handrails or footrails that help crew members move about the *spacecraft*.

mock test An operational test of a complete *launch vehicle* or *rocket* system performed without actually firing the rocket *engine*(s).

mock-up A full-sized replica or dummy of something, such as a *spacecraft*, often made of some substitute material, such as wood, and sometimes incorporating actual functioning pieces of equipment, such as *engines*.

mode A functioning position or arrangement that allows for the performance of a given task. For example, we can say that a *spacecraft* has gone from an operational mode to a *rendezvous* (or standby) mode.

mode Any of the various stationary vibration patterns of which an elastic body is capable.

model atmosphere 1. Any theoretical representation of the *atmosphere*, particularly of vertical temperature distribution.

mode of vibration A characteristic pattern assumed by a system undergoing vibration in which the motion of every particle is *simple harmonic* with the same *frequency;* also called "mode shape."

moderator A material, such as ordinary water, heavy water or graphite, used in a *nuclear reactor* to slow down high-velocity *neutrons*, thus increasing the likelihood of further *fission*. Compare this term with reflector. *See:* Absorber, thermal neutrons

mode shape *See:* Mode of vibration

modification complete Date when modification of existing facilities is finished. *Certification* by the site activation office completes this *milestone*.

modulating Term applied to a control system or device in which the controlled variable is proportional to a sensed *parameter* and is continuously variable within the regulated range.

modulation The variation in the value of some parameter characterizing a periodic *oscillation*. Specifically, the variation of some characteristic (e.g., amplitude, *frequency,* phase, etc.) of a *radio wave,* called the "carrier wave", in accordance with the instantaneous values of another wave, called the "modulating wave".

Variation of amplitude is called *amplitude modulation* variation of frequency is *frequency modulation;* and variation of phase is *phase modulation.* The formation of very short bursts of a carrier wave separated by relatively long periods during which no carrier wave is transmitted is called *pulse modulation.*

module 1. In the *Space Transportation System* a pressurized manned laboratory suitable for conducting science, applications and technology activities.

2. A self-contained unit of a *launch vehicle* or *spacecraft* that serves as a building block for the overall structure. It is common aerospace practice to refer to the module by its primary function, for example, the "command module" in the *Apollo* expeditions to the *Moon.*

3. A one-package assembly of functionally related electronic parts, usually a "plug-in" unit arranged to function as a system or subsystem; a *"black box."* *See:* Spacelab

molecular effusion The passage of *gas* through a single opening in a plane wall of negligible thickness where the largest dimension of the *hole* is smaller than the *mean free path.*

molecular flow The flow of gas through a duct under conditions such that the *mean free path* is greater than the largest dimension of a transverse section of the duct.

molecular flux The net number of gas molecules crossing a specified surface in unit time, those having a *velocity* component in the same direction as the normal to the surface at the point of crossing being counted as positive and those having a velocity component in the opposite direction being counted as negative.

molecular weight The weight of a given *molecule* expressed in *atomic weight units.*

molecule A group of atoms held together by chemical forces. The atoms in the molecule may be identical, as in H_2, S_2, and S_6, or different, as in H_2O and CO_2. A molecule is the smallest unit of matter which can exist

by itself and retain all its chemical properties. Compare this term with *atom* and *ion.*

moment [Symbol M] A tendency to cause
rotation about a point or *axis,* as of a control surface about its hinge or of an airplane
about its *center of gravity;* the measure of
this tendency, equal to the product of the
force and the perpendicular distance
between the point of axis of rotation and
the line of action of the force.

Moment of Inertia [Symbol: I] Of a body
about an *axis,* the Σmr^2, where m is the
mass of a *particle* of the body and r its distance from the axis.

momentum Quantity of motion.

Linear momentum is the quantity
obtained by multiplying the *mass* of a body
by its linear speed. Angular momentum is
the quantity obtained by multiplying the
moment of inertia of a body by its angular
speed.

The momentum of a system of *particles* is
given by the sum of the moments of the
individual particles which make up the
system, or by the product of the total mass
of the system and the *velocity* of the *center
of gravity* of the system.

The momentum of a continuous medium
is given by the integral of the velocity over
the mass of the medium, or by the product
of the total mass of the medium and the
velocity of the center of gravity of the
medium.

momentum separation Stratification of
flow in a *combustion chamber* (especially in
a gas generator) as a result of the high relative *velocity* of the hot core gases.

momentum thickness Thickness of the
potential flow with a momentum equal to
that lost in the boundary layer as a result of
wall shear forces.

monocoque Term applied to a structure in
which the stressed outer skin carries all or a
major portion of the torsional and bending
stresses.

monopropellant Liquid propellant that
decomposes exothermally to produce hot
gas; e.g., hydrogen peroxide and hydrazine.

monitor 1. An instrument that measures the
level of *ionizing radiation* in an area.
See: radiation detection instrument, radiation monitoring.
2. To observe, listen in on, keep track of, or

exercise surveillance over by any appropriate means, as, to monitor radio signals;
to monitor the flight of a *rocket* by radar; to
monitor a landing approach.

monochromator A device to isolate a single color or spectral *wavelength* from a
spectrum.

monochromatic Pertaining to a single
wavelength or, more commonly, to a narrow band of wavelengths.

monocular A small telescope, like binoculars but for one eye only.

Moon The Earth's only natural satellite and
closest celestial neighbor. It has a diameter
of 3476 km and is 384,400 km from Earth.
The lunar crust and mantle are quite thick,
extending inward to more than 800 km.
Beneath that is a partly molten zone extending to within 700 km of the center. The
question of whether there is a core remains
unresolved. The surface consists of high-
lands composed of alumina-rich rocks that
formed from a globe-encircling molten sea,
and marl made up of volcanic melts which
surfaced about 3.5 billion years ago.

motion The act, process, or instance of
change of position. Also called movement,
especially when used in connection with
problems involving the motion of one craft
relative to another.

Absolute motion is motion relative to a
fixed time frame for reference. Actual
motion is motion of a craft relative to the
Earth. Apparent or relative motion is
change of position as observed from a
reference point which may itself be in
motion. Diurnal motion is the apparent
daily motion of a *celestial body.* Direct
motion is the apparent motion of a *planet*
eastward among the *stars,* retrograde
motion, the apparent motion westward
among the stars. Motion of a celestial body
through space is called space motion, which
is composed of two components: proper
motion, that component perpendicular to
the line of sight; and radial motion, that
component in the direction of the line of
sight.

motion, apsidal Advance or regression of
the *line of apsides. Earth oblateness* causes
the major axis to precess forward, for
orbital inclinations less than i = 63.5 de.,
while for higher inclinations, the motion is

retrograde.

motion, nodal   Advance or regression of a *line of nodes* along a reference plane (usually equatorial or ecliptic plane). *Earth oblateness* causes a regression of the nodes, decreasing with increasing *orbital inclination* until the nodes are stationary for *polar orbits.*

Multimission Modular Spacecraft (MMS)
The *NASA Goddard Space Flight Center* (GSFC) in Greenbelt, Md has designed the multimission modular spacecraft (MMS) to serve a wide variety of missions to be launched from the *Space Shuttle.* The MMS concept is to create *spacecraft* for particular missions from a family of standard functional modules. From the start of the U.S. Space Program spacecraft have been designed to accommodate precisely the shape, size, and weight of their specific missions. [See: spacecraft] Consequently, these spacecraft were made with customized components and fixed structures which could not be reused. Of course, this design approach was only logical, since an on-orbit retrieval/refurbishment capability—as now provided by the Shuttle—did not previously exist. To eliminate the costly and time-consuming design, development, production and procurement activities associated with conventional, uniquely integrated spacecraft, NASA investigated new approaches to spacecraft fabrication. These efforts were driven by the same philosophy and target dates which governed the Shuttle program, and compatibility with *STS* was a major design requirement. The multimission modular spacecraft resulted from these study efforts.

The MMS concept employs standard modules to perform basic spacecraft functions, such as *attitude* control, data handling, and power supply. To satisfy specific mission requirements, these standard subsystem modules are integrated with and complemented by standard options and

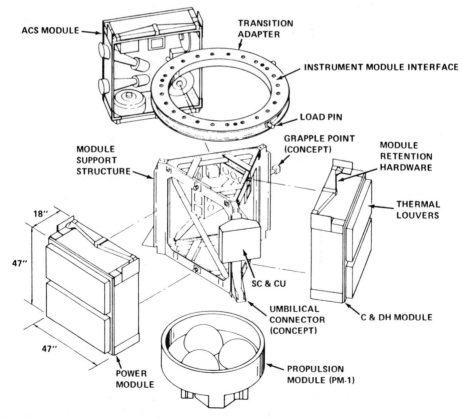

Multimission Modular Spacecraft Components.   (Drawing courtesy of NASA.)

mission-unique equipment into a particular MMS spacecraft. Since this approach results in simplified interfaces between spacecraft components and shortened development times, the MMS represents a major resource savings concept.

The MMS can be used in low-Earth-orbit (LEO) or in *geosynchronous orbit* (GEO) in support of a wide range of remote sensing and research missions. [See Table (1)] Although not directly classitifed as an STS element, the MMS is a *NASA payload carrier* fully compatible with the launch environments and other interface requirements of the Shuttle. The MMS can also be used with expendable *launch vehicles*, including the *Delta* 2910 and 3910 series. The MMS can either be serviced on orbit by the Shuttle, or retrieved and returned to Earth for refurbishment and relaunch.

Table (2) summarizes the general MMS performance capabilities.

Table 1 Candidate MMS Missions
- Weather Monitoring
- Marine and Ocean State Monitoring
- Earth Resource Monitoring and Commodity Prediction
- Water Resource and Pollution Monitoring
- Space Science Experiments
- Biomedical Research
- Communications Research

Table (2) MMS General Performance Capabilities

PAYLOAD WEIGHT CAPABILITY
— for Shuttle launches in excess of 4,536 kg (10,000 lbm) (limited by the payload configuration)
— for expendable launch vehicles 1,814 kg (4,000 lbm)

OPERATING ALTITUDES
— Low-Earth-Orbit (all inclinations), 500 km (270 n mi) to 1,600 km (864 n mi)
— Geosynchronous Orbit

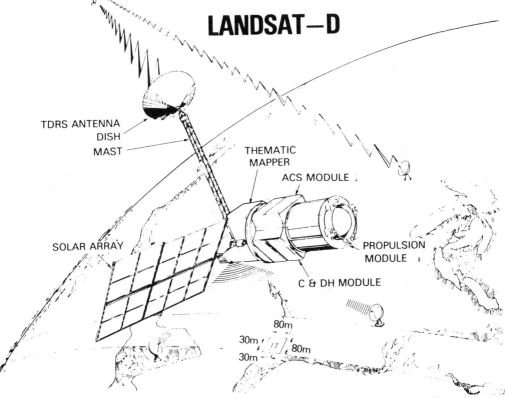

LANDSAT−D

TDRS ANTENNA
DISH
MAST

THEMATIC MAPPER

ACS MODULE

SOLAR ARRAY

PROPULSION MODULE

C & DH MODULE

80m
30m
80m
30m

Typical communication links for an MMS mission in LEO. (Here a LANDSAT-D spacecraft).   (Drawing courtesy of NASA.)

TYPES OF MISSIONS
— Earth, solar, or stellar-pointed missions, or special purpose missions
— LEO or GEO orbits: inertial pointed or payload pointed

LAUNCH VEHICLE
— fully Shuttle, Delta, Atlas or Titan compatible
— IUS launched
— Shuttle on-orbit serviced or retrieved

LIFE EXPECTANCY/REDUNDANCY
— minimum life expectancy of two years
— MMS is designed to have no single-point failure that would prevent resupply or retrieval by the Shuttle.

multiple-degree-of-freedom system   A mechanical system for which two or more *coordinates* are required to define completely the position of the system at any instant.

multiplexer   A mechanical or electrical device that permits time-sharing of a circuit by two or more simultaneous signals.

multiplexing   The simultaneous transmission of two or more signals within a single transmission path or channel. Three basic methods of multiplexing involve the separation of signals by time division, *frequency* division and phase division.

multiplication factor (or constant) [Symbol: K]   The ratio of the number of *neutrons* present in a *reactor* in any one neutron generation to that in the immediately preceding generation. *Criticality* is achieved when this ratio is equal to one. The "infinite" multiplication factor is the ratio in a theoretical system from which there is no leakage, that is, a reactor of infinite size; for an actual reactor (from which leakage does occur), the term effective multiplication factor, which is the ratio based on neutrons available after leakage, is commonly used. *See:* generation time, leakage, neutron, reactivity

multipropellant   A *rocket propellant* consisting of two or more substances fed separately to the *combustion chamber*.

multi-stage compressor   An *Axial flow compressor* having two or more, usually more than two, stages of rotor and stator *blades;* a radial-flow compressor having two

or more *impeller wheels,* Also called a multiple stage compressor.

multistage launch   A launch that uses several stages to boost the payload into *orbit.* After the first-stage *booster* used its *fuel,* it is jettisoned and the secondary booster is fired. When the second-stage fuel is gone, that booster is jettisoned, and so on. Such multistage launching allows very high payload *velocities.*

multistage rocket   A vehicle having two or more rocket units, each firing after the one in back of it has exhausted its *propellant.* Normally, each unit, or stage, is *jettisoned* after completing its firing. Also called a multiple-stage rocket or, infrequently, a step rocket.

N

nacelle   An enclosed portion of an aircraft, not part of the fuselage proper, that is used for housing cargo, the engine, etc.

nadir   Point on the *celestial sphere* that is vertically below the observer, or 180° from the *zenith.* On the celestial sphere the nadir, the center of the Earth and the zenith can be visualized as points on a straight line, with the nadir lying directly beneath an observer (i.e., 90° below the horizon) and the zenith directly overhead.

Napierian base   The logarithmic base, e.

natural circulation reactor   A *nuclear reactor* in which the *coolant* (usually water) is made to circulate without pumping, that is, by natural *convection* resulting from the different densities of its cold and reactor-heated portions.

natural frequency   1. The frequency of *free oscillation* of a system. For a multiple-degree-of-freedom system, the natural frequencies are the frequencies of the normal *modes* of the vibration.
2. The undamped *resonance frequency* of a physical system. The system may be mechanical, pneumatic or electrical.

natural radiation, natural radioactivity

See: Background radiation

Natural uranium *Uranium* as found in nature, containing 0.7% of ^{235}U, 99.3% of ^{238}U, and a trace of ^{234}U. It is also called normal uranium.

See: uranium

nautical mile A unit of distance used principally in navigation. For practical navigation it is usually considered the length of 1 minute of any great circle of the Earth, the meridian being the great circle most commonly used. Also called sea mile.

1 naut mi = 1.852 km = 1.15 statute miles

Navier-Stokes equations The equations of motion for a *viscous fluid* that may be written

$$\frac{dN}{dt} = -\frac{1}{\rho} \nabla P + F + v\nabla^2 N + \frac{1}{3} v\nabla(\nabla \cdot N)$$

where p is the pressure; ρ is the density; t is the time; F is the total external force; N is the fluid velocity; and v is the kinematic viscosity. For an *incompressible fluid,* the term in $\nabla \cdot N$ (divergence) vanishes and the effects of viscosity then play a role analogous to that of temperature in *thermal conduction* and to that of density in simple diffusion.

Solutions of the Navier-Stokes equations have been obtained only in a limited number of special cases. The equations are derived on the basis of certain simplifying assumptions concerning the *stress tensor* of the *fluid;* in one dimension they represent the assumption referred to as the Newtonian friction law.

navigation The practice or art of directing the movement of a craft from one point to another.

Navigation usually implies the presence of a human, a navigator, aboard the craft. Compare this term with *guidance.*

See: (Orbiter) guidance, navigation and control systems

navigational planets The four *planets* commonly used in celestial surface and air navigation: *Venus, Mars, Jupiter* and *Saturn.*

Navstar Global Positioning System [GPS]

The Navstar Global Positioning System (GPS), a space-based radio navigation network, is currently undergoing full-

Fig. 1 The Navstar Global Positioning System Satellites will broadcast navigation signals to land, sea and aerospace forces. (Illustration courtesy of the U.S. Air Force.)

scale engineering development at the *Air Force Systems Command's Space Division* (SD) in Los Angeles, Calif. This multiservice program will provide precise navigation coverage for users throughout the world by the late 1980s. (See Fig. 1). In the fully operational system, *satellites* circling the globe every 12 hours will beam continuous navigation signals to *Earth.* With proper equipment, a user can process the signals and determine his position within tens of feet (i.e. several meters), his velocity within a fraction of a mile per hour (i.e. less than a kilometer per hour), and the time within a millinth of a second.

To demand this information, a GPS user only needs to push a few buttons. A user set will then automatically select four satellites most favorably located, lock onto their navigation signals, and compute the user's position, velocity, and time. User sets are being developed for integration with aircraft, land vehicles and ships, and lightweight backpack unit is under production and test for use by ground troops. Promising airborne and ground tests of these various user sets have been under way for more than three years at the Army's Yuma Proving Ground in Yuma, Ariz.

Overall, the Navstar program is being developed in three phases. During the initial concept validation phase, six satellites were positioned in two orbital planes to provide periodic three-dimensional coverage at Yuma and other test sites. This satellite constellation is being maintained in the second phase for the Trident Improved Accuracy Program and GPS. In the third phase, a three-plane, fully operational system of satellites will transmit continuous three-dimensional navigation signals to users around the world. The satellites will be tracked, controlled, and monitored by personnel at the Navstar Master Control Station to be located within the continental United States.

Navstar GPS equipment is being developed so it can perform many of the functions now collectively accomplished by a number of separate navigation systems. A single Navstar receiver aboard an *aircraft,* for example, could achieve or surpass the capabilities of up to five individual pieces of equipment. Further, Navstar could do so at a lower equipment cost.

Navstar GPS has the following:

ACCURACY:Position: Within 16 meters (52 feet) in three dimensions. Velocity: Within 0.1 meter/second (0.3 foot/ second) in three dimensions. Time: Within a millionth of a second.

APPLICATIONS: Precision weapons delivery; en route navigation for space, air, land, and sea vehicles; aircraft runway approach; photomapping; geodetic surveys; aerial rendezvous/refueling; tactical missile navigation system updating; air traffic control; range instrumentation and safety, as well as search and rescue operations.

Five widely separated monitor stations will passively track all satellites on view and accumulate ranging data from the navigation signals. This information will be processed at a Navstar Master Control station for use in satellite orbit determination and systematic error elimination. The control station will also act as the two-way communications link with the satellites. Through these links, the satellites' computers will be updated with information so users will receive optimum mission performance. In addition, the communications link will provide for telemetry, tracking, and command and control functions.

nebula   A cloud of *interstellar* gas or dust. It can either be seen as a dark hole or band against a brighter background, called a dark nebula, or as a luminous patch of light, called a bright nebula. Using data obtained from the *NASA Ames Research Center* Kuiper Airborne *Observatory* scientists created an infrared map of the "Swan Nebula," also known as M17. Located in the southern sky in the constellation Sagittarius, this region contains a cloud of *ionized,* interstellar gas excited by recently formed *stars.* Astronomers call this type of cloud an *HII* (ionized hydrogen) *region.* The M17 HII region is bounded on the southwest by a *molecular* cloud. The far *infrared* radiation is brightest just adjacent to the HII region, the result of thermal emission from dust grains, most of which are located outside the ionized gas. The core of the molecular cloud is heated primarily by infrared radiation from dust within and next to the HII region. The energy flow appears to be from the HII region into the molecular cloud, illustrating the intimate association between an HII region and the molecular cloud out of which it formed.

needle valve   Valve with a long tapered rod (pintle) for gradual opening or closing of the valve throat when translated.

negative altitude
See: altitude

negative feedback
See: feedback

negative g   In designating the direction of *acceleration* on a body, the opposite of positive g, for example, the effect of flying an outside loop in the upright seated position.

negative gain   Term for a control circuit in which an increase in regulated *pressure* causes an amplified decrease in control pressure.

negative temperature coefficient   1. The partial derivative of any physical variable with respect to *temperature,* when the value of the variable decreases as temperature increases.

2. The decrease in *reactivity* of a *nuclear reactor* with increase in temperature.
See: temperature coefficient of reactivity

Increasing temperature within the reactor increases the average neutron energy. Since the cross section of the fissionable material decreases with increased neutron energy, the net effect of increased temperature is to decrease the number of fissions.

negatron A negative *electron.*

neptunium series (sequence) The series of *nuclides* resulting from the *radioactive decay* of the man-made nuclide, neptunium-237. Many other man-made nuclides decay into this sequence. The end-product of the series is stable bismuth-209, which is the only nuclide in the series that occurs in nature.

Neptune Eighth major planet from the Sun, with a diameter of 49,500 km. Neptune has a very deep atmosphere in which methane has been detected. Hydrogen and helium are believed to be the planet's principal constituents. Inward from its thin outer atmosphere, Neptune is made up of progressively denser layers of gases and then liquids, principally hydrogen, that surround a layer of ice and an iron-silicate core. Neptune has two known satellites: Triton and Nereid. Triton seems to be orbiting so close to the planet that it appears in danger of being torn apart by Neptune's gravity.

net positive suction head [abbr. NPSH] The difference, at the pump inlet, between the head due to total fluid pressure and the head due to propellant vapor pressure, expresses in feet of the propellant being pumped; this is the head available to suppress cavitation in the pump.

net positive suction pressure (NPSP) The *pressure* needed at the *Space Shuttle Main Engine* pump inlets to cause the pumps to work properly.
See: External Tank

net present value Equal to the discounted net cash flow at a given rate of return or interest minus the original investment.

network 1. A combination of *electrical elements.*
2. A group of parts or systems combined to provide a closed information loop, i.e., one that provides for inquiry or command, response, and interpretation of response in relation to inquiry or command.
head due to propellant vapor pressure, expressed in feet of the propellant being

pumped; this is the head available to suppress cavitation in the pump.

neutral Without an electrical charge; neither positive nor negative.

neutral buoyancy Overall density of an object that is almost equal to that of water. A tendency to remain stationary when submerged at depths under the water.
See: buoyancy

neutral buoyancy simulator Objects in space, being *weightless,* behave a little like *neutral buoyancy* objects in water. *Astronauts* train by moving the objects underwater, thereby simulating activities of weightlessness. Fig (1) shows *Astronaut* Anna L. Fisher wearing a *pressurized spacesuit,* which allows her to float freely for a few minutes in a water immersion facility (WIF). The water immersion facility, located at the Johnson Space Center, provides one of two ways in the *Earth*'s *atmosphere* to simulate the *weightlessness* of space. The other technique, which provides a brief period of weightlessness, is the use of an aircraft flying a parabolic curve.
See: buoyancy

neutrino [symbol ν (nu)] An electrically neutral *elementary particle* with a negligible mass. It interacts very weakly with matter and hence is difficult to detect. It is produced in many nuclear reactions, for example, in *beta decay,* and has high penetrating power; neutrinos from the *Sun* usually pass right through the *Earth.*
See: cosmic rays, neutron, nuclear reaction

neutron [Symbol n] An uncharged *elementary particle* with a mass slightly greater than that of the *proton,* and found in the *nucleus* of every *atom* heavier than hydrogen. A free neutron is unstable and decays with a half-life of about 13 minutes into an electron proton, and neutrino. Neutrons sustain the fission *chain reaction* in a *nuclear reactor.*

neutron capture The process in which an atomic *nucleus* absorbs or captures a *neutron.* The probability that a given material will capture neutrons is measured by its neutron capture *cross section,* which depends on the energy of the neutrons and on the nature of the material.
See: capture, nuclear reaction, radiative capture

neutron density The number of *neutrons* per cubic centimeter in the *core* of a *reactor*. *See:* flux

neutron flux
See: flux

neutral density filter A *filter* that attenuates or reduces all regions equally; it is not spectrally selective (no color).

neutron economy The degree to which *neutrons* in a *reactor* are used for desired ends instead of being lost by leakage or nonproductive absorption. The ends may include propagation of the *chain reaction,* converting *fertile* to *fissionable* material, producing *isotopes,* or research.

newton [Symbol: N] The *SI unit* of *force.* It is defined as the force that provides a one kilogram *mass* with an *acceleration* of one meter per second per second.

Newtonian friction law The statement that the tangential force (i.e., the force in the direction of the flow) per unit area acting at an arbitrary level within a *fluid* contained between two rigid horizontal plates, one of which is motionless and the other of which is in steady motion, is proportional to the shear of the fluid motion at that level. Mathematically, the law is given by

$$\tau = \mu(\partial u / \partial z)$$

where τ is the tangential force per unit area, usually called the shearing stress; μ is a constant of proportionality, called the dynamic viscosity; and $\partial u / \partial z$ is the shear of the fluid flow normal to the resting plate. Also called Newton formula for the stress.

Newtonian mechanics The system of mechanics based upon *Newton laws of motion* in which *mass* and *energy* are considered as separate, conservative, mechanical properties, in contrast to their treatment in relativistic mechanics.

Newtonian speed of sound An approximation to the *speed of sound* a in an ideal (perfect) gas given by the relation

$$a^2 = p/\rho$$

where p is *pressure* and ρ is *density.*

Newton derived this expression by assuming the propagation of sound to be an isothermal process. It leads to values about 16% below those observed.

Newtonian telescope A *reflecting telescope* in which a small plane mirror reflects the convergence beam from the objective to an eyepiece at one side of the telescope. After the second reflection the rays travel approximately perpendicular to the longitudinal *axis* of the telescope.

Newton's law of gravitation Every particle of matter in the universe attracts every other particle with a force, F, acting along the line joining the two particles, proportional to the product of the masses m_1 m_2 of the particles and inversely proportional to the square of the distance r between the particles, or

$$F = \frac{Gm_1m_2}{r^2}$$

where G = universal gravitational constant;

$$G = [6.6732 \pm 0.0031] \times 10^{-11} \ Nm^2/kg^2$$

Newton's laws The *three laws of motion* and the law of gravitation published in 1687, explaining almost all the motions of *planets* and *satellites* with high accuracy.

Newton's laws of motion A set of three fundamental postulates forming the basis of the mechanics of rigid bodies, formulated by Newton in 1687.

The "first law" is concerned with the principle of *inertia* and states that if a body in motion is not acted upon by an external *force*, its *momentum* remains constant (law of conservation of momentum). The "second law" asserts that the rate of change of momentum of a body is proportional to the force acting upon the body and in the direction of the applied force. A familiar statement of this is the equation

$$F = ma$$

where F is vector sum of the applied forces, m   the mass, and a the vector acceleration of the body. The "third law" is the principle of action and reaction, stating that for every force acting upon a body there exists a corresponding force of the same magnitude exerted by the body in the opposite direction.

nitrile butadiene rubber (NBR) Insulation found inside the motor case of the *Space Shuttle Solid Rocket Booster.* *See:* Solid Rocket Booster

nitrogen cycle The exchange of nitrogen between animals and plants, in which plants convert urea or nitrates to protein, animals digest protein and excrete its nitrogen contant as urea, which is taken up again by plants.

no bleed Term applied to a control circuit in which there is no flow of gas through the pilot valve under steady-state conditions.

noctilucent Luminous during the night.

noctilucent clouds *Clouds* of unknown composition which occur at great heights, 75 to 90 kilometers. They resemble thin cirrus, but usually with a bluish or silverish color, although sometimes orange to red, standing out against a dark night sky. Sometimes called luminous clouds.

These clouds have been seen rarely, and then only during twilight, especially with the sun between 5° and 13° below the horizon.

nodal circle Pattern of vibration *nodes* that forms a circle.

nodal diameter Pattern of vibration nodes that forms a diametrical line.

node 1. One of the two points of intersection of the *orbit* of a *planet*, planetoid, or *comet* with the *ecliptic*, or of the orbit of a *satellite* with the plane of the orbit of its *primary*.

That point at which the body crosses to the north side of the reference plane is called the *ascending* node; the other, the *descending node*. The line connecting the nodes is called line of nodes. Also called nodal point. *See:* regression of the nodes.

2. A point, line or surface in a *standing wave* where some characteristic of the wave field has essentially zero *amplitude*.

3. A terminal of any branch of a *network* or a terminal common to two or more branches of a network.

no-fire limit Maximum current, power, voltage or capacitance that can be applied to the firing circuit of an *electroexplosive device* without firing the device.

noise 1. Any undesired *sound*. By extension, noise is any unwanted disturbance within a useful frequency band, such as undesired electric waves in a transmission channel or device.

When caused by natural electrical discharges in the atmosphere, noise may be called "static."

2. An erratic, intermittent or statistically random *oscillation*.

3. In electrical circuit analysis, that portion of the unwanted *signal* which is statistically random, as distinguished from hum, which is an unwanted signal occurring at multiples of the power-supply frequency.

noncavitating valve Flow-control valve that meters the flow of liquid propellant without *cavitation* occuring in its operating range.

nonconformance A condition of any article or material or service in which one or more characteristics do not conform to requirements. Includes failures, discrepancies, defects, and malfunctions.

noncondensable gas A gas whose temperature is above its critical *temperature*, so that it cannot be liquified by increase of pressure alone.

nondestructive testing Testing to detect internal and concealed defects in materials using techniques that do not damage or destroy the items being tested. *X-rays*, isotopic radiation and ultrasonics are frequently used.

nondimensional parameter Any parameter of a problem which has the dimensions of a pure number, usually rendered so deliberately.

nonequilibrium composition Exhaust gas chemical composition resulting from incomplete chemical reaction of the products of combustion in the exhaust gas.

nonimpinging injector An *injector* used in rocket engines that employs parallel streams of *propellant* usually emerging normal to the face of the injector. In this injector, mixing is usually obtained by turbulence and diffusion. The *V-2* used a nonimpinging injector.

nonlinear damping Damping due to a *damping force* that is not proportional to *velocity*.

nonlinear distortion *Distortion* caused by a deviation from a proportional relationship between specified measures of the output and input of a system.

nonmodulating Term for a control system or device in which the controlled variable (flow, pressure, etc.) cycles between limits;

sometimes called a "bang-bang" system.

nonpositive-displacement pump   Pump in which the *fluid pressure* is raised by alternately adding to and then diffusing the kinetic energy of the fluid; examples of this type of pump are the axial-flow, Barske, centrifugal-flow, drag, Pitot, and Tesla.

normal   1. Equivalent to usual, regular, rational or standard conditions.
2. Perpendicular.
   A line is normal to another line or a plane when it is perpendicular to it. A line is normal to a curve or curved surface when it is perpendicular to the tangent line of plane at the point of tangency.

normal distribution   The fundamental frequency distribution of statistical analysis. A continuous variate x is said to have a normal distribution or to be normally distributed if it possesses a density function $f(x)$ which satisfies the equation

$$f(x) = \frac{1}{\sigma\sqrt{2\pi}} e - (x - \mu)^2 / 2\sigma^2$$

where μ is the arithmetic mean (or first moment) and σ is the standard deviation. Also called *Gaussian distribution*.

normal emittance   *Emittance* in a direction perpendicular to the surface or in a small solid angle whose axial ray is perpendicular to the surface.

normalize   1. To change in scale so that the sum of squares, or the integral of the squares of the transformed quantity is unity. *See:* orthogonal functions.
2. To transform a random variable so that the resulting random variable has a normal distribution.
3. In computer operations, to adjust the exponent and coefficient of a floating-point result so that the coefficient is in the prescribed normal range. Also called standardize.

normally closed valve   Powered valve that returns to a closed position on shutoff or on failure of the actuating energy or signal.

normally open valve   Powered valve that returns to an open position on shutoff or on failure of the actuating energy or signal.

normal mode of vibration   A *mode of free vibration* of an undamped system. In general, any composite motion of a vibrating system can be analyzed into a summation of its normal modes. Also called natural mode, characteristic mode and eigenmode.

normal positions   Positions assumed by the elements in a fluid-system component when no operating forces are applied.

"normal" propellants   *Bipropellants* that derive *thermal energy* primarily from *oxidizer-fuel* reactions.

normal shock wave   A *shock wave* perpendicular, or substantially so, to the direction of flow in a *supersonic flow*. Sometimes shortened to "normal shock."

nozzle   1. Carefully shaped aft portion of the thrust chamber that controls the expansion of the exhaust products so that the thermal energy produced in the combustion chamber is efficiently converted into kinetic energy, thereby imparting thrust to the vehicle.
2. Convergent passage in a pump or turbine that directs fluid into or leads it away from the impeller or turbine wheel.
3. That part of a *rocket thrust chamber* assembly in which the combustion gases are accelerated to high *velocity*.
See: Space Shuttle Main Engines, Solid Rocket Booster; Rocket

nozzle assembly   A part of the *Space Shuttle Main Engine* that provides the maximum possible thrust efficiency by allowing continued expansion of the combustion gases coming from the *main combustion chamber*.
See: Space Shuttle Main Engine

nozzle-contraction area ratio   The ratio of the cross-sectional area for gas flow at the *nozzle* inlet to that at the *throat*.

nozzle efficiency   The efficiency with which a *nozzle* converts *potential energy* into *kinetic energy*. It is commonly expressed as the ratio of the actual change in kinetic energy to the ideal change at the given *pressure* ratio.

nozzle exit area   The cross-sectional area of a *rocket nozzle* measured at the nozzle exit that is available for gas flow.

nozzle-expansion area ratio   The ratio of the cross-sectional area available for gas flow at the exit of a *nozzle* to the cross-sectional area available at the *throat*.
See: expansion ratio

nozzle extension   Nozzle structure that is added to the main nozzle in order to increase expansion area ratio or provide a change in nozzle construction.

nozzle throat   That portion of a *nozzle* with the smallest cross section.

nozzle thrust coefficient [symbol C_F]   A measure of the amplification of a *thrust* due to gas expansion in a particular nozzle as compared with the thrust that would be exerted if the *chamber pressure* acted only over the *throat area*. Also called "thrust coefficient."

nuclear battery   A radioisotopic generator.

nuclear-electric rocket engine   A rocket engine in which a *nuclear reactor* is used to generate electricity that is used in an electric propulsion system.
See: space nuclear power

nuclear emulsion   A very thick photographic emulsion used in the study of *cosmic rays* and other energetic particles. The path of the particle through the thick emulsion is recorded in three dimensions.

nuclear energy   The energy liberated by a nuclear reaction (*fission* or *fusion*) or by *radioactive decay*.
See: decay; radioactive; nuclear explosive; nuclear reactor

nuclear explosive   An explosive based on *fission* or *fusion* of atomic nuclei.

nuclear fission   *See:* fission, fusion

nuclear fuel   *Fissionable material* of a reasonably long life, used or usable in producing energy in a *nuclear reactor*.

nuclear power plant   Any device, machine or assembly that converts *nuclear energy* into some form of useful power, such as mechanical or electrical power. In a nuclear electric power plant, heat produced by a *reactor* is generally used to make steam to drive a *turbine* that in turn drives an electric generator.

nuclear radiation   The emission of *neutrons* and other particles from an atomic *nucleus* as the result of *nuclear fission* or *nuclear fusion*.

nuclear reaction   A reaction involving a change in an atomic *nucleus,* such as *fission, fusion, neutron capture* or *radioactive decay*, as distinct from a chemical reaction, which is limited to changes in the electron structure surrounding the nucleus. Compare this term with *thermonuclear reaction*.

nuclear weapons   A collective term for *atomic bombs* and *hydrogen bombs*. Any weapons based on a nuclear explosive.

nucleate boiling   Formation and breaking away of bubbles from active bubble sites (nuclei) on a submerged heated surface; the rising bubbles stir the liquid so that heat transfer from the surface to the liquid is much greater than that due to normal convection.

nucleon   A constituent of an atomic *nucleus*, that is, a *proton* or a *neutron*.

nucleonics   The science and technology of nuclear energy and its applications.

nucleus   1. The central portion of an atom, consisting of protons and neutrons bound together by a strong nuclear force. It is positively charged. The total number of protons plus neutrons is the mass number. 2. The small, permanent portion of a comet.

nuclide   A general term applicable to all atomic forms of the elements. The term is erroneously used as a synonym for "isotope", which properly has a more limited definition. Whereas isotopes are the various forms of a single element (hence are a family of nuclides) and all have the same *atomic number* and number of *protons*, nuclides comprise all the isotopic forms of all the elements. Nuclides are distinguished by their *atomic number, atomic mass* and energy state. Compare this term with *element* and *isotope*.

nuclide designation   In accordance with the recommendations of the International Unions of Pure and Applied Chemistry, the following designations, illustrated in figure 1, are used for *nuclides*:

The *mass number* of a nuclide is placed as a superscript to the left of the symbol for the chemical element of the nuclide, rather than to its right, as was formerly done; for example, ^{14}N rather than N^{14} for nitrogen-14.

The *atomic number* is placed as a left subscript; for example $^{14}_6C$ for carbon-14 or $^{235}_{92}U$ for uranium-235.

The state of *ionization* is shown as a right superscript; for example, Ca^{++} or SO_4^-.

The number of *neutrons* in the *nucleus* is shown as a right subscript; for example, $^{40}_{20}Ca_{20}$ for the isotope of calcium-40 containing 20 protons (its atomic number, shown as the left subscript) and 20 neutrons (right subscript) in its nucleus.

Excited states are shown either as part of the left superscript or, sometimes, the right superscript; for example, ^{110m}Ag or $^{110}Ag^m$ indicates an excited state of a silver-110 nucleus; He* indicates an excited state of a helium atom.

nutation   1. The oscillation of the axis of any rotating body, as a *gyroscope* rotor. 2. Specifically, in astronomy, irregularities in the precessional motion of the *equinoxes* because of varying positions of the *Moon* and, to a lesser extent, of other *celestial bodies* with respect to the *ecliptic*.

Because of nutation, the Earth's axis nods like a top, describing a slightly wavy circle about the ecliptic pole. The maximum displacement is about 9.21 seconds (constant of nutation) and the period of a complete cycle is 18.60 *tropical years* (period of Moon's node, nutation period).

O

object glass
See: objective

objective   The *lens* or combination of lenses which receives light rays from an object and refracts them to form an image in the focal plane of the eyepiece of an optical instrument, such as a *telescope*. Also called object glass.

objective prism   A small prism placed in front of the *object-glass* of a *telescope*, producing small-scale stellar scale *spectra* in the *field of view*.

oblateness   the degree of flattening of an oblate spheroid.

oblique shock wave   A *shock wave* inclined at an *oblique* angle to the direction of flow in a *supersonic flow* field. Sometimes shortened to oblique shock. Compare this term with *normal shock*.

obliquity of the ecliptic [Symbol: ϵ]   The angle between the plane of the *ecliptic* (the plane of the *Earth's orbit*) and the plane of the *celestial equator*.

The obliquity of the ecliptic is computed from the following formula: 23 degrees 27 minutes 08.26 seconds − 0.4684(t − 1900) seconds, where t is the year for which the obliquity is desired.

observed   In astronomy and navigation, pertaining to a value which has been measured in contrast to one which is computed.

occlusion   Specifically, the trapping of undissolved *gas* in a solid during solidification.

occultation   The disappearance of a *celestial body* or astronomical object behind another body of larger apparent size.

When the *Moon* passes between the observer and a *star*, the star is said to be occulted. The three associated terms, occultation, *eclipse*, and *transit* are exemplified by the motions of the satellites of Jupiter. An eclipse occurs when a satellite passes into the shadow cast by the planet; an occultation occurs when a satellite passes directly behind the planet, so that it could not be seen even if it were illuminated; and a transit occurs when a satellite passes between the observer and the planet, showing against the disk of the planet.

octave   The interval between any two frequencies having the ratio of 1.2.

The interval in octaves between any two frequencies is the logarithm to the base 2 (or 3.322 times the logarithm to the base 10) of the frequency ratio.

ocular   Pertaining to or in relation to the eye.

oculogyral illusion   The apparent movement of an image in *space* in the same direction as that in which one seems to be turning when the semicircular canals (in the ears) are stimulated.

O/F   Ratio of mass flow rate of *oxidizer* to mass flowrate of *fuel* at the time of combustion.

offgassing   The emanation of volatile matter of any kind from materials.

ohmic heating   In *plasma physics*, the *energy* imparted to charged particles as they respond to an electric field and make collisions with other particles.

omni   1. A prefix meaning all, as in omnidirectional.
2. Short for omnirange.

OMS delta-V kit The auxiliary *propellant* tanks that can be added to the basic *orbital maneuvering subsystem* (OMS). Each tank can be used to increase orbital velocity by 152 m/sec (500 ft/sec). This tank is an optional *Space Shuttle flight kit*.

OMS primary thrusters Components of the *orbital maneuvering system* (OMS) consisting of the *bipropellant rocket engines* of the *Orbiter reaction control system* that are used for normal *translation* and *attitude* control.

on-off Term referring to a system or device in which full-stroke actuation or deactuation occurs in response to input signals.

opaque plasma A *plasma* through which an *electromagnetic wave* cannot propagate and is either absorbed or reflected.

In general, a plasma is opaque for frequencies below plasma frequency. The fact that a plasma is opaque over a certain frequency range will change the radiation properties within that frequency range. Any radiation emitted within the volume of the plasma is quickly absorbed. In this opaque region, therefore, the plasma can only radiate from its surface.

open-cycle reactor system A *nuclear reactor* system in which the *coolant* passes through the reactor core only once and is then discarded. Compare this term with *closed-cycle reactor system*.
See: nuclear rocket

open loop Term referring to an electrical or mechanical system in which the response of the output to the input is scheduled or preset; there is no *feedback* of the output for comparison and corrective adjustment.

operating cycles The cumulative number of times an *item* completes a sequence of activation and return to its initial state; e.g., a switched-on/switched-off sequence, a valve-opened/valve-closed sequence, tank-pressurized/depressurized, or *dewar cryogenic* exposure/drain.

operating life The maximum operating time/cycles that an *item* can accrue before replacement or refurbishment without risk of degradation of performance beyond acceptable limits.

operating parameter sensitive item Any item which has a limited life due to variances in its operating parameters (i.e., drift rate in gyro mechanisms) which may not be directly related to operating or calendar time.

operating pressure Nominal pressure to which the fluid-system components are subjected under steady-state conditions in service operations.

operational bioinstrumentation system [OBS] The operational bioinstrumentation system (OBS) provides an amplified electrocardiograph (ECG) *analog* signal from any two designated *crewmembers* on board the *Shuttle* to the Shuttle *avionics* where it is transmitted to the ground in *real time* or stored on tape for dump or postflight return. The major components of the system are a battery-operated signal conditioner, batteries, cables, and *electrodes*. The operational bioinstrumentation system is used on all flights during prelaunch, *launch,* and entry. Onorbit use is limited to *extravehicular activity* (EVA) unless *intravehicular activity* (IVA) use is requested by the flight surgeon.

For flights in which the crewmembers wear *pressure suits,* the OBS is routed through the crewmembers' constant-wear garments and ejection seats and restrained to the pressure suit and seat. For *shirtsleeve flights,* the OBS is routed through the crewmember's constant-wear garment [See: Shuttle crew equipment] and flight suit and restrained to the flight suit and seat. For all flights, the OBS interfaces with one or two of the five biomedical jacks. Three electrodes are placed on the skin in the standard OBS configuration (one sternal, one right chest, and one lateral lead position). The electrodes detect the moving electric field generated by heart muscle depolarization and repolarization, and this *electric field* is amplified and conditioned by the signal conditioner.

operational deflections Deflections imposed on a structure during engine operation or flight (e.g., by thrust vector gimballing, thermal differential, flight accelerations and mechanical vibration).

operational pressure transients Rises in operating pressure (due to water hammer, rapid startup, or shutdown) with sufficient duration to be felt as loads on the system or structure.

operator   Any device that causes an *actuator* to function.

opposition   1. The situation of two *celestial bodies* having either celestial longitudes or *sidereal hour angles* differing by 180°. The term is usually used only in relation to the position of a planet or the moon from the *Sun.* Compare this term with conjunction. 2. The situation of the periodic quantities differing by half a *cycle.*

optical air mass   A measure of the length of the path through the *atmosphere* to sea level traversed by light rays from a *celestial body,* expressed as a multiple of the path length for a light source at the *zenith.*

optical path   1. A *line of sight.*
2. The path followed by a ray of *light* through an optical system.

optical pyrometer   A device for measuring the *temperature* of an incandescent radiating body by comparing its *brightness* for a selected *wavelength* interval within the *visible spectrum* with that of a standard source; a monochromatic radiation pyrometer.

Temperatures measured by optical pyrometer are known as brightness temperatures and except for *black bodies* are less than the true temperature.

optical thickness   Specifically, in calculations of the transfer of *radiant energy*, the mass of a given absorbing or emitting material lying in a vertical column of unit cross-sectional area and extending between two specific levels. Also called optical depth.

If z_1 and z_2 are the lower and upper limits, respectively, of a layer in which the variation of a density ρ of some absorbing or emitting substance is given as a function of height z, then the quantity

$$\int_{z_1}^{z_2} \rho(z)\, dz$$

is called the optical thickness of that substance within that particular layer.

orange peel   Surface roughening that occurs when a metal of coarse grain is stressed beyond its elastic limit; the grain pattern formed resembles the outer surface of an orange.

orbit   1. In nuclear energy the region occupied by an *electron* as it moves about the *nucleus* of an *atom.*

2. In astronomy and space science the path followed by a *satellite* around an astronomical body, such as the *Earth* or *Moon.*

When a body in space is moving around a *primary body*, such as the Earth, under the influence of gravitational force alone, its path is called its **orbit**. If a *spacecraft* is traveling along a closed path with respect to the primary body, its orbit will be a circle or an *ellipse.* Perfectly circular orbits are not achieved in practice. However, the ellipse approaches a circle when the *eccentricity* becomes small. For example, the *planets,* with the *Sun* as the primary body, follow nearly circular orbits. When a satellite makes a full trip in orbit around its primary body, it is said to complete a **revolution,** and the time required is called its **period,** or **period of revolution.**

If an observer is measuring the period, the value measured will depend upon the observer's point of reference. If the observer were located, or example, far out in space, he could see the orbit of the spacecraft against the background of *fixed stars* and determine the satellite's period by timing the interval between successive passages over some point in that background. These measurements would then indicate the time needed to make one complete transit of the ellipse and would be called the *sidereal* period of revolution. The word sidereal means "of or relating to the stars." The sidereal period is not affected in any way by the rotation of the primary body (e.g., the Earth) under the satellite.

If, on the other hand, the observer were not far out in space, but rather standing on the equator with the satellite in a low Earth orbit moving directly east above the equator, the observer would use his own position as the reference point for measuring the satellite's period. Then, when the satellite has passed through one complete ellipse, it will actually be behind the observer because the Earth's rotation will have carried him a distance eastward. The satellite will be over the observer again only after it has traveled an *additional* distance eastward. The satellite's period as now measured by the observer will obviously be greater than the sidereal period; it is called the *synodic* period. The word synodic refers to "a meeting or conjunction." At the beginning of the synodic period, the position of the spacecraft over the observer results in a certain grouping or meeting of the Earth, spacecraft and Sun. At the end of the period the spacecraft will be over the observer again, and this same grouping will be repeated.

In practice very few satellites are placed in equatorial orbits (except for *geosynchronous* missions). Most orbits are inclined at an angle to the equator (*inclination*), as shown in figure 1. In the case of an inclined orbit, the spacecraft will not make successive passes over the observer. The observer moves with the Earth on a circle in a plane parallel to the plane of the equator, while the spacecraft moves through an ellipse in a plane inclined to the plane of the equator. Thus, the point at which the spacecraft passes over the observer's longitude changes with each pass. For an inclined orbit the time elapsing between two consecutive passes over the reference longitude is called the *synodic period*.

In the early aerospace operations the practice arose of referring to the synodic periods simply as "revolutions" and the sidereal periods as "orbits." Thus, for example, "*Gemini* VII" astronauts Borman and Lovell completed 206 revolutions and 220 orbits during their 14-day mission.

If the spacecraft is orbiting in an easterly direction (the same direction as the Earth's rotation), the orbit is said to be *posigrade*, and the synodic period is greater than the

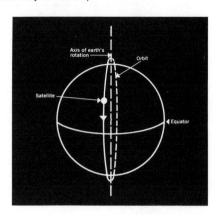

Fig. 2 A satellite in polar orbit. (Drawing courtesy of NASA.)

sidereal period. If the spacecraft is revolving in a westerly direction (opposite to the Earth's rotation), the orbit is said to be *retrograde*. In this case the eastward motion of an observer on Earth causes him to meet the satellite before it has completed a full transit of the ellipse, and the synodic period is therefore less than the sidereal period. Vehicles in retrograde orbits require a greater expenditure of energy to attain *orbital velocity*, since the launch vehicle must overcome the velocity of rotation of the Earth. If a spacecraft is placed in a *polar orbit* (see fig. 2), the synodic and sidereal periods are equal.

The sidereal period can easily be computed using the equation:

$$P = 2\pi\sqrt{(a^3)/(GM)}$$

where   a is the average radius of the orbit (i.e., the semimajor axis of the ellipse)

M is the mass of the Earth

G is the *constant of universal gravitation*

The value of "a" is found by averaging the *apogee* and *perigee* distances as measured from the center of the Earth. The apogee is the farthest distance of the spacecraft from Earth, while the perigee is the nearest distance.

orbital Taking place in *orbit*, as orbital refueling, orbital launch, or pertaining to an orbit as orbital plane.

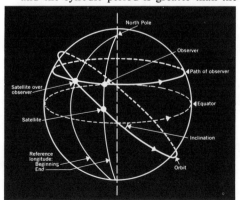

Fig. 1 Relative positions of a satellite and an observer at the beginning and end of the synodic period. (Drawing courtesy of NASA.)

orbital flight test (OFT)   One of the developmental flights of the *Space Shuttle* system.

orbital maneuvering system (OMS)   Two orbital maneuvering engines, located in external pods on each side of the aft fuselage of the *Orbiter structure*, provide *thrust* for *orbit* insertion, orbit change, orbit transfer, *rendezvous* and *de-orbit*. These maneuvering engines can provide a velocity change of 305 m/sec (1,000 ft/sec) when the *Orbiter* carries a *payload* of 29,500 kg (65,000 lb$_m$). A portion of this velocity change capacity is used during the final ascent to orbit. The 10,900 kg (24,000 lb$_m$) of usable *propellant* is contained in two pods, one on each side of the aft fuselage. Each pod contains a high-*pressure* helium storage bottle, the tank pressurization regulators and controls, a *fuel* tank, an *oxidizer* tank and a pressure-fed *regeneratively cooled rocket engine*.

There are also provisions for including up to three sets of auxiliary tanks (called orbital maneuvering system kits), each of which permits an additional 150 m/sec (500 ft/sec) velocity change, to achieve an overall velocity change capability of 755 m/sec (2,500 ft/sec). These OMS kits are located in the *payload bay* and use the same type of propellant tank, helium bottles and pressurization system components as the OMS pods.

Each OMS engine develops a *vacuum* thrust of 27 kilonewtons (6,000 lb$_f$) and uses a *hypergolic* propellant combination of nitrogen tetroxide and monomethyl hydrazine. These propellants are burned at a nominal oxidizer-to-fuel mixture ratio of 1.65 and a chamber pressure of 860 kN/m^2 (125 psia).

The OMS engine is designed for 100 *missions (flights)* with a service life of 10 years and is capable of sustaining 1000 starts and 15 hours of cumulative firing time. Each engine is 196 cm (77 in.) long and weighs 118 kg (260 lb$_m$). *Pitch* and *yaw* electromechanical *actuators* (attached to the vehicle structure at the forward end of the *combustion chamber*) *gimbal* the engine. The controller for the actuators is mounted in the pod structure.

The major components of the OMS engine are the platelet injector; the fuel regeneratively cooled combustion chamber; the radiation-cooled nozzle extension, an 80 percent *engine bell* that extends from an area ratio of 6:1 where it attaches to the chamber to an expansion ratio of 55:1 at the exit plane; a series-redundant *ball-type* bio-*propellant* valve, which is opened by regulated nitrogen gas supplied by a pneumatic actuation assembly; the thrust mount and gimbal assembly; and the propellant lines.

orbital maneuvering system (OMS) kits   Auxiliary *propellant* tanks, each of which provides an additional 150 m/sec (500 ft/sec) *velocity* change capacity to the maneuvering engines of the *Orbiter* vehicle for such operations as *orbit* change, orbit transfer, *rendezvous* and *de-orbit*. Up to three OMS kits can be located in the *payload bay* to achieve an overall velocity change capability of 755 m/sec (2,500 ft/sec) for the *orbital maneuvering system*. *See:* orbital maneuvering system

orbital motion   Continuous motion in a closed path such as a *circle* or an *ellipse*. *See:* orbit

orbital period   The interval between successive passages of a *satellite* or *spacecraft* through the same point in its *orbit*. Often called period.

orbital velocity   The average *velocity* at which a *satellite, spacecraft*, or other orbiting body travels around its *primary*.

Orbiter avionics system   The Shuttle *avionics* system controls, or assists in controlling, most of the *Space Shuttle* systems (see Fig. 1 and Fig. 2). Its functions include guidance, navigation, control and electrical power distribution for the *Orbiter*, the *External Tank* and the *Solid Rocket Boosters*. In addition the avionics control the communications equipment and can control *payloads*. Orbiter avionics automatically determine vehicle status and operational readiness and provide sequencing and control for the External Tank and Solid Rocket Boosters during launch and ascent. Automatic vehicle flight control can be used for every *mission* phase except *docking*. Manual control is also available at all times and can be used at the option of the crew.

The avionics equipment is arranged to

Avionics

FORWARD CABIN AREA
   Star trackers
   Inertial measurement unit

FLIGHT DECK
   Manual controls
   Indicators
   Displays
   Inertial optical unit

DRIVERS AND ACTUATORS
   Aerosurfaces
   Propulsive elements

AFT AVIONICS BAYS
   Rate gyros
   Aerosurface servo amplifier
   Reaction jet OMS driver (aft)
   Multiplexer-demultiplexer units

FORWARD AVIONICS BAYS
   Tacans
   Radar altimeters
   MSBLS receivers
   Air data transducer assembly
   RCS jet driver (forward)
   General-purpose computers
   Mass memories
   Multiplexer-demultiplexer units
   UHF receiver
   Rendezvous sensor
   Electronics
   Accelerometers
   One-way Doppler extractor

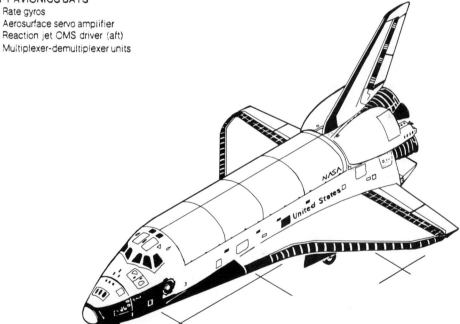

Fig. 1   A summary of the Orbiter's avionics system.   (Drawing courtesy of NASA.)

facilitate checkout, access and replacement with minimal disturbance to the other sub-systems. Almost all the electrical and electronics equipment is installed in three areas of the Orbiter: the flight deck, the forward avionics equipment bays and the aft avionics equipment bays (see Fig. 3).

The avionics are designed with *redundant hardware* and *software* to withstand multiple failures. The Space Shuttle avionics system consists of more than 200 electronic "*black boxes*" connected to a set of five computers through common communications lines called "data *buses*." The electronic black boxes offer dual or triple redundancy for every function. The status of individual avionics components is checked by a performance monitoring computer program. The status of critical vehicle functions—such as the position of the *hatches* and exterior doors, the separation mechanisms for the External Tank and Solid Rocket Boosters, and excessive temperatures for certain locations—is continuously monitored and displayed for the crew. The computer programs necessary to accomplish the different avionics functions are stored in tape *mass memories* and transferred to the computer memories as needed. The most detailed levels of the programs are called "principal functions," of which there are approximately 300. Principal functions are grouped together to create major *modes* concerning specific flight aspects, such as first-stage ascent, orbital stationkeeping and landing.

The avionics system is closely interrelated with three other Orbiter systems: the *guidance, navigation and control system;* the *controls and displays system;* and the *communications and data systems.*

The avionics system is segregated into subsystems by function. The guidance, navigation and control subsystem—in conjunction with the displays and controls, computers, and navigational aids—provides guidance, navigation and automatic (or manual) control of the Shuttle in all flight phases.

The *S-band phase modulated* (PM) *microwave* communications system can be used to provide two-way *Doppler* tone ranging (ground to Orbiter to ground). The *Doppler shift* is determined on the ground and used to compute a revised vehicle *state vector.* (A state vector is a mathematical representation of a vehicle's position in three-dimensional space and time—that is, motion. The state vector is used for *trajectory* computations.) The Doppler shift is then transmitted to the *spacecraft* computer for onboard computational use. A one-way Doppler system on the Orbiter obtains ranging information by comparing the S-band carrier *frequency* shift to an onboard frequency standard. This information may

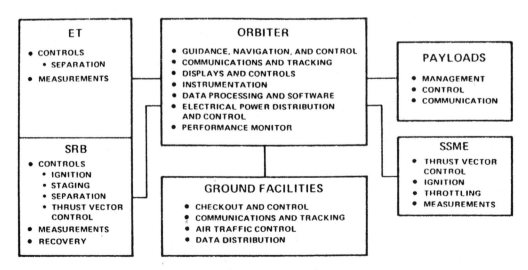

Fig. 2   Schematic of the Orbiter's avionics system.   (Drawing courtesy of NASA.)

also be used for vehicle position computations. A *rendezvous radar* is used for detecting, acquiring and tracking *active* and *passive targets*.

Three navigational aids are used during entry and landing. The *tactical air navigation* system determines the *bearing* and slant range from the Orbiter to a ground station (of known position with respect to the landing site) for fairly long ranges. The microwave scanning beam landing system determines the elevation, *azimuth* angles and range of the Orbiter from the landing site for fairly close ranges. The *radar altimeter* provides *absolute altitude* with respect to the runway from 762 meters (2,500 ft.) to touchdown. These navigational data are used to update the Orbiter computers during the landing phase of a *flight*.

The Orbiter uses four separate communications systems: S-band phase modulated (PM), S-band *frequency modulated* (FM), *Ku-band* and *ultrahigh frequency* (UHF). The S- and Ku-bands are microwave systems. The S-band PM and Ku-band systems provide for the transmission and reception of voice, engineering and scientific data and commands. The Ku-band provides for television transmission to the ground. The S-band FM equipment provides for the transmission of engineering and scientific data and television. The S-band payload system provides for sending voice and commands to *detached payloads* and receiving voice and data from detached payloads. The UHF system provides two-way (amplitude modulated) voice for air-to-ground, orbital and *extravehicular astronaut* communications. A hardwired audio system provides communications from point to point within both the Orbiter and the *payload bay*. Closed-circuit television is provided by cameras on the *remote manipulator system* (RMS) to aid in RMS control and by fixed cameras in the payload bay for observation of payloads.

The Orbiter displays and controls subsystem allows the crew to supervise, control and monitor all Orbiter systems. It includes the display and control panels, the spacecraft and the *RMS hand controllers*, the *cathode ray tube* displays, the keyboards and assoicated electronics, the timing displays, the caution-and-warning system, and the spacecraft lighting. It also provides the caution-and-warning and status information on the External Tank, the Solid Rocket Boosters and the payload systems.

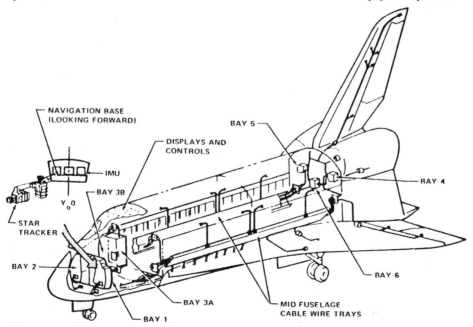

Fig. 3   Installation configuration of the Orbiter's avionics system.   (Drawing courtesy of NASA.)

The multifunction display system consists of a keyboard unit for manual control and data entry, a television display unit that displays *alphanumeric* and graphic information on a 12.7-cm by 17.8-cm (5-in. by 7-in.) screen, and a display electronics unit that stores and processes display data. Two keyboards are located in the forward flight station for use by the *pilot* and the *crew commander,* and a third is located in the aft mission *station* for use by the *mission* and *payload specialists.* Three television display units are also located in the forward flight station, with a fourth in the *mission specialist station.*

The electrical power distribution and control subsystem controls, conditions and distributes electrical power throughout the vehicle. The power system can be controlled through Orbiter avionics and is capable of performing specified power switching and sequencing of devices that must operate in a time-critical priority sequence. It also provides power to *attached payloads.*

The engine interface unit provides a two-way redundant *digital* interface for main engine commands. Each of the three engine interface units connects the main computers with the three *Space Shuttle Main Engine* controllers. Main engine functions such as initiation, ignition, gimbaling, thrust throttling and shutdown are internally controlled by the main engine controllers through inputs from the guidance equations computed in the Orbiter general-purpose computers.

The distinguishing characteristic of payload data is that it will not necessarily be standardized for Orbiter computers. To accommodate the various forms of payload data, the payload data interleaver integrates payload data into the Orbiter avionics so that the data can be transmitted to the ground. This applies to attached payloads; *free-flying satellites* are expected to have independent data transmission systems.

The Orbiter data processing subsystem handles data processing, data transfer, data entry and data display in conjunction with the operations of the Orbiter avionics system. The data processing system consists of the following: (1) 5 general-purpose computers for computation and control, (2) 2

magnetic-tape mass memories for large-volume bulk storage, (3) time-shared serial-*digital* data *buses* (essentially party lines) to accommodate the data traffic between the computers and other Orbiter subsystems, (4) 19 *multiplexer-demultiplexer* units to convert and format data at various subsystems, (5) 3 remote engine interface units to command the Orbiter main rocket engines and (6) 4 multifunction television display systems for the crew to monitor and control the Orbiter and payload systems.

Each general-purpose computer is a modified IBM AP-101 microprogram-controlled computer. The computer has a 106,496-word (36 *bits* to the word) memory. The *Apollo* Command Module computer had a central memory of 38,912 words (16 bits to the word).

As part of the *fail-safe* design of the avionics system, four of the five computers are arranged as a redundant group during critical flight operations, such as launch/ascent or entry/landing. In this mode the four computers are linked as a voting set, with each one capable of being used as the flight control computer and with each one checking on the other three. The crew can select which of the four computers is in control. The fifth computer is used for the backup flight control system, which would control the Shuttle should all four voting computers fail. Each of the four computers in the redundant set synchronizes itself to the other three computers 440 times each second. In this way the computer set is able to achieve a high degree of *reliability.* During non-critical flight periods one computer is used for guidance, navigation and control tasks and another for systems management. The remaining three can be used for payload management or can be deactivated. During critical phases of a mission, each of the five computers in the system performs approximately 325,000 operations each second using *floating-point arithmetic.* The *Shuttle crew* can ask more than 1,000 questions of the system and have information displayed as alphanumeric symbols, as graphs, or as a combination of the two (including moving and flashing characters or symbols) on any of the four television display sets.

In addition to the central memory stored in the computers themselves, 34 million bits of information (34 megabits) are also stored in two magnetic-tape devices. Critical programs and data are loaded in both tape machines and are protected from erasure. Normally, one *mass memory* unit is activated for use, and the other is held in reserve for operation if the primary unit fails. However, it is possible to use both units simultaneously on separate data buses or to have both communicate with separate computers.

The data bus network consists of digital data signal paths between the computers and the avionics subsystems and secondary channels between the *telemetry* system and the units that collect instrumentation data. This system is also fail-safe. The data transfer technique uses time-division *multiplexing* with *pulse-code modulation*. In this system, data channels are multiplexed together, one after the other, and information is coded on any given channel by a series of *binary* pulses corresponding to discrete information. There are 24 data buses on the Orbiter, and an additional 28 buses connect the Orbiter avionics with the Solid Rocket Boosters and the External Tank.

The multiplexer-demultiplexer units are used in numerous remote locations of the Orbiter to handle the functions of serial data time multiplexing-demultiplexing associated with the digital data buses and for signal conditioning. They act as translators, putting information on or taking it off the data buses.

All Orbiter and payload data are time-tagged with coordinated *universal time* and mission-elapsed time generated by the master timing unit. This device also supplies synchronizing signals to other electronic circuits, as required, and to the computers.

Orbiter communications and data systems

Flexible systems that provide adequate capability to accommodate the operational and scientific requirements of the wide range of *Space Shuttle payloads*. The *Orbiter* systems (see Fig. 1) consist of radiofrequency systems, a general-purpose computer system, special processors for interfacing between payloads and radiofrequency systems, a television system and

tape recording systems. The supporting ground systems include the *ground space tracking and data network*, the *tracking and data relay satellite system* (TDRSS), the *Mission Control Center* and the *Payload Operations Control Centers*.

The **Orbiter radiofrequency systems** and data services are summarized in table 1. Included are an *S-band phase modulation* (PM) transmitter/receiver, a *Ku-band* transmitter/receiver, two independent S-band frequency modulation (FM) transmitters, an S-band payload interrogator transmitter/receiver and a Ku-band *rendezvous radar*.

The **S-band phase modulation transmitter/receiver** normally operates in phase lock with the uplink carrier. The system can operate on either of two sets of frequencies. The antenna system consists of four selectable *omnidirectional antennas*, one located in each quadrant of the Orbiter. One mode of the system is compatible with the ground tracking system, and the other mode is compatible with the *satellite* tracking system. The S-band PM transmitter accepts *real-time* operational data at either of two rates: 96 or 192 *kilobits* per second (kbps). The low data rate consists of 64 kbps of *telemetry* and one 32-kbps *digital* voice channel; the high data rate consists of 128 kbps of telemetry and two digital voice channels of 32 kbps each. The data received by the S-band PM receiver are either 32 or 72 kbps. The 32-kbps rate consists of 8 kbps of command and a single 24-kbps voice channel; the 72-kbps rate consists of 8 kbps of command and two 32-kbps voice channels. The ground station mode, but not the satellite tracking mode, provides a ranging capability by modulation of multiple tones on a 1.7 *megahertz subcarrier* in the signal received from the ground station. The received subcarrier is routed to the transmitter and back to the ground station. The S-band PM system is available for use in all phases of the *mission* and will be available on all flights of the Orbiter.

The **Ku-band system** provides *duplex* communications between the Orbiter and the *Mission Control Center* through the tracking satellite when antenna line of sight

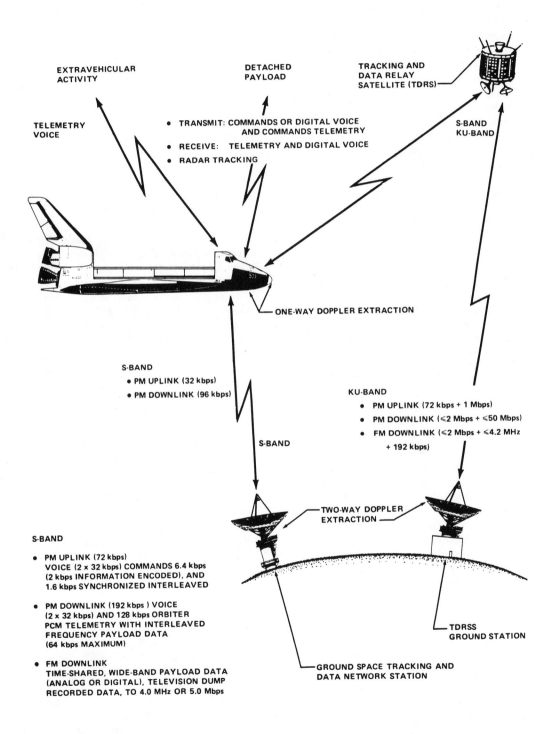

EXTRAVEHICULAR
ACTIVITY

DETACHED
PAYLOAD

TRACKING AND
DATA RELAY
SATELLITE (TDRS)

TELEMETRY
VOICE

S-BAND
KU-BAND

- TRANSMIT: COMMANDS OR DIGITAL VOICE
  AND COMMANDS TELEMETRY
- RECEIVE: TELEMETRY AND DIGITAL VOICE
- RADAR TRACKING

ONE-WAY DOPPLER EXTRACTION

S-BAND
- PM UPLINK (32 kbps)
- PM DOWNLINK (96 kbps)

KU-BAND
- PM UPLINK (72 kbps + 1 Mbps)
- PM DOWNLINK (≤2 Mbps + ≤50 Mbps)
- FM DOWNLINK (≤2 Mbps + ≤4.2 MHz
  + 192 kbps)

S-BAND

TWO-WAY DOPPLER
EXTRACTION

S-BAND

- PM UPLINK (72 kbps)
  VOICE (2 x 32 kbps) COMMANDS 6.4 kbps
  (2 kbps INFORMATION ENCODED), AND
  1.6 kbps SYNCHRONIZED INTERLEAVED

- PM DOWNLINK (192 kbps) VOICE
  (2 x 32 kbps) AND 128 kbps ORBITER
  PCM TELEMETRY WITH INTERLEAVED
  FREQUENCY PAYLOAD DATA
  (64 kbps MAXIMUM)

- FM DOWNLINK
  TIME-SHARED, WIDE-BAND PAYLOAD DATA
  (ANALOG OR DIGITAL), TELEVISION DUMP
  RECORDED DATA, TO 4.0 MHz OR 5.0 Mbps

TDRSS
GROUND STATION

GROUND SPACE TRACKING AND
DATA NETWORK STATION

Fig. 1   Space Shuttle orbital communications and tracking links.   (Drawing courtesy of NASA.)

exists. The system uses a *high-gain antenna* deployable from inside the *payload bay*. Use of this system is limited to orbital operations when the *payload bay doors* are open. The Ku-band transmitter accepts data in either of two modes, with three channels of data in each mode. In the first mode, channel 1 provides 192 kilobits per second of real-time operational data identical to the format transmitted by the S-band PM system in the high data rate mode. Channel 2 provides: (1) real-time *attached-payload* data up to 2 *megabits* per second (Mbps), (2) playback of operational data from the record/playback system or (3) playback of payload data from the record/playback system. Channel 3 provides real-time attached-payload digital data, or playback thereof, at rates up to 50 Mbps. In the second mode, channels 1 and 2 are the same as in the first mode, but channel 3 provides: (1) television video to 4.5 megahertz *bandwidth*, (2) attached- or *detached-payload analog* data to 4.5 megahertz bandwidth or (3) attached- or detached-payload digital data up to 4 Mbps.

The Ku-band receiver link includes the same 72-kbps data described for the S-band PM system high data rate mode. Additionally, a capability is provided for a text and graphics channel at 144 kbps, for a total received data rate of 216 kbps. The primary Ku-band system and the rendezvous radar system are partly integrated, with a common antenna and front-end electronics. For Ku-band acquisition between the Orbiter and the tracking satellite, the broad-beam Orbiter S-band PM signal will be acquired by the tracking satellite and used to point the satellite high-gain Ku-band antenna. The Orbiter Ku-band antenna will then acquire and automatically track the satellite Ku-band signal.

The *S-band payload interrogator* is a transmitter/receiver designed to communicate with detached payloads. The payload interrogator and its companion *avionics* packages will be available in the latter part of the *orbital flight test program*. The interrogator/transmitter operates on any one of the selectable channels and carries command data to the detached payloads at rates up to 2 kbps.

The *S-band frequency modulation transmitter* provides a single wide-band data channel from the Orbiter to the ground stations. Inputs to the transmitter are selectable and can be any of the following: (1) real-time or playback main engine (i.e., *Space Shuttle Main Engine*) data consisting of three channels at 60 kbps each, (2) playback of operational or payload data at rates up to 1.024 Mbps, (3) real-time payload data up to 5 Mbps or 4-megahertz bandwidth and (4) television video. The transmitter uses one of the two omnidirectional antennas located on the upper and lower hemispheres of the Orbiter.

The Orbiter data system interfaces with the radiofrequency systems through a series of special-purpose processors: the pulse-code-modulation (PCM) master unit, the network signal processor, the Ku-band signal processor and the frequency modulation signal processor.

The *payload signal processor* serves as an interface between the payload and the payload interrogator for demodulation of data from detached payloads. The processor transmits command data to attached payloads and to the payload interrogator for transmission to detached payloads. Provision is made in mass memory for changing the processor configuration to handle different data streams and bit rates.

Space Transportation System payload data will not always be formatted and standardized for Orbiter computers. Attached payload data are fed by the *payload data interleaver* into the avionics system for *downlink* to the ground. *Free-flying payloads* are expected to have independent data downlink systems.

Two 14-track recorders are provided with up to 80 minutes capacity each. Tape speeds of 15 cm/sec to 305 cm/sec (6 in./sec to 120 in./sec) in 14 steps are available, although no more than 4 steps are available on any one mission. One recorder is assigned as the operations recorder and the other as the payload recorder.

A closed-circuit television system (see Fig. 2) is scheduled for use during orbital operations, including crew compartment activities and out-the-window observations. Typical examples of crew compartment activities are crew operations, hardware inspection and experimentation; typical examples of out-the-window observations include the *Earth*, payloads and observations of the exterior of the Orbiter. The television system is composed of the camera, the power cable, the monitor, the monitor cable, the lens assemblies (a six-to-one 25-mm zoom lens and a three-to-one 9-mm wide-angle zoom lens) and the camera bracket. Also included are the video interface unit, console monitors and a videotape recorder. The videotape recorder will be used for *remote manipulator system* operations. Television operations involving ground commands and downlink are restricted to S-band coverage during the early flights. For later flights the Ku-band downlink will also be available.

The onboard television system consists of a video switcher capable of receiving camera inputs from as many as 10 cameras. The various video camera outputs may be transmitted to Earth by S-band FM or Ku-band as well as switched to two onboard console television monitors. The basic Space Shuttle television camera is a monochrome (one-color) system with changeable lens configurations that allow for various size lenses and for lenses that contain a field sequential color wheel. The capability exists for future addition of a videotape recorder.

The tracking and data relay satellite system will consist of two *geosynchronous* operational satellites separated by 130°, a third in-orbit spare satellite and a single *ground station* located at White Sands, New Mexico. The system will be a commercial service operated by Western Union. The TDRSS will be capable of supporting two Orbiters simultaneously and other Earth-orbital free-flying satellites in multiaccess *modes*. It will also provide radiofrequency interfaces to the Orbiter Ku-band system and to the S-band PM system. At standard flight altitudes of 185 km to 370 km (100 nm to 200 nm), the tracking satellite will normally provide communications for 40 percent to 90 percent of the day, compared to a typical ground network communications contact of 7 percent to 30 percent of the day.

Data from the tracking satellite ground station will be carried over two diversely routed duplex 1.544-Mbps lines. One will be routed to the Mission Control Center at the *Johnson Space Center* (JSC) through the *Goddard Space Flight Center* (GSFC), and the other will go directly to the Mission Control Center. Some consideration has been given to expanding these services to 6.3-Mbps rates to accommodate limited payload data streams. A single duplex 224-kbps line between GSFC and the Mission Control Center will be provided for *backup* communication. The ground station will provide a short-duration recording capability as protection for circuit outages.

Both real-time and playback operation telemetry will be transmitted in real time on both circuits. Metric data and digital voice will be transmitted in addition to telemetry. Data from the Mission Control Center will include commands, digital voice and digital text/graphics. Domestic communications satellite services are envisaged for real-time remoting of high-data rate payload data streams; video lines will be provided for remoting of mission television. The GSFC-JSC communications links provided for the tracking satellite data will also be used for ground station data transmission.

The Mission Control Center at the Johnson Space Center in Houston, Texas, will perform flight control. Additionally, the control center will provide for monitoring and control of specific payloads assigned to JSC. As the Space Shuttle becomes operational, the emphasis of the Mission Control Center will shift from basic systems monitoring to payload monitoring, mission management and multiple flight support.

Three Payload Operations Control Centers have been identified to interface with the Mission Control Center during the Space Shuttle Program. These are at the Goddard Space Flight Center, at the Jet Propulsion Laboratory and at the Johnson Space Center. The payload center at GSFC will be responsible for control of Earth-orbital free-flying payloads, the one at JPL for control of deep-space payloads and the one at JSC for control of Shuttle-attached payloads.

Payload data interleaved in the Orbiter

operational telemetry downlink will be routed to the Mission Control Center. The control center will process the data, extract the payload data, format them for output and transmit them together with selected Orbiter data to the appropriate payload center. *Ephemeris*, command-verification and ground-systems data will be transmitted with the telemetry data. In addition to the interleaved data, some payloads will transmit directly to the ground or to satellite tracking systems; these data will be routed directly to the appropriate payload center.

When payloads are attached to the Orbiter or are operating detached from the Orbiter through the payload interrogator link, commands from the payload centers will be formatted into the Orbiter command format and *uplinked* to the payload by the Orbiter systems. The control center will perform message error checks on payload center commands and provide command verifications to the payload centers. A capability to remote Orbiter-to-ground voice from the payload centers will be provided in coordination with the Mission Control Center use of this system. After completion of the Space Shuttle mission, deep-space and Earth-orbital free-flying payloads will be controlled from the appropriate payload center independent of the Mission Control Center.

The text and graphics hardcopy system provides the crew with an onboard method of reproducing hardcopies of ground-generated data. These data include such information as *crew activity plans, maneuver pads*, general messages, schematics, photographs, crew procedures and *trajectory* data. The text and graphics hardcopy system is composed of a Polaroid camera system that provides a photographic copy of data displayed on an onboard television tube. A maximum of 100 pages of text data can be uplinked and stored in the Orbiter mass memory for television display and hardcopy.

The software stored in and executed by the Orbiter general-purpose computers is the most sophisticated and complex set of programs ever developed for *aerospace* use. The programs are written to accommodate almost every aspect of Shuttle operations, including vehicle checkout at the manufacturer's plant; flight *turnaround* activity at the *Kennedy Space Center;* prelaunch and final countdown; and navigation, guidance and control during the ascent, orbital, entry and landing phases and during *abort* or other contingency mission phases. In-flight programs monitor the status of vehicle subsystems; provide consumables computations; control the opening and closing of the *payload bay doors*; operate the remote manipulator system; perform fault detection and reporting; provide for payload monitoring, commanding, control and data acquisition; provide *antenna* pointing for various communications systems; and provide backup guidance, navigation and control for the ascent, orbital, entry and landing phases and for aborts. These primary computer programs are written so that they can be executed by a single computer or by all computers executing an identical program in the same time frame. This multicomputer mode is used for critical flight phases, such as launch, ascent, entry and aborts.

The Orbiter computer programs are written in a two-level hierarchy. The first level is the system software group, consisting of three sets of programs: (1) the flight computer operating program (the executive), which controls the processors, monitors key system parameters, allocates computer resources, provides for orderly program interruptions for higher priority activities and updates computer memory; (2) the user interface program, which provides instructions for processing crew commands or requests; and (3) the system control program, which *initializes* each computer and arranges for multicomputer operation during critical flight periods. The system software group programs also tell the computers how to perform and how to communicate with the other equipment. The second level of memory groups is the applications processing software. This group contains specific software programs for guidance, navigation and control; systems management; payload operations; and vehicle checkout.

The two program groups are combined to

Fig. 2   Orbiter closed-circuit television system.   (Drawing courtesy of NASA.)

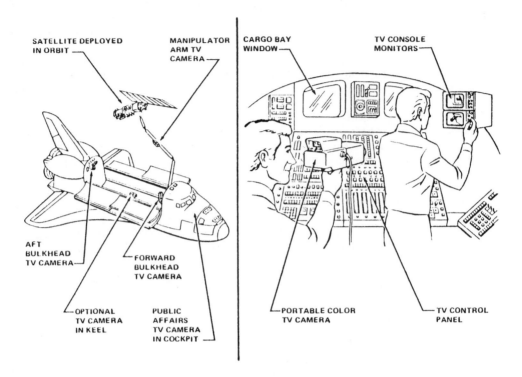

form a memory configuration for a specific mission phase. The guidance, navigation and control programs contain functions required for launching, flying into orbit, maneuvering in orbit and returning to an *Earth* landing. The systems management programs handle data management, performance monitoring, and special and display control processing. The payload processing programs contain instructions for control and monitoring of Orbiter payload systems. This set of instructions can be revised depending on the nature of the payload. The vehicle checkout program contains instructions for loading the memories in the main engine computers and for checking the instrumentation system. This program also aids in vehicle subsystem checkout and in ascertaining that the crew displays and controls perform properly. It is also used to update *inertial measurement unit state vectors.*

Orbiter Crew Accommodations The *Orbiter* cabin is designed as a combination working and living area [See: Figs. 1 and 2].

The pressurized crew compartment has a volume of 71.5 cubic meters (2525 cubic feet) and contains three levels. The flight deck contains the displays and controls used to pilot, monitor, and control the *Orbiter,* the integrated Shuttle vehicles, and the mission *payloads.* Seating for as many as four crewmembers can be provided. The mid deck contains passenger seating, the living area, an *airlock,* the galley, sleeping compartments, the toilet, and *avionics* equipment compartments. An aft hatch in the airlock provides access to the *payload bay.* The lower deck contains the environmental control equipment and is readily accessible from above through removable floor panels. Located outside the crew module in the payload bay are provisions for a *docking module* and a *transfer tunnel* with an adapter to allow crew and equipment transfer for docking, *Spacelab,* and *extravehicular* operations. [See also: Orbiter structure; Orbiter avionics system; Orbiter environmental control and life support system.]

The daily routine for crewmembers

Fig. 1   Summary of Orbiter crew accomodations and equipment.   (Drawing courtesy of NASA.)

ORBITER ACCOMMODATIONS
Seats, restraints, and mobility aids
Egress systems
Flight data file
Sighting aids
Photographic equipment
Window shades and filters
Stowage areas
Food systems and equipment
Sleeping accommodations
Crew hygiene systems and accommodations
Housekeeping equipment
Airlock

CREW EQUIPMENT
Survival equipment
Medical kits
Radiation instrumentation
Operational bioinstrumentation
Crew clothing
Space suit assembly

aboard flights of the *Space Transportation System* will vary according to crew assignment but each member will follow a detailed schedule each day. Time is allotted for each person for sleep, personal hygiene, work, meal preparation, and eating as well as routine Orbiter subsystem housekeeping. A 24-hour time period is normally divided into an 8-hour sleep period and a 16-hour awake period for each crewmember. Adjustments will be made to the daily schedule when specific flight activities require. One-or two-shift operations will depend on the number

of crewmembers and the specific flight and operational requirements.

Two fixed sleeping bags and two alternate locations are located in the mid deck for crewmembers (See: Fig. 3]. During a one-shift operation, all crewmembers sleep simultaneously; concurrent sleep periods are scheduled for two-shift operations. If all crewmembers are sleeping simultaneously, at least one will wear a communication headset to ensure reception of ground calls and Orbiter caution-and-warning alarms. The headset is connected to a nearby communication outlet in the mid deck. All crew-members wear noise-suppressing earplugs while sleeping. Forty-five minutes is allo-cated for each crewmember to prepare for bed and another 45 minutes is set aside when they wake to wash and get ready for the day ahead. Three 1-hour meal periods are scheduled for all onboard the Orbiter. This hour includes actual eating time and the time required to clean up. Breakfast, lunch, and dinner are scheduled as close to routine hours as possible. Dinner will be scheduled at least 2 to 3 hours before crew-members begin preparations for sleep. A galley and dining/work area will be located in the mid deck. The galley area includes a food preparation center, food and equip-ment storage, hot and cold water dis-pensers, food trays, an oven, a water heater, and waste storage. A 20-minute food prepa-ration period is required 30 minutes before the meal. (Orbiter "food system" is de-scribed in detail later on in this section).

In addition to time scheduled for sleep and meals, each crewmember has house-keeping tasks that require from 5 to 15 min-utes of his time at given intervals through-out the day. These include cleaning the waste compartment, dumping excess water, replacing the carbon dioxide scrubbing can-

isters (see "Housekeeping System" further on in this section), purging the *fuel cells*, giving daily status reports to the ground contollers, and aligning the *inertial measurement unit* (the device that directs the vehicle *attitude* in space). A 15-minute period is also set aside at the end of each day for one of the crewmembers to put the garbage out.

Seats, restraints, and mobility aids are provided in the Orbiter to enable the crew to perform all tasks safely and efficiently and to provide them with proper body posi-tioning. These devices include operational seats; foot restraint platforms/shoes; work/dining table and portable desk; in-flight restraints such as *Velcro*, snaps, *bungees*, tethers, and sleep restraints; and mobility aids and devices. Foot restraints, handholds, and mobility aids are also provided in the *Spacelab*.

The commander and pilot operational seats, which replace the ejection seats used in the *orbital flight tests*, provide proper body positioning so that control of the vehi-cle can be maintained throughout the flight. These seats provide comfortable support and proper body positioning during launch, entry, and orbital flight.

The *mission* and *payload specialist* seats are similar to the commander and pilot

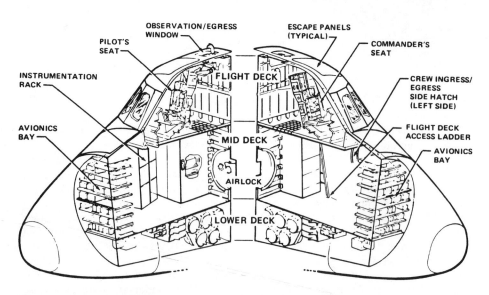

Fig. 2   Orbiter crew station   (Drawing courtesy of NASA.)

Fig. 3   Locations of the sleeping bags in the Orbiter crew cabin.   (Drawing courtesy of NASA.)

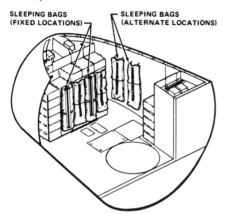

SLEEPING BAGS (FIXED LOCATIONS)   SLEEPING BAGS (ALTERNATE LOCATIONS)

seats and are required to provide support and restraint during launch and entry or during high-acceleration maneuvers. During flight, these seats may be removed and stowed. Restraints used in conjunction with the seats consist of two shoulder harnesses and a lapbelt.

The adjustable foot restraints in the Orbiter and the Spacelab are used for orbital operations only. An adjustable foot restraint platform in the Orbiter is provided for use at the onorbit station. The restraint platform is required for crewman optical alignment sight (COAS) operations and may be required for all out-the-window operations for small crewmembers. Suction-cup foot restraint shoes are provided for securing crewmen to various crew station locations and to the foot restraint platform.

The mid-deck work/dining table accommodates as many as four persons simultaneously for dining; it is also used as a workbench with appropriate retention devices for orbital operations. It is stowed during launch and enty and unstowed as required during orbital operations. As many as four portable work desks are supplied to provide portable work surfaces for retaining material and as an alternate means of supporting the food trays.

In-flight restraints such as Velcro, snaps, straps, and bungees are provided to assist in securing various in-flight equipment. Mobility aids and devices are provided for all passageways within the crew module, the docking module, the airlock, and the Spacelab. These devices include handholds, footholds, handrails, and ladders and are permanently attached to the Orbiter.

Adjustable seats for the commander and pilot allow movement of the seats in both the fore/aft and up/down directions for launch and orbit. The specialist seats are mounted to the flight deck and mid deck. Restraints are built into the seat system and controls are provided to lock and unlock the seat back for tilt change. The specialist seats can be removed and installed without tools by using the quick-disconnect fitting on each seat leg. These seats can be folded and stowed in the mid deck.

The foot restraint platform used at the aft-flight-deck onorbit stations can be preadjusted from 7.6 to 30.5 centimeters (3 to 12 inches) in height as desired by the crewman. Laced and zippered shoes are worn by the crewmen for the entire flight. Two suction cups installed on a plate attached to the shoe are used to secure the crewmen to the deck of the Orbiter and the Spacelab during orbital operations.

Mobility aids and devices are located in the Orbiter for movement of the flightcrew before launch and during orbital flight. These devices consist of (1) handholds for *ingress* and *egress* to and from crew seats for launch and landing configurations, (2) handholds in the primary interdeck access opening for ingress and egress for launch and landing configurations, (3) a platform in front of the airlock for ingress and egress to and from the mid-deck seats when the Orbiter is in the launch configuration, and (4) an interdeck access ladder for egress from the flight deck to the mid deck after landing. Additional mobility aids are provided on the aft mid-deck manufacturing access panel when the airlock is removed.

Egress provisions are those pieces of *hardware* that have been incorporated into the Orbiter to provide the crew with egress capabilities under emergency conditions. These provisions include the escape panel system, the side hatch egress bar, thermal aprons, the descent device, the personal egress air pack (PEAP), the lifevest, and survival equipment. The primary mode of emergency egress is through the side hatch; the escape panel system provides the secondary egress route [See: Fig. 4] The thermal aprons, descent devices, and egress bar assist the crew during egress. The lifevest and the survival equipment assist the crew after egress.

The escape panel system consists of the ejection escape panels and the left-hand overhead window (LOW) escape panel. The LOW escape panel will replace the ejection escape panels when the operational seats replace the ejection seats. The side hatch egress bar helps stabilize the crewmember's drop to the ground during egress through the primary route. The thermal aprons provide thermal protection from the exterior surfaces of the Orbiter during egress. The descent devices provide a controlled descent from the escape panels on the top of the Orbiter down the side of the Orbiter to the ground. The descent device is restrained to the crewmember by means of the integrated harness on the pressure suit or the egress sling. The personal egress air pack provides a regulated supply of air for egress and escape in a contaminated atmosphere. The lifevest provides flotation for the crewmembers in case of bailout. The vest is not flown when the pressure suit and integrated harness are flown.

The survival equipment consists of the ejection seat kit and the Orbiter survival kit. The seat kit provides the equipment needed for survival for the commander and pilot individually after ejection. The Orbiter survival kit provides survival equipment for up to seven crewmembers for 48 hours after landing at a remote site. The Orbiter survival kit will be flown when the ejection seats are replaced by the operational seats or when more than two crewmembers fly before the replacement of the ejection seats.

The escape panel system is normally activated by the crew from inside the vehicle with the console *jettison* T-handle located in the center console. In the event the crew is partly or completely disabled, ground rescue forces can activate the system from outside the vehicle with the T-handle located behind an access door above the leading edge of the right wing. This access door is clearly marked with a rescue arrow and instructions. To prevent inadvertent activation of the system, the console T-handle has a safing pin that is removed for launch and entry. The LOW escape panel has a *pyrotechnic* system that jettisons the panel for exit from the Orbiter.

The side *hatch* egress bar is permanently attached to the side hatch and serves as a handhold for jumping off the side hatch. It reduces the drop to the ground from the inboard side of the hatch from 3 meters (10.5 feet) to approximately 1 meter (3.5 feet). For egress out the side hatch, the crewmember deploys the egress bar, a 3.18-centimeter (1.25-inch) diameter bar with a knurled handle at the end. The bar swings out for use.

Thermal aprons are protective covers used by the crewmember to drape over external surfaces of the vehicle that he may contact during egress. The temperature of the outer hatch structure is 425 to 450 K (150° to 175° C or 300° to 350°F) and the thermal protection tiles on top of the Orbiter are approximately 365 K (90°C or 200°F). The apron has attach points to anchor the apron to predetermined attach points at the egress exit.

The descent device permits emergency egress through the ejection seat panels or the right-hand overhead escape panel at a controlled rate of descent. Each descent device consists of a descent line, a tether, a descent control assembly, and a deployment bag. The descent line is a 0.97-centimeter (0.38-inch) diameter nylon line with a preformed loop on one end to anchor it. The tether is attached to the lower end of the descent control by a swivel fitting and connects to the crewmember's integrated harness or egress sling by a snap fitting. The descent control assembly allows the crewmember to lower himself to the ground at a controlled rate of descent. The deployment bag houses the stowed descent line and prevents line entanglement before deployment.

The commander and pilot each have a descent device in a stowage bag located on the inboard side of the commander's seat rail support structure. The two descent devices are anchored to the seat rail support structure by clevises. Other crewmembers

PRIMARY EGRESS ROUTE

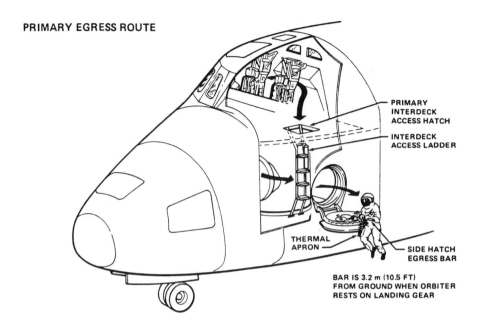

PRIMARY
INTERDECK
ACCESS HATCH

INTERDECK
ACCESS LADDER

THERMAL
APRON

SIDE HATCH
EGRESS BAR

BAR IS 3.2 m (10.5 FT)
FROM GROUND WHEN ORBITER
RESTS ON LANDING GEAR

SECONDARY EGRESS ROUTE

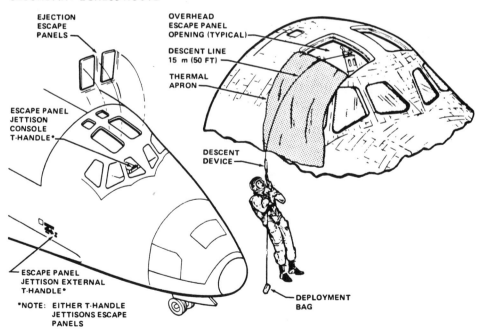

EJECTION
ESCAPE
PANELS

OVERHEAD
ESCAPE PANEL
OPENING (TYPICAL)

DESCENT LINE
15 m (50 FT)

THERMAL
APRON

ESCAPE PANEL
JETTISON
CONSOLE
T-HANDLE*

DESCENT
DEVICE

ESCAPE PANEL
JETTISON EXTERNAL
T-HANDLE*

*NOTE: EITHER T-HANDLE
JETTISONS ESCAPE
PANELS

DEPLOYMENT
BAG

Fig. 4   Shuttle egress routes (postlanding with ejection seats installed.)   (Drawing courtesy of NASA.)

will have their own descent devices. When the ejection seats are flown, the descent device is used for emergency egress through the ejection escape panels; when the operational seats are flown, it will be used for emergency egress through the LOW escape panel.

The personal egress air pack provides air for egress (normal and emergency) for the commander and pilot only and is worn when the atmosphere inside or outside the Orbiter is contaminated or not known to be safe. The lifevest provides 24 hours of flotation for a maximum-sized crewmember. It is designed so that the performance of emergency tasks will not be degraded when the vest is worn. The lifevests will be stowed on the seats within easy reach of the crewmembers. The Orbiter survival kit provides land and sea survival capability for seven crewmembers for 48 hours. The kit is packaged in a single container sized to be deployed through the side hatch or the LOW escape panel by a single crewmember. The kit contains an eight-man liferaft with a carbon dioxide inflation assembly, a mooring lanyard assembly, two oral inflation tubes, a bellows pump, a bailing bucket, and a sea anchor; signaling equipment consisting of a personal distress signal kit, two smoke/illumination flares, a Sun mirror, two radio beacons with spare batteries, and two sea dye markers; and other equipment consisting of a two-part individual survival kit, a survival blanket, a survival knife assembly, and a desalter bag and chemical packets.

The Flight Data File (FDF) is a flight reference data file that is readily available to crewmen within the Orbiter. It consists of the onboard complement of documentation and related crew aids and includes (1) FDF documentation, such as procedural checklists (normal, backup, and emergency procedures), malfunction procedures, crew activity plans, schematics, photographs, cue cards, star charts, *Earth* maps, and crew notebooks; (2) FDF stowage containers; and (3) FDF ancillary equipment, such as tethers, clips, tape, and erasers. The Flight Data File is similar for all flights in quantity and stowage locations. The baseline stowage volume is sufficient to contain all FDF items for all Orbiter configurations except the *pallet-mounted payload.* In this case, a larger Flight Data File, and consequently additional locker space, is required because all payload operations are performed in the Orbiter.

FDF items are used throughout the flight—from prelaunch use of the Ascent Checklist through crew egress use of the Entry Checklist. Packaging and stowage of Flight Data Files are accomplished on an individual flight basis. FDF items will be stowed in five types of stowage containers: lockers, the flight-deck module, the commander's and pilot's seat-back FDF assemblies, the mid-deck FDF assembly, and the map bag. The portable containers are stowed in a mid-deck modular locker for launch and entry. If the flight carries a *Spacelab module,* all Spacelab books are stowed for launch in a portable container on the mid deck and transferred in flight to a location in the Spacelab. The FDF stowage is flexible and easily accessible.

Sighting aids include all items used to aid Shuttle crew and passenger visibility within and outside the crew module. Sighting aids include the crewman optical alignment sight (COAS), the payload bay door COAS (PLBD COAS), binoculars, and adjustable mirrors.

The COAS is used onorbit to provide (1) range and range rate during *rendezvous* and *docking,* (2) a fixed *line-of-sight attitude* reference for verification of *inertial measurement unit* (IMU) and *star tracker* performance, (3) the capability for backup IMU alignment, and (4) a backup attitude check before any major burn. The PLBD COAS provides the capability to check PLBD thermal/vibrational misalignment onorbit. The COAS and PLBD COAS must be activated for use and deactivated after use. The binoculars can be used to view objects in the payload bay and remote to the Orbiter. The adjustable mirrors provide rear and side visibility to each crewmember for assessing Orbiter and *External Tank* separation. The mirrors are also used to check controls and to display statusing and man/seat interfaces.

The COAS is used in the right-hand overhead window to provide range and range rate for rendezvous/docking, for IMU backup alignment, and for Z-axis sighting. The COAS is used in the left-hand forward window for stationkeeping/tracking and miscellaneous Orbiter alignment tasks and as a backup location for IMU alignment. The COAS is a *collimator* device similar to an aircraft gunsight. The COAS weighs approximately 1 kilogram (2.5 pounds), is 24 centimeters (9.5 inches) long, and requires Orbiter 115 ± 5 volt alternating-current power. The COAS consists of a lamp with an intensity control, a *reticle*, a barrel-shaped housing, a mount, a combiner assembly, and a power cable. The reticle consists of a 10° circle, vertical and horizontal crosshairs with 1° marks, and an elevation scale on the right side of −10° to 31.5°.

The PLBD COAS is used in the aft window during onorbit operations to determine the amount of door deflection and to determine that the payload bay doors have closed. The 10 by 40 binoculars are a space-modified version of the commercial Leitz Trinovid binoculars. This unit is noted especially for its small size, high magnification, wide field of view, and rugged sealed construction. The adjustable mirrors are approximately 8 by 13 centimeters (3 by 5 inches) and weigh less than 0.5 kilogram (1 pound). The mirrors (one each for the *commander* and *pilot*) are mounted on the forward flight deck vertical handholds. The mirror is mounted on a ball and can be adjusted by rotating it to the desired position.

The Orbiter windows are designed to provide external visibility for *entry*, landing, and orbital operations. Vision requirements for *atmospheric* flight require that the *flightcrew* be provided with forward, left, and right viewing areas. Orbital missions require visibility for rendezvous, docking, and payload handling. Potentially, these large areas of transparency will expose the crew to *Sun* glare; therefore, window shades and filters are provided to preclude or minimize exposure when desirable during orbital operations. The window shade and filter use locations are identified in Fig. 11. All shades and filters are for onorbit use. Shades are provided for all windows, while

filters are supplied for the aft station and overhead windows only. The overhead window shades are installed preflight for the launch and are also used during reentry. The aft window filters are installed preflight for launch and are stowed for entry. The window shades and filters are stowed on the *mid deck* in the *galley area*. [See: Orbiter structure]

The forward-station window shades (W-1 through W-6) are fabricated from *Kevlar* epoxy glass fabric with silver and *Inconel*-coated *Teflon* tape on the outside surface and paint on the inside surface. When the shade is installed next to the inner window pane, a silicone rubber seal around the periphery deforms to prevent light leakage. The shade is held in place by the shade installation guide, the hinge plate, and the *Velcro* keeper. The overhead window shades (W-7 and W-8) are the same as the forward shades except the rubber seal is deleted and the shade is sealed and held in place by a separate seal around the window opening, a hinge plate and secondary frame, and a Velcro retainer. The overhead window filters are fabricated from Lexan and are used interchangeably with the shades. The aft window shades (W-9 and W-10) are the same as the overhead window shades except that a 1.6-centimeter (0.63-inch) wide strip of *Nomex* Velcro hook has been added around the perimeter of the shade. The shade is attached to the window by pressing the Velcro strip to the pile strip around the window opening. The aft window filters are the same as the overhead window filters except for the addition of the Velcro hook strip. The filters and shades are used interchangeably. The side-hatch window cover is permanently attached to the window frame and is hinged to allow opening and closing.

There camera systems—16, 35, and 70 millimeter—will be used by the Orbiter crews to document activities inside and outside the Orbiter. All three camera systems are used to document onorbit operations; the 16-millimeter camera is also used during the launch and landing phases of the flight.

All 16-millimeter camera equipment is the same configuration as that used during the *Apollo, Skylab*, and *Apollo-Soyuz Test*

Project missions and the Shuttle *Approach and Landing Test.* The only exception is the camera mount. The 16-millimeter camera is a motion-picture-type camera with independent shutter speeds and frame rates. The camera can be operated in one of three modes: pulse, cine, or time exposure. In the pulse mode, the camera operates at a continuous frame rate of 2, 6, or 12 frames per second. In the cine mode, the camera operates at 24 frames per second. In the time-exposure mode, the first switch actuation opens the shutter and the second actuation closes it. The camera uses 43-meter (140-foot) film magazines and has 5-, 10-, and 18-millimeter lenses.

The 35-millimeter camera system is the same type as used in previous *United States manned space flights* with the exception of a new flash unit. The camera is a motorized battery-operated Nikon camera with reflex viewing, through-the-lens coupled light metering, and automatic film advancement. The camera has three automatic (electrically controlled) modes of operation—single exposure, continuous, and time—plus the standard manual mode. The 35-millimeter camera uses an f/1.4 lens.

The 70-millimeter camera system is composed of the same equipment as used during the Apollo-Soyuz Test Project. It is a modified battery-powered motor-driven single-reflex Hasselblad camera and is provided with 80- and 250- millimeter lenses and film magazines. Each magazine contains approximately 80 exposures. The 70-millimeter camera has only one mode of operation, automatic; however, there are five automatic-type camera functions from which to select. The camera has a fixed viewfinder that provides through-the-lens viewing.

Crew equipment onboard the Orbiter will be stowed in lockers with insertable trays. The trays can be adapted to accommodate a wide variety of soft goods, loose equipment, and food. The lockers are interchangeable and attach to the Orbiter with screw fittings. The lockers can be removed or installed in flight by crewmen. There are two sizes of trays: a half-size tray (two of which fit inside a locker) and a full-size tray. Approximately 4.2 cubic meters (150 cubic feet) of

stowage space is available, almost 95 percent of which is on the mid deck.

The lockers will be made of either epoxy- or polyimide-coated Kevlar honeycomb material joined at the corners with aluminum channels. Inside dimensions are appoximately 25 by 43 by 50 centimeters (10 by 17 by 20 inches). The honeycomb material is approximately 0.64 centimeter (0.25 inch) thick and was chosen for its strength and light weight. The lockers contain about 0.06 cubic meter (2 cubic feet) of space and can hold up to 27 kilograms (60 pounds-mass) Dividers will be used in the trays to provide a friction fit for zero-g retention. This will reduce the necessity for straps, bags, Velcro snaps, and other cumbersome attach devices previously used. Soft containers will be used in Orbiter spaces too small for the fixed lockers. The trays will be packed with gear in such a way that no item covers another or, if it does, the gear is of the same type. This method of packing will reduce the confusion usually associated with finding and maintaining a record of loose equipment.

Stowage areas in the Orbiter crew compartment are located in the forward flight deck, the *aft flight deck,* the mid deck, the *equipment bay,* and the *airlock module.*

In the forward flight deck, the Flight Data File is located to the right of the pilot's seat. A fire extinguisher is located forward of the file below the pilot's window. Special provisions for the orbital flight test phase are attached to both the commander's and the pilot's seats. These provisions include the emergency survival kit, parachutes, the portable oxygen system, and (behind the commander's seat) the emergency egress kit. In the aft flight deck, stowage lockers are located below the rear payload control panels in the center of the deck. Container modules can be mounted to the right and left of the payload control station. Since these side containers are interchangeable, they may not be carried on every mission, depending on any payload-unique installed electronic gear. In the mid deck, container modules can be inserted in the forward *avionics bay.* Provisions for 33 containers are available in this area. In addition, there is an area to the right side of the airlock module where nine containers can be attached. *Lithium hydroxide (LiOH) canisters* will be stowed below the mid deck in the equipment bay. The airlock module will be used to stow equipment directly related to *extravehicular activity*, such as the *extravehicular mobility unit (space suit/* backpack).

The Orbiter is equipped with food, food stowage, and food preparation and dining facilities to provides each crewman with three meals per day plus snacks and an additional 96 hours of contingency food. The food supply and food preparation facilities are designed to accommodate flight variations in the number of crewmen and flight durations ranging from two crewmen for 1 day to seven crewmen for 30 days.

The galley, which is located in the cabin working area, [See: Fig. 5] is modular and can be removed for special missions. In addition to cold and hot water dispensers, it will be equipped with a pantry, an oven, food serving trays, a personal hygiene station, a water heater, and auxiliary equipment storage areas. The oven will be a forced-air *convection* heater with a maximum temperature of 355 K (82°C or 180F). There are no provisions for food freezers or refrigerators.

The food consists of individually packaged serving portions of dehydrated, thermostabilized, irradiated, intermediate moisture, natural form, and beverage foods. The food system relies heavily on dehydrated food, since water is a byproduct of the *fuel cell* system onboard the Orbiter. Off-the-shelf thermostabilized cans, flexible

SPACE SHUTTLE TYPICAL MENU[a]

DAY 1	DAY 2	DAY 3	DAY 4
Peaches (T)	Applesauce (T)	Dried peaches (IM)	Dried apricots (IM)
Beef patty (R)	Beef jerky (NF)	Sausage (R)	Breakfast roll (I) NF)
Scrambled eggs (R)	Granola (R)	Scrambled eggs (R)	Granola w/blueberries (R)
Bran flakes (R)	Breakfast roll (I) (NF)	Cornflakes (R)	Vanilla instant breakfast (B)
Cocoa (B)	Chocolate instant breakfast (B)	Cocoa (B)	Grapefruit drink (B)
Orange drink (B)	Orange-grapefruit drink (B)	Orange-pineapple drink (B)	
Frankfurters (T)	Corned beef (T) (I)	Ham (T) (I)	Ground beef w/
Turkey tetrazzini (R)	Asparagus (R)	Cheese spread (T)	pickle sauce (T)
Bread (2) (I) (NF)	Bread (2) (I) NF)	Bread (2) (I) (NF)	Noodles and chicken (R)
Bananas (FD)	Pears (T)	Green beans and broccoli (R)	Stewed tomatoes (T)
Almond crunch bar (NF)	Peanuts (NF)	Crushed pineapple (T)	Pears (FD)
Apple drink (2) (B)	Lemonade (2) (B)	Shortbread cookies (NF)	Almonds (NF)
		Cashews (NF)	Strawberry drink (B)
		Tea w/lemon and sugar (2) (B)	
Shrimp cocktail (R)	Beef w/barbeque sauce (T)	Cream of mushroom soup (R)	Tuna (T)
Beef steak (T) (I)	Cauliflower w/cheese (R)	Smoked turkey (T) (I)	Macaroni and cheese (R)
Rice pilaf (R)	Green beans w/mushrooms (R)	Mixed Italian vegetables (R)	Peas w/butter sauce (R)
Broccoli au gratin (R)	Lemon pudding (T)	Vanilla pudding (T) (I)	Peach ambrosia (R)
Fruit cocktail (T)	Pecan cookies (NF)	Strawberries (R)	Chocolate pudding (T) (R)
Butterscotch pudding (T)	Cocoa (B)	Tropical punch (B)	Lemonade (B)
Grape drink (B)			

[a]Abbreviations in parentheses indicate type of food: T = thermostabilized, I = irradiated, IM = intermediate moisture, FD = freeze dried, R = rehydratable, NF = natural form, and B = beverage.

Table 1   Space Shuttle typical Menu.   (Table courtesy of NASA.)

pouches, and semirigid plastic containers will be used for food packaging.

The menu will be a standard 6-day menu instead of the personal-preference type used in previous programs. The menu will consist of three meals each day plus additional snacks and beverages Table 1. The daily menu will be designed to provide an average energy intake of 3000 calories for each crewmember. The food system also includes a pantry of foods for snacks and beverages between meals and for individual menu changes [See: Table 2]

Food preparation activities will be performed by one crewman 30 to 60 minutes before mealtime. The crewman will remove the selected meal from the storage locker,

reconstitute those items that are rehydratable, place the foods to be heated into the galley oven, and assemble other food items on the food trays. Meal preparation for a crew of seven can be accomplished by one crewmember in about a 20-minute period. Utensils and trays are the only items that require cleaning after a meal. Cleaning will be done with sanitized "wet wipes" that contain a quaternary ammonium compound. As on the *Skylab* flights, the crewmembers will use regular silverware.

Two basic systems for sleeping are available on the Orbiter: sleeping bags, available on the first flights, and rigid sleep stations, available on later flights. After the rigid sleep stations are available, the Orbiter can

SPACE SHUTTLE FOOD AND BEVERAGE LIST

Foods[a]

Applesauce (T)	Chicken and noodles (R)	Peaches, dried (IM)
Apricots, dried (IM)	Chicken and rice (R)	Peaches, (T)
Asparagus (R)	Chili mac w/beef (R)	Peanut butter
Bananas (FD)	Cookies, pecan (NF)	Pears (FD)
Beef almondine (R)	Cookies, shortbread (NF)	Pears (T)
Beef, corned (I) (T)	Crackers, graham (NF)	Peas w/butter sauce (R)
Beef and gravy (T)	Eggs, scrambled (R)	Pineapple, crushed (T)
Beef, ground w/pickle sauce (T)	Food bar, almond crunch (NF)	Pudding, butterscotch (T)
Beef jerky (IM)	Food bar, chocolate chip (NF)	Pudding, chocolate (R) (T)
Beef patty (R)	Food bar, granola (NF)	Pudding, lemon (T)
Beef, slices w/barbeque sauce (T)	Food bar, granola/raisin (NF)	Pudding, vanilla (R) (T)
Beef steak (I) (T)	Food bar, peanut butter/granola (NF)	Rice pilaf (R)
Beef stroganoff w/noodles (R)	Frankfurters (Vienna sausage) (T)	Salmon (T)
Bread, seedless rye (I) (NF)	Fruitcake	Sausage patty (R)
Broccoli au gratin (R)	Fruit cocktail (T)	Shrimp creole (R)
Breakfast roll (I) (NF)	Green beans, french w/mushrooms (R)	Shrimp cocktail (R)
Candy, Life Savers, assorted flavors (NF)	Green beans and broccoli (R)	Soup, cream of mushroom (R)
Cauliflower w/cheese (R)	Ham (I) (T)	Spaghetti w/meatless sauce (R)
Cereal, bran flakes (R)	Jam/jelly (T)	Strawberries (R)
Cereal, cornflakes (R)	Macaroni and cheese (R)	Tomatoes, stewed (T)
Cereal, granola (R)	Meatballs w/barbeque sauce (T)	Tuna (T)
Cereal, granola w/blueberries (R)	Nuts, almonds (NF)	Turkey and gravy (T)
Cereal, granola w/raisins (R)	Nuts, cashews (NF)	Turkey, smoked/sliced (I) (T)
Cheddar cheese spread (T)	Nuts, peanuts (NF)	Turkey tetrazzini (R)
Chicken a la king (T)	Peach ambrosia (R)	Vegetables, mixed Italian (R)

Beverages

		Condiments
Apple drink	Instant breakfast, vanilla	Barbeque sauce
Cocoa	Lemonade	Catsup
Coffee, black	Orange drink	Mustard
Coffee w/cream	Orange-grapefruit drink	Pepper
Coffee w/cream and sugar	Orange-pineapple drink	Salt
Coffee w/sugar	Strawberry drink	Hot pepper sauce
Grape drink	Tea	Mayonnaise
Grapefruit drink	Tea w/lemon and sugar	
Instant breakfast, chocolate	Tea w/sugar	
Instant breakfast, strawberry	Tropical punch	

[a]Abbreviations in parentheses indicate type of food: T = thermostabilized, I = irradiated, IM = intermediate moisture, FD = freeze dried, R = rehydratable, and NF = natural form.

Table 2   Space Shuttle Food and Beverage List.   (Table courtesy of NASA.)

Fig. 5   One proposed Orbiter galley configuration.   (Drawing courtesy of NASA.)

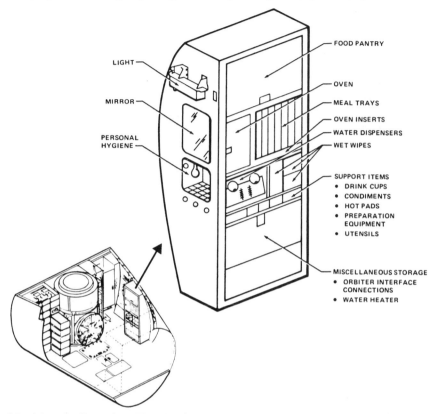

fly with either configuration. The sleeping bags are the same as the *Apollo* sleep restraint. For early flights, they will be constructed of *Beta* material perforated for *thermal* comfort and will be modified to include attach fittings for Shuttle installation. On later flights, the perforated Beta material will be replaced by perforated Nomex for better temperature control. Light masks and earplugs are provided with all sleeping bags. Four rigid sleep stations [See: Fig. 6] are provided: three horizontal stacked units and one vertical unit. Each station is provided with a sleep pallet, a sleep restraint, personal stowage, a light, ventilation inlet and outlet, and overhead light shields. The backs of the sleep restraints face toward the deck in the middle and top stations (stations 2 and 3); in the bottom sleep station (station 1), the back of the sleep restraint faces toward the overhead. This arrangement is intended to provide a 66-centimeter (26-inch) clearance between the bottom sleeping pallet and the

Orbiter floor for underfloor stowage access.

The waste collection system is an integrated multifunctional system used to collect, process, and store solid and liquid wastes. The system is used the same as a standard facility and performs the following general functions: (1) collecting, storing, and drying fecal wastes, associated toilet paper, and emesis-filled bags; (2) processing wash water from the personal hygiene station; (3) processing urine; (4) transferring the collected *fluids* to the waste storage tanks in the waste management system; and (5) venting the air and *vapors* from the wet trash container and stowage compartment. The waste collection system accommodates both male and female crewmembers and consists of the commode assembly, urinal assembly valving, instrumentation, interconnecting plumbing, mounting framework, and restraints. The waste collection system is located on the mid deck of the Orbiter in a 74-centimeter (29-inch) wide compartment immediately aft of the side hatch.

The personal hygiene system provides for the hygienic needs of the individual crewmembers with the following equipment: a personal hygiene station, personal hygiene kits, towels and washcloths, and a tissue dispenser. When the galley is not flown, the personal hygiene station is located on the left side of the mid deck and provides ambient and chilled water through a flexible line to a water dispenser. No drain is provided. When the galley is flown, the personal hygiene station is located on the left side of the mid deck and provides ambient and hot water plus a drain.

The galley with its built-in personal hygiene station will not be available on early Shuttle flights. In the interim, the personal hygiene needs of the crew will be met by the water dispenser. The personal hygiene station consists basically of the hygiene water valve and water gun, which is a part of the water dispenser. The hygiene water valve is a manually operated squeeze valve that provides ambient water at 291 to 308K (18° to 35°C or 65° to 95°F). A 22.9- by 30.5-centimeter (9- by 12-inch) *Mylar* mirror is provided that mounts on the waste collection system door by means of Velcro for use onorbit. The mirror is stowed for launch and entry in a modular stowage locker. The personal hygiene station for later flights is part of the galley [See: Fig. 7] In addition to ambient water from the

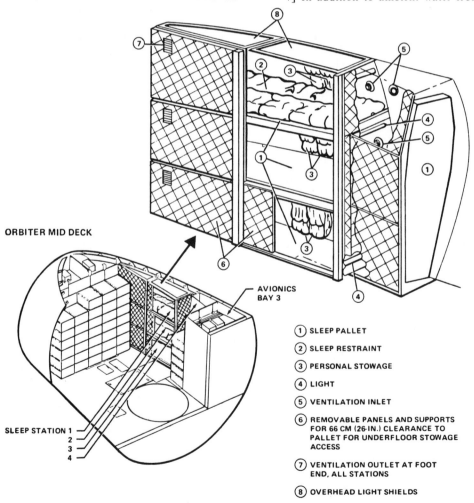

ORBITER MID DECK

AVIONICS BAY 3

SLEEP STATION 1
2
3
4

1 SLEEP PALLET

2 SLEEP RESTRAINT

3 PERSONAL STOWAGE

4 LIGHT

5 VENTILATION INLET

6 REMOVABLE PANELS AND SUPPORTS FOR 66 CM (26-IN.) CLEARANCE TO PALLET FOR UNDERFLOOR STOWAGE ACCESS

7 VENTILATION OUTLET AT FOOT END, ALL STATIONS

8 OVERHEAD LIGHT SHIELDS

Fig. 6   Rigid sleep station provisions on the Orbiter.   (Drawing courtesy of NASA.)

Fig. 7   The inorbit locations of the Space Shuttle's personal hygiene equipment. (Flights with the galley).   (Drawing courtesy of NASA.)

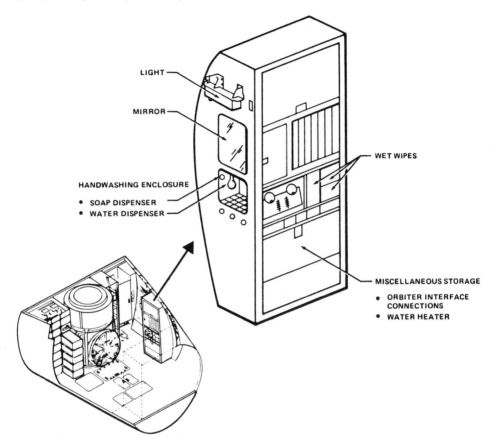

Orbiter water system, the galley water heater provides hot water. The personal hygiene station includes a hand washing enclosure, a mirror, a light, a soap dispenser, and controls for water dispensing, draining, and temperature adjustment. Fluid waste from the personal hygiene station is discharged into the waste collection system. Fluid wastes are low sudsing and relatively free of particulates to be compatible with the operation of the waste collection system fan separators. A mixing *valve* at the personal hygiene station adjusts water temperature, a water valve activates the water ejection, and an airflow valve connects the personal hygiene station to the waste collection system.

A personal hygiene kit is furnished for each crewman to provide for dental hygiene, hair care, nail care, shaving, etc. A typical personal hygiene kit contains a toothbrush, toothpaste, dental floss, and an antichap lipstick; a comb and brush; nail clippers; shaving cream, a safety razor and blades or a mechanical (windup) shaver, and a styptic pencil; skin emollient; soap; and stick deodorant. Pockets, loops, and Velcro are provided in the personal hygiene kit containers to maintain articles in an orderly manner and to permit efficient removal, use, and replacement of components in flight. The maximum weight of the kit will be 2.3 kilograms (5 pounds). A quantity of towels sufficient to support one crewman for 7 days is packaged together to form a towel assembly. Each crewmember is provided with seven cotton washcloths and three cotton towels. The washcloths are 30.5 by 30.5 centimeters (12 by 12 inches) and the towels are 46.6 by 68.8 centimeters (16 by 27 inches). The tissues are paper and are absorbent, multi-ply, and low-linting. The quantities are the same as for the towels.

The housekeeping system is used for cleaning the Orbiter crew station in orbit. The major components of the system are cleaning equipment, carbon dioxide absorbers, trash containers, and replacement parts.

The three trash accumulators, for both wet and dry trash, *interface* with the Orbiter in their use locations and are stowed for launch and entry. The wet trash containers are connected to the waste management system through a vent hose. Cleaning materials are stowed in lockers. The carbon dioxide absorbers (*lithium hydroxide canisters*) are stowed in a mid-deck floor stowage compartment and are periodically placed in the *environmental control system* downstream from the fans. Used absorbers are stowed in the original container. The vacuum cleaner is stowed in a modular locker and is powered by the Orbiter electrical power system.

The cleaning operations performed in the Orbiter include cleaning the waste collection system urinal and seat, the dining area and equipment, floors and walls (as required), the personal hygiene station, and the *air filters*. The materials and equipment available for cleaning operations are biocidal cleanser, disposable gloves, general-purpose wipes, and a vacuum cleaner. The biocidal cleanser is a liquid biocidal detergent formulation in a container approximately 5 centimeters (2 inches) in diameter and 15 centimeters (6 inches) long, with a built-in bladder, dispensing valve, and nozzle. The cleanser is sprayed on the surface to be cleaned and wiped clean with dry general-purpose wipes. The cleanser is used for periodic cleansing of the waste collection system urinal and seat, the dining area

and equipment, and the personal hygiene station. It will also be used, as required, to clean walls and floors. Disposable plastic gloves are worn while using the biocidal cleanser. General-purpose wipes are dry wipes used to spread the biocidal cleanser and to dry the cleaned surface; they are also used for general-purpose cleaning. The vacuum cleaner is provided for general housekeeping and for cleaning the Orbiter cabin air filter and the *Spacelab* environmental control system filters.

Carbon dioxide absorbers.—The carbon dioxide absorbers are *lithium hydroxide canisters* that are placed in the *Environmental Control and Life Support System (ECLSS)* to filter carbon dioxide from the recirculated air. The absorbers are stowed in a mid-deck floor compartment. They are removed from stowage and inserted into the environmental control system according to the number of people onboard. The absorber changeout schedule is shown in Table 3. The system will operate with two active absorber cartridges that are changed on a rotating basis. The nominal carbon dioxide *partial pressure* in the *Orbiter cabin* is 666 N/m^2(5.0 mmHg) within a range of 0 to 1013 N/m^2 (0 to 7.6 mmHg). The lithium hydroxide bed of the absorber cartridge is sized to maintain the carbon dioxide partial pressure within this range when changed out as scheduled in Table 3.

The trash management operations include routine stowage and daily collection of wet and dry trash such as expended wipes, tissues, and food containers. Wet trash includes all items that could *offgas*. The equipment available for trash management includes trash bags, trash bag liners, wet trash containers, and the stowable wet trash vent hose. Three trash bags are located in the crew compartment. Each bag contains a disposable trash bag liner. Two

CARBON DIOXIDE ADD RATE/REPLACEMENT TIME

Number in crew	Carbon dioxide add rate, kg/hr (lb/hr)	Absorber cartridge replacement time, hr	Cartridge alternating changeout time, hr
4	0.160 (0.352)	24	12
7	.279 (.616)	11	5.5
10	.399 (.880)	6.4	3.2

Table 3   Orbiter's carbon dioxide obsorber add rate/replacement time.   (Table courtesy of NASA.)

bags will be designated for dry trash and one for wet trash. At a scheduled time each day, the dry trash bag liner will be removed from its trash bag. The trash bag liner will be closed with a strip of Velcro and stowed in an empty locker. For long-duration flights when more than 0.23 cubic meter (8 cubic feet) of wet trash is expected, the wet trash bag liners will be removed at a scheduled time each day and placed in a wet trash container. The container is then closed with a zipper and the unit is stowed. If expansion due to wet trash *offgassing* is evident, the container is connected to a vent in the waste management system for overboard venting.

The wet trash container is made of airtight fabric and is closed with a seal-type slide fastener. The container has a volume of approximately 0.02 cubic meter (0.7 cubic foot) and has an air inlet valve on one end and a quick disconnect on the other end. The container is attached to the waste management vent system at a point beneath the commode, enabling air to flow through the wet trash container and then overboard. Attachment is made through a 104-centimeter (41-inch) long vent hose filter. When the container is full, it is removed and stowed in a modular locker. A 0.2-cubic-meter (8-cubic-foot) wet trash stowage compartment is available under the mid-deck floor. Each day, the wet trash bag liners will be removed from the trash bags and stowed in the wet trash stowage compartment, which is vented overboard. If the compartment becomes full, the wet trash bag liners will be stowed in wet trash containers.

The *airlock* in the Shuttle Orbiter accommodates *astronaut* extravehicular operations without the necessity for cabin decompression or for decompression of an attached pressurized manned payload in the Orbiter *payload bay*. The airlock [See: Fig. 8] is a modular cylindrical structure 160 centimeters (63 inches) inside diameter by 210 centimeters (83 inches) long with two D-shaped pressure-sealing hatches and a complement of airlock support systems. Access to the airlock from the crew compartment and from the airlock to the payload bay is provided by the two *hatches* located on opposite sides of the airlock.

The airlock is removable and can be

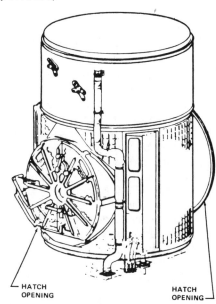

Fig. 8 The Orbiter airlock. (Drawing courtesy of NASA.)

HATCH OPENING

HATCH OPENING

installed in one of three different Orbiter locations, depending on the payload carried. The *baseline* location is inside the crew compartment, allowing maximum use of the payload bay volume. The airlock may also be rotated 180° and positioned in the payload bay, still attached to the aft cabin *bulkhead*. For a habitable payload mission such as *Spacelab*, the airlock may be positioned on top of a pressurized *tunnel adapter*, which connects the cabin with the pressurized payload. The airlock provides stowage during flight for the *extravehicular mobility unit*. During *extravehicular activity (EVA)*, the airlock supplies oxygen, cooling water, communications, and power to the crewmen.

The primary structure of the airlock is composed of machined aluminum sections welded together to form a cylinder with hatch mounting flanges. The upper cylindrical section and the bulkheads are made of nonvented aluminum *honeycomb*. Two semicylindrical aluminum sections are welded to the airlock primary structure to house the *ECLSS* and *avionics* support equipment. Each semicylindrical section has three feedthrough plates for plumbing and cable routings from the Orbiter to support the airlock subsystems.

The airlock hatches permit the *EVA crewmember* to transfer from the Orbiter crew compartment to the payload bay. Both airlock hatches open toward the primary pressure source, the *Orbiter cabin*. Each hatch opening has a clear passageway 101.6 centimeters (40 inches) in diameter, with one flat side that reduces the minimum dimension of the hatch opening to 91.4 centimeters (36 inches).

The airlock depressurization system is designed to discharge an airlock volume of 3.8 cubic meters (133 cubic feet), assuming two suited crewmembers are in the airlock, at a rate of 620 *pascals*/sec in 6 minutes. Following the EVA and hatch closing, the airlock volume is repressurized by using the equalization valve on the airlock/cabin hatch. Each airlock has two pressure equalization valves that are operable from both sides of the hatch. Four repressurization modes are available to the crewmember following an EVA. One mode is considered normal, two are classified as emergency modes because of time constraints on either the space suit or the crewmember, and the fourth is for rescue operations.

Orbiter environmental control and life support system (ECLSS)   A system that provides a comfortable *shirtsleeve environment* (289 K to 305 K or 16 C to 32 C or 61 F to 90 F) for the *Orbiter* crew and a conditioned thermal enviornment (heat controlled) for the electronic components. The ECLSS bay, which includes air-handling equipment, *lithium hydroxide* canisters, water circulation pumps, and supply and waste water, is located in the mid-deck of the Orbiter and contains the pressurization system, the air-revitalization system, the active thermal control system, and the water and waste management system. The systems interact to pressurize the crew compartment with a breathable mixture of oxygen and nitrogen while keeping toxic gases below harmful levels, controlling *temperature* and *humidity*, cooling equipment, storing water for drinking and personal hygiene and *spacecraft* cooling, and processing crew waste in a sanitary manner.

The electrical power system provides the ECLSS with drinking water, oxygen for the cabin *atmosphere* and power to run the *fans, pumps* and electrical circuits. In return the ECLSS provides heat removal for the electrical power system. The ECLSS is the prime heat-removal system onboard the Orbiter and thus interfaces with all other Shuttle systems.

The pressurization system provides a mixed-gas atmosphere of oxygen and nitrogen at sea level *pressure* (101.4 kN/m^2 or 14.7 psia). The air-revitalization system controls the *relative humidity* (between 35 percent and 55 percent) and the carbon dioxide and carbon monoxide levels. It also collects heat by circulating the cabin air and then transfers the heat to the active thermal control system, which collects excess heat and transports it to the exterior of the spacecraft through water and *Freon* circulating loops. The water and waste management system provides the basic *life support* functions for the *Shuttle crew*.

Pressurization helps provide structural stability and aids in the transfer of thermal energy (heat) from crew and equipment. The Orbiter is pressurized with 21 percent oxygen and 79 percent nitrogen, which provide a breathable atmosphere comparable to that of the normal *Earth* environment. The spacecraft cabin compartment is normally maintained at sea level pressure and is kept at this pressure by a regulator. In an emergency this regulator can be turned off, and another regulator will maintain the cabin at 55kN/m^2 (8.0 psia).

The pressurization system is composed of two oxygen systems, two nitrogen systems and one emergency oxygen system. Oxygen comes from the same *cryogenic* storage tanks that supply the electrical power system. The cryogenic oxygen pressure is controlled by heaters to 575.7 kN/m^2 (835 psia to 852 psia), and the oxygen is delivered in gaseous form to the oxygen ECLSS supply valve. When the oxygen system supply valve is opened, the oxygen is permitted to flow through a restrictor that is basically a *heat exchanger*, where the oxygen is warmed before passing through the regulators.

The nitrogen system has four storage tanks: two 23 kg (50 lb$_m$) 22,750 kN/m^2 (3,300 psi) tanks for each of the two redun-

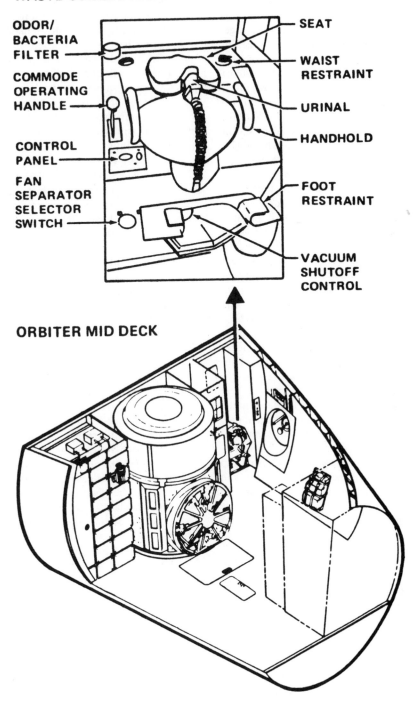

WASTE COLLECTOR

ODOR/ BACTERIA FILTER

COMMODE OPERATING HANDLE

CONTROL PANEL

FAN SEPARATOR SELECTOR SWITCH

SEAT

WAIST RESTRAINT

URINAL

HANDHOLD

FOOT RESTRAINT

VACUUM SHUTOFF CONTROL

ORBITER MID DECK

The Orbiter's waste collection system.   (Drawing courtesy of NASA.)

dant systems, nitrogen system 1 and nitrogen system 2. The storage tanks are constructed of filament-wound *Kevlar* fiber with a titanium liner and are located in the forward *payload bay*. The gaseous nitrogen tank is filled to a pressure of 22,750 kN/m^2 (3,300 psi) before *lift-off*. Motor control valves control the 22,750 kN/m^2 (3,300 psi) inlet to the 1,380 kN/m^2 (200 psi) regulator. Nitrogen arrives at the oxygen and nitrogen control valve at 1,380 kN/m^2 (200 psi). These pressures—oxygen at 690 kN/m^2 (100 psi) and nitrogen at 1,380 kN/m^2 (200 psi)—are important to the operation of the two-gas pressurization system.

There are numerous openings in the crew compartment to provide for proper airflow for pressure-control purposes. A negative relief valve will open when a pressure differential of 1.4 kN/m^2 (0.2 psi) is detected. This ensures that if the pressure in the compartment is lower than that outside the pressure vessel, air will flow into the cabin. The inlets (orifices) are sealed with tethered caps, which are dislodged by a negative pressure of 1.4 kN/m^2 (0.2 psi).

The isolation and vent valves provide for flow from the inside to the outside of the pressure vessel, but they can reverse the flow. These valves enable the crew to vent the cabin if the need arises. The maximum flow at 13.8 kN/m^2 (2 psi) is 408 kg/hr (900 lb$_m$/hr). The cabin relief valves, when activated, provide protection against pressurization of the vehicle above 107 kN/m^2 (15.5 psi). If activated the valves will open and flow to a maximum of 68 kg/hr (150 lb$_m$/hr).

At approximately 1 hour and 26 minutes before launch, the cabin is pressurized by *ground support equipment* to approximately 110 kN/m^2 (16 psi) for a leak check. The isolation and vent valves are then opened, and the spacecraft is allowed to return to 105.5 kN/m^2 (15.3 psi). The vent valve is then closed.

Air circulation is provided by two cabin fans, only one of which will be used at a time. Each fan turns at 11,200 rpm (revolutions per minute), propelling the air (up to 635 kg/hr or 1,400 lb$_m$/hr) to the lithium hydroxide canisters that cleanse it. The canisters contain a mixture of activated char-

coal, which removes the odor, and lithium hydroxide, which removes the carbon dioxide. The canisters are changed on a scheduled basis. Heated cabin air is passed through the cabin heat exchanger and the excess heat is transferred to the water coolant loop. Additional thermal energy (heat) is collected from the electronics. Two cabin temperature controllers maintain the cabin temperature set by the crew, but only one controller is used at a time. During flight the controller senses the temperature in the supply and return air *ducts* and controls the temperature between 291 K and 300 K (18 C and 27 C or 65 F and 80 F).

Crew compartment humidity is controlled as air is drawn across coldplates in the cabin heat exchanger. Temperature change as the cabin air passes over these coldplates causes *condensation*. Centrifugal fans separate the water and air; the air is returned to the cabin, and the water is forced into the waste tank. This system can remove up to 1.8 kg (4 lb$_m$) of water each hour. Two fans located in the *avionics* bay provide the required airflow for the enclosed air circulation system. Each avionics bay has a heat exchanger to cool the air as it is circulated.

As the active thermal control system circulates cabin air, it collects excess heat from the crew and the flight deck electronics; two water circulation systems, two Freon circulation systems, the space *radiators*, the *flash evaporators* and the ammonia boilers transfer the collected thermal energy (heat) into space. Two complete and separate water coolant loops (called loops 1 and 2) flow side-by-side throughout the same area of the spacecraft collecting excess heat from the avionics bays, cabin windows and various other components. The only difference in the two systems is that loop 1 has two pumps, whereas loop 2 has a single pump. Between 500 kg and 590 kg (1,100 lb$_m$ and 1,300 lb$_m$) of water pass through each loop each hour. Both loops will be used during lift-off, ascent and entry. Loop 2 alone will be used for orbital operations.

The circulating coolant water collects heat from the cabin heat exchanger and from coldplates throughout the cabin area. When the water passes through the Freon interchanger, the heat is passed to the Freon coolant loops and is then delivered to the radiators, the primary heat-rejection system onboard the Orbiter. Two complete and identical Freon coolant loops, each with its own pump package and accumulator, transport the thermal loads to the radiator system. The primary Freon coolant loop also cools the spacecraft *fuel cells.* Because Freon can be toxic under certain conditions, it does not flow through the cabin compartment. Instead the Freon coolant loop collects heat from various coldplates outside the crew compartment. Heat collected by the active thermal control system is finally ejected from the spacecraft by the space radiators and flash evaporators.

The space radiators, which consist of two deployable and two fixed panels on each payload bay door, are active during orbital operations when the payload bay doors are opened. The radiator panels contain 111 square meters (1,195 ft²) of effective heat-rejection area. Each panel is 3 m (10 ft) wide and 4.6 m (15 ft) long. The aft radiator panels are single-sided, whereas the forward panels are double-sided, allowing heat rejection from both sides of the panel. Each panel contains parallel tubes through which the Freon heat loop fluid passes. There are 68 tubes in the forward panels and 26 in each of the aft panels. More than 1.5 kilometers (1 mile) of Freon tubing is in the radiator panels. This system is designed to have a heat-rejection capability of 5,480 kJ/hr (5,200 Btu/hr) during ascent (with the payload bay doors closed) and 23 kJ/hr (21.5 Btu/hr) during orbital operations (that is, with the payload bay doors open).

The flash evaporator and the ammonia boilers provide for the transfer of waste heat from the Freon heat transport loop to water, using the *latent heat* capacity of water. This system functions during Orbiter operations at altitudes above 43 km (140,000 ft) during ascent and above 30.5 km (100,000 ft) during entry.

Water for Shuttle crew use (e.g., food preparation, drinking and personal hygiene) onboard the Orbiter is furnished as a by-product of the fuel cell operation. The drinkable water generated by this operation is fed into the storage tanks at the flow rate of approximately 3.2 kg/hr (7 lb$_m$/hr). The water temperature is about 283 K (10 C or 50 F). When the water tanks are full, a fuel cell relief valve automatically dumps the excess water overboard. Bacteria are controlled by a microbial check valve located in the supply line between the fuel cells and the potable water supply tank. Two 75-kg (165-lb$_m$) water tanks are provided as a supply source for the flash evaporators, and one tank is isolated to provide potable water to the galley.

The waste collection system (WCS) is an Earthlike commode for Orbiter crew members. It is an integrated multifunctional device used to collect and process biowastes as well as wash water from the personal hygiene station and cabin condensate water from the cabin heat exchanger. The waste collection system accommodates both male and female crew members and consists of the commode assembly, the urinal assembly valving, instrumentation, interconnection plumbing, the mounting framework and restraints. It is located on the mid-deck of the Orbiter in a compartment immediately aft of the side hatch. The unit is approximately 69 cm by 69 cm by 74 cm (27 in. by 27 in. by 29 in.) and has two major independent and interconnected assemblies: the urinal, designed to handle fluids, and the commode, designed to handle solid biowastes. Two privacy curtains of Nomex cloth are attached to the inside of the compartment door and serve to isolate the WCS compartment and the galley personal hygiene station from the rest of the mid-deck.

Designed for use both in the weightless space environment and in the Earth's atmosphere, the waste collection system is used in the same manner as the facilities onboard jet airliners. It differs from conventional bathroom commodes in that separate receptacles for the collection of liquid and solid body wastes are built into the seat. In space high-velocity airstreams compensate for Earth's *gravity* to force waste matter into the respective chambers. Airstreams also assist in the operation of a water-flush mechanism for cleaning after each use. The waste matter is vacuum dried, stored and chemically treated to prevent odor and bacterial growth. Like toilets on airliners the facility will be serviced when the Orbiter lands back on Earth.

Foot restraints, a waist restraint and handholds are part of the WCS. The foot restraints are simple toeholds attached to a fold-up step. The handholds are multipurpose and can be used for positioning or actual stabilization at the option of the user. The waist restraint encircles the user and is attached on either side. The restraint has a negator spring take-up reel that removes slack in the belt and exerts an evenly distributed downward force on the user to ensure an adequte seal between the user and the seat.

The Orbiter fire detection and suppression system consists of two related but not physically connected subsystems: the detection system and the suppression system. There are four portable Freon (1301) fire extinguishers and three fixed extinguishers onboard the Orbiter. Nine early warning smoke detectors sense any significant increase in the gaseous or particulate products of combustion within the crew compartment or avionics bays. Should a significant increase occur, a signal activates a siren and illuminates a warning light on the fire detection and suppression control panel.

Two smoke detectors are located in each of the three avionics bays. When smoke is detected appropriate signals displayed on the fire detection and suppression control panel warn the crew to take necessary action. If the sensors detect smoke in the avionics bay, the crew initiates the appropriate system, which will rapidly extinguish the fire. One Freon 1301 extinguisher is mounted in each avionics bay. Each of these extinguishers has 1.6 kg (3.5 lb$_m$) of Freon 1301 in a pressure vessel 20.3 cm (8 in.) long and 11.4 cm (4.5 in.) in diameter.

If sensors detect smoke anywhere in the crew compartment or behind the electrical panels, the crew will use one of the portable extinguishers, which are 33.5 cm (13.2 in.) long and can be used as a backup for the avionics bay extinguishers. The portable extinguishers have a tapered nozzle that can be placed into fire holes on the electrical panels to extinguish fires behind the panels.

Freon 1301 (bromotrifluoromethane) is one of the most effective chemical fire suppressants. It operates by breaking the chemical chain reaction in a flame rather than by smothering it. A relatively small amount of the chemical is sufficient to suppress a fire.
See: Orbiter power generation systems, Orbiter structure

Orbiter hatches   The *Space Shuttle Orbiter* hatches consist of the side hatch, the airlock hatch, the tunnel adapter hatch and the docking module hatch.
See: Orbiter structure

Orbiter mechanical subsystems   *Orbiter* subsystems with electrical and hydraulic *actuators* that operate the *aerodynamic* control surfaces, landing/*deceleration* system, *payload bay* doors, deployable radiators, and *payload* retention and handling subsystems. Orbiter/*External Tank* propellant disconnects and a variety of other mechanical and *pyrotechnic* devices comprise the balance of the mechanical subsystems.
See: Orbiter structure

Orbiter power generation systems (PGS)
The *Space Shuttle Orbiter* has one system to supply electrical power and another to supply hydraulic power. Electrical power is generated by three *fuel cells* that use *cryogenically* stored hydrogen and oxygen reactants. Hydraulic power is derived from three independent hydraulic pumps, each driven by its own hydrazine-fueled *auxiliary power unit* and cooled by its own boiler. (See fig. 1.)

The power reactant storage and distribution system cotains the *cryogenic* oxygen and hydrogen reactants that are supplied to the fuel cells and the oxygen that is supplied to the *environmental control and life support system*. The oxygen and hydrogen are stored in double-walled, vacuum-jacketed *Dewar*-type spherical tanks in a *supercritical condition*—97 K (−176 C or −285 F) for oxygen and 22 K (−251 C or −420 F) for hydrogen. In this supercritical condition the hydrogen or oxygen takes the form of a cold, dense, high-pressure gas that can be expelled, gauged (quantity-measured) and controlled under *zero-g* conditions. Automatic controls, activated

ELECTRICAL SYSTEM

- POWER GENERATION SYSTEM
- POWER REACTANT STORAGE AND DISTRIBUTION SYSTEM

FUEL CELL POWERPLANT
2-KILOWATTS MINIMUM,
7-KILOWATTS CONTINUOUS,
12-KILOWATTS PEAK PER FCP
15-MINUTE DURATION ONCE
EACH 3 HOURS

FUEL CELL POWERPLANT SYSTEM
14-kilowatt continuous/24-kilowatt peak
27.5 to 32.5 volts direct current

REACTANT STORAGE
5508 megajoules (1530 kilowatt-hours)
mission energy
950 megajoules (264 kilowatt-hours)
abort/survival energy
51 kilograms (112 pounds) oxygen for ECLSS
42 kilograms (92 pounds)
hydrogen per tank
354 kilograms (781 pounds)
oxygen per tank

TOTAL
LOADED
QUANTITY

OXYGEN TANKS
0.035-CUBIC-METER (12.3-CUBIC-FOOT) CAPACITY
7240 kN/m^2 (1050 PSIA) MAXIMUM PRESSURE

HYDROGEN TANKS
0.665-CUBIC-METER (23.5-CUBIC-FOOT) CAPACITY
2310 kN/m^2 (335 PSIA) MAXIMUM PRESSURE

UMBILICAL SERVICE

HYDRAULIC SYSTEM

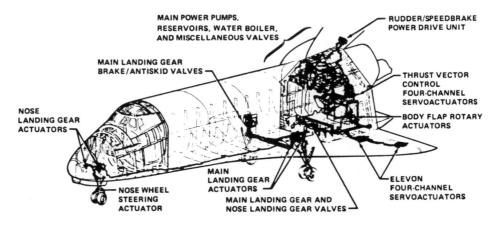

Fig. 1   The Orbiter power generation systems. Four hydrogen-oxygen fuel cells supply all the electrical power for the Orbiter during all mission phases, and three hydrazine-fueled turbines drive pumps to provide hydraulic pressure.   (Drawing courtesy of NASA.)

by *pressure*, energize tank heaters and thus add thermal energy (heat) to the reactants to maintain pressure during depletion. Each tank has relief valves to prevent overpressurization form abnormal operating conditions. *Redundancy* is provided by having two components for each major function or by providing manual override of the automatic controls.

The distribution system consists of filters, check valves and shutoff valves. The valves and plumbing are arranged so that any tank can be used to supply any subsystem, or any tank can be isolated in case of failure. In addition the distribution system can be isolated into two halves by using valves provided for that purpose, so that a distribution system failure can be tolerated.

Each oxygen tank can store 354 kg (781 lb$_m$) of oxygen, and each hydrogen tank can store 42 kg (92 lb$_m$) of hydrogen. The hydrogen tank is 115.6 cm (45.5 in.) in diameter, and the oxygen tank is 93.5 cm (36.8 in.) in diameter. The basic system consists of four oxygen and four hydrogen tanks, with additional tanks available to take care of added requirements. The oxygen and hydrogen tanks are shown in figure 2.

The fuel cell system produces the electrical power required by the Orbiter during all *mission* phases. It is composed of three units located in the Orbiter's mid-body. Each fuel cell power plant is composed of a single stack of 64 cells divided electrically into parallel connected substacks of 32 cells each. Electrical power is produced by the chemical reaction of hydrogen and oxygen, which are supplied continuously as needed to meet output requirements. A by-product of this reaction is drinkable water (needed and used by the *Shuttle crew*). Each fuel cell

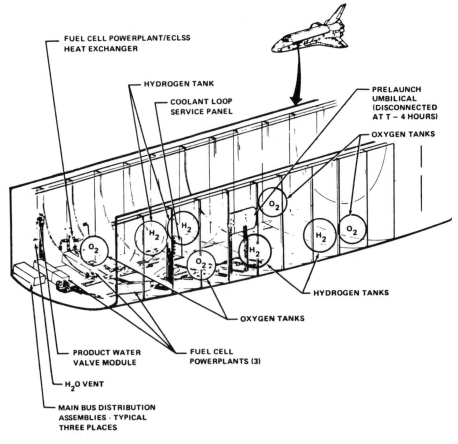

Fig. 2   The Orbiter's electric power system.   (Drawing courtesy of NASA.)

is connected to one of the three independent electrical buses.

During peak and average power loads, all three fuel cells and buses are used; during minimum power loads only two fuel cells are used, but they are interconnected to the three buses. In this case the third fuel cell is placed on standby but can be reconnected instantly to support higher electric loads. Alternately the third fuel cell is shut down under the condition of a 278 K (4.4 C or 40 F) minimum temperature requirement and can be reconnected within 15 minutes to support higher load demands.

The electrical power requirements of a *payload* will vary throughout a *Shuttle mission*. During the 10-minute launch-to-orbit mission phase and the 30-minute de-orbit-to-landing phase, when most of the experiment hardware is in a standby mode or completely turned off, 1,000 watts average to 1,500 watts peak are available from the Orbiter. During payload equipment operation in *orbit*, the capability exists to provide as much as 7,000 watts maximum average to 12,000 watts peak for major energy-consuming payloads. For a seven-day mission payload, 180 megajoules (50 kilowatt-hours) of electrical energy are available. *Mission kits* containing consumables for 3,060 megajoules (850 kilowatt-hours) each are also available.

The Space Shuttle fuel cells will be serviced between flights and reflown until each has accumulated 5,000 hours of on-line service. Although the Shuttle fuel cells are no larger than those used during the *Apollo Program*, each has six times the output of the Apollo fuel cells.

The Shuttle power plant is a single stack of 64 cells divided electrically into two parallel connected substacks of 32 cells each. An accessory section containing components for reactant management, thermal control, water removal, electrical control and monitoring is located at one end of the stack. The entire accessory section can be separated from the cell stack for *maintenance*. The interface panel, which is part of the accessory section, provides (1) fluid connections for hydrogen and oxygen supply and purge, water discharge, and coolant entry and outlet and (2) electrical con-

nectors for power output, control instrumentation, power input and preflight verification data. A complete Orbiter fuel cell power plant measures 35.6 cm by 43.2 cm by 101.6 cm (14 in. by 17 in. by 40 in.) and weighs approximatley 92 kg (202 lb$_m$).

Each cell consists of an *anode*, a *cathode* and a matrix containing the *catalyst* of potassium hydroxide. Magnesium separator plates provide rigidity, electron transfer paths (the means to distribute hydrogen and oxygen to the cells), and water and waste heat removal. Electrons produced by the reaction flow through the separator plates to the power takeoff point, where the power feeds into the electrical system.

The electrical control unit (ECU) consists of a start/stop control and the isolation and control relays. The heater group is composed of the end cell and the sustaining and startup heaters and their control switches. The pump motors are the only components that draw significant power during normal operation. The ECU starts and stops the power plant independent of any outside support equipment. When the operator starts the system, the hydrogen pump separator, the coolant pump and the heaters are activated; when the fuel cell stack reaches operating temperature, the startup heater cycles off automatically. A coolant pressure interlock prevents coolant overheating. Power for the startup and sustaining heaters comes from the fuel cell power plant direct current power bus. Each heater is controlled by a spearate switch. The sustaining heaters keep fuel cell stack temperatures at the proper level during sea level operations at low-power settings, and the end-cell heaters keep end-cell temperatures near those of the rest of the stack. Power for driving the hydrogen and oxygen valves comes from the vehicle-essential direct current control bus, and pump motors draw 115-volt 400-hertz three-phase alternating current (ac) power from the Orbiter ac bus.

The Orbiter flight control system *actuators*, main engine *thrust-vector-control* actuators and utility actuators are powered by the three *auxiliary power units* (APUs). This differs from most commercial and military aircraft in that these applications typically use main engine shaft power to drive

Fig. 3   The Orbiter's hydraulic system.   (Drawing courtesy of NASA.)

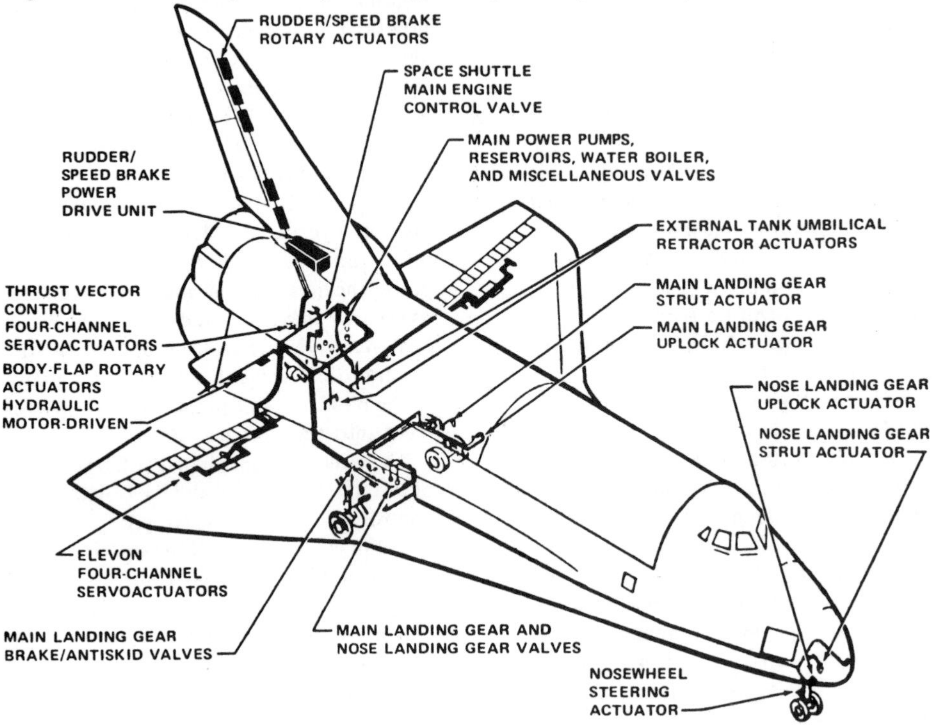

the hydraulic pump. The Shuttle auxiliary power units and hydraulic system (see fig. 3) are sized so that any two of the three systems can perform all flight control functions.

The auxiliary power units are started before launch and used for engine control during boost. The systems are shut down during orbital operation; they are restarted five minutes before the de-orbit burn and continue to operate until approximately five minutes after landing. The circulation pumps are operated after landing to circulate the hydraulic fluid through the water *heat exchangers* to dissipate the postlanding *heat soakback*.

The main pumps—the heart of the hydraulic system—are capable of 238-liter/min (63-gal/min) output at the minimum power engine speed. A depressurization valve on each pump reduces the output pressure, and therefore the input *torque* requirement, during *bootstrap* startup of the power units. The hydraulic fluid is supplied to the appropriate subsytems at a

nominal 20,684 kN/ m^2 (3,000 psig) through thermally insulated titanium lines. Suction pressure is provided by a bootstrap reservoir pressurized by the main pump during APU operation and by a gaseous nitrogen accumulator when the APU is not operating.

Cooling for the hydraulic fluid and the APU lubrication oil is provided by a water-spray heat exchanger located in the return line of each system. Fluid circulation for thermal control during in-orbit *coldsoak* is provided by an electric-motor-driven circulation pump in each hydraulic system.

The auxiliary power system (see fig. 4), consisting of three independent 103-kilowatt (138-horsepower) *turbine* engines, converts the chemical energy of liquid hydrazine into mechanical shaft power, which drives hydraulic pumps that provide power to operate aerosurfaces, main engine thrust-vector-control and engine valves, landing gear and other system actuators. The APU is unique in that it is not an auxiliary power system but actually drives the hydraulic system pumps. Each APU weighs 39 kg (85 lb$_m$). The auxiliary power system provides primary hydraulic power during the launch and entry phases of a Shuttle flight.

The turbine engines are started several minutes after termination of *thrust* from the Orbiter main engines. All three units are required to be operating in order to maintain thrust vector control for the three *Space Shuttle Main Engines*, because the vector control system for each main engine is connected to a single power unit. If shaft power from a power unit is prematurely interrupted during launch, it may become necessary to *abort* the normal launch cycle and either return to the launch site or seek a lower Earth orbit and return to a landing site near the end of the first orbit. Conversely, the entry functions that require APU power are arranged such that power can be supplied from more than one unit. Therefore, a satisfactory entry can be accomplished with only two of the three APUs operating. This is called *"fail-safe"* redundancy.

Approximately 134 kg (295 lb$_m$) of hydra-

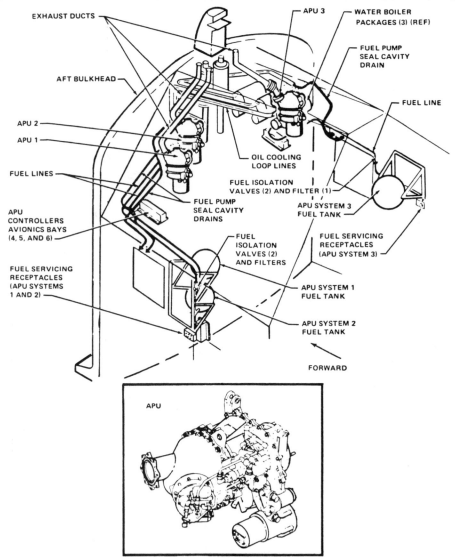

Fig. 4   The Orbiter's auxiliary power unit system.   (Drawing courtesy of NASA.)

zine is stored in a fuel tank in each unit and is pressurized with helium to no more than 2,550 kN/m^2 (370 psig), providing 91 minutes of APU operation. The hydrazine pressure is boosted to approximately 10,342 kN/m^2 (1,500 psig) by the APU fuel pump, which is driven from the unit's *gearbox*. The high-pressure hydrazine is routed from the fuel pump outlet to the gas generator valve module, which controls the flow of hydrazine to the *gas generator*. When the valves in the gas generator valve module are open,

hydrazine flows into the gas generator. The APU controller automatically cycles these valves to maintain the proper turbine speed of 72,000 rpm.

The gas generator contains a bed of granular catalyst that decomposes the hydrazine into a hot gas of approximately 1,200 K (927 C or 1,700 F) and 6,900 1/m^2 to 8,300 kN/m^2 (1,000 psi to 1,200 psi). These hot gases are expanded to a lower pressure, accelerated to a high velocity and directed at the turbine blades by a *nozzle*. The

Purge, Vent, and Drain System

PURGE SUBSYSTEM (PREFLIGHT AND POSTFLIGHT)

Circulates conditioned gas during launch preparations to remove contaminants and toxic gases and maintain specified temperature and humidity

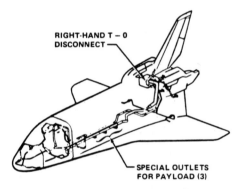

VENT SUBSYSTEM (ALL PHASES)

Allows unpressurized areas to depressurize during ascent and repressurize during descent and landing

DRAIN SUBSYSTEM (PREFLIGHT AND POSTFLIGHT)

Removes accumulated water and other fluids

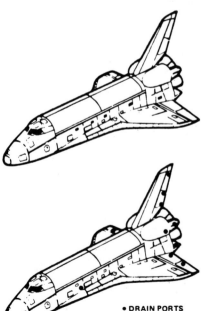

Fig. 1   Functions of the Orbiter purge, vent and drain system.   (Drawing courtesy of NASA.)

Fig. 2   Details of the Orbiter purge, vent and drain system.   (Drawing courtesy of NASA.)

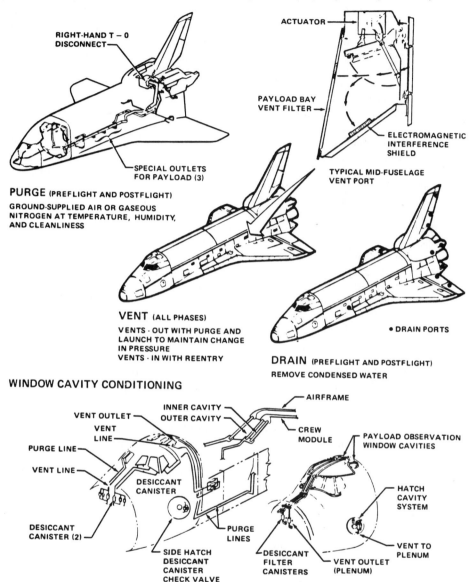

RIGHT-HAND T – 0
DISCONNECT

ACTUATOR

PAYLOAD BAY
VENT FILTER

ELECTROMAGNETIC
INTERFERENCE
SHIELD

SPECIAL OUTLETS
FOR PAYLOAD (3)

TYPICAL MID-FUSELAGE
VENT PORT

PURGE (PREFLIGHT AND POSTFLIGHT)
GROUND-SUPPLIED AIR OR GASEOUS
NITROGEN AT TEMPERATURE, HUMIDITY,
AND CLEANLINESS

VENT (ALL PHASES)
VENTS - OUT WITH PURGE AND
LAUNCH TO MAINTAIN CHANGE
IN PRESSURE
VENTS - IN WITH REENTRY

• DRAIN PORTS

DRAIN (PREFLIGHT AND POSTFLIGHT)
REMOVE CONDENSED WATER

WINDOW CAVITY CONDITIONING

INNER CAVITY
OUTER CAVITY

AIRFRAME

VENT OUTLET
VENT
LINE

CREW
MODULE

PAYLOAD OBSERVATION
WINDOW CAVITIES

PURGE LINE

VENT LINE

DESICCANT
CANISTER

HATCH
CAVITY
SYSTEM

DESICCANT
CANISTER (2)

PURGE
LINES

SIDE HATCH
DESICCANT
CANISTER
CHECK VALVE

DESICCANT
FILTER
CANISTERS

VENT OUTLET
(PLENUM)

VENT TO
PLENUM

OUTER CAVITY SYSTEM

INNER CAVITY SYSTEM

momentum of the high-velocity gas is transferred to the blades of the turbine wheel as the direction of the gas is changed during its passages through the blades. The torque applied to the blades by the high-velocity gas is transferred to the gearbox through the turbine shaft.
See: abort, Orbiter structure

Orbiter purge, vent and drain (PVD) system   A system that removes the gases and

liquids that accumulate in the unpressurized spaces of the *spacecraft* and prevents ice buildup in the ground disconnect between the *Orbiter* and the *External Tank*. (See figs. 1 and 2.) During launch preparations on the *pad*, the purge subsystem circulates conditioned gas (air, gaseous nitrogen or a mixture of the two) through the forward fuselage, *payload bay*, tail group and *orbital maneuvering system* pods to remove contaminants or toxic gases as well as to maintain proper *temperature* and *humidity* levels.

The vent subsystem allows spacecraft cavities to depressurize during ascent through a series of 18 vents and outlets in the fuselage skin. These vent ports also allow repressurization during descent and landing. Electromechanical *actuators* open and close the vent ports.

Accumulated water and other fluids in the Orbiter drain through *limber holes* (much like limber holes in frames along the keel of a boat) to the lowest point for removal. Additional tubing and connections allow draining of those compartments that cannot drain through limber holes. A separate subsystem modulates pressure in the cavities between window panes and prevents fogging or frosting of windows in flight. On the launch pad a ground-based *mass spectrometer* samples purged gases from the Orbiter to determine whether hazardous levels of explosive or toxic gases are accumulating.

See: Orbiter structure

Orbiter structure   The Orbiter structure consists of the *forward fuselage* (upper and lower forward fuselage and the *crew module*), the wings, the *mid fuselage*, the *payload bay doors*, the *aft fuselage* and the *vertical stabilizer*. Most of the Orbiter structures are constructed of conventional aluminum. A cutaway view of the Orbiter appears in Figure 1 while Orbiter dimensions are given in Figure 2.

The "cockpit," living quarters and *experiment* oeprator's station are located in the forward fuselage. This area houses the pressurized crew module and provides support for the *nose section, nose gear*, and the *nose gear wheel well and doors*.

The forward fuselage (see Fig. 3) is of conventional *aircraft* construction with type 2024 aluminum alloy skin-*stringer* panels, frames and *bulkheads*. The panels are composed of single-curvature stretch-formed skins with riveted stringers spaced approximately 8cm to 13 cm (3 in. to 5 in.) apart. The frames are riveted to the skin-stringer panels. The spacing between the major frames is 76.2 cm to 91.4 cm (30 in. to 36 in.). The forward bulkhead is constructed of

flat aluminum and formed sections (upper), riveted and bolted together, and a machined section (lower). The bulkhead provides the interface fitting for the nose section, which contains large machined beams and struts. The two nose landing gear doors are constructed of aluminum alloy honeycomb and have *aerodynamic* seals. The forward fuselage skin has structural provisions for the installation of antennas and air data sensors.

The forward *reaction control system* (RCS), which is constructed of aluminum, houses the RCS engines and tank and is attached to the forward fuselage at 16 attach points.

The 71.5 cubic meter (2,525 cubic foot) crew station module (see Fig. 4) is a three-section pressurized working, living and stowage compartment in the forward portion of the Orbiter. It consists of the *flight deck*, the *mid-deck*/equipment bay and an *airlock*. Outside the aft bulkhead of the crew module in the *payload bay*, a *docking module* and a transfer tunnel with an adapter can be fitted to allow crew and equipment transfer for docking, *Spacelab* and *extravehicular* operations. The two-level crew module has a foward flight deck, with the *commander's* seat positioned on the left and the *pilot's* seat on the right.

Therefore, the flight deck (see Fig. 5), designed in the usual pilot/copilot arrangement, permits the vehicle to be piloted from either seat and permits one-man emergency return. Each seat has manual flight controls, including rotation and *translation* hand controllers, rudder pedals and speed-brake controllers. The flight deck seats four. The in-orbit displays and controls at the aft end of the flight deck/crew compartment are shown in figure 6. The displays and controls on the left are for operating the Orbiter, and those on the right are for operating and handling the *payloads*.

More than 2,020 separate displays and controls are located on the flight deck. These include toggle switches, circuit breakers, rotary switches, push buttons, thumbwheels, metered and mechanical readouts, and separate indicating lights. Approximately three times more displays and controls are onboard the Orbiter than were

Fig. 1   Space Shuttle Orbiter structures.   (Drawing courtesy of NASA.)

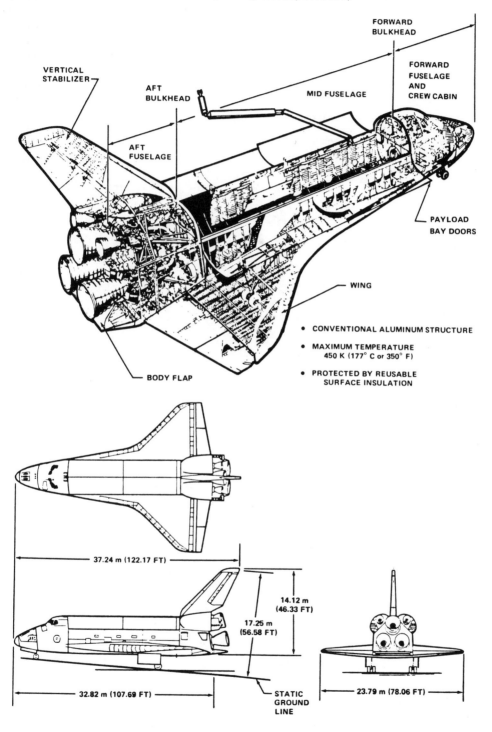

- CONVENTIONAL ALUMINUM STRUCTURE
- MAXIMUM TEMPERATURE
  450 K (177° C or 350° F)
- PROTECTED BY REUSABLE
  SURFACE INSULATION

Fig. 2   Dimensions of the Orbiter vehicle.   (Drawing courtesy of NASA.)

onboard the *Apollo Command Module*.

The payload-handling portion of the in-orbit station contains displays and controls required to manipulate, deploy, release and capture payloads. Displays and controls are provided at this station to open and close payload bay doors; to deploy *radiators;* to deploy, operate and stow manipulator arms; and to operate payload-bay-mounted lights and television cameras. Two closed-circuit television monitors display the payload bay video pictures for monitoring payload manipulation operations. The *rendezvous* and *docking* portion of the in-orbit station contains displays and controls needed to execute Orbiter *attitude/transla-*

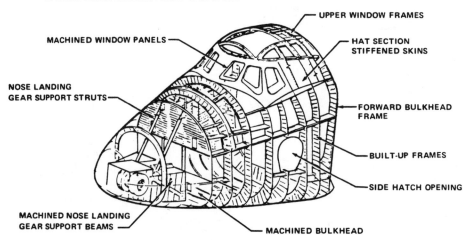

- INTEGRALLY MACHINED ALUMINUM SKIN-STRINGER PANELS
- WELDED CONSTRUCTION
- MECHANICALLY ATTACHED
  - FRAMES
  - BEAMS
  - FLOORS
  - EQUIPMENT SUPPORT

FLIGHT-DECK FLOOR

SIDE HATCH

AVIONICS BAY PARTITION

MID-DECK FLOOR

(a) Crew module.

- ALUMINUM CONSTRUCTION
- RIVETED SKIN/STRINGER/FRAME STRUCTURE

UPPER WINDOW FRAMES

MACHINED WINDOW PANELS

HAT SECTION STIFFENED SKINS

NOSE LANDING GEAR SUPPORT STRUTS

FORWARD BULKHEAD FRAME

BUILT-UP FRAMES

SIDE HATCH OPENING

MACHINED NOSE LANDING GEAR SUPPORT BEAMS

MACHINED BULKHEAD

(b) Structure.

Fig. 3   The Orbiter's forward fuselage.   (Drawing courtesy of NASA.)

tion maneuvers for terminal-phase rendezvous and docking. Rendezvous radar displays and controls and *cross-pointer* displays of *pitch and roll* angles are provided at this station as well as translation and rotation hand controllers, flight control mode switches and attitude direction indicators.

The *mission station* is located aft of the pilot's station on the right side and has displays and controls for Orbiter-to-payload interfaces and payload subsystems. An auxiliary caution-and-warning display at this station alerts the crew to critical malfunctions in the payload systems. The station manages Orbiter subsystem functions that do not require immediate access and in-orbit housekeeping. A *cathode-ray tube* (i.e., a television screen) display and keyboard are located at this station for monitoring payloads and Orbiter subsystems. Payload conditions during ascent and entry can also be displayed at the forward flight stations by caution-and-warning and television displays.

Six pressure windshields, two overhead windows and two rear-viewing payload bay windows are located on the upper flight deck of the crew module, and a window is located in the crew entrance/exit *hatch* located in the mid-deck, or mid section, of the crew module. These six windshields, providing pilot visibility, are the largest pieces of glass ever produced with the optical quality required for "see-through" viewing. Each of the six outer windshields is constructed of silica glass for high optical quality and thermal shock resistance. The construction of the overhead windows is identical to that of the windshields with the exception of the center pane, which is made of tempered aluminosilicate instead of fused silica glass. The rear-viewing payload bay windows consist of two panes of tempered aluminosilicate glass, because no thermal (outer) window is required. The side hatch (mid-deck), which is 101.6 cm (40 in.) in diameter, has a clear-view window 25.4 cm (10 in.) in diameter in its center. The window consists of three panes of glass identical in construction to the Orbiter's six windshields.

The-mid deck (see Fig. 7) contains provisions and stowage facilities for four crew sleep stations. Stowage for the *lithium hydroxide canisters* and other gear, the *waste management system,* the *personal hygiene station* and the work/dining table is also provided in the mid-deck.

The nominal maximum *Shuttle crew* size is seven. The mid-deck can, however, be reconfigured by adding three rescue seats in

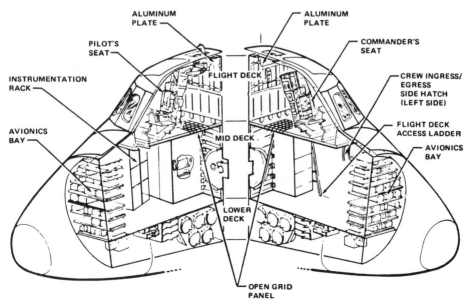

Fig. 4   Layout of the Orbiter's crew module.   (Drawing courtesy of NASA.)

Fig. 5   The Orbiter's flight deck crew cabin arrangement.   (Drawing courtesy of NASA.)

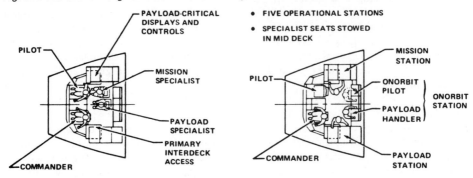

(a) Launch/entry.

(b) Onorbit.

place of the modular stowage and sleeping provisions. The Orbiter's seating capacity will then accommodate a rescue flight crew of three and a maximum rescued crew size of seven.

Access to the mid-deck from the flight deck is through two 66-cm by 71-cm (26-in. by 28-in.) interdeck access hatches and, from the exterior, through the Orbiter side hatch. A ladder attached to the port (left) interdeck hatch permits easy ground entry by the crew and ground crew from the mid-deck to the flight deck. The airlock allows passage to the payload bay. Environmental control equipment and additional stowage space are located below the mid-deck.

Expended lithium hydroxide canisters and wet trash are also stowed below the mid-deck floor. Removable floor panels provide access to the environmental control equipment.

The airlock (se Fig. 8) provides access for *extravehicular activity* (EVA). It can be located in one of several places: inside the Orbiter crew module in the mid-deck area mounted to the aft bulkhead, outside the cabin also mounted to the aft bulkhead or on top of a *tunnel adapter* that can connect the pressurized Spacelab module with the Orbiter cabin. A docking module can also serve as an EVA airlock. The airlock, two *space suits,* expendables for two six-hour

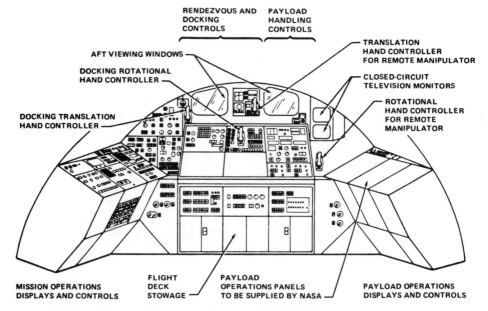

Fig. 6   The Orbiter's aft flight deck configuration.   (Drawing courtesy of NASA.)

payload EVAs and one contingency or emergency EVA, and mobility aids such as handrails enable the crew to perform a variety of tasks outside the pressurized crew cabin.

The airlick is cylindrical, with an inside diameter of 160 cm (63 in.) and a length of 211 cm (83 in.). It allows two crew members room for changing space suits. The hatches are D-shaped, with the flat side of the D making the minimum clearance 91.4 cm (36 in.). The shape, size and location of the hatches allow the two Shuttle crew members to move a package 46 cm by 46 cm by 127 cm (18 in. by 18 in. by 50 in.) through the airlock.

The transfer tunnel and the tunnel adapter (see fig. 9) provide for transfer of the crew and equipment between the Spacelab and the crew module. The tunnel has flexible elements and a number of segments to accommodate different flight locations. The tunnel mates with the tunnel adapter at the forward end of the payload bay for operations outside the crew module and for rescue when a docking module is not carried. Electrical and *fluid* interface lines between the Orbiter and the Spacelab extend along the exterior of the tunnel. The tunnel adapter has two access hatches: one on top for access to an airlock and the other on the aft end for access to the payload bay. For operations outside the crew module, the airlock will be placed on top of the tunnel adapter.

For Shuttle missions requiring direct docking of two vehicles, a docking module can be substituted for the airlock and installed on the tunnel adapter. The docking module is extendable and provides an airlock function for EVA for two crew members when extended or for one crew member when retracted.

The mid-fuselage structure (see Fig. 10) interfaces with the forward fuselage, the aft fuselage and the wings and, in addition to forming the payload bay of the Orbiter, supports the payload bay doors, hinges and tie-down fittings· the forward *wing glove;* and various Orbiter system components.

The forward and aft ends of the mid fuselage join the bulkheads of the forward and aft fuselage. The length of the mid

Fig. 1   The Orbiter's mid-deck crew cabin arrangement.   (Drawing courtesy of NASA.)

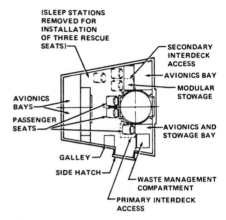

(a) Launch/entry.

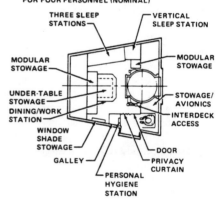

(b) Onorbit.

fuselage is 18.3 m (60 ft), the width is 5.2 m (17 ft) and the height is 4.0 m (13 ft). The mid fuselage weighs approximately 6,124 kg (13,502 lb$_m$).

There are 12 main-frame assemblies that stabilize the mid-fuselage structure. These assemblies consist of vertical side and horizontal elements. The side elements are machined, whereas the horizontal elements have machined flanges with baron/aluminum tube trusses. The boron/aluminum tubes (tubular struts) have diffusion-bonded titanium end fittings that provide substantial weight savings. The upper portion of the mid fuselage consists of the sill and door *longerons.* The machined sill longerons are not only the primary body-bending elements but also serve to support the longitudinal loads from payloads in the cargo bay. There are 13 payload bay door longerons and associated backup structure.

The payload bay doors (see Fig. 11) consist of a left-hand and a right-hand door that are hinged to the mid fuselage and latched at the forward and aft fuselage and at the top centerline of each door. The doors provide an opening for payload deployment and retrieval and serve as structural support for the Orbiter radiators. The payload bay is not pressurized. The doors are 18.3 m (60 ft) long and 4.6 m (15 ft) in diameter and are constructed of graphite/epoxy composite material. The doors weigh 1,480 kg (3,264 lb$_m$). Each door supports four radiator panels. The two forward radiator panels on each door can be tilted, but the two aft radiators remain fixed. When the doors are opened the tilting radiators are unlatched and moved to the proper position. This allows *thermal radiation* (heat rejection) from both sides of the panels, whereas the four aft radiator panels radiate from the upper side only.

The controls for operating the payload bay doors and positioning the radiator panels are located at the aft station on the flight deck. The crew has the capability of selecting automatic or manual operation. In the automatic mode all sequences are performed automatically after the proper switch is selected and activated, whereas the manual mode allows the crew to select latch group opening and closing sequences.

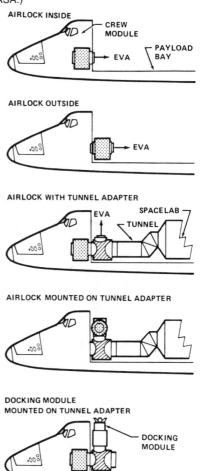

Fig. 9   The Orbiter's airlock/tunnel adapter/docking module configurations.   (Drawing courtesy of NASA.)

Structural attach points for payloads are located at 9.9-cm (3.9-in.) intervals along the tops of the two Orbiter mid-fuselage main longerons. Some payloads may not be attached directly to the Orbiter but to *payload carriers* that are attached to the Orbiter. The *Inertial Upper Stage*, Spacelab and Spacelab *pallet*, or any specialized cradle for holding a payload are typical carriers.

The *remote manipulator system* (RMS) is a 15.2-m (50-ft) long articulating arm that is remotely controlled from the Orbiter's flight deck. (See Fig. 12) The elbow and wrist movements of the RMS permit payloads to

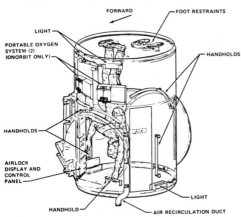

Fig. 8   The Orbiter's airlock.   (Drawing courtesy of NASA.)

be grappled for deployment out of the payload bay attach points or to be retrieved and secured for return to *Earth*. Because the RMS can be operated from the shirtsleeve environment of the cabin, an EVA maneuver is not required.

The standard remote manipulator is mounted with its "shoulder" on the left main longeron (facing forward); a second manipulator can be mounted on the right longeron for handling certain types of payloads. A television camera and lights near the outer end of the RMS arm permit the operator to see on television monitors what his "hands" are doing. Payloads will carry markings and alignment aids to help the RMS operator maneuver payloads. The RMS operator has a 62° *field-of-view* out the two aft windows on the flight deck and 80° through the two overhead observation windows (see Fig. 13). Three floodlights are located along each side of the payload bay.

The aft fuselage (see Fig. 14) is approximately 5.5 (18 ft) long, 6.7 m (22 ft) wide and 6.1 m (20 ft) high. It carries and interfaces with the *orbital maneuvering system/ reaction control system* pod (left and right sides), wing aft spar, mid fuselage, *Space Shuttle Main Engines* (SSMEs), heat shield, body flap and vertical stabilizer. The aft fuselage provides for the following: (1) the load path to the mid-fuselage main longerons; (2) main wing spar continuity across the forward bulkhead of the aft fuselage; (3) structural support for the body flap; (4) structural housing around all internal systems for protection from the operational environment (*pressure, thermal* and *acoustics)*; and (5) controlled internal pressure venting during flight.

The aft bulkhead separates the aft fuselage from the mid fuselage and is composed of machined and beaded sheet metal aluminum segments. The upper portion of the

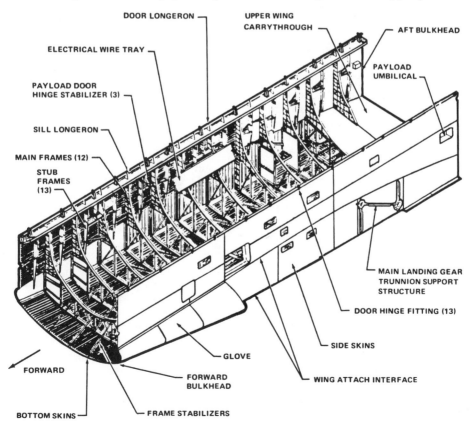

Fig. 10   The Orbiter's mid-fuselage structure.   (Drawing courtesy of NASA.)

Fig. 11   The Orbiter's payload bay doors.   (Drawing courtesy of NASA.)

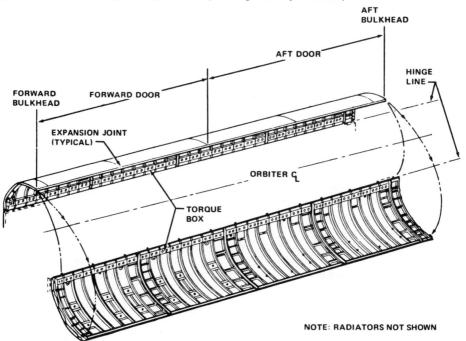

bulkhead attaches to the front spar of the vertical *fin*. The internal *thrust* structure carries loads from the Space Shuttle Main Engines. This structure includes the SSME load reaction truss structure, engine interface fittings and the SSME *actuator* support structure.

The body flap (see Fig. 15) is an aluminum structure consisting of ribs, spars, skin panels and a trailing-edge assembly. The main upper and lower and the forward lower honeycomb skin panels are joined to the ribs, spars and honeycomb trailing-edge assembly with structural fasteners. Addition of the removable upper forward honeycomb skin panels completes the body flap structure. This structure, which is covered with reusable surface insulation on its mold-line surfaces, is attached to the lower aft fuselage by four rotary actuators. With its aerodynamic and thermal seals, the body flap provides the Shuttle pitch trim control and thermally shields the main engines during re-entry.

The wing (see Fig. 16) is the aerodynamic lifting surface that provides conventional *lift* and control for the Orbiter. The wing consists of the wing glove; the intermediate section, which includes the main landing gear well; the *torque box;* the forward spar for mounting the leading-edge structure *thermal protection system;* and the wing/ *elevon* interface, the elevon seal panels and the elevons. The wing is constructed of conventional aluminum alloy with a multirib and spar arrangement with skin-stringer stiffened covers or honeycomb skin covers. Each wing is approximately 18.3 m (60 ft) long at the fuselage intersection, with a maximum thickness of 1.5 m (5 ft).

The forward wing glove is an extension of the basic wing and aerodynamically blends the wing leading edge into the mid fuselage. The forward wing box is of a conventional design of aluminum multiribs and aluminum truss tubes with sheet metal caps. The upper and lower wing skin panels are constructed of stiffened aluminum.

The intermediate wing section consists of the same conventional design of aluminum multiribs with aluminum truss tubes. The upper and lower skin covers are made of aluminum honeycomb construction. A portion of the lower wing surface skin panel comprises the main landing gear door, which is constructed of aluminum with machined hinge beams and hinges. The recessed area in the door is for tire clearance. The intermediate section houses the main landing gear compartment and carries a portion of the main landing gear loads.

The wing torque box area incorporates the conventional aluminum multirib truss arrangement with four spars, which are constructed of corrugated aluminum to minimize thermal loads. The forward closeout beam is constructed of aluminum honeycomb and provides the attachment for the leading-edge structure thermal protection system. The rear spar provides the attachment interfaces for the elevons, hinged upper seal panels and associated hydraulic and electrical system components. The upper and lower wing skin panels are constructed of aluminum stiffened skin.

The elevons provide flight control during atmospheric flight. The two-segment elevons are of conventional aluminum multirib and beam construction with aluminum honeycomb skins. Each segment is supported by three hinges. Flight control systems are attached along the forward extremity of the elevons. The upper leading edge of each elevon incorporates rub strips, which are of titanium/*Inconel* honeycomb construction and are not covered with thermal protection material. The rub strips provide the sealing surface for the elevon seal panels. A tension bolt splice along the upper surface and a shear bolt splice along the lower surface are used to attach the wing to the fuselage.

The vertical stabilizer (see Fig. 17) consists of a structural fin, the rudder/speed brake (the rudder splits in half for speed-brake control) and the systems for positioning the rudder/speed-brake control surface. The vertical stabilizer structure consists of a torque box of aluminum integral stringers, web ribs and two machined aluminum spars. The lower trailing-edge area of the fin that houses the rudder/speed-brake power drive unit has aluminum honeycomb skin. The fin is attached to the forward bulkhead of the aft fuselage by ten bolts. The rudder/speed brakes consist of conventional

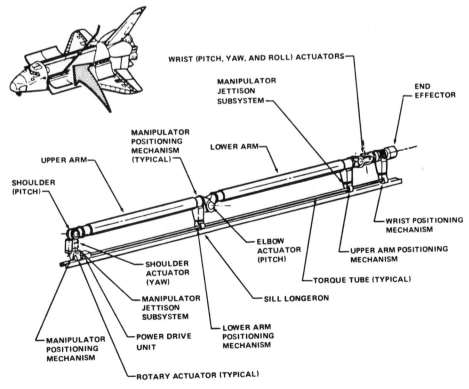

Fig. 12   The Orbiter's remote manipulator arm.   (Drawing courtesy of NASA.)

Fig. 13   The remote manipulator system operator's visual observation of payloads.   (Drawing courtesy of NASA.)

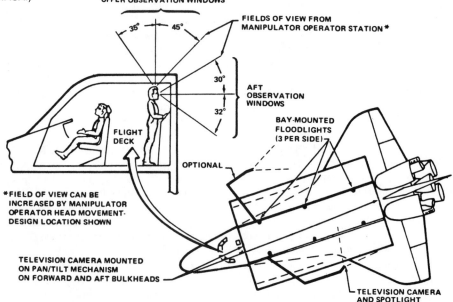

aluminum ribs and spars with aluminum honeycomb skin panels and are attached to the vertical stabilizer by rotating hinges. The rudder/speed-brake assembly is divided into upper and lower sections. Each section splits to both sides of the fin in the speed-brake mode.

The Orbiter mechanical subsystems (see Fig. 18) with electrical and hydraulic actuators operate the aerodynamic control surfaces, landing/*deceleration* system, payload bay doors, deployable radiators, and payload retention and handling subsystems. Orbiter/*External Tank* propellant disconnects and a variety of other mechanical and *pyrotechnic* devices comprise the balance of the mechanical subsystems.

Aerodynamic control surface movement is provided by hydraulically powered actuators that position the elevons and by hydraulically powered drive units that position the body flap and combination rudder/speed brake. Three redundant 20,684 KN/m^2 (3,000 psi) systems supply

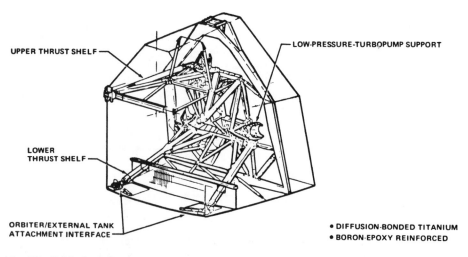

Fig. 14   The Orbiter's aft fuselage thrust structure.   (Drawing courtesy of NASA.)

Fig. 15   The Orbiter's body flap.   (Drawing courtesy of NASA.)

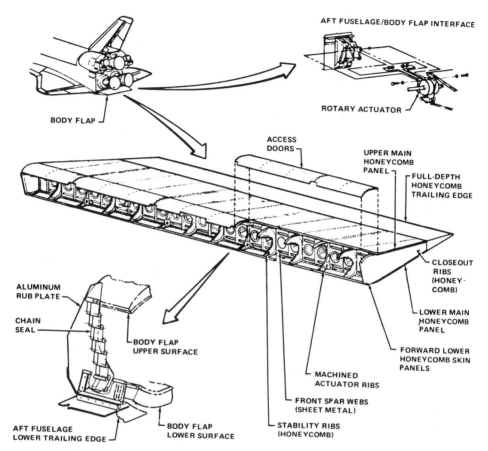

AFT FUSELAGE/BODY FLAP INTERFACE

ROTARY ACTUATOR

BODY FLAP

ACCESS DOORS

UPPER MAIN HONEYCOMB PANEL

FULL-DEPTH HONEYCOMB TRAILING EDGE

CLOSEOUT RIBS (HONEY-COMB)

LOWER MAIN HONEYCOMB PANEL

FORWARD LOWER HONEYCOMB SKIN PANELS

ALUMINUM RUB PLATE

CHAIN SEAL

BODY FLAP UPPER SURFACE

MACHINED ACTUATOR RIBS

FRONT SPAR WEBS (SHEET METAL)

AFT FUSELAGE LOWER TRAILING EDGE

BODY FLAP LOWER SURFACE

STABILITY RIBS (HONEYCOMB)

the necessary hydraulic power. Elevon seal panels that provide an aerodynamic/ thermal seal between the upper leading edges of the elevon and the upper wing surfaces are actuated by push/pull rod mechanisms attached directly to the elevons.

The fully retractable tricycle landing gear is designed to provide safe landings at speeds up to 409 km/h (254 mph). The shock struts are of conventional aircraft design with dual wheels/tires. Braking is accomplished by special lightweight carbon-lined beryllium brakes with anti-skid protection.

The payload bay doors, deployable radiators, vent doors (forward fuselage, payload bay and wing, aft payload bay and aft fuselage), star tracker doors and separation-system closeout doors are operated by electromechanical actuators that must per-

form reliably after severe environmental exposure during ascent and entry and during orbital operations. The payload bay doors, when closed and latched, are part of the Orbiter structure and react to fuselage *torsional loads.*

The payload retention subsystem includes remotely controlled retention latches that hold down or release the payload items but do not transmit Orbiter stresses, such as bending, to the payloads. The payload-handling subsystem consists primarily of remotely controlled manipulator arms (one arm is normally installed; a second is optional) that can move the paylads in or out of the payload bay while in *orbit.*

The Orbiter landing gears are arranged in a conventional tricycle configuration consisting of a nose landing gear and left and right main landing gears. The nose landing gear retracts forward and up into the

forward fuselage; the main landing gears
retract forward and up into the wings. Each
landing gear is held in the retracted position
by an uplock hook. The landing gears are
extended by releasing the uplock hooks
hydraulically, thus enabling the landing
gear to free fall, assisted by springs and
hydraulic pressure. The landing gear doors
(two doors for the nose gear and one door
for each main landing gear) open through a
mechanical linkage attached to the landing
gears. The landing gears will reach the fully
extended position within a maximum of ten
seconds and are locked in the down position
by spring-loaded *bungees*. The nose landing
gear door opening and gear extension are
also assisted by a pyrotechnic actuator to
ensure gear deployment if adverse air load
conditions occur.

The nose landing gear and the main land-

ing gears contain a shock strut that is the
primary source of shock attenuation during
Orbiter landing impact. The shock struts
are conventional "pneudraulic" (gaseous
nitrogen/hydraulic fluid) shock absorbers.
A floating diaphragm within the shock strut
separates the gaseous nitrogen as it
disperses throughout the hydraulic fluid in
zero-g conditions. The separation is required to assure proper shock-strut performance at Orbiter touchdown.

For orbital missions the landing gears are
retracted, locked and checked out before
launch. Throughout the mission, until re-
entry is completed and preparations to land
are initiated, the landing system remains
essentially dormant. A landing gear "down"
command removes the hydraulic isolation
and initiates gear extension. After touchdown brake application by the crew deceler-

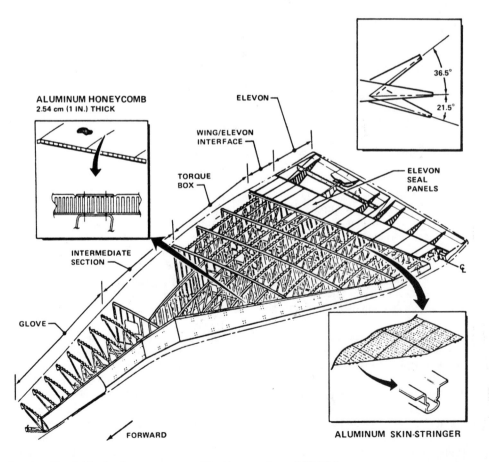

Fig. 16   The Orbiter's wing structure.   (Drawing courtesy of NASA.)

Fig. 17   The Orbiter's vertical stabilizer.   (Drawing courtesy of NASA.)

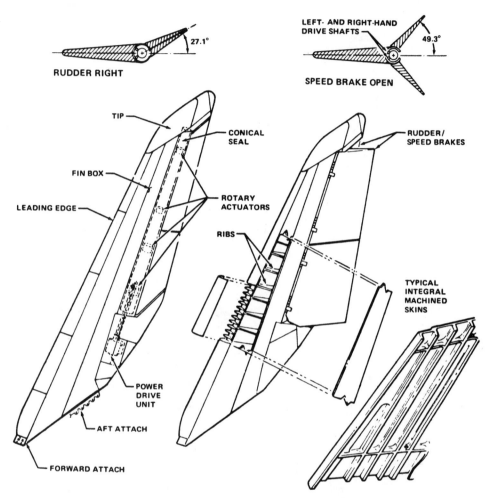

ates the Orbiter, with nose-wheel steering supplementing the directional control provided by aerodynamic forces on the rudder. Both the nose gear and the wing-mounted main gears are equipped with dual wheels, and both retract forward, thereby maintaining free-fall extension capabilities. Each of the four main wheels is equipped with a carbon-lined beryllium brake and fully modulated skid-control system.

The nose landing gear consists of two wheel and tire assemblies. The nose gear is steerable when it is extended and there is weight on it. The nose-gear tires have a rated static load of 105 kilonewtons (23,700 lb_f) and an inflation pressure of 2,070 kN/m^2 (300 psi). The nose-gear tire life is

five normal landings. Each main and nose-gear tire is rated landings at approximately 409 km/h (254 mph) with a maximum Orbiter load.

The Orbiter separation and pyrotechnic subsystem operates as follows. The External Tank and tank umbilical plates are securely attached to the External Tank support during ascent. At tank separation the umbilical plates—containing the propellant lines, electrical connectors and vent lines—are released by energizing three *solenoids*. This release opens the hooks holding the umbilical plates together, and the plates are separated by a retract system. After the umbilical separation signal is received, the External Tank is separated from the Orbiter

Fig. 18   The Orbiter's mechanical subsystems.   (Drawing courtesy of NASA.)

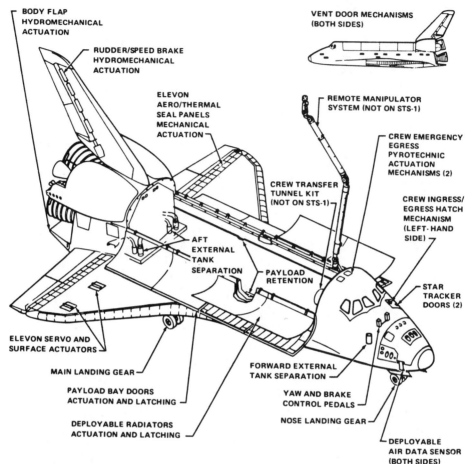

by simultaneously releasing the forward and aft structural ties.

The crew enters the crew compartment through the side hatch (see Fig. 19), which also serves as the primary emergency escape route. This hatch is on the left-hand side of the vehicle, forward of the wing, and is hinged to open outward. It is 101.6 cm (40 in.) in diameter and contains 28 latches, an actuator mechanism with provisions for interior/exterior operation, a viewport window-hatch coverplate, a ratchet assembly, an *egress* bar and access doors.

The actuator mechanism is equipped with a handcrank on the crew module side for interior operation by the crew. A series of over-center latches driven by a series of links from the actuator are used to hold the hatch in the closed sealed position. Two separate operations are required before the

actuator will actuate (open) the latches. The actuator handle must first be unlocked and then rotated counterclockwise to unlatch (clockwise to latch) the hatch. The actuator is capable of being driven manually either from within the crew module or from outside. The ratchet assembly is used to hold the hatch open while the Orbiter is on the *launch pad.*

For emergency egress the crew will deploy the egress bar and descend. A hatch T-handle tool is provided for exterior ground operation and is used to operate the latch mechanism after removing the access door.

The side hatch has a window 25.4 cm (10 in.) in diameter. The window cavity is covered by three panes (two inboard pressure panes and one outboard thermal pane).

The airlock hatch will be used primarily

for extravehicular operations but will also be used for stowage or retrieval of gear in the airlock and for transfer to and from the Spacelab or a docked vehicle. The airlock can be located either in the payload bay or in the mid-deck of the Orbiter. The airlock has two universal hatches, except when it is located in the crew station mid-deck. Here the inner hatch will be a stowable hatch, and the outer hatch will be a universal hatch. The hatches are closed before flight and are opened for crew movement between the crew cabin, payload bay and Spacelab. The inner airlock hatch is opened during the first day of activity and normally remains open for any contingency. The inner hatch will be closed before re-entry.

The tunnel adapter, when in use, is located on the payload bay side of the manufacturing access panel. The adapter allows access to the airlock and the tunnel when both are in the payload bay at the same time. The tunnel connects the adapter and the Spacelab. When the tunnel adapter is onboard the Orbiter, the hatches will be closed for launch or re-entry and opened for movement between the crew cabin and the tunnel or between the crew cabin and the airlock or docking module. The docking module hatches remain closed except for transfer to and from a docked vehicle or for extravehicular operations.

See: Apollo Program, External Tank, extravehicular activity, radiation heat transfer

Orbiter thermal protection system (TPS)

The overall economic feasibility of a reusable *Space Transportation System* hinges on protecting the *Space Shuttle Orbiter*— which will experience widely varying thermal and *aerodynamic* environments typical of both *aircraft* and *spacecraft*—in a way that does not require significant refurbishment between trips into space. The thermal protection system is designed to limit the *temperature* of the Orbiter's aluminum and graphite epoxy structures to a nominal value of 450 K (177 C or 350 F) during

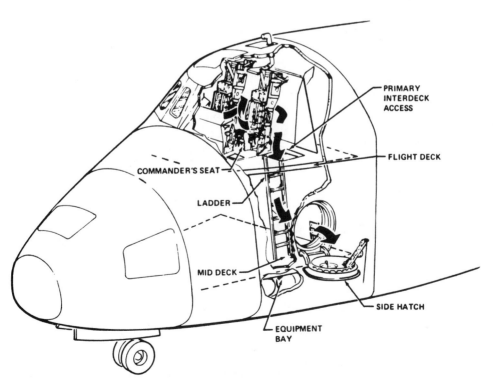

Fig. 19   Orbiter crew cabin ingress/egress.   (Drawing courtesy of NASA.)

ascent and entry. Maximum surface temperatures during entry vary from 1,783 K (1,510 C or 2,750 F) on the wing leading edge to less than 589 K (316 C or 600 F) on the upper fuselage. The thermal protection system must also endure exposure to non-heating environments during prelaunch, launch, inorbit. landing and turnaround operations similar to those encountered by conventional aircraft. It must also sustain the mechanical forces induced by deflection of the *airframe* as it responds to the same external environment. The system is designed to withstand 100 ascents and entries with a minimum of refurbishment and *maintenance.*

The thermal protection system consists of materials applied externally to the Orbiter that maintain the airframe outer skin within acceptable temperature limits. (See Fig. 1.)

Internal insulation, heaters and purging facilities are used to control interior compartment temperatures. The Orbiter thermal protection system is a passive system consisting of the following four materials, selected for weight efficiency and stability at high temperatures.

(1) Coated *reinforced carbon-carbon* (RCC) for the nose cap and wing leading edges, where temperatures exceed 1,533 K (1,260 C or 2,300 F).

(2) *High-temperature reusable surface insulation* (HRSI), for areas where maximum surface temperatures reach 922 K to 978 K (649 C to 704 C or 1,200 F to 1,300 F).

(3) *Low-temperature reusable surface insulation* (LRSI), for areas where surface temperatures reach 644 K to 922 K (371 C to 649 C or 700 F to 1,200 F).

(4) *Flexible reusable surface insulation*

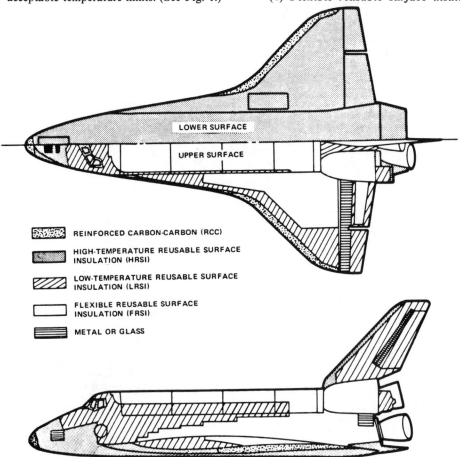

LOWER SURFACE

UPPER SURFACE

REINFORCED CARBON-CARBON (RCC)

HIGH-TEMPERATURE REUSABLE SURFACE INSULATION (HRSI)

LOW-TEMPERATURE REUSABLE SURFACE INSULATION (LRSI)

FLEXIBLE REUSABLE SURFACE INSULATION (FRSI)

METAL OR GLASS

Fig. 1   Orbiter thermal protection system.   (Drawing courtesy of NASA.)

(FRSI) (coated *Nomex* felt), for areas where the surface temperature does not exceed 644 K (371 C or 700 F).

The reinforced carbon-carbon is an all-carbon composite that consists of layers of graphite cloth contained in a carbon matrix formed by *pyrolysis*. To prevent oxidation at elevated temperatures, the outer graphite-cloth layers are chemically converted to silicon carbide. RCC covers the nose cap and wing leading edge areas of the Orbiter, which receive the highest thermal (heating) loads.

The high-temperature reusable surface insulation consists of approximately 20,000 tiles located predominantly on the lower surface of the vehicle. The tiles nominally

measure 15 cm by 15 cm (6 in. by 6 in.) and vary in thickness from 1.3 cm to 8.9 cm (0.5 in. to 3.5 in.), depending on local heating. The high-temperature material is composed of a low-density high-purity silica fiber insulator made rigid by ceramic bonding. Each tile is bonded to a strain isolator pad made of Nomex fiber felt, and the total composite is directly bonded to the vehicle.

The low-temperature reusable surface insulation consists of approximately 7,000 tiles applied to the upper wing and fuselage side surfaces of the Orbiter in the same manner as the HRSI. These tiles nominally measure 20 cm by 20 cm (8 in. by 8 in.) and vary in thickness from 0.5 cm to 2.5 cm (0.2 in to 1 in.). The low-temperature tiles are

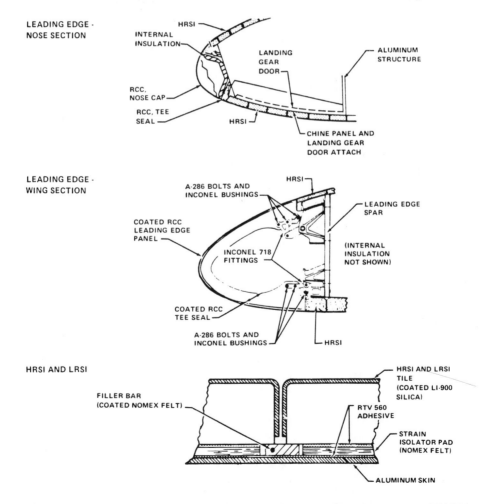

Fig. 2   Typical Orbiter interfaces with the thermal protection system.   (Drawing courtesy of NASA.)

the same as the high-temperature tiles except that the coating has a different optical pigment for obtaining low solar *absorbance* and high solar *emittance*. These two sets of tiles cover approximately 70 percent of the Orbiter.

The basic raw material for the tiles is a high-purity silica fiber that was selected for its low *thermal conductivity*, low thermal expansion and high-temperature stability. The reusable tiles can easily be repaired. Coating scratches can be repaired in place by spray techniques and torch firing, and small gouges or punctures can be cored out and replaced with standard-sized plugs. Complete tiles can be removed and replaced in about 45 hours.

The flexible reusable surface insulation consists of 0.9-m by 1.2-m (3-ft by 4-ft) sheets of Nomex felt that are directly bonded to the structure. Before installation the Nomex felt, which varies in thickness from 0.41 cm to 1.62 cm (0.16 in. to 0.64 in.), is coated with a silicone elastomeric film to waterproof it and to give the surface the desired optical properties. The Nomex felt is applied to the upper parts of the *payload bay doors*, the sides of the fuselage and the upper wing.

Typical interfaces with the thermal protection system are shown in Figure 2. The leading edge subsystem to HRSI transition is shown for the nose and outer wing leading edge. A typical joint for the two tile systems is also shown. The tiles are bonded to the aluminum skin with RTV 560 (a silicone resin cement) with a strain isolator pad in between. Filler bars are installed under the tiles at the intertile gaps. In areas where high surface *pressure* gradients would cause crossflow of *boundary-layer* air within the intertile gaps, tile-gap fillers are provided to minimize heating within the gaps. These gap fillers are fabricated using a silica fiber cloth cover with an alumina fiber filler.

A passive (non-active) thermal control system helps maintain spacecraft systems and components at specified temperature limits. This system uses available spacecraft heat sources and heat sinks supplemented by insulation blankets, thermal coatings and thermal isolation methods. Heaters are provided on components and systems where passive thermal control techniques are not adequate to maintain required temperatures. The insulation blankets are of two basic types: fibrous bulk and multilayer (see Fig. 3). The bulk blankets are made of 32-kg/m^3 (2-lb$_m$/ft^3) density fibrous material with a sewn cover of reinforced double-goldized *Kapton*. The cover material has numerous small holes for venting purposes. Goldized tape is used for cutouts, patching and reinforcements. Tufts are used throughout the blankets to minimize billowing during venting. The multilayer blankets are constructed of alternate layers of perforated double-goldized Kapton reflectors and *Dacron* net separators for a total of 20 reflector layers, with the two cover halves counting as two layers. The covers, tufting and goldized tape are similar to those of the bulk blankets. *See:* Orbiter structure

ordnance ring   The component of the foward assembly of the *Space Shuttle Solid Rocket Booster* that connects the frustum with the forward skirt and contains a linear-shaped *pyrotechnic* charge that cuts the frustum and forward skirt apart. *See:* Solid Rocket Booster

ordnance train   A network of small explosive charges.

"O"-ring   Sealing ring with a circular cross section, which may be either hollow or solid; generally made of *elastomer* or plastic but can be metal.

O-ring squeeze   Compression of the *O-ring* cross section between opposite surfaces of a gland.

orthogonal   Orignally, at right angles; later generalized to mean the vanishing of a sum (or integral) of products.

The cosine of the angle between two vectors V_1 and V_2 with respective components x_1, y_1, z_1, and x_2, y_2, z_2 is proportional to the sum of products $x_1x_2 + y_1y_2 + z_1z_2$. Hence if the vectors are perpendicular, the latter sum

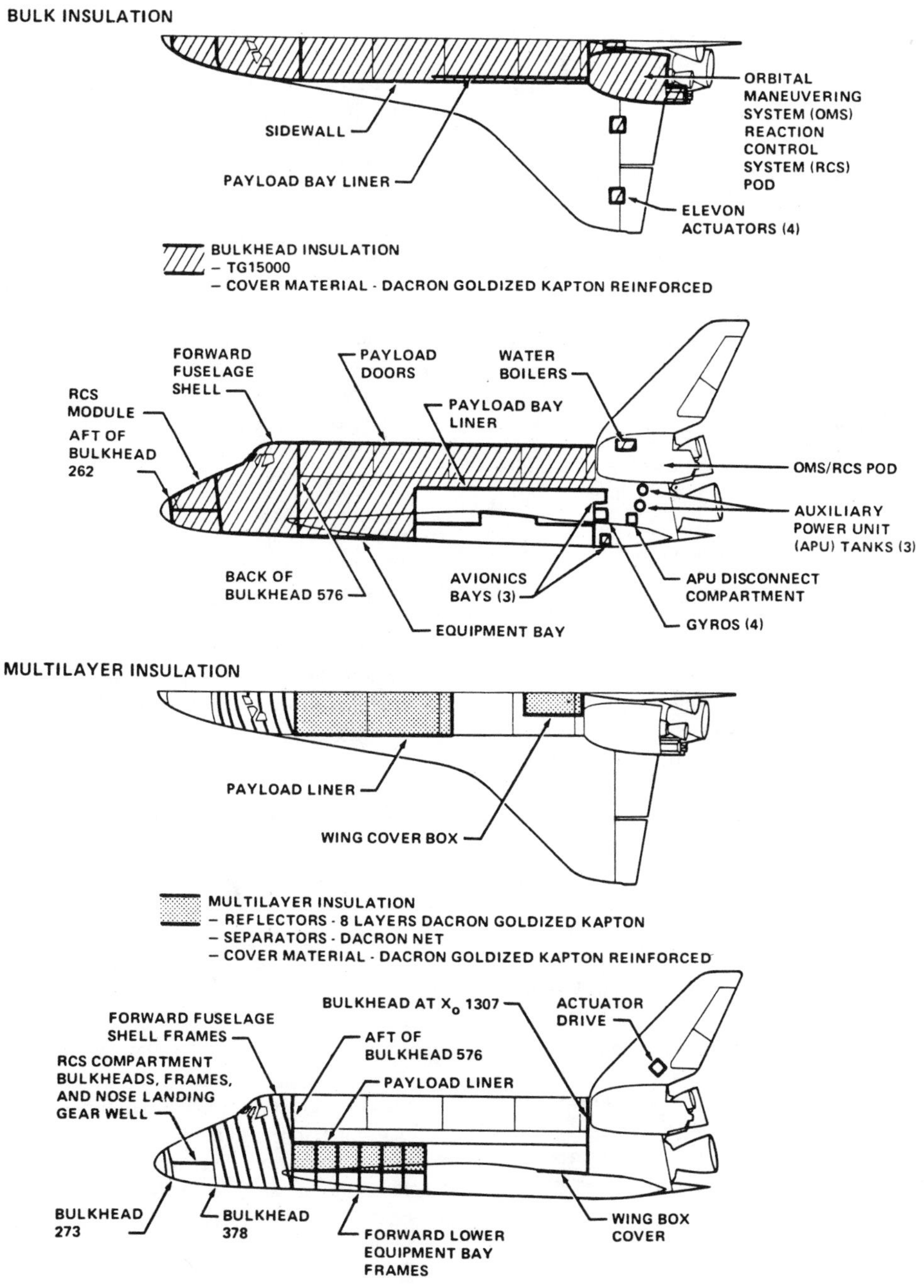

BULK INSULATION

SIDEWALL

PAYLOAD BAY LINER

ORBITAL MANEUVERING SYSTEM (OMS) REACTION CONTROL SYSTEM (RCS) POD

ELEVON ACTUATORS (4)

BULKHEAD INSULATION
– TG15000
– COVER MATERIAL - DACRON GOLDIZED KAPTON REINFORCED

RCS MODULE AFT OF BULKHEAD 262

FORWARD FUSELAGE SHELL

PAYLOAD DOORS

WATER BOILERS

PAYLOAD BAY LINER

OMS/RCS POD

AUXILIARY POWER UNIT (APU) TANKS (3)

APU DISCONNECT COMPARTMENT

GYROS (4)

BACK OF BULKHEAD 576

AVIONICS BAYS (3)

EQUIPMENT BAY

MULTILAYER INSULATION

PAYLOAD LINER

WING COVER BOX

MULTILAYER INSULATION
– REFLECTORS - 8 LAYERS DACRON GOLDIZED KAPTON
– SEPARATORS - DACRON NET
– COVER MATERIAL - DACRON GOLDIZED KAPTON REINFORCED

BULKHEAD AT X_o 1307

ACTUATOR DRIVE

FORWARD FUSELAGE SHELL FRAMES

AFT OF BULKHEAD 576

RCS COMPARTMENT BULKHEADS, FRAMES, AND NOSE LANDING GEAR WELL

PAYLOAD LINER

BULKHEAD 273

BULKHEAD 378

FORWARD LOWER EQUIPMENT BAY FRAMES

WING BOX COVER

Fig. 3   The Orbiter's passive thermal control system.   (Drawing courtesy of NASA.)

equals zero. For this reason, any two series of numbers $(x_1, x_2, \ldots x_n)$ and $(y_1, y_2, \ldots y_n)$ is said to be orthogonal if

$$\Sigma \; x_i y_i = 0$$

orthogonal functions A set of functions, any two of which, by analogy to orthogonal *vectors*, vanish if their product is summed by integration over a specified interval.

For example, f(x) and g(x) are orthogonal in the interval x = b if

$$\int_a^b f(x) \; g(x) \; dx = 0$$

The functions are also said to be normal if

$$\int_b^a [f(x)]^2 \; dx = \int_a^b [g(x)^2] \; dx = 1$$

The most familiar examples of such functions, many of which have great importance in mathematical physics, are the sine and cosine functions between zero and 2π.

ortho hydrogen
See: cryogenics

oscillation 1. Fluctuation or vibration on each side of a mean value or position.
2. Half an oscillatory scycle, consisting of a fluctuation or vibration in one direction; half a vibration.
3. The variation, usually with time, of the magnitude of a quantity with respect to a specific reference when the magnitude is alternately greater and smaller than the reference.

oscillator An electronic device producing radio waves of a given frequency. Crystal oscillators give a highly accurate frequency.

oscillatory combustion Unstable combustion in the *rocket engine* or motor, characterized by *pressure oscillations* in either transverse or axial modes.

oscilloscope 1. An instrument for producing a visual representation of *oscillations* or changes in an electric current.
2. Specifically, a *cathode-ray* oscilloscope.

outgassing Release of *gas* from a material when it is exposed to an ambient pressure lower than the vapor pressure of the gas; generally refers to the gradual release of gas from enclosed surfaces when an enclosure is vacuum pumped.

output 1. The yield or product of an activity furnished by man, machine, or a system.
2. Power or *energy* delivered by an engine, generator, etc.
3. The electrical signal coming out of a *transducer* and is a function of the applied stimulus. Compare this term with *input*.

overexpanding nozzle A *nozzle* in which the *fluid* is expanded to a lower *pressure* than the external pressure. An overexpanding nozzle has an exit area larger than the optimum.

overexpansion Expansion of *nozzle* exhaust *gas* to a pressure lower than the ambient *pressure*.

overpressure The transient pressure over and above the *atmospheric pressure* caused by a *shock wave* from a *nuclear* explosion.

overstress A value of any stress parameter in excess of the upper limit of the normal working range or in excess of rated value.

overstroke Displacement of an operating component that exceeds the maximum allowable displacement.

oxidizer Material whose main function is to supply oxygen or other oxidizing materials for *deflagration* of a *solid propellant* or combustion of a *liquid fuel*.

P

pack concentration The method of applying a coating by packing the article to be coated in a powder mixture and heating to reaction temperature.

packing fraction The difference between the actual mass of a *nuclide* and the nearest whole number, divided by the *mass number*, A; or (M-A)/A. An equivalent statement is that it is the mass defect divided by the mass number. It is positive for most nuclides with mass number less than 12 and more than 180, which therefore tend to be less stable, and negative for most other nuclides, which tend to be more stable.

pad The platform from which a rocket vehicle is launched.
See: launch pad.

pad degree Water sprayed upon certain *launch pads* during the launch of a *rocket* so as to reduce the temperatures of critical parts of the pad or the rocket.

pallet An unpressurized platform designed for installation in the *Orbite cargo bay* for mounting instruments and equipment that require direct space exposure.
See: Spacelab

pallet train More than one *pallet* rigidly connected together to form a single unit.
See: Spacelab

parabola An open curve all points of which are equidistant from a fixed point, called the *focus,* and a straight line.

The limiting case occurs when the point is on the line, in which case the parabola becomes a straight line.

parabolic Pertaining to, or shaped like, a parabola.

parabolic orbit An *orbit* shaped like a parabola; the orbit representing the least eccentricity (that of 1) for escape from an attracting body.

parabolic effect A reflecting surface having a *cross section* along the *axis* in the shape of a *parabola.*

Parallel rays striking the reflector are brought to a focus point, or if the source of the rays is placed at the focus, the reflected rays are parallel.

parabolic reflector A reflecting surface having the *cross section* along the *axis* in the shape of a *parabola.*

Parallel rays striking the reflector are brought to a focus at a point, or if the source of the rays is placed at the focus, the reflected rays are parallel.

parachute recovery system A component of the *Space Shuttle Solid Rocket Booster.* This system (pilot, drogue and main parachutes) is designed to permit the recovery of the spent (used) *Solid Rocket Boosters* by gently returning them to an ocean impact area where they can be recovered for refurbishing and subsequent reuse.

paraglider A flexible-winged, kite-like vehicle designed for use in a recovery system for *launch vehicles* or as an *entry vehicle.*

para hydrogen
See: cryogenics

parameter 1. In general any quantity of a problem that is not an independent variable. More specifically, the term is often used to distinguish, from dependent variables, quantities which may be assigned more or less arbitray values for purposes of the problem at hand.
2. In statistical terminology, any numerical constant derived from a population or a probability distribution. Specifically, it is an arbitrary constant in the mathematical expression of a probability distribution. For example, in the distribution given by
$$f(x) = \alpha e^{-az}$$
the constant α is a parameter.
3. In celestial mechanics, the *semi-latus rectum.*

parameterization The representation, in a mathematical model, of physical effects in terms of perhaps oversimplified *parameters,* rather than realistically requiring such effects to be in consequences of the dynamics of the system.

Parameterization is often used in systems analysis to determine the effect on the system of changing one parameter while holding other parameters constant.

parametric equations A set of equations in which the *independent variables* or coordinates are each expressed in terms of a parameter.

For example, instead of investigating $y = f(x)$ or $F(x,y) = 0$ it is often advantageous to express both x and y in terms of a parameter u: $x = g(u)$; $y = G(u)$. The parameter may or may not have a useful geometric or physical interpretation.

parasitic capture Any *absorption* (as in a *nuclear reactor*) of *neutrons* in reactions which do not cause further *fission* or the production of new *fissionable material.* In a reactor, the process is undesirable.
See: capture, neutron economy

parent A *radionuclide* that upon *radioactive decay* or disintegration yields a specific nuclide (the daughter), either directly or as a later member of a radioactive series.

parking orbit An *orbit* of a *spacecraft* around a *celestial body,* used for assembly of components or to wait for conditions favorable for departure from the orbit.

part   One of the constituents into which a thing may be divided. Applicable to a major *assembly, subassembly* or the smallest individual piece in a given thing.

partial derivative   The ordinary *derivative* of a function of two or more variables with respect to one of the variable, the others being considered constants. If the variables are x and y, the partial derivatives of (x,y) are written $\partial f/\partial x$ and $\partial f/\partial y$.

partial pressure   The *pressure* exerted by a designated component or components of a gaseous mixture.

particle   A minute constituent of *matter,* generally one with a measurable mass. The primary particles involved in radioactivity are alpha particles, beta particles, *neutrons,* and protons. Compare this term with *antiparticle* and *photon.*

particle accelerator   An *accelerator.*

pascal [symbol Pa]   Named after the French scientist Blaise Pascal (1623–62), this is the *SI unit* of *pressure.* It is defined as the pressure that results from a *force* of one *newton* (N) acting uniformly over an area of one square meter. This can be expressed as follows:

$$1 \text{ Pa} = 1 \text{ N}/\text{m}^2$$

pass   1. A single circuit of the *Earth* by a *spacecraft* or *satellite.* Passes start at the time the satellite crosses the equator from the southern hemisphere into the northern hemisphere (the ascending node).
See: orbit.
2. The period of time the satellite is within *telemetry* range of a *ground station.*

passivation   The formation of a compact and continuous corrosion-resistant film on the surface of metal exposed to air (natural passivation) or to a chemical solution (artificial passivation).

passive   Containing no power sources to augment output power, e.g. passive electrical network, passive reflector. Applied to a device that draws all its power from the *input signal.* Compare this term with active.

passive homing   The homing of an *aircraft* or *spacecraft* wherein the craft directs itself toward the *target* by means of electromagnetic waves or signals transmitted or radiated by the target. Compare this term with *active homing.*

passive homing guidance   *Guidance* in which a craft or missile is directed toward a destination by means of natural radiations or signals from the destination.

passive target   An orbital *payload* or *satellite* that is stabilized along its three axes and is detected, acquired and tracked by means of *electromagnetic* energy reflected from the surface (or skin) of the target.

path   1. Of a *satellite* or *spacecraft,* the projection of the *orbital* plane on the *Earth's* surface, the locus of the satellite *subpoint.*
Since the Earth is turning under the satellite, the path of a single orbital pass will not be a closed curve. Path and track are used interchangeably. On a cylindrical map projection, the path is a sine-shaped curve.
2. Of a *meteor,* the projection of the *trajectory* on the *celestial sphere,* as seen by the observer.
3. *flightpath.*

payload   Originally, the revenue-producing portion of an aircraft's load, such as passengers, cargo and mail. Then, by extension, the term "payload" became applied to what a *rocket* carries over and above what is necessary for the operation of the vehicle during flight. In *Space Transportation System* terminology the payload is defined as the total complement of specific instruments, space equipment, support hardware and consumables carried in the *Orbiter* (but not included as part of the basic Orbiter payload support) to accomplish a particular activity in space.

payload assist module   A system designed to provide the necessary injection velocities to place spacecraft into their required orbits above the low altitude parking orbit attainable from the Space Transportation System.

payload carrier   One of the major classes of standard *payload* carriers certified for use with the *Space Shuttle* for low-cost payload operations. The payload carriers are classified as either habitable *modules (Spacelab)* or attached but uninhabitable module(s) (Spacelab *pallet, free flyer, satellite*(s), the *inertial upper stage* or any specialized cradle for holding a payload in the *Orbiter cargo bay)*.

payload-handling subsystem (PHS) A subsystem that consists primarily of remotely controlled manipulator arms (one arm, the *remote manipulator system,* is normally installed; a second arm is optional) that can move *payloads* in and out of the payload bay while the *Orbiter* is in *orbit*. *See:* Orbiter structure, remote manipulator system

payload mass ratio [Symbol: ζ] Of a *rocket* or *launch vehicle,* the ratio of the effective *propellant* mass m_p to the initial vehicle mass m_o or

$$\zeta = m_p/m_o$$

Also called mass ratio.

payload model A projected list of possible future *Shuttle payloads* that can be used as a reference base for planning purposes.

payload retention subsystem (PRS) A subsystem that includes remotely controlled retention latches that hold down or release *payload* items but do not transmit *Orbiter* stresses, such as bending, to the payload. *See:* Orbiter structure

payload retrieval mission An option to revisit an orbiting *payload* to capture it and return it to Earth by the *Space Shuttle.*

payload specialist The non-career astronaut who flies as a *Space Shuttle* passenger and is responsible for attaining the *payload/experiment* objectives. He or she is the onboard scientific expert in charge of payload/experiment operations. The payload specialist has a detailed knowledge of the payload instruments and their subsystems, operations, requirements, objectives and supporting equipment. As such, he or she is either the *principal investigator* conducting his or her experiment in orbit or the direct representative of the principal investigator. Of course, the payload specialist must also be knowledgeable about certain basic *Orbiter* systems, such as food and hygiene accommodations, life support systems, hatches, tunnels, and caution and warning systems.

NASA is currently investigating appropriate policy and processes for selecting payload specialists for those non-NASA payloads to be flown aboard the Shuttle. In NASA Management Instruction 7100, entitled "Payload Specialist for Non-NASA Payloads," the payload specialist is identified as an individual selected to operate assigned payload elements on a specific *Space Transportation System* (STS) *flight* or *mission.* Furthermore, the *payload sponsor* is identified as the individual or organization who funds the development and flight of a payload.

The *Space Transportation System* has been developed to expand man's capabilities to take advantage of the benefits of space in a routine, resource-efficient manner. Part of this capability involves the opportunity for non-career astronaut investigators to conduct in-orbit payload/experiment work. It is therefore current NASA policy to provide these selected individuals the opportunity to perform as payload specialists aboard Shuttle flights—conducting experiments, making direct observations and engaging in scientific activities that support mission objectives. As currently envisioned by NASA payload sponsors who contract for the full payload on a particular flight will be allowed to select and use two payload specialists (subject, of course, to the approval of the NASA Administrator). Payload sponsors who contract for at least one-half of the payload of a particular Shuttle flight will be allowed to select and use one payload specialist (on a space available basis), subject, again, to the NASA Administrator's approval.

Payload sponsors who have contracted for at least one-half the payload on a particular Shuttle flight may request resumes of available *mission specialists* who might have the required qualifications to operate their payloads or experiments; or the payload sponsors can propose their own payload specialist(s). If the sponsors choose to nominate their own payload specialists, the nominee(s) must then be approved by the NASA Administrator or his designee. If the payload sponsors have contracted for less than one-half the payload on a Shuttle flight, and if this payload is determined by NASA to be sufficiently unique and complex to require a payload specialist, then the NASA Director of STS Operations will first propose available mission specialists who have qualifications appropriate for the operation of the specific experiment. If none of the available mission specialists satisfies the payload sponsor's requirements, then the sponsor will be requested to nominate a payload specialist, who must subsequently be approved by the NASA Administrator.

As part of the overall flight preparation, NASA will identify and negotiate those times when the payload specialist(s) must be available for *flight independent training.* Flight or mission independent training, which is conducted at the *Johnson Space Center* in Houston, Texas, is the flight familiarization training that is required for every Shuttle mission. During this training the payload specialist will learn how to operate certain Orbiter systems, such as food and hygiene facilities and hatches, and will develop proficiency in the normal and emergency procedures that are required for safe crew operations. NASA will also establish other mission tasks to be accomplished during a flight and will allocate those tasks among the available personnel on that flight. Consequently, a payload specialist may be required to participate in in-orbit activities not related to his or her payload.

Payload specialists selected by the payload sponsor or nominated by NASA must be certified in writing by (1) passing a NASA Class III physical examination (or its military equivalent) at a qualified flight medicine examination facility (2) obtaining a statement of competence from the payload sponsor and (3) obtaining a statement of successful completion of flight independent training *completion,* and acceptance of flight readiness by the *commander* of the Shuttle flight on which the payload is to be flown.

For NASA or NASA-related payloads, the actual selection of payload specialists is made by a committee known as the Investigator Working Group (IWG), which consists of principal investigators or their representatives who have been selected for a particular mission. As currently planned, this committtee formulates the technical and professional requirements for payload

specialist (s), and, from the names proposed by the principal investigators, nominates and evaluates potential candidates and makes the final selection of the best candidates. The payload specialists themselves will be drawn from the scientific and technical community—domestic or foreign—having a specific interest in the particular mission.

The payload specialist is responsible for managing and operating the payload elements/experiments assigned to him or her. He or she will be an expert in payload/experiment design and operation and in onboard decisions about payload/experiment operations. The payload specialist will also be subject to the authority of the mission specialist and operate in compliance with mission rules and *Payload Operations Control Center* directives.

Since the Shuttle is capable of carrying more payloads than personnel to operate payloads, the payload specialist will most likely also be cross-trained as necessary to assist the mission specialist or other payload specialists in payload/experiment operations. However, the payload specialist may not be required to manage payloads/experiments outside his or her area of expertise. Under certain circumstances, the payload specialist may actually become responsible for all experiments on board. He or she may also operate various payload support systems on either the Orbiter or *Spacelab* in conjunction with payload/experiment operation. These support systems include an *instrument pointing system,* data management and command systems and scientific airlocks. However, the responsibility for in-orbit management of Orbiter systems and attached payload support systems as well as for *extravehicular activity* and payload manipulation using the *remote manipulator system* resides with the basic Shuttle crew.

Payload specialists come from many sources. They may be employed by universities, private industry, government agencies (either U.S. or foreign) or even be self-employed. Other groups planning to fly payload specialists, such as the *European Space Agency,* are developing similar screening and selection processes in close coordination with NASA.

Creative use of the payload specialist in the Shuttle era will greatly accelerate the rate at which future space technology and advanced space systems directly improve the quality of life for all on Earth. Building upon roles and functions tested in early Shuttle missions, STS users of the next decade will routinely exploit the payload specialist position in ways not now imagined!

See: Shuttle crew, mission specialist, humanization of space, payload specialist training

payload specialist training Process by which a *payload specialist* is trained for a *Space Transportation System* (STS) flight. It consists of *flight-independent training* and *flight-dependent training.*

Flight-independent training for a payload specialist involves learning the Shuttle functions that any crewperson must be able to perform effectively during a flight. A total of 124 hours of flight-independent training is required. Payload specialists who have flown before will be required to take a proficiency examination and repeat any such training that is deemed necessary.

Flight-dependent training is divided into two categories: (1) *payload discipline training* and (2) training necessary to support STS/*payload* integrated operations. The first category, payload discipline training, consists of training related to the individual *experiment*/payload. This type of training

includes use of the *payload sponsor*'s research facilities, prototype or engineering development payload hardware and possibly experimental flight hardware. However, there may be specific limitations on the use of flight hardware for training purposes. Payload sponsors must therefore carefully evaluate the unique training requirements for their individual payloads/experiments, since crew payload training is ultimately the responsibility of such sponsors. Payload discipline training should occur in a time frame that is compatible with the overall *Shuttle crew* and payload specialist training schedules. It may start as early as two years before a Shuttle flight.

The second category of flight-dependent training, that necessary to support STS/ payload integrated operations, consists of integrated simulations involving the entire flight crew and ground-based *flight operations support team.* These integrated simulations, which take place in one or more of the *Johnson Space Center* (JSC) *training facilities,* also involve the appropriate *Payload Operations Control Center.* Approximately 115 hours will be devoted to this type of training.

The training in support of STS/payload integrated operations occurs in residence at JSC, takes place at the end of the payload specialist training program and lasts for two months at nearly full time. About half this time is devoted to formal classroom and *simulator* training. The remaining time at

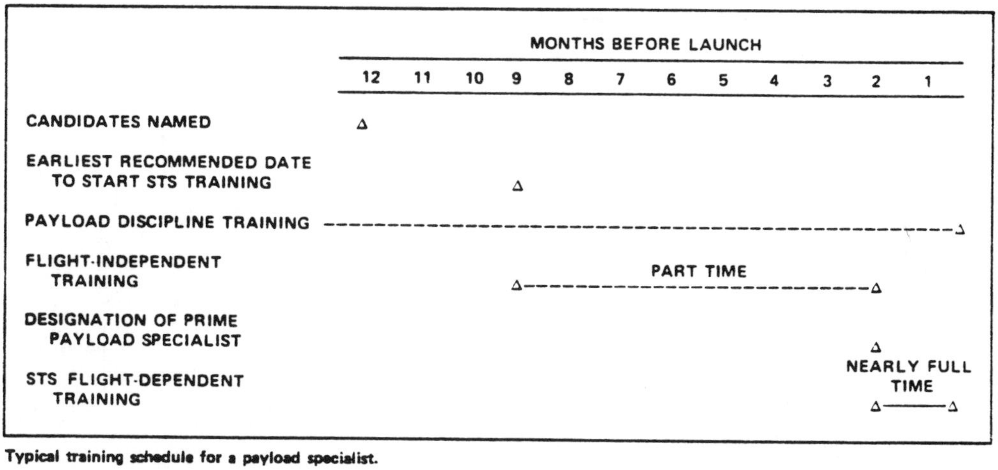

Typical training schedule for a payload specialist.

Fig. 1   Typical payload specialist training schedule.   (Table courtesy of NASA.)

JSC can be used for STS/payload *flight plan* integration and reviews, development of flight/mission rules, meetings regarding various flight techniques and reviews of flight requirements implementation. For more complex multidiscipline payloads, this type of integrated training may require more than two months.

Figure 1 shows a typical training schedule for a payload specialist, including both flight-independent and flight-dependent training. Although the schedule in figure 1 covers a one-year period, for some payloads the sponsor may want to evaluate and screen the payload specialist for a longer time period before the Shuttle flight.

Figure 2 shows the typical activities involved in payload specialist training. (The figure lists the approximate hours required for each type of training, but not for each individual activity. An "X" after an activity indicates that some time is required in the facility listed, and a "C" indicates coordinated training with at least one Shuttle crew member present.) In general, a payload spe-

cialist schedule for an *"Orbiter*-only" flight is required to undergo approximately 180 hours of training, while one scheduled for a flight with a *Spacelab* pressurized crew *module* is required to undergo about 239 hours.

See: payload specialist

payload support equipment (PSE) The *Shuttle* flight equipment needed to support the *payload,* such as caution and warning, data recording, controlled functions and instrumentation.

payload system The combination of a *spacecraft,* instrument *module* and other *mission*-unique equipment, such as a solar array and antenna system, that can be used together to meet a specified mission objective.

pebble bed reactor A *nuclear reactor* in which the *fissionable fuel* (and sometimes also the *moderator*) is in the form of packed or randomly placed pellets, which are cooled by *gas* or liquid.

Peclet number [N_{Pe}] 1. A *nondimensional number* arising in problems of heat transfer

Orbiter habitability

Training type	Facility				Total
	Class	ORB 1-g or SLS	Water tank	Launch pad	
Shuttle Program orientation	X				
STS systems overview	X				
Space-flight physiology	X				
Crew systems					
Ingress/egress	X	X			C
Habitability	X	X			
Stowage	X	X			
Emergency/survival	X	X	C	C	
Medical	X	X			
Crew station activation/ deactivation	X	C			
Hours (approximate)	50	45	1	4	100

Orbiter systems

Training type	Facility			Total
	Class	CSTA	ORB 1-g	
Orbiter systems	X		X	
Guidance, navigation, and control/ software	X	X		
Ground-support network	X			
Hours (approximate)	15	4	4	23

Spacelab systems

Training type	Facility		Total
	Class	SLS	
Electrical power distribution subsystem	X		
Environmental control subsystem	X		
Common payload support equipment	X	X	
Command and data management subsystem	X	X	
Instrument pointing subsystem	X	X	
Hours (approximate)	8	13	21

Spacelab habitability (module only)

Training type	Facility		Total
	Class	SLS	
Spacelab crew systems emergency/safety	X	X	
Hours (approximate)	2	3	5

Integrated crew/ground simulations

Training type	Facility		Total
	ORB 1-g	SMS or SLS	
Ascent		C	
Entry		C	
On-orbit operations	C	C	
Hours (approximate)	10[a]	48	48

[a]The Orbiter 1-g trainer is used in conjunction with the simulators for transfer; therefore this figure is part of the simulator total.

Orbiter phases

Training type	Facility (classroom only)
Orbiter phase training: crew activity plan and data management	X
Onboard pointing coordination	X
Hours (approximate)	10

Spacelab phases

Training type	Facility (SLS and SMS)
Activation and checkout	C
Orbital operations	C
Deactivation	C
Hours (approximate)	32

Fig. 2   Payload specialist training activities.   (Table courtesy of NASA.)

in *fluids*. It is defined as the ratio of *heat convection* to *heat conduction*.

$$N_{Pe} = \frac{\text{heat convection}}{\text{heat conduction}}$$

$$= \frac{\rho\, C_p V l}{k}$$

where ρ is the mass density, C_p is the *specific heat* at constant pressure, V is the (fluid) *velocity*, l is the *characteristic length,* k is the *heat conductivity.*
2. A nondimensional number arising in mass transfer problems. It is defined as the ratio of bulk mass transfer to diffusive mass transfer.

$$N_{pe} = \frac{\text{bulk mass transfer}}{\text{diffusive mass transfer}}$$

$$= \frac{l V}{D}$$

where l is the characteristic length, V is the (fluid) veolocity, D is the mass diffusivity.

pendulum A mass suspended from a fixed point so that it can swing back and forth freely.

percussion primer Mechanical initiator in which the impact of a firing pin against an anvil ignites an impact-sensitive initiating charge.

perfect fluid In simplifying assumptions, a *fluid* chiefly characterized by lack of *viscosity* and, usually, by incompressibility. Also called an ideal fluid inviscid fluid.

perfect gas
See: ideal gas

perfect radiator
See: black body

perfect vacuum
See: absolute vacuum

perforation The central cavity of a *propellant grain*

peri A prefix meaning near as in *perigee.*

periapsis The orbital point nearest the center of attraction.
See: orbit

periastron
See: apastron

pericynthian That point in the *trajectory* of a vehicle which is closest to the *Moon.*

perifocus (periapsis) An orbiting body's point of nearest approach to the surface of the central body.

See: orbit; orbital elements

perigee That orbital point nearest the *Earth* when the Earth is the center of attraction.
    That orbital point farthest from the Earth is called *apogee.* Perigee and apogee are also used referring to *orbits* of *satellites,* especially artificial satellites, around any planet or satellite, thus avoiding coinage of new terms for each *planet* and *Moon.*

perigee propulsion A programmed-thrust technique for *escape* from a *planet* that uses intermittent applications of *thrust* or *perigee* (when vehicle *velocity* is high) and coasting periods.

permanent magnetism *Magnetism* that is retained for long periods without appreciable reduction, unless the magnet is subjected to demagnetizing force.

permeability 1. Of a magnetic material, the ratio of the *magnetic induction* to the *magnetic-field intensity* in the same region.
2. The ability to permit penetration or passage. In this sense, the term is applied particularly to substances which permit penetration or passage of fluids.
See: permeability coefficient

permeability coefficient The steady-state rate of flow of *gas* through unit area and thickness of a solid barrier per unit pressure differential at a given temperature. Also called permeability.

permeability The index to the resistance of a given penetrable barrier to passage (permeation) of a given gas.

permeation As applied to gas flow through solids, the passage of *gas* into, through, and out of a solid barrier having no holes large enough to permit more than a small fraction of the gas to pass through any one hole. The process always involves *diffusion* through the solid and may involve various surface phenomena, such as *sorption, dissociation, migration* and *desorption* of the gas *molecules.*

permissible dose The amount of *radiation* that may be received by an individual within a specified period with expectation of no harmful result to himself.

perturbation 1. Any departure introduced into an assumed *steady state* of a system, or a small departure from the nominal path such as a desired *trajectory.*
2. Specifically, a disturbance in the regular

motion of a *celestial body,* the result of a force additional to that which causes the regular motion, specifically, a gravitational force.

phase   A solid, liquid, or gaseous homgeneous form existing as a distinct part of a heterogeneous system.

phase lock   The technique of making the phase of an *oscillator* signal exactly follow the phase of a reference signal by comparing the phases of the two signals and using the resultant difference to adjust the *frequency* of the oscillator.

phase modulation [abbr. PM]   Angle modulation in which the angle of a sine-wave carrier is caused to depart from the carrier angle by an amount proportional to the instantaneous value of the modulating wave.

Combinations of phase and frequency modulation are commonly referred to as frequency modulations.

perilune   In a *lunar orbit* the point nearest to the *Moon.*

perimeter   The outer boundary of the flow area at a given station in a *solid-propellant rocket.*

period
See: orbit

period of revolution
See: orbit

phase space   The sum of the three dimensions of ordinary space and the three dimensions of velocity space.

phosphor   A *luminescent* substance, a material capable of emitting light when stimulated by *radiation.*
See: scintillation

photocathode   A cathode that releases *electrons* in proportion to light *flux* shining on it.

photochemical reaction   A chemical reaction that involves either the *absorption* or *emission* of *radiation.*

photo densitometer   An instrument to measure the transparency of photographs.

photodissociation   The dissociation (splitting) of a *molecule* by the absorption of a *photon.* The resulting components may be *ionized* in the process (photoionization).

photoelectric   Pertaining to the *photoelectric effect.*

photoelectric cell   A transducer that converts *electromagnetic radiation* in the *infrared, visible* and *ultraviolet* regions into electrical quantities such as *voltage, current* or *resistance.* Also called photocell.
See: photoelectric effect

photoelectric effect   The *emission* of an *electron* from a surface as the surface absorbs a *photon* of *electromagnetic radiation.* Electrons so emitted are termed photoelectrons.

The effectiveness of the process depends upon the surface metal concerned and the *wavelength* of the radiant energy to which it is exposed. Cesium, for example, will emit electrons when exposed to visible *radiation.* The energy of the electron produced is equal to the energy of the incident photon minus the amount of work needed to raise the electron to a sufficient energy level to free it from the surface. The resultant energy of the electron, therefore, is proportional to the *frequency* (i.e., inversely proportional to the wavelength) of the incident radiation.

photoelectron   An *electron* that has been ejected from its parent *atom* by interaction between that atom and a high energy *photon.*

Photoelectrons are produced when *electromagnetic radiation* of sufficiently short *wavelength* is incident upon metallic or other solid surfaces (*photoelectric effect)* or when *radiation* passes through a *gas.*

photoionization   The *ionization* of an *atom* or *molecule* by the collision of a high-energy *photon* with the *particle.*
See: photoelectron, photodissociation

photometer   An instrument for measuring the *intensity* (brightness) of *light* or the relative intensity of a pair of lights.

If the instrument is designed to measure the intensity of light as a function of *wavelength,* it is called a spectrophotometer. Photometers may be divided into two classes: photoelectric photometry in which a photoelectric cell is used to compare electrically the intensity of an unknown light with that of a standard light, and visual photometers in which the human eye is the sensor.

photometry   The measurement of the *intensity* of *light.*

photon   According to the quantum theory of *radiation,* the elementary quantity, or quantum, of radiant energy. It is regarded as a discrete quantity having momentum equal to $h\mu/c$, where h is the Planck constant, μ is the *frequency* of the radiation, and c is the speed of light in a vacuum. The photon is never at rest, has no electric charge and no magnetic moment, but does have a spin moment. The energy of a photon (the unit quantum of energy) is equal to $h\mu$.

photo polymerization   The reaction between two or more chemicals which results from *light* (e.g., sunlight).

photosphere   The intensely bright visible portion of the *Sun,* radius 695,000 kilometers. Other stars have a similar surface.

phototube   An *electron tube* that contains a *photocathode* and has an *output* depending on the total photoelectric emission from the irradiated area of the *photocathode.*

physical acceleration   The *acceleration* experienced by a human or an animal test subject in an accelerating vehicle.

pickoff   A sensing device, used in combination with a gyroscope in an automatic pilot or other automatic apparatus, that responds to angular movement to create a signal or to effect some type of control.

pickup   1. A device that converts a sound, scene, or other form of intelligence into corresponding electric signals (e.g., a microphone, a television camera or a phonograph pickup).
2. The minimum current, *voltage,* power or other value at which a *relay* will complete its intended function.
3. *Interference* from a nearby circuit or electrical system.

piggyback experiment   An experiment which rides along with the primary experiment on a space-available basis without interfering with the mission of the primary experiment.

pile   Old term for *nuclear reactor.* This name was used because the first reactor was built by piling up graphite blocks and natural uranium.

pilot   The second in command of a *Space Shuttle flight.* He or she assists the *commander* as required in conducting all phases of the *Orbiter* flight. The pilot has such authority and responsibilities as are delegated to him or her by the commander, as

for example during two-shift orbital operations. The pilot will normally operate the *remote manipulator system* (RMS) at the *payload handling station,* using the RMS to deploy, release or capture *payloads.* At this station, the pilot can open and close the *cargo (payload) bay* doors; deploy the cooling radiators; deploy, operate and stow the manipulator arm; and operate the lights and television cameras in the payload bay. The pilot is also the second crewperson for *extravehicular activities.* The commander or pilot will be available to perform specific payload operations, if appropriate and at the discretion of the *Space Transportation System user.*
See: Shuttle crew, commander, mission specialist

pilot circuit   Flow-control elements (orifices, diaphragms, springs, etc.) that, in combination with a *pilot valve,* control the operation of a larger valve.

pilot parachute   A small parachute used to pull a *drogue* or main parachute from stowage.
See: Solid Rocket Booster

pilot valve   A low-capacity valve that amplifies a low-power control signal to operate a larger valve.

pinch effect   In controlled *fusion* experiments, the effect obtained when an electric current, flowing through a column of *plasma,* produces a *magnetic field* that confines and compresses the plasma.

pinion   A toothed wheel with a small number of teeth designed to mesh with a larger wheel or rack.

pion   An *elementary particle* (contraction of pi-meson). The mass of a charged (positive or negative) pion is about 273 times that of an *electron*; that of an electrically neutral pion is 264 times that of an electron.
See: meson

Pioneer   A series of NASA probes designed to explore the solar system. The probes, launched from the late 1950s through the mid 1970s, explored the Van Allen radiation belt and monitored solar activity. Pioneer 10 and 11 made extensive observations of Jupiter and Saturn.

Pioneer-Venus probes   Two probes, launched in 1978, to make extensive observations of Venus. Information from the probes significantly altered astronomers' theories on the makeup of the planet.

pip   Signal indication on the scope (e.g. *oscilloscope*) screen of an electronic instrument, produced by a short sharply peaked pulse of *voltage*. Also called blip.

pitch   1. Angular motion of a vehicle about a lateral axis passing through its midpoint or *center of gravity* and perpendicular to the longitudinal axis.
2. Distance between corresponding points on adjacent teeth of a gear or on adjacent blades on a turbine wheel, as measured along a prescribed arc, the pitchline.

pitchover   The programmed turn from the vertical that a *rocket* or *launch vehicle* (under power) takes as it describes an arc and points in a direction other than vertical.

pitting   Creation of surface voids by mechanical erosion, chemical corrosion, or cavitation.

plages   Clouds of calcium or hydrogen vapor that show up as bright patches on the surface of the *photosphere* of the *Sun*.

planetary aberration   A displacement in the apparent position of a *planet* in the *celestial sphere* due to the relative movement of the observer and the planet.
See: aberration

planetary boundary layer   That layer of the *atmosphere* from a *planet*'s surface to the *geostrophic wind level*. Above this layer lies the "free atmosphere." Also called atmospheric boundary layer.

planetary circulation   1. The system of large-scale disturbance in a planet's *troposphere* when viewed on a hemispheric or world-wide scale.
2. The mean or time-averaged hemispheric circulation of a planetary *atmosphere*; also called general circulation.

planet fall   The landing of a *spacecraft* or spaceship on a *planet*.

planetoid   An *asteroid* or *minor planet*.

plane wave   A wave in which the *wave fronts* are everywhere parallel planes normal to the direction of *propagation*.

Planet   One of the nine large, nonluminous bodies in the Solar System. The distinction between a planet and a satellite may not be clearcut, except in that a satellite orbits around a planet.

plasma   An electrically neutral gaseous mixture of positive and negative *ions*. Sometimes called the "fourth state of matter," since it behaves differently from *solids, liquids* and *gases*. High temperature plasmas are used in controlled *fusion* experiments. *See:* charged particle

plasma engine   A reaction engine using magnetically accelerated *plasma* as propellant. A plasma engine is a type of electrical engine.

plasma frequency   The natural frequency for motion of *electrons* in a *plasma*. The plasma frequency

$$f = \sqrt{Ne^2/\pi m}$$

where e is charge on the electron; m is mass of the electron; and N is number of electrons per cubic centimeter.

plasma generator   1. A machine, such as an electric-arc chamber, that will generate very high heat fluxes to convert neutral *gases* into *plasma*.
2. A device which uses the interaction of a plasma and electrical field to generate a current.

plasma jet   A *magnetohydrodynamic rocket engine* in which the ejection of plasma generates thrust.
See: electric propulsion

plasma physics   The study of the properties of *plasmas*.

plasma rocket   A *rocket* using a *plasma engine*. Also called electromagnetic rocket.

plasma sheath   1. The boundary layer of charged *particles* between a *plasma* and its surrounding walls, electrode, or other plasmas. The sheath is generated by the interaction of the plasma with the boundary material. Current flow may be in only one direction across the sheath (single sheath), in both directions across the sheath (double sheath), or when the plasma is immersed in a magnetic field, may flow along the sheath surface at right angles to the magnetic field (magnetic current sheath).
2. An envelope of ionized gas that surrounds a body moving through an atmosphere at hypersonic velocities. The plasma sheath affects transmission, reception, and diffraction of radio waves; thus it is

important in operational problems of *spacecraft,* especially during *reentry.*

plastic High-molecular-weight material that while usually firm and hard (although often flexible) in its finished state is at some stage in its manufacture soft enough to be formed into a desired shape by application of heat or pressure or both.

plastic deformation Permanent distortion of material under applied stress great enough to strain the material beyond its elastic limit; also called plastic flow, permanent strain and permanent distortion.

plasticity The tendency of a loaded body to assume a deformed state other than its original state when the load is removed.

plates The six continental masses and the sea floors adjacent to them on *Earth*; subplates are segments of these plates. Plate tectonics is a study of the motion of plates and subplates.

plug nozzle A annular nozzle that discharges exhaust gas with a radial inward component; a truncated aerospike.

Pluto Ninth planet in the Solar System with a diameter of about 6400 km (approx. 4000 mi.) or roughly half of Earth's. The planet remains largely a mystery. It is covered with methane ice. Considering the covering of ice, it is believed that Pluto's diameter could be less than our Moon's which is about 3475 km (roughly 2160 mi.). Pluto has one satellite: Charon.

plutonium [Symbol: Pu] A heavy radioactive, man-made, metallic element with *atomic number* 94. Its most important *isotope* is *fissionable* plutonium-239, produced by *neutron* irradiation of uranium-238. It is used for *nuclear reactor fuel* and in weapons.

pneumatic Operated, moved or effected by *gas* used to transmit *energy.*

pneumatic control assembly (PCA) A component of the *Space Shuttle Main Engine* that does the following: (1) controls ground-supplied gaseous nitrogen used for engine prestart *purges* and vehicle-supplied helium for the operational purge, (2) controls the *oxidizer bleed valve* and the *fuel bleed valve,* and (3) provides for emergency shutdown control of the *main propellant valves* in the event of electrical power loss to the engine.

See: Space Shuttle Main Engine

pod An enclosure, housing, or detachable container of some kind, as an engine pod.

pogo A term coined to describe the longitudinal (vertical) dynamic *oscillations* (vibration) generated by the interaction of the *(launch) vehicle* structural dynamics with the *propellant* and the engine combustion process. The name appears to have originated from the similarity of this bumping phenomenon to the up-and-down bumping motion of a pogo stick.

pogo suppressor A device within the *Space Shuttle Main Engine* that absorbs any vibration of the vehicle's structure during the engine combustion process.

See: Space Shuttle Main Engine; pogo

point-source flow A concept in flowfield analysis in which flow is considered to originate at a point from which it diverges outward uniformly in all directions.

poison 1. (in nuclear engineering) Any material of high absorption *cross section* that absorbs neutrons unproductively and hence removes them from the *fission* chain reaction in a reactor, decreasing its *reactivity.*
2. (in general) Any material that interferes with catalytic action.

Poisson distribution A one-parameter discrete frequency distribution giving the probability that n points (or events) will be (or occur) in an interval (or time) x, provided that these points are individually independent and that the number occurring in a subinterval does not influence the number occurring in any other nonoverlapping subinterval. It has the form

$$f(n,x) = e^{-\sigma x}(\sigma x)^n/n!$$

The mean and variance are both σx, and σ is the average density (or rate) with which the events occur. When σx is large, the Poisson distribution approaches the *normal distribution.* The binomial distribution approaches the Poisson when the number of events n becomes large and the probability of success P becomes small in such a way that $nP \rightarrow \sigma x$.

Poisson equation 1. The partial differential equation

$$\nabla^2 \varphi = f$$

where ∇ is the Laplacian operator; φ is the scalar function of position and f is a given function of the independent space variables. For a special case, F = 0, the Poisson equation reduces to the Laplace equation.
See: relaxation method
2. The relationship between the *temperature* T and *pressure* p of a perfect gas undergoing an *adiabatic* process; given by

$$T = \text{constant} \times p$$

where μ is the *Poissant constant.*

This equation defines a family of process lines, called isentropes or dry adiabatics, each of which represents the changes of state possible in a fluid with a constant value of *entropy.*

Poissant constant (symbol μ)   The ratio of the *gas content* to the *specific heat* of a gas at constant pressure.
See: Poisson equation.

polar coordinates   In a plane, a system of curvilenar coordinates in which a point is located by its distance r from the origin (or pole) and by the angle θ which a line (radius vector) joining the given point and the origin makes with a fixed reference line, called the polar axis. The relations between rectangular Cartesian coordinates and polar coordinates are

$$x = r \cos \theta, \ y = r \sin \theta, \ r^2 = x^2 + y^2$$

where the origins of the two systems coincide and the polar axis coincides with the X-axis.

polar orbit   The *orbit* of an *Earth satellite* that passes over or near the Earth's poles.

polybutadiene acrylic acid acrylonitrile terpolymer (PBAN)   A type of solid rocket motor *propellant* (polymer binder).
See: Solid Rocket Booster

polychromator   A device that produces colors from a source of white light; synonomous with spectrograph.

polyisocyanurate foam insulation   The thermal protection material applied over the *External Tank's* oxygen tank, intertank and hydrogen tank. This insulation primarily reduces the boiloff rate of the *cryogenic propellants;* it also eliminates ice formation on the outside of the External Tank due to the extremely cold propellants

within.
See: External Tank

polytropic atmosphere   A model atmosphere in *hydrostatic equilibrium* with a nonzero *lapse rate.* The vertical distribution of *pressure* and *temperature* is given by

$$p/P_0 = (TK_0) \ g/R\gamma$$

where p is the pressure; T is the Kelvin temperature; g is the acceleration of gravity; R is the gas constant for the gases concerned; and γ is the environmental lapse rate. The subscipt 0 denotes values at the *planet's* surface.

Polytropic process   A *thermodynamic* process in which changes of *pressure* p and *density* p are related according to the formula

$$P^{-\lambda} = P_0 P_0^{-\lambda}$$

where λ is a constant and the subscript 0 denotes initial values of the variables. Therefore, pressure and *temperature* are similarly related:

$$P/P_o = (T/T_o)k$$

where k is the coefficient of polytropy. For isobaric processes, k = 0; for isoteric processes k = l; for adiabatic processes k = c_p/R, where c_p is the specific heat at constant pressure and R is the gas constant.

In meteorology this formula is applied to individual gas parcels and should be distinguished from that for a *polytropic atmosphere,* which describes a distribution of pressure and temperature in space.

pool reactor   A *nuclear reactor* in which the *fuel elements* are suspended in a pool of water that serves as the *reflector, moderator,* and *coolant.* Popularly called a swimming pool reactor, it is usually used for research and training.

poppet   A valving element (ball, cone, or disk) that moves perpendicularly to and from its seat to control flow.

poppet valve   A mushroom- or tulip-shaped valve commonly used in inlets and exhausts.

poppet valve   A valve constructed to control flow by translating a poppet to or from a seat in the valve housing; translation of the poppet away from the seat can result in essentially orifice flow.

popping   Sudden, short-duration surges of

pressure in a *combustion chamber.*

posigrade A posigrade course correction adds to *spacecraft* speed; a *retrograde* correction slows it down.
See: orbit.

posigrade rocket An auxiliary *rocket* that fires in the direction in which the vehicle is pointed, used for example in separating two stages of a vehicle.

position 1. A point in space.
2. A point defined by stated or implied *coordinates,* particularly one on the surface of the *Earth.*
3. *altitude*
4. A crew member's station aboard an *aircraft, spacecraft* or the *Space Shuttle.*

position vector
See: vector

positive acceleration 1. *Acceleration* such that speed increases.
2. Accelerating *force* in an upward sense or direction, e.g., from bottom to top, seat to head, etc.; acceleration in the direction that this force is applied.

positive displacement pump Pump in which *fluid* is forced into a high-*pressure* region by reducing the column of a chamber that is momentarily sealed off from a low-pressure region.

positive feedback *Feedback* which results in increasing the amplification.

positive G *Acceleration* experienced in a downward, or head-to-feet direction; also known as eyeballs down.
See: physiological acceleration

positive gain A term for a control circuit in which an increase in regulated pressure causes an amplified increase in control pressure.

positron [Symbol: β^+] An *elementary particle* with the mass of an *electron* but charged positively. It is the "antielectron." It is emitted in some *radioactive* disintegrations and is formed in *pair production* by the interaction of high-energy *gamma rays* with matter.
See: antimatter

potential [Symbol: ϕ] A function of space, the *gradient* of which is equal to a *force.* In symbols, F $= -\nabla\phi$, where F is the force; ∇ is the del-operator; and ϕ is the potential. A *force* that may be so expressed is said to be

conservative, and the work done against it in motion from one given equipotential surface to another is independent of the path of the motion.
See: Gibbs function, potential energy.

potential core The length of unmixed flow in a *gas generator.*

potential energy *Energy* possessed by a body by virtue of its position in a *gravity* field in contrast with *kinetic energy,* that possessed by virtue of its motion.

potential flow Flow with the effects of *viscosity* not considered.

potentiometer An instrument for measuring differences in *electric potential* by balancing the unknown voltage against a variable known voltage. If the balancing is accomplished automatically, the instrument is called a "self-balancing potentiometer"; also, a variable electric resistor.

pot life The length of time a polymerizing *fluid* can be held or worked before setting up to a gel or solid.

power 1. [Symbol: *P*] (in physics or thermodynamics) Rate of doing *work.*
2. Luminous intensity.
3. The number of times an object is magnified by an optical system, such as a telescope. Usually called magnifying power
4. The result of multiplying a number by itself a given number of times, as the third power of a number is its cube; the superscript which indicates this process as in $2^3 = 2 \times 2 \times 2.$

power density The rate of heat generated per unit volume of a *nuclear reactor core.*

power plant 1. The complete assemblage or installation of *engine* or engines with accessories (induction system, cooling system, ignition system, etc.) that generates the motive power for a self-propelled vehicle or vessel such as an *aircraft, rocket,* etc.
2. An engine or engine installation regarded as a source of *power.*

power reactor A nuclear *reactor* designed to produce useful *power,* as distinguished from reactors used primarily for research or for producing *radiation* or *fissionable materials.* Compare this term with production reactor and research reactor

power series An infinite series of increasing powers of the variable, of the form

$$\overset{\infty}{\Sigma}\, a_n x^n = a_0 + a_1 x + a_2 x^2 \ldots + a_n x_n$$

Both the variable and the coefficients may take on complex values. The totality of values of x for which a power series is convergent is called the interval of convergence of the series.

power spectrum   The square of the amplitude of the (complex) *Fourier* coefficient of a given periodic function. Thus if f(t) is periodic with period Tm, its F o u r i e r coefficients are

$$F\,(n) = 1/T \!\int_0^T f(t) e^{-in\omega t} dt$$

where $\omega = 2\pi/T$ and the power spectrum of f (t) is $[F(n)]^2$. Here n takes integral values and the spectrum is discrete. The total *energy* of the periodic function is infinite, but the *power,* or energy per unit period, is finite.

Poynting-Robertson effect   The gradual decrease in *orbital velocity* of a small *particle* such as a micrometeorite in orbit about the Sun due to the *absorption* and reemission of *radiant energy* by the particle.

Prandtl number [Pr]   1. In heat transfer, the *dimensionless parameter* expressing the ratio of *momentum diffusivity* to thermal diffusivity.

$$Pr = \frac{\text{momentum diffusivity}}{\text{thermal diffusivity}}$$

$$Pr = \frac{Cp\,\mu}{k}$$

where Cp is the *specific heat* at constant pressure

   μ is the absolute *viscosity*
   k is the heat conductivity

2. In mass transfer, the dimensionless parameter expressing the ratio of momentum diffusivity to mass diffusivity

$$Pr = \frac{\text{momentum diffusivity}}{\text{mass diffusivity}}$$

$$Pr = \frac{\mu}{\rho D}$$

where μ is the absolute viscosity
   ρ is the density
   D is the mass diffusivity

preamplifier   An *amplifier,* the primary function of which is to raise the output of a low-level source to an intermediate level so that the signal may be further processed without appreciable degradation in the signal-to-noise ratio of the receiver.

preburners   Each *Space Shuttle Main Engine* has *fuel* and *oxidizer* preburners that provide hydrogen-rich hot gases at approximately 1,030 K (760 C or 1,400 F). These gases drive the fuel and oxidizer high-pressure turbopumps.
See: Space Shuttle Main Engine

precession   Change in the direction of the *axis* of rotation of a spinning body, as a gyro, when acted upon by a *torque.*
   The direction of motion of the axis is such that it causes the direction of spin of the gyro to tend to coincide with that of the impressed torque. The horizontal component of precession is called drift, and the vertical component is called topple.

precision   The quality of being exactly or sharply defined or stated. A measure of the precision of a representation is the number of distingusihable alternatives from which it was selected, that is sometimes indicated by the number of *significant digits* it contains.

recombustion chamber   In a *rocket,* a chamber in which the *propellants* are ignited and from which the burning mixture expands torchlike to ignite the mixture in the main *combustion chamber.*

predisperser   A secondary *diffraction grating* used to preselect a portion of a *spectrum* to be dispersed by the main diffraction grating.

predissociation   A process by which a *molecule* that has absorbed *energy* separates into constituents before it loses energy by *radiation.*
See: dissociation.

preflight   Occurring before *launch vehicle liftoff.*

preload   The mechanical load applied to components in an assembly at the time of assembly to ensure dimensional accuracy and proper operation; bolt or nut torque, and spring force, for example, are means for providing preload.

pressurant   *Gas* that provides *ullage pressure* in a *propellant* tank.

pressure (symbol p)   A *thermodynamic*

property that two systems have in common when they are in mechanical *equilibrium.* Pressure can be defined as the normal component of *force* per unit area exerted by a *fluid* on a boundary. According to **Pascal Principle** the pressure at any point in a fluid at rest in the same in all directions. (Blaise Pascal was the French mathematician, physicist and philosopher who discovered this *hydrostatic* principle in the 1650's.)

The fundamental equation of hydrostatics and *aerostatics* is:

$$\frac{dp}{dz} = -\rho g$$

where p is the pressure
    z is the height or depth
    P is the fluid density
    g is the acceleration due to gravity at the surface of the Earth 9.81 m/sec² (32.2 ft/sec²)

Assuming the density (ρ) is constant, we can compute the hydrostatic pressure of a liquid from the following relationship:

$$p(z) = -\rho g z$$

In aerostatics density appears as a variable (i.e., a function of height in the *atmosphere*), and consequently, pressure as a function of height cannot be evaluated until the density is specified in some manner. For example, air is frequently regarded as an **ideal gas** obeying the following relationship:

$$p = \rho R T$$

where p is pressure
    ρ is density
    R is the specific gas constant
    T is temperature (absolute)
Consequently, the fundamental law of aerostatics can be rewritten as follows, assuming temperature is a constant (i.e., $T = T_o$):

$$\frac{dp}{dz} = -[p/(RT_0)] g$$

or

where p_0 is now the atmospheric pressure at $z = 0$. This relationship, known as Halley's Law, states that pressure diminishes exponentially with height for an *isothermal* (constant temperature) atmosphere. Since the

Earth's atmosphere in reality exhibits temperature changes with altitude, Halley's Law is useful only when limited thicknesses of the atmosphere are considered.

In engineering terminology the total pressure exerted on a boundary wall is called the **absolute pressure,** whereas the pressure exerted on a surface or boundary by the atmosphere is called the **atmospheric pressure.** Atmospheric pressure varies with location and elevation on the Earth's surface and is essentially the result of the "weight of the air" at some particular location. Atmospheric pressure is often called **barometric pressure.** One reference value for atmospheric pressure is the standard atmosphere (symbol atm), which is defined as the pressure produced by a column of mercury (Hg) exactly 760 mm in height at 273.15 K (0 C) temperature and under standard gravitational acceleration (g).

1 standard   = 1.013×10^5 N/m² (*pascals*)
atmosphere = 1.013 bar
             = 29.92 inches of Hg at 0 C
             = 14.696 lb$_f$/in.² (psi)

$$p(z) = p_0 e - [-[92]/_{[RT_0]}]$$

Gauge pressure (p_{gauge}) is used to designate the difference between absolute pressure (p_{abs}) and atmospheric pressure (p_{atm}) in a particular system. This relationship can be shown as follows:

$$p_{gauge} = p_{abs} - p_{atm}$$

It is normally measured with an instrument that has atmospheric pressure as a reference.

A *negative gauge pressure* (p_{vac}) occurring when the atmospheric pressure is greater than the absolute pressure, is called a *vacuum.* It can be shown by the following equation:

$$p_{vac} = p_{atm} - p_{abs}$$

The *SI unit* of pressure is the pascal (1 Pa = 1 N/m^2). In high-vacuum technology pressure is often expressed in atmospheres, while in *meteorology* pressure is measured in millibars (1 bar = 10^5 Pa). In chemical technology pressure is often expressed in terms of millimeters of mercury.

pressure-actuated seal   A seal designed such that the *pressure* of the *fluid* being sealed activates or increases the sealing action.

pressure altimeter   An *altimeter* that utilizes the change of *atmospheric pressure* with height to measure *altitude*. It is commonly an aneroid altimeter. Also called barometric altimeter.

pressure broadening   The process whereby the width of the lines in an *emission spectrum* or *absorption spectrum* of a gaseous radiative medium is increased due to perturbations of the energy states by collisions of the *molecules* or *atoms* within the *gas*. The extent of this line-broadening effect is directly proportional to the number of impacts experienced by the emitter or absorber per unit time, and hence is proportional to the *pressure*. Compare this term with *Doppler broadening*.

pressure divider   A term applied to a *pneumatic* or *hydraulic* circuit in which an intermediate *pressure* between two flow restrictors in series is used as a control signal.

pressure fed   A term for a propulsion system in which *tank ullage pressure* expels the *propellants* from the tanks into the combustion chamber of the engine.
See: pump fed

pressure-laden sequence   A method to effect *fail-safe* engine starts by sequencing the operation of *rocket engine* control valves; the sequencing is achieved by vent mechanisms on the control system or propellant feed system or both that are triggered by pressure changes.
See: Space Shuttle Main Engine.

pressure overshoot   The maximum or relative-maximum point that occurs on the pressure -time curve, as in *ullage pressurization* or engine ignition.

pressure ratio   1. Ratio of *combustion chamber pressure* to ambient pressure.
2. Ratio of *turbine* inlet pressure to turbine outlet pressure.

pressure recovery   The conversion of *velocity head* to *pressure head* in a *fluid* system.

pressure regulator   A *pressure* control valve that varies the volumetric flowrate through itself in response to a downstream pressure signal so as to maintain the downstream pressure nearly constant.

pressure surface   The concave surface of a *pump* or *turbine blade*; along this surface, the *fluid pressures* are highest.

pressure thrust   In rocketry, the product of the cross-sectional area of the exhaust *jet* leaving the *nozzle exit* and the difference between the *exhaust pressure* and the *ambient pressure*.

pressure vessel   A strong-walled container housing the core of most types of power *reactors*; it usually also contains *moderator, reflector,* thermal shield and control rods.

pressure wave   1. In *meteorology,* a short-period of *oscillation* of *pressure* such as that associated with the propagation of sound through the *atmosphere*. These waves are usually recorded on sensitive microbarographs capable of measuring pressure changes of amounts down to 10^{-4} millibar. Typical values for the *period* and *wavelength* of pressure waves are ½ to 5 seconds and 100 to 1500 meters, respectively.

2. A wave or periodicity that exists in the variation of *atmospheric pressure* on any scale, usually excluding normal diurnal and seasonal trends. Such waves can persist for an indefinite length of time only if they coincide approximately with the free oscillations of the atmosphere. Waves of a period longer than that associated with the passage of large-scale weather disturbances are difficult to isolate, since they usually have such a small *amplitude* that they can be extracted from the data only by means of precise statistical methods.

pressurization   The sequence of operations that increases the *ullage pressure* to the desired level some time before the main sequence of *propellant* flow and engine firing; in *launch vehicles,* prepressurization occurs prior to liftoff.

pressurization system   The set of *fluid*-system components that provides and maintains a controlled gas pressure in the *ullage* space of the *launch vehicle* propellant tanks.
See: Orbiter environmental control and life support system

pressurized   Containing air, or other *gas,* at

a pressure that is higher than the pressure outside the container.

pressurized crew module
See: Orbiter structure

pressurized habitable environment Any space *module* in which a person may perform activities in a *shirtsleeve environment*.

prestage A step in the action of igniting a large *liquid rocket* taken prior to the ignition of the full flow, and consisting of igniting a partial flow of propellants into the *thrust chamber*.

prevalve(s) Device(s) located in the *Space Shuttle Main Engine* (SSME) that keeps the flow of *liquid hydrogen* and *liquid oxygen* from the *external tank* from entering the SSME.
See: Space Shuttle Main Engine

primary body The *celestial body* about which a *satellite* or other body orbits, or from which it is escaping, or towards which it is falling. The primary body of the *Moon* is the *Earth*; the primary body of the Earth is the *Sun*.

primary leakage Leakage from the upstream side to the downstream side of a *fluid*-system component.

primary scattering Any scattering process in which *radiation* is received at a detector, after having been scattered just once; to be distinguished from multiple scattering.

primary seal The seal intended to limit *primary leakage*.

prime charge In an *electroexplosive device*, the material in contact with the bridge wire the electrical heating of which causes the material to *deflagrate*.

prime meridian The zero meridian of *longitude* passing through Greenwich, England, adopted as a standard of east-west reference.

primer Material applied to surfaces of *solid rocket motor* cases, insulation, or liners to enhance bond strengths.

prime vertical The vertical circle normal to the *celestial meridian* and therefore intersecting the horizon at its east and west points.

primitive atmosphere The *atmosphere* of a *celestial body* as it existed in the early stages of its formation; specifically, the *Earth*'s atmosphere of 3 billion or more years ago, thought to consist of water vapor, carbon

dioxide, methane, and ammonia gas.

principal planets The larger bodies revolving about the *Sun* in nearly circular orbits. The known principal *planets,* in order of their distance from the Sun, are: *Mercury, Venus, Earth, Mars, Jupiter, Saturn, Uranus, Neptune,* and *Pluto.*

probable error [Symbol: p_e] In statistics, that value e_p for which there exists an even *probability* (0.5) that the actual error exceeds e_p. The probable error p_e is 0.6745 times the *standard deviation* σ.

probability The chance that a prescribed event will occur, represented as a number P in the range $0 \leq P \leq 1$. The probability of an impossible event is zero and that of an inevitable event is unity.

Probability is estimated empirically by relative frequency, that is, the number of times the particular event occurs divided by the total count of all events in the class considered.

probability integral The classical form of the definite integral of the special *normal distribution* for which the *mean* N = O and *standard deviation* $\sigma = 1/\sqrt{2}$. Geometrically, the probability integral equals the area under this density curve between $-z$ and z, where z is an arbitrary positive number. Often denoted by the symbol er*f z* (read error function of *z*) the probability integral is defined thus:

$$\text{er} f z \equiv \frac{2}{\sqrt{\pi}} \int_0^2 e^{-x^2} dx$$

Also called error function, er*f*.

probe 1. Any device inserted in an environment for the purpose of obtaining information about the environment.

2. In geophysics, a device used to make a *sounding*.

3. Specifically, an instrumented vehicle moving through the *upper atmosphere* or *space* or landing upon another *celestial body* in order to obtain information about the specific environment.

In this sense, almost any instrumented *spacecraft* can be considered a probe. However, earth *satellites* are not usually referred to as probes. Also, almost any instrumented *rocket* can be considered a probe. In practice, rockets which attain an *altitude* of less than 1 earth radius (6437km) [4000 miles]

are called *sounding rockets,* those which attain an altitude of more than 1 earth radius are called probes or space probes. Spacecraft that enter into orbit aound the *Sun* are called *deep-space probes.* Spacecraft designed to pass near or land on another celestial body are often designated lunar probe, Martian probe, Venus probe, etc.
4. Specifically, a slender device or apparatus projected into a moving *fluid,* as for measurement purposes

processability   The measure of relative ease with which a *component, assembly,* or *system* can be produced with state-of-the-art techniques.

procurement/fabrication complete   The date when all procurement and fabrication for a particular facility has been finished. *Certification* of acceptance by the site activation office completes this *milestone.*

production reactor   A *nuclear reactor* designed primarily for large-scale production of *plutonium-239* by *neutron* irradiation of *uranium-238.* Also a reactor used primarily for the production of *radioactive isotopes.*

progressive burning   The condition in which *thrust,* pressure, or burning surface increases with respect to time or to *web* burned.
See: solid propellant rocket

progressivity ratio   Ratio of final to initial burning surface of a *solid propellant.*

project   A scheduled undertaking, within a *program,* which may involve the research and development, design, construction, and operation of system and associated *hardware* or hardware only, to accomplish a scientific or technical objective.

projectile   1. Any object, especially a *missile,* fired, thrown, launched, or otherwise projected in any manner, such as a bullet, a guided rocket missile, a sounding rocket, a pilotless airplane, etc.
2. Originally, an object, such as a bullet or artillery shell, projected by an applied external force.

prominences   Jets of luminous matter ejecting from the Sun's *chromosphere* for thousands of kilometers.
See: Sun

prompt criticality   The state of a *nuclear*

reactor when the *fission chain reaction* is sustained solely by *prompt neutrons,* that is, without the help of *delayed neutrons.*
See: criticality

prompt neutrons   *Neutrons* that are emitted immediately following nuclear *fission,* as distinct from *delayed neutrons,* which are emitted for some time after fission has occurred. Prompt neutrons comprise more than 99% of fission neutrons. Compare this term with *delayed neutrons.*

prompt radiation   *Radiation* produced by the primary *fission* or *fusion* process, as distinguished from the radiation from *fission products,* their *decay chains* and other delayed reactions.

proof pressure   *Pressure* that a pressurized component must sustain and still function satisfactorily; proof pressure is the maximum limit pressure multiplied by the proof-test *safety factor* and is the reference from which the pressure levels for acceptance testing are established.

proof test   *Pressure* test to prove the structural integrity of a component or assembly without exceeding allowable stresses or producing any permanent deformation.

propellant   A material, such as a *fuel,* an *oxidizer,* an *additive,* a *catalyst* or any compound or mixture of these, carried in a *rocket vehicle* that releases energy during combustion and thus provides *thrust* to the vehicle. Propellants are commonly in either liquid or solid form.
See: liquid propellant rocket; solid propellant rocket; Space Shuttle Main Engine; Solid Rocket Booster(s)

propellant feed system   A component of the *Space Shuttle Main Engine* that includes four turbopumps, two of which are low-pressue and two high-pressure. There is one of each for the *liquid hydrogen fuel* and *liquid oxygen oxidizer.*

propellant mass fraction [Symbol: ζ]   Of a *rocket,* the ratio of the effective *propellant* mass m_p to the initial vehicle mass m_o or

$$\zeta = m_p/m_o$$

Also called mass ratio, propellant mass ratio.

propellant stratification   Non-isothermal temperature distribution in the propellant.

proportional counter tube An *electron tube* that produces an electric pulse which is proportional to the *energy* of a *photon* entering the tube.

proprioceptive stimulation Stimulation originating within the deeper structures of the body (muscles, tendons, joints, etc.) for sense of body position and movement and by which muscular movements can be adjusted with a great degree of accuracy and equilibrium can be maintained.

propulsion system A vehicle system that includes the engines, tanks, lines, and all associated equipment necessary to provide the propulsive force as specified for the vehicle.

propulsive efficiency [Symbol: $\eta\rho$] The efficiency with which *energy* available for propulsion is converted into *thrust* by a *rocket engine*.

$$\eta\rho = (2u/c)/[1 + (u/c)^2]$$

where u is the absolute vehicle *velocity*, and c is the effective exhaust velocity with respect to the vehicle. Propulsive efficiency is a maximum when u = c.

protective clothing Special clothing worn by a *radiation* or hazardous material worker to prevent contamination of his body or his personal clothing.

protective survey An evaluation of the *radiation* hazards incidental to the production, use or existence of radioactive materials or other sources of radiation under a specific set of conditions.

protection Provisions to reduce exposure of persons to radiation or toxic substances. For example, protective barriers to reduce external *radiation* or measures to prevent inhalation of *radioactive materials* or toxic substances.

proton An *elementary particle* with a single positive electrical charge and a *mass* approximately 1837 times that of the *electron*. The *nucleus* of an ordinary or light hydrogen *atom*. Protons are constituents of all nuclei. The *atomic number* (Z) of an atom is equal to the number of protons in its nucleus.

proton binding energy
See: binding energy

proton storm The *flux* of *protons* sent into space by a *solar flare.*

proton synchrotron A type of particle *accelerator* for producing *beams* of very high energy *protons* (in the GeV range).

protoplanet Any of a *star's planets* as it emerges during the process of *accretion* or the formative period of the star system (e.g., the *Solar System*).

protostar A *star* in the making. Specifically, the stage in a young star's evolution after it has separated from a gas cloud, but prior to its sufficient collapse to support *thermonuclear reaction.*
See: fusion

protosun The *Sun* as it emerges in the formation of the *Solar System.*
See: protostar

prototype 1. Of any mechanical device, a production model suitable for complete evaluation of mechanical and electrical form, design and performance.
2. The first of a series of similar devices.
3. A physical standard to which replicas are compared, as the prototype kilogram.
4. A *spacecraft* or component that has passed or is undergoing tests, qualifying the design for fabrication of flight units or components thereof.

proving stand A *test stand* for *reaction engines,* especially *rocket engines.*

provisioning activity The actions of the provisioning team of the *Space Transportation System* Projects Office that is responsible for selecting provisioned items and determining requirements for them.

psychomotor ability Of or pertaining to muscular action ensuing directly from a mental process, as in the coordinated manipulation of *aircraft* or *spacecraft* controls.

psychophysical quantity A physical measurement, as a *threshold,* dependent on human attributes or perception.

pulse 1. A variation of a quantity whose value is normally constant; this variation is characterized by a rise and a decay, and has a finite duration.

The word pulse normally refers to a variation in time; when the variation is in some other dimension, it should be so specified, such as space pulse.

This definition is so broad that it covers almost any transient phenomenon. The only

features common to all pulses are rise, finite duration and decay. It is necessary that the rise, duration and decay be of a quantity that is constant (not necessarily zero) for some time before the pulse and has the same constant value for some time afterwards. The quantity has a normally constant value and is perturbed during the pulse. No relative time scale can be assigned.
2. An electrical signal arising from a single event of *ionizing radiation.*
3. An *aerospace medicine* term describing the intermittent change in the shape of an artery due to an increase in the tension of its walls following the contraction of the heart. The pulse is usually counted at the wrist (radial pulse), but may be taken over any artery that can be felt.

pulse amplifier   An *amplifier* designed specifically to amplify the intermittent signals of a *radiation detection instrument,* incorporating appropriate pulse-shaping characteristics.

pulsed reactor   A type of research *nuclear reactor* with which repeated short, intense surges of power and *radiation* can be produced. The *neutron flux* during each surge is much higher than could be tolerated during a steady-state operation.

pulse height   The measure of the strength or signal amplitude of a *pulse* delivered by a detector; frequently measured in *volts.*

pulse height analyzer   An electronic *circuit* that sorts and records *pulses* according to height or *voltage.*

pulse height selector   A *circuit* designed to select and pass *voltage pulses* in a certain range of *amplitudes.*

pulsejet   A pulsejet engine.

pulsejet engine   A type of compressorless *jet engine* in which a combustion takes place intermittently, producing *thrust* by a series of explosions, commonly occurring at the approximate *resonance frequency* of the engine. Often called a pulsejet. The German V-1 rocket contained a pulsejet engine.

pump   A machine for transferring *mechanical energy* from an external source to the *fluid* flowing through it, the increased *energy* being used to lift the fluid or increase the fluid pressure.

pump fed   Term for a *propulsion system* that incorporates a *pump* that delivers pro-

pellant to the *combustion chamber* at a *pressure* greater than the tank *ullage* pressure.

purge   To rid a line or tank of residual *fluid;* especially of *fuel* or *oxidizer* in the tanks or lines of a *rocket* after a test firing or simulated test firing.

pyrogen   Small *rocket* motor used to ignite a larger rocket motor.

pyrolysis   Chemical decomposition of a material by heat.

pyrometer   An instrument for the measurement of *temperatures*; generally applied to instruments measuring temperatures above 600°C (1112°F).

pyrometry   High-temperature *thermometry,* the technique of remote (at a distance) measurement of *temperatures,* generally above 600°C (1112°F).

pyrophoric fuel   A *fuel* that ignites spontaneously in *air.* Compare this term with *hypergolic propellants.*

pyrotechnics   *Igniters* (other than *pyrogens*) in which solid explosives or energetic propellant-like chemical formulations are used as the heat-producing material.

Q

Q-band   A *frequency band* (used in *radar*) extending approximately from 36 to 46 *gigahertz.*

quadrant   An ancient *astronomical* instrument for measuring *altitudes.* Considered the most important instrument used by astronomers from antiquity to the invention of the telescope [by Galileo in 1609], it was a quarter circle (90° graduated arc) with an attached sighting device capable of measuring the *apparent* positions of *celestial bodies.*

quadrature   1. The position of a *planet* or the *Moon* when at 90° (right angles) to the *Sun* as seen by an observer on *Earth.* More formally, an *elongation* of 90°, usually specified as "east" or "west" in accordance with the direction of the celestial body from the Sun. The Moon, for example, is at quadration at first and last quarters. Compare this term with *conjunction*.

2. The situation of two periodic quantities (having the same frequency and waveworm) differing by a quarter of a cycle [i.e., differ in phase by 90°].

quality factor [symbol: Q] A measure of the sharpness of *resonance* or frequency selectivity of a resonant vibratory system having a single degree of freedom, either mechanical or electrical.
See: vibration.

quantitation A number that best represents the statistical results of biological measurements involving many individuals.

quantity/distance Term for a system for specifying safe distances for locations of buildings for processing or storing *propellant* or propellant ingredients (liquid or solid).

quantization The process of converting from continuous values of information to a finite number of discrete values.

quantum [plural: quanta] The quantity of energy possessed by a *photon*. According to *quantum theory*, it is equal to the *frequency (ν)* of the radiation and Planck's constant (h).

$$E = h\mu$$

where $h = 6.626 \times 10^{-34}$ joule-seconds. Consequently, a photon carries a quantum of *electromagnetic energy*.

quantum discontinuity The discontinuous absorption or emission of *energy* associated with a *quantum jump.*

quantum electrodynamics A study of *electromagnetic* interactions based on relativistic quantum mechanical theory.
See: quantum theory

quantum electronics The fields of *electronics* involving the application of *quantum mechanics* to the production and amplification of power in solid crystals (generally at frequencies from 10^9 to 10^{11} hertz).

quantum jump The transition from one fixed or stationary *atomic* or *molecular* state to another, accompanied by the absorption of emission of energy ("*quanta*").

quantum mechanics The mathematically-sophisticated physical theory that emerged from Planck's original *quantum theory*. It deals with the mechanics of atomic systems (and related physical systems), using quantities that can be measured. Quantum mechanics has developed in several, essentially equivalent, mathematical formulations: wave mechanics, matrix mechanics, and relativistic quantum mechanics.

De Broglie's theory (Louis de Broglie—a French physicist [presented in 1924]), that a *particle* can also be regarded as a *wave,* was the founding concept for "wave mechanics." It is based on the *Schrodinger wave equation* which describes the wave properties of matter and it relates a system's energy to a *wave function.* Generally, it has been observed that the stable or stationary states of a system (e.g., an *atom* or *molecule*) can only possess certain "allowed" wave functions (called eigenfunctions) and certain "allowed" values of energy (called eigenvalues).

Another form of quantum mechanics, called matrix mechanics, was formulated by Max Born (a German-born British scientist) and Werner Heisenberg (a German physicist). They represented observable physical quantities such as *momentum, energy* and spatial position by matrices (see: matrix). Heisenberg's uncertainty principle—that the product of: the uncertainty in the measured value of momentum ($\triangle p$) and the uncertainty in the value of position ($\triangle x$) is of the same *order of magnitude* as *Planck's constant (h), i.e.,* $\triangle p_1 \times \triangle x \sim h$, played a key role in the development of matrix mechanics. This formulation of quantum mechanics involves the concept, therefore, that a measurement on the system disturbs the system itself, at least to some extent. Appropriate matrix elements (depicting momentum and position) are connected with the transition probabilities between various system states. In essence, however, as shown by Schrodinger (an Austrian physicist) wave mechanics and matrix mechanics are just different mathematical formulations of the same basic principles.

The final formulation of quantum mechanics, is called relativistic quantum mechanics. In 1925 Wolfgang Pauli (an Austro-Swiss physicist) postulated the *spin* of the *electron*—giving an electron in an atom a fourth *quantum number.* Then in 1928 Paul Dirac (a British physicist) extended quantum mechanic principles so that they also satisfied the principles of *relativity.* Consequently, the properties of spin could be obtained from the relativistic Schrodinger equation.

quantum theory   The theory, first stated in 1900 by the German physicist Max Planck, that all *electromagnetic radiation* is emitted and absorbed in *"quanta"* or discrete energy packets versus continuously. Each quantum of energy has a magnitude hμ, where h is the Planck constant [a quantity having the dimensions of energy x time, h = 6.626 × 10⁻³⁴ joule-seconds] and μ is the *frequency.* That is,

$$E = h\mu$$

In his theory Planck postulated that an *oscillation* could change its energy only in discrete packets or an integral number of quanta [i.e., by hμ, 2hμ, 3hμ....nhμ]. This quantum of energy concept was used by Albert Einstein in 1905 to explain the *photoelectric effect,* assuming that light was radiated in quanta or *photons.* In 1913, the Danish physicist, Niels Bohr, developed his theory of atomic *spectra,* by assuming that an *atom* can only exist in certain energy states and that photons (light) are emitted or absorbved when the atom changes from one energy state to another. Bohr proposed that the *angular momentum* of an orbiting *electron* can only assume or equal an integral number of units (i.e., nh/2π where n = 0, 1, 2, 3....and h is the Planck constant). In quantum theory, therefore, physical quantities are "quantized"—that is, certain restrictions are imposed on these physical quantities restricting their values to a number of discrete values. Quantum changes in energy are, subsequently, small and observable, only on the atomic scale. Classical mechanics, however, can adequately describe the behavior of large scale systems or oscillations. The original quantum theory applied "quantum of energy" concepts to problems in classical mechanics. More precise developments and refinements, involving the formulation of a new system of mechanics, resulted in the creation of *quantum mechanics* in the 1920s.

Finally, the extension of quantum mechanics to include Einstein's *special theory of relativity* resulted in the formulation of "relativistic quantum mechanics."

quartz [Symbol: SiO₂]   Commonly crystalline; fused to form a transparent glasslike substance (silica) with a melting point of about 1973 K (1700°C).

Quasi-stellar objects [abbr. QSO]   Stars that are powerful radio sources. Many are not visible and have been detected with *radio telescopes.* Visible *quasars* exhibit extreme *red shifts* in their *spectra.*

quench   To limit or stop the electrical discharge in an *ionization* detector.

quincunx   Geometrical pattern on which four items are spaced equally around a central item.

R

---

rack and pinion   Mechanical linkage for operation of a rotary valve in which the actuator shaft incorporates a straight sided rack to drive a pinion gear attached to the rotary shaft.

rad   (Acronym for radiation absorbed dose.) The basic unit of absorbed dose of ionizing radiation. A dose of one rad means the absorption of 100 ergs of radiation energy per gram [0.01 *joule* per kilogram] of absorbing material. Compare this term with rem and roentgen.
See: absorbed dose

radar   (radio detection and ranging) A transmitter that sends a radio pulse toward an object and measures the time interval until the reflected (echo) pulse comes back. The time interval gives the range (distance) of the object.

radar altimeter [RA or R/A]
See: Orbiter avionics system; Orbiter guidance, navigation and control system.

radar altitude   The altitude of an *aircraft, aerospace vehicle* or *spacecraft* as determined by a *radio altimeter*; essentially, the actual distance from the nearest terrain feature.

radar range 1. The distance from a *radar* to a target as measured by the radar.
2. The maximum distance at which a radar set is effective in detecting targets.

radar range equation The relation between the maximum range R_{max} at which a point target is detectable and the properties of the radar and the target

$$R_{max} = [(PA^2\lambda^2\sigma)/(4\pi)^3 \int_{min})]_{1/4}$$

where P is the transmitted power of the radar; λ is its wavelength; σ is the scattering cross section of the target; A is the antenna gain and χ_{min} is the threshold signal.
2. The maximum distance at which a radar set is effective in detecting targets.

radar reflectivity The measure of the efficiency of a *radar target* in intercepting and returning a *radar signal*. It depends upon the size, shape, aspect and the dialectric properties at the surface of the target. It includes the effects of not only reflection but also scattering and diffraction.

radar reflector A device capable of or intended for reflecting radar signals.

radial-burning Term applied to a *solid-propellant grain* that burns in the *radial* direction, either outwardly (e.g., an internal-burning grain) or inwardly (e.g., an internal-external burning tube or rod and tube).

radial equilibrium Flow in an annular passage in which there is no *radial velocity* component; i.e., the *fluid* pressure forces in the radial direction are in equilibrium with the *centrifugal* forces.

radial motion Motion along a radius.

radial velocity 1. Movement of a *celestial body* toward or away from an observer; it is considered positive if receding, and negative if approaching.
2. In general, the *velocity* at which two objects (e.g., two *spacecraft*) approach or recede from one another.
3. In radar applications, that *vector* component of the velocity of a moving *target* that is directed away from or toward the ground station.

radian A unit of angle (and a supplementary *SI unit*). One radian is the angle subtended at the the center of a circle by an arc equal in length to a radius of the circle.

$$\text{one radian} = \frac{360°}{2\pi} \approx 57° 17'45''$$

$$\approx 57.296°$$

$$2\pi \text{ radius} = 360°$$

radiance In *radiometry,* a measure of the intrinsic radiant *intensity* emitted by a *radiator* in a given direction. It is the irradiance (radiant flux density) produced by radiation from the source upon a unit surface area oriented normal to the line between source and receiver, divided by the solid angle subtended by the source at the receiving surface. It is assumed that the medium between the radiator and receiver is perfectly transparent; therefore, radiance is independent of attenuation between source and received. Radiance is measured in watts per *steradian* per square meter.

radiant 1. Pertaining to the emission or the measurement of electromagnetic radiation. Compare this term with *luminous.*
2. In astronomy, the apparent location on the *celestial sphere* of the origin of the luminous *trajectories* of *meteors* seen during a *meteor shower.*

For convenience, the common meteor showers are named for the *constellations* of *stars* in which their radiants appear.
3. In describing auroras, a projected point of intersection of lines drawn coincident with auroral streamers; that is, the point from which the *aurora* seems to originate.

radiant energy 1. The *energy* of any type of *electromagnetic radiation.* Also called radiation.

radiant energy density The instantaneous amount of *radiant energy* contained in a unit volume of propagating medium.

radiant energy thermometer An instrument which determines the *black-body temperature* of a substance by measuring its thermal *radiation.*

The substance need not be thermally black over the whole *spectrum,* since it is possible to limit the measurement to those frequencies where it is black.

radiant flux [Symbol: I] The rate of flow of *radiant energy*; the total power emitted or received by a body in the form of *electro-*

magnetic radiation. Typically measured in watts.

radiant temperature The *temperature* obtained by use of a total radiation *pyrometer* when sighted upon a nonblack body.

This is always less than the true temperature.

radiation The emission and propagation of *energy* through matter or space by means of *electromagnetic disturbances* which display both wave-like and particle-like behavior; in this context the "particles" are known as *photons.* Also, the energy so propagated. The term has been extended to include streams of fast-moving *particles (alph* and *beta* particles, free *neutrons, cosmic radiation,* etc.). Nuclear radiation is that emitted from atomic *nuclei* in various nuclear reactions, including alpha, beta, and gamma radiation and neutrons.
See: electromagnetic, radiation, ionizing radiation, quantum

radiation accidents Accidents resulting in the spread of *radioactive* material or in the exposure of individuals to *radiation.*

radiation burn Radiation damage to the skin. Betz burns result from skin contact with or exposure to emitters of *beta particles.* Flash burns result from sudden thermal radiation.

radiation chemistry The branch of chemistry that is concerned with the chemical effects, including decomposition, of energetic *radiation* or particles on matter. Compare this term with *radiochemistry*

radiation constants
See: Planck's Law

radiation cooling Cooling of a *combustion chamber* or *nozzle* in which *heat* loss by *radiation* balances heat gained from the combustion products, and the chamber or nozzle wall thereby operates in *thermal equilibrium.*

radiation damage A general term for the harmful effects of *radiation* on matter.

radiation dectection instruments Devices that detect and record the characteristics of *ionizing radiation.*

radiation dose The amount of *radiation* absorbed by a material, system or living tissue in a given amount of time.
See: rad

radiation dosimetry The measurement of

the amount of *radiation* delivered to a specific place or the amount of radiation that was absorbed there.

radiation illness An acute organic disorder that follows exposure to relatively severe doses of *ionizing radiation.* It is characterized by nausea, vomiting, diarrhea, blood cell changes, and in later stages by hemorrhage and loss of hair.

radiation laws 1. The four physical laws that together, fundamentally describe the behavior of *black-body radiation:* (a) the *Kirchoff law* is essentialy a *thermodynamic* relationship between emission and absorption of any given *wavelength* at a given temperature; (b) the *Planck law* describes the variation of intensity of a black-body radiation at a given temperature, as a function of wavelength; (c) the *Stefan-Boltzmann law* relates the time rate of *radiant energy* emission from a black body to its absolute temperature; (d) the *Wien law* relates the wavelength of maximum intensity emitted by a black body to its absolute temperature.
2. All the more inclusive assemblage of empirical and theoretical laws describing all manifestations of radiative phenomena.

radiation medicine That branch of medicine dealing with the effect of *radiation,* specifically high-energy radiation such as X-rays, gamma rays, and energetic *particles* on the body and with the prevention or cure of pysiological injuries resulting from such radiation.

radiation monitoring Continuous or periodic determination of the amount of *radiation* present in a given area.

radiation pressure *Pressure* exerted upon a body or surface exposed to *electromagnetic radiation.* The magnitude of any radiation pressure effect is directly proportional to the *radiant energy density,* and is very small by most standards. However, radiation pressure has a perceptible effect on the *orbit* of *spacecraft* and *satellites,* especially those with a large reflecting surface, such as *Echo.* For *diffuse* radiation, the total *normal* radiation pressure exerted on the body or surface is $E_d/3$, where Ed is the total energy density of the incident radiation.

radiation shielding Reduction of *radiation*

by interposing a shield of absorbing material between any radioactive source and a person, laboratory area, or radiation-sensitive device.

See: absorber, shield

radiation sickness
See: radiation illness

radiation source Usually a man-made, sealed source of *radioactivity* used in radiotherapy, radiography, as a power source for batteries, or in various types of industrial gauges. Machines such as *accelerators,* and radioisotopic generators and natural radionuclides may also be considered as sources.

radiation standards Exposure standards, permissible concentrations, rules for safe handling, regulations for transportation, regulations for industrial control of radiation, and control of radiation, and control of radiation exposure by legislative means.

radiation sterilization Use of radiation to cause a plant or animal to become sterile, that is, incapable of reproduction. Also the use of radiation to kill all forms of life (especially bacteria) in food, surgical sutures,etc.

radiative capture A nuclear capture process whose prompt result is emission of *electromagnetic radiation* only, as when a *nucleus* captures a *neutron* and emits *gamma rays.*

radiator 1. Any source of *radiant energy,* especially *electromagnetic radiation.*
2. A device that dissipates the heat from something, not necessarily by radiation only.
See: orbiter structure

radiators (Orbiter)
See: Orbiter structure

radio 1. A prefix denoting *radioactivity* or a relationship to it, or a relationship to radiation.
2. Communication by *electromagnetic waves,* without a connecting wire.
3. Pertaining to *radiofrequency,* as in radio wave.

radioactive Exhibiting *radioactivity* or pertaining to radioactivity.

radioactive cloud A mass of air and vapor in the atmosphere carrying radioactive debris from a nuclear explosion.
See: atomic cloud

radioactive contamination 1. Deposition of radioactive material in any place where it may harm persons, spoil experiments, or make products or equipment unsuitable or unsafe for some specific use.
2. The presence of unwanted radioactive matter.
3. Radioactive material found on the walls of vessels in used-fuel processing plants, or radioactive material that has leaked into a reactor coolant. Often referred to only as contamination.

radioactive series A succession of *nuclides,* each of which transforms by *radioactive* disintegration into the next until a *stable nuclide* results. The first member is called the *parent,* the intermediate members are called *daughters,* and the final stable member is called the end product.
See: decay, radioactive

radioactive source A radiation source.

radioactive standard A sample of radioactive material, usually with a long *half-life,* in which the number and type of radioactive *atoms* at a definite reference time is known. These are used in calibrating radiation measuring equipment or for comparing measurements in different laboratories. Compare this term with *radiation source.*

radioactive tracer A small quantity of *radioactive isotope* (either with carrier or carrier-free) used to follow biological, chemical or other processes, by detection, determination or localization of the radioactivity.

radioactivity The spontaneous decay or disintegration of an unstable atomic *nucleus,* usually accompanied by the emission of *ionizing radiation,* such as *alphaparticles, beta particles* and *gamma rays.* The radioactivity, frequently shortened to "activity," of natural and artificial radioisotopes decreases exponentially with time, governed by the fundamental relation:

$$N = N_0\, e\text{-}\lambda t$$

where N is the number of radionuclides at time (t), N_0 is the number of radionuclides at the start of the count (i.e., at $t = 0$), λ is the decay constant, t is time.

The decay constant (λ) is related to the *half-life* ($T_{1/2}$) of the radioisotope by the equation:

$$\lambda = \frac{\ln^2}{T_{1/2}} = \frac{0.69315}{T_{1/2}}$$

The half-lives for different radioisotopes vary in value from 10^{-8} seconds to 10^{10} years.

Natural radioactivity is simply the spontaneous disintegration of nautrally occurring *radioisotopes*. Radioisotopes which do not normally occur in nature, but are made in *nuclear reactors* or *accelerators* are called artificial radioactivity.

radiobiology   The body of knowledge and the study of the principles, mechanisms and effects of ionizing *radiation* on living matter.

radiochemistry   The body of knowlege and the study of the chemical properties and reactions of *radioactive materials*. Compare this term with *radiation chemistry*.

radioecology   The body of knowledge and the study of the effects of *radiation* on species of plants and animals in natural communities.

radioelement   An element containing one or more *radioactive isotopes*; a radioactive element.

radio energy   *Electromagnetic radiation* of a *wavelength* greater than about 1000 micrometers (10^{-6} meters) (0.01 centimeter). The high-frequency end of the radio-energy spectrum is known as *microwave radiation*.

radio frequency (RF)   In general, a *frequency* at which coherent *electromagnetic radiation* of energy is useful for communication purposes; specifically, a frequency above approximately 10^4 *hertz* and below approximately 3×10^{12} hertz.

radiogenic   Of radioactive origin; produced by radioactive transformation.
See: decay, radioactive, transmutation

radiography   The use of *ionizing radiation* for the production of shadow images on a photographic emulsion. Some of the rays (*gamma rays* or *X-rays*) pass through the subject, while others are partially or completely absorbed by the more opaque parts of the subject and thus cast a shadow on the photographic film.

radioisotope   A *radioactive isotope*. An unstable isotope of an element that decays or disintegrates spontaneously, emitting *radiation*. More than 1300 natural and arti-

ficial radioisotopes have been identified.
See: decay, radioactive

radiology   The science which deals with the use of all forms of *ionizing radiation* in the diagnosis and the treatment of disease.

radioluminescence   Visible light caused by *radiations* from *radioactive* substances; an example is the glow from luminous paint containing radium and crystals of zinc sulfide, which give off light when struck by alpha particles from the radium.
See: luminescence

radiolysis   The dissociation (or decomposition) of *molecules* by *radiation*. Example: A small proportion of water in a *nuclear reactor core* disociates into hydrogen and oxygen during operation of the reactor.

radio meteor   A *meteor* that has been detected by the reflection of a radio signal from the meteor trail of relatively high *ion* density (ion column).

Such an ion column is left behind a meteoroid when it reaches the region of the *upper atmosphere* between about 80 and 120 kilometers, although occasionally radio meteors are detected at higher altitudes. The maximum reflection occurs when the column is perpendicular to the line of the transmitter-receiver.

radiometry   The science of measurement of *radiant energy*.

radiomutation   A permanent, transmissable change in form, quality or other characteristic of a cell or offspring from the characteristics of its parent, due to *radiation* exposure.

radionuclide   A *radioactive nuclide*.

radioresistance   A relative resistance of cells, tissues, organs, or organisms to the injurious action of *radiation*. Compare this term with *radiosensitivity*.

radiosensitivity   A relative susceptibility of cells, tissues, organs or organisms to the injurious action of *radiation*. Compare this term with *radioresistance*.

radiosonde   A balloon-borne device, used for the simultaneous measurement and transmission of meteorological data. It consists of a radio transmitter and a meteorograph (a compact instrument package that records *temperature, pressure, relative humidity*, etc.). The device is carried aloft into the *Earth's atmosphere* by a balloon and subsequently transmits radio signals containing appropriate meteorological data.

radiospectrum
See: radiofrequency

radiotherapy   Radiation therapy.

radio waves   *Electromagnetic waves* of *wavelength* between 1 millimeter and several thousand kilometers and frequencies between 300 gigahertz and a few kilohertz. The higher frequencies are used for *spacecraft* communications.
See: radio frequency

radium [Symbol: Ra]   A radioactive metallic element with atomic number 88. As found in nature, the most common isotope has an atomic weight of 226. It occurs in minute quantities associated with *uranium* in pitch blend, carnotite and other minerals; the uranium decays to radium in a series of *alpha* and *beta* emissions. By virtue of being an alpha- and gamma- emitter, radium is used as a source of *luminescence* and as a *radiation source* in medicine and *radiography*.

ram air   Air entering an airscoop or air inlet as a result of the high-speed forward movement of a vehicle.

ramjet (engine)   A reaction propulsion (jet) engine that has a specially shaped tube or *duct* open at both ends into which *fuel* is fed at a controlled rate. The air needed for combustion is shoved or "rammed" into the duct and compressed by the forward motion of the vehicle/engine assembly. This "rammed" air passes through a *diffuser* and is mixed with *fuel* and burned. The combustion products are then expanded in a *nozzle*. The ramjet engine cannot operate under static conditions. The duct geometry depends on whether or not the ramjet is to operate at *supersonic* velocities.

ramping   Opening or closing of a valve at a controlled rate to achieve a desired flow-vs-time relation.

random   Eluding precise prediction, completely irregular.

In connection with probability and statistics, the term random implies collective or long-run regularity; thus a long record of the behavior of a random phenomenon presumably gives a fair indication of its general behavior in another long record, although the individual observations have no discernible system of progression. Compare this term with *stochastic.*

random noise   An *oscillation* whose instantaneous *amplitudes* occur, as a function of time, according to a normal (Gaussian) curve. Also called Gaussian noise, random Gaussian noise.

random vibration   Vibration characterized by a wide continuous band of multiple *frequencies.*

range   1. The difference between the maximum and minimum of a given set of numbers; in a *periodic* process it is twice the amplitude, i.e., the wave height.
2. The distance between two objects, usually an observation point and an object under observation.
3. A maximum distance attributable to some process, as in visual range or the range of a rocket.
4. An area in and over which rockets are fired for testing, as for example the Eastern Test Range.

range rate   The rate at which the distance from the measuring equipment to the target or signal source being tracked is changing with respect to time.

rare earths   A group of 15 chemically similar metallic elements, including Elements 57 through 71 on the *Periodic Table* of the *Elements,* also known as the *Lanthanide series.*

rarefied gas dynamics   The study of the phenomena related to the molecular or non-molecular nature of gas flow at densities where

$$\lambda/1 > 0.01$$

when λ is molecular mean free path and 1 is a characteristic dimension of the flow field.

Flow with $\lambda/1 > 0.01$ is called molecular flow.
Flow with $\lambda/1 < 0.01$ is called continuum flow.
Flow with $\lambda/1 \approx 0.01$ to .1 is called slip flow.
Flow with $\lambda/1 \approx 0.1$ to 10 is called transition flow.
Flow with $\lambda/1 > 10$ is called free molecule flow.

Slip flow and transition flow are not always distinguished from each other. The value 1 is sometimes used instead of 10 as the boundary value for transition flow and free molecule flow.

raster   The pattern followed by the *electron-beam* exploring element scaning the screen of a television transmitter or receiver.

raster line   One line of a *raster,* or scanning pattern.

rate gyro   A *single-degree-of-freedom gyro* having primarily elastic restraint of its spin axis about the output axis. In this gyro an output signal is produced by *gimbal* angular displacement, relative to the base, that is proportional to the angular rate of the base about the input *axis.*

rate integrating gyro   a *single-degree-of-freedom gyro* having primarily viscous restraint of its spin axis about the output axis. In this gyro an output signal is produced by *gimbal* angular displacement, relative to the base, that is proportional to the integral of the singular rate of the base aobut the input axis.

rate of decay   1. Of a sound, the time rate at which the sound pressure level (or other stated characteristic) decreases at a given point and at a given time. A commonly used unit is the decibel per second.
2. Of a *radioactive nuclide,* the number of nuclei of that nuclide changing (or disintegrating) per unit of time. It is usually expressed as the instantaneous rate of decay by $=dN/dt$ where N is the total number of the state nuclides present at a given time t. *See:* radioactivity.

raw data   Data that is in a form ready for processing.

rawin   A measurement of wind direction and speed at *altitude* by *radar* tracking of a balloon-borne target.

rawinsonde   A combination of *raob* and *rawin;* an observation of *temperature, pressure,* relative humidity, and winds-aloft by means of *radiosonde* and radio direction finding equipment or *radar* tracking.

ray   1. An elemental path of radiated *energy;* or the energy following this path. It is perpendicular to the *phase fronts* of the *radiation.*
See: incident ray, reflected ray, refracted ray.
2. One of a series of lines diverging from a common point, as radii from the center of a circle.
3. A long, narrow, light-colored streak on the lunar surface originating from a *crater.* Rays range in length to over 150 kilometers and usually several radiate from the same crater, like spokes of a wheel.

Rayleigh atmosphere   An idealized *atmosphere* consisting of only those particles, such as *molecules,* that are smaller than about one-tenth the *wavelength* of all *radiation* incident upon that atmosphere. In such an atmosphere, simple *Rayleigh scattering* would prevail.

This model atmosphere is amenable to reasonably complete theoretical treatment, and therefore has often served as a useful starting point in descriptions of the opitcal properties of actual atmospheres. The *polarization* of skylight, for example, exhibits almost none of the complexities found in the real atmosphere.

Rayleigh flow   Steady frictionless flow in a constant area *duct* with heat being added or removed.

Rayleigh Number [Symbol: N_{Ra}]   The nondimensional ratio between the product of *buoyancy forces* and heat *advection* and the product of *viscous* forces and heat *conduction* in a *fluid.* It can be expressed as:

$$N_{Ra} = \frac{\text{Gravity}}{\text{Thermal diffusivity}}$$
$$= \frac{C_p \rho^2 g_1^3 \beta \, \Delta T}{\mu k}$$

where Cp is the specific heat at constant pressure, ρ is the (mass) density, g is the acceleration of gravity, l is the characteristic length, β is the volume expansion coefficient wiht temperature, ΔT is the temperature difference, μ is the absolute viscosity, κ is the termal (heat) conductivity. The Rayleigh Number equals the product of the *Grashof* and *Prandtl* numbers. It is a critical parameter in the theory of thermal instability.

Rayleigh scattering
See: scattering

reaction   1. Term in *pump* and *turbine* design for the ratio of static *headrise* in the rotor to static headrise in the stage.
2. Response of vehicle to the *thrust* of the vehicle engines.

3. Chemical activity between substances (e.g., propellant and contacting surfaces). *See:* Newton's Laws

reaction-control jets Small propulsion units on a *spacecraft* used to rotate it or to *accelerate* it in a specific direction.

reaction engine An engine that develops *thrust* by its reaction to ejection of a substance from it; specifically, such an engine that ejects a jet or stream of gases created by the burning of *fuel* within the *engine*.

A reaction engine operates in accordance with Newton third law of motion, i.e., to every action (force) there is an equal and opposite reaction. Both *rocket engines* and *jet engines* are reaction engines. Also called reaction motor.

reaction propulsion *Propulsion* by reaction to a *jet* or jets ejected from one or more *reaction engines*.

reaction time In *human engineering*, the interval between an input signal (physiological) or a stimulus (psychophysiological) and the response elicited by the signal.

reaction turbine A type of *turbine* having rotary blades shaped so that they form a ring of *nozzles*, the turbine being rotated by the reaction of the *fluid* ejected from between the blades. Compare this term with *impulse turbine*.

reactivity A measure of the departure of a *nuclear reactor* from *critically*. It is about equal to the effective multiplication factor (keff) minus one and is thus precisely zero at criticality. If there is excess reactivity (positive reactivity), the reactor is *supercritical,* and its power will rise. Negative reactivity (*subcriticality*) will result in a decreasing power level.

reactor
See: nuclear reactor

reactor core In a *nuclear reactor* the region containing the *fissionable material*.

readout 1. The action of a *transmitter* sending data either instantaneously with the acquisition of the data or by playing of a magnetic tape upon which the data have been recorded.
2. The data transmitted by the action described in sense 1.
3. In *computer* operations, to extract information from storage.

readout station A recording or receiving radio station at which data are received from a *transmitter* in a *probe, satellite* or other *spacecraft*.

real time Time in which reporting on or recording of an event is simultaneous with the events; essentially, "as it happens."

real-time data Data presented in usable form at essentially the same time the event occurs.

receiver 1. The initial component or sensing element of a measuring system.
2. An instrument used to detect the presence of and determine the information carried by *electro-magnetic radiation*. A receiver includes circuits designed to detect, amplify, rectify and shape the incoming *radiofrequency* signals received at the *antenna* in such a manner that the information-containing component of this received energy can be delivered to the desired indicating or recording equipment.

receiver gain
See: gain

reciprocating engine An *engine*, especially an internal combustion engine, in which a piston or pistons moving back and forth work upon a crankshaft or other device to create rotational movement.

recognition The psychological process in which an observer so interprets the visual stimuli he receives from a distant object that he forms a correct conclusion as to the exact nature of that object.

Recognition is a more subtle phenomenon than the antecedent step of detection, for the latter involves only the simpler process of interpreting visual stimuli to the extent of concluding that an object is present at some distance from the observer.

recombination The process by which a positive and a negative *ion* join to form a neutral *molecule* or other neutral *particle*, also a process by which radicals or dissociations species join to form molecules.

Recombination is applied both to the simple case of capture of free electrons by positive atomic or molecular ions, and also to the more complex case of neutralization of a positive small ion by a negative small ion or a similar (but much more rare) neutralization of large ions.

Recombination is, in general, a process accompanied by emission of *radiation*. The light emitted from the channel of a lightning stroke is recombination radiation as is *airglow*. The much less concentrated recombinations steadily occurring in all parts of the atmosphere where ions are forming and disappearing does not yield observable radiation.

The rate at which electrons, small ions, and large ions recombine is a function of their respective mobilities and of their concentrations.

recombination coefficient   A measure of the specific rate at which oppositely charged *ions* join to form neutral particles (a measure of ion *recombination*).

recombination energy   The energy released as heat or light when two oppositely charged *ions* join to form a neutral *atom* or *molecule,* or two dissociated atoms combine to form a stable molecule.

reconnaissance satellites
See: U.S. Air Force role in space

recoverable   Of a *rocket* vehicle or one of its parts, so designed or equipped as to be located after flight and recovered with or without damage.
See: solid rocket boosters

recovery   The procedure or action that occurs when the whole of a *satellite,* or a satellite instrumentation package, or other part of a *rocket vehicle* is recovered after a launch; the result of this procedure.
See: Space Shuttle mission profiles

recovery vessels
See: Solid Rocket Booster retrieval and processing

recrystallization   1. In metals, the change from one crystal structure to another, as occurs on heating or cooling through a critical *temperature.*
2. The formation of a new strain-free grain structure from that existing in *cold-worked* metal, usually accomplished by heating.

rectifier   1. A device for obtaining a pure sample of a substance from a mixture (e.g., krypton from liquid air). It depends on the difference in the boiling points of the various constituents to achieve separation.
2. In electrical engineering, a device that converts alternating current to direct current, by permitting *current* flow in one direction only.

recurring training   The process of maintain-

ing or updating the knowledge and skills a person needs for *certification.* The courses used in this type of training are *(Shuttle) flight-independent.*

recycle   In a *countdown:* To stop the count and to return to an earlier point in the countdown, as in "we have recycled, now at T minus 2 hours counting."

recycling   The reuse of *fissionable material,* after it has been recovered by chemical processing from spent or depleted reactor fuel, reenriched and then refabricated into new fuel elements.

red fuming nitric acid [abbr. RFNA]   Concentrated nitric acid (HNO_3) in which nitrogen dioxide (NO_2) has been dissolved. It is used as an *oxidizer* in *liquid propellant rockets.*

redline   1. Term denoting a critical value for a *parameter* or a condition that, if exceeded, threatens the integrity of a system, performance of a vehicle, or success of a mission.
2. To establish a critical value as described in 1.

redundancy (of design)   The existence of more than one means for accomplishing a given task, where all means must fail before there is an overall failure to the system. Parallel redundancy applies to systems where both means are working at the same time to accomplish the task, and either of the systems is capable of handling the job itself in case of failure of the other system. Standby redundancy applies to a system where there is an alternative means of accomplishing the task that is switched in by a malfunction sensing device when the primary system fails.

redundant   Incorporating multiple identical components to achieve increased *reliability.*

redundant design   Design in which more than one unit is available for the performance of a given function, so that failure of a unit will not cause failure of the system or *abort* the mission.

Reech number   A *nondimensionl* number expressing the ratio of *gravity* force to *inertia* force, that is,

$$\frac{\text{Reech}}{\text{Number}} = \frac{\text{gravity force}}{\text{inertia force}} = \frac{gl}{V^2}$$

where g is the acceleration of gravity, l is

the characteristic length, V is the velocity.
It is the reciprocal of the *Froude number.*

reefing line A rope used to restrict the deployment of a parachute.

reentry The event occurring when a *spacecraft, aerospace vehicle* or other object comes back into the sensible atmosphere after being rocketed to higher altitudes; the action involved in this event. Also called entry.

reentry body That part of a *space vehicle* that reenters the *atmosphere* after flight above the *sensible atmosphere.*

reentry nose cone A *nose cone* designed especially for *reentry,* consisting of one or more chambers protected by an outer shield.

reentry trajectory That part of a *rocket*'s *trajectory* that begins at *reentry* and ends at target or at the surface of the *Earth.*

If the rocket is unguided at reentry, its reentry trajectory is *ballistic* in character.

reentry window The area at the limits of the *Earth*'s *atmosphere* through which a *spacecraft* or *aerospace vehicle* in a given *trajectory* can pass to accomplish a successful *reentry.*

reference signal In *telemetry,* the signal against which data-carrying signals are compared to measure differences in time, phase, *frequency,* etc.

reflectance [symbol ρ] The ratio of the radiant *flux* reflected by a body to that incident upon it.

For an *opaque* body, the sum of the reflectance and the absorptance for the incident radiation is unity $\rho + \alpha = 1$.

reflected wave 1. A *shock wave, expansion wave* or *compression wave* reflected by another wave incident upon a wall or other boundary.
2. In electronics, a *radio wave* reflected from a surface or object.

reflecting telescope A *telescope* which collects light by means of a concave mirror.

reflection The process whereby a surface of discontinuity turns back a portion of the incident *radiation* into the medium through which the radiation approached.
See: albedo, reflectivity

When the scale of the irregularities on the reflecting surface is small compared to the *wavelength,* specular reflection (also called

mirror reflection) results; if the irregularities are large compared to the wavelength, diffuse reflection occurs. The process of reflection is not affected by wavelength except as the relative scale of the irregularities of the surface change with wavelength. The fraction of the incident radiation reflected does depend on wavelength because of the selective nature of the *absorptivity* and *transmissivity.* The idealized white body is a total reflector: a *black body* reflects none of the incident radiation.

The laws of specular reflection are: (first law) the reflected ray lies in the same plane as the incident ray and the normal to the surface at the point of incidence; and (second law) the angle of reflection equals the angle of incidence, both measured from the normal to the surface.

reflectivity 1. A measure of the fraction of *radiation* reflected by a given surface; defined as the ratio of the *radiant energy* reflected to the total that is incident upon that surface.

The reflectivity of any given substance is, in general, a variable strongly dependent upon the *wavelength* of the incident radiation. The reflectivity of a given surface for a specified broad spectral range, such as the visible spectrum or the solar spectrum, is referred to as the *albedo.*
2. In *thermal radiation,* a property of a material, measured as the *reflectance.*

reflector 1. In general, any object that reflects incident *energy*; usually it is a device designed for specific reflection characteristics.
2. In an *antenna,* a parasitic element located in a direction other than the general direction of the major lobe of radiation.
3. A layer of material immediately surrounding a *nuclear reactor core* which scatters back or reflects into the core many *neutrons* that would otherwise escape. The returned neutrons can then cause more *fissions* and improve the neutron economy of the reactor. Common reflector materials are graphite, beryllium and natural uranium.

refraction The bending of *electromagnetic* energy, such as light and *radio waves,* at the point where the medium through which they are passing experiences changes in density, composition or other properties. For a change in medium density the electromagnetic *ray* undergoes a smooth bending over a finite distance. At a density continuity between two media, the *refractive index* changes through an interfacial layer that is thin compared to the *wavelength*; thus the refraction is abrupt, essentially discontinuous.

refractive index [Symbol: n] The ratio of c (the *velocity* of *light* in a vacuum) to the velocity of light through a transparent substance. Each such substance lowers the velocity of light slightly and by a different amount—low-density air has a different refractive index from high-density air and from water, glass, sulfuric acid, etc.
See: refraction

refracting telescope A telescope that collects light by means of a lens or system of lenses. Also called refractor.

refractory A material, usually *ceramic,* that resists the action of heat, does not fuse at high temperatures, and is very difficult to break down.

refractory metal A metal with melting point above 2200°C (4000°F). Usually refers to columbium, molybdenum, tantalum or tungsten.

regenerator A device used in a *thermodynamic* process for capturing and returning to the process heat that would otherwise be lost.

regression Erosion of the surface of a material.

regressive burning Condition in which *thrust, pressure,* or burning surface decreases with time or with *web* burned.

regulating rod A *nuclear reactor control rod* used for making frequent fine adjustments in *reactivity.* Compare this term with *shim rod.*

regulator Flow-control device that adjusts the *pressure* and controls the flow of *fluid* to meet the demands of a *liquid-propellant rocket* system.

reheating The addition of heat to a *working fluid* in an engine after a partial expansion.

reimbursement schedule The prelaunch timetable on which a *Space Transportation System user* pays NASA the costs associated with his or her *Shuttle mission.*

reinforced carbon-carbon (RCC) material

See: Orbiter thermal protection system

relative biological effectiveness [abbr. RBE] A factor used to compare the biological effectiveness of different types of *ionizing radiation.* It is the inverse ratio of the amount of absorbed radiation, required to produce a given effect, to a standard (or reference) radiation required to produce the same effect.
See: absorbed dose, rad, rem

relative humidity (symbol ϕ) This can be defined as (1) the ratio of the actual mass of vapor (m_v) to the mass of vapor required to produce a *saturated mixture* (m_{sat}) at the same *temperature* (T); (2) the ratio of the actual *partial pressure* of the vapor (p_v) in a mixture to the *saturation pressure* of the vapor (p_g) at the temperature of the mixture. If the vapor behaves like an *ideal* (perfect) *gas,* the relative humidity can be expressed as follows:

$$\phi = \frac{m_v}{m_{sat}} = \frac{p_v}{p_g}$$

relative temperature scale The Celsius (formerly centigrade) scale and the Fahrenheit scale are relative temperature scales.
See: absolute temperature scale

relaxation time 1. In general, the time required for a system, object or fluid to recover to a specified condition or value after disturbance.
2. Specifically, the time taken by an exponentially decaying quantity to decrease in *amplitude* by a factor of $1/e = 0.3679$.

relief valve Pressure relieving device that opens automatically when a predetermined *pressure* is reached.

rem (Acronym for roentgen equivalent man) The unit of dose of any *ionizing radiation* that produces the same biological effect as a unit of *absorbed dose* of ordinary *X-rays.* The RBE dose (in rems) = RBE × absorbed dose (in rads).
See: relative biological effectiveness

remaining body That part of a rocket or vehicle that remains after the separation of a *fallaway section* or *companion body.*

   In a *multistage rocket,* the remaining body diminishes in size successively as each section or part is cast away and successively becomes a different body.

remedial action Action to correct a non-conforming article or material.

remote control Control of an operation from a distance, especially by means of *telemetry* and electronics; a controlling switch, lever, or other device used in this kind of control; as in remote-control armament, remote-control switch, etc.

remotely piloted vehicle(s) [abbr. RPV] An *aircraft* or *aerospace vehicle* the pilot of which does not fly on board but rather controls it at a distance (i.e. remotely) from a manned aircraft, aerospace vehicle or ground station.

remote manipulator system (RMS) A 15.2-meter (50-ft) long articulating arm that is remotely controlled from the aft flight deck of the *Orbiter*. The elbow and waist movements of the RMS permit *payloads* to be grappled for deployment out of the cargo *bay* attach points or to be retrieved and secured for return to *Earth*. Because the RMS can be operated from the shirtsleeve environment of the Orbiter cabin, an extra-vehicular activity is not required.

The standard remote manipulator is mounted with its "shoulder" on the left main *longeron* for handling certain types of payloads. A television camera and lights near the outer end of the RMS arm permit the operator to see on television monitors what his "hands" are doing. Payloads will carry markings and alignment aids to help the RMS operator maneuver them. The RMS operator has a 62° field of view out the two aft windows on the Orbiter's flight deck and an 80° view through the two overhead observation windows. Three floodlights are located along each side of the payload bay.
See: Orbiter structure

rendezvous The close approach of two or more *spacecraft* in the same *orbit,* so that docking can take place. These objects meet at a preplanned location and time with essentially zero relative velocity. A rendezvous would be involved, for example, in the construction, servicing or resupply of a *space station,* or when the *Space Shuttle Orbiter* performed on-orbit repair/servicing of a satellite.
See: Space Shuttle mission profiles

rendezvous radar [RR]
See: Orbiter avionics system

rep (Acronym for roentgen equivalent physical) An obsolete unit of absorbed dose of any *ionizing radiation,* with a magnitude of 93 ergs per gram. It has been superseded by the *rad.*

repeatablity Capability of a component or assembly to operate in the same way and in the same time each time it is actuated.

repressurization Sequence of operations during the vehicle flight that utilizes an on-board *pressurant* supply to restore the *ullage pressure* to the desired level after a burn period.

research reactor A *nuclear reactor* primarily designed to supply *neutrons* or other *ionizing radiation* for experimental purposes. It may also be used for training, materials testing, and production of *radioisotopes.*

reseat pressure *Pressure* at which a flow-control device (e.g., a valve) will close and shut off flow as specified.

residual In *celestial mechanics* and *trajectory* analysis, the deviation between an observed and a computed value, usually in the sense observed minus computed.

residual hazard A hazard that is not counteracted by development or utilization of special procedures.

residual load Of a vehicle, the sum of the *payload,* all items directly associated with the payload, and other relatively fixed weights of the overall vehicle; calculated as the difference between gross weight and the sum of *propellant,* tank, structure and power-plant weights.

resilience The property of a material that enables it to return to its original shape and size after deformation; e.g., it is this property of a sealing material that makes it possible for a seal to maintain sealing pressure despite wear, misalignment, or out-of-round conditions.

resolution 1. The ability of a film, a lens, a combination of both, or an optical system to render barely distinguishable a standard pattern of black and white lines.

When the resolution is said to be 10 lines per millimeter, it means that the pattern whose line plus space width is 0.1 millimeter is barely resolved, the finer patterns are not resolved, and the coarser patterns are more clearly resolved.

2. In *radar,* the minimum angular separation at the *antenna* at which two targets can be distinguished; or the minimum range at which two targets at the same *azimuth* can be separated.

3. Of a *gyro,* a measure of response to small changes in input; the maximum value of the minimum input change that will cause a detectable change in the output for inputs greater than the threshold, expressed as a percent of one half the input range.

resonance   1. The phenomenon of amplification of a *free wave* or oscillation of a system by a forced wave or oscillation of exactly equal period. The forced wave may arise from an impressed force upon the system or from a boundary condition. The growth of the resonant amplitude is characteristically linear in time.

2. Of a system in forced oscillation, the condition which exists when any change, however small, in the frequency of excitation causes a decrease in the response of the system.

3. In nuclear physics, the phenomenon whereby *particles* such as *neutrons* exhibit a very high interaction probability with nuclei at specific kinetic energies of the particles. *Cross sections* for neutron *capture* and *scattering,* for example, exhibit peaks at these so-called resonance energies and have relatively low values between the peaks.

resonance frequency   A *frequency* at which *resonance* exists. Also called resonant frequency.

response time   In a control flow device, the interval from receipt of signal to completion of the commanded action, a total comprised of electrical delay, plus pneumatic or hydraulic control system delay plus travel time for the movable element.

restart   Specifically, the act of firing a *stage* of a *rocket* after a previous powered flight and a coast phase in a *parking orbit.*
See: orbit

rest mass [Symbol: m₀]   According to *relativistic* theory, the mass that a body or object has when it is at absolute rest. Mass increases when the body is in motion according to

$$m = m_0/\sqrt{1 - (v^2/c^2)}$$

where m is its mass in motion; m_0 is its rest mass; v is the body's speed of motion; and c is the speed of light. Newtonian physics, in contrast with relativistic physics, makes no distinction between rest mass and mass in general.

restricted propellant   A *solid propellant* having only a portion of its surface exposed for burning, the other surfaces being covered by an *inhibitor.*

restricted surface   Surface of a *solid propellant grain* that is prevented from burning by the use of *inhibitors.*

restriction inlet   Flow path of reduced cross section or an orifice through which *gas* flows into a control-pressure region; gas flows into the control-pressure region through a throttling valve positioned by the pressure sensor.

restrictor   1. In *solid-propellant rockets,* a layer of fuel containing no *oxidizer,* or of noncombustible material, adhered to the surface of the propellant so as to prevent burning in that region.

2. Discrete flow resistance in a *fluid* flow passage; usually an orifice.

reticle   A system of lines, dots, cross hairs or wires in the focus of the eyepiece of an optical instrument that serves as a reference.

retort   Vessel used in an oven or furnace to close the work being heat treated in a controlled atmosphere.

retrieval   The process of using the *remote manipulator system* and/or other handling aids to return a *captured payload* to a stowed or berthed position. No payload is considered retrieved until it is fully stowed for safe return or berthed for repair and maintenance.

retrofire   To ignite a *retrorocket.*

retrofit   The addition to or modification of a *spacecraft, aerospace vehicle,* launch vehicle, etc.....after it has become operational (i.e., been used for its intended purpose or mission).

retrograde   In reverse or backward direction.
See: retrograde motion

retrograde motion   1. Motion in an *orbit* opposite to the usual orbital direction of *celestial bodies* within a given system. Specifically, of a *satellite,* motion in a direction opposite to the direction of rotation of the *primary.*

2. The apparent motion of a *planet* westward among the stars.

retrograde orbit
See: orbit

retroreflection   *Reflection* wherein the reflected rays return along paths parallel to those of their corresponding incident rays.

retroreflector   Any instrument used to cause reflected rays to return along paths parallel to those of their corresponding incident rays. For example, three mirrors perpendicular to each other (like the inside corner of a box). These mirrors reflect any entering light ray back on itself.

retrorocket   A small *rocket engine* on a *satellite, spacecraft,* or *aerospace vehicle* used to produce a retarding *thrust* or *force* that opposes the object's forward motion. This action reduces the system's velocity.

retrothrust   *Thrust* used for a braking maneuver; a *reverse thrust.*
See: retrorocket

reusable surface insulation (RSI)
See: Orbiter thermal protection system

reverberation   1. The persistence of *sound* in an enclosed space, as a result of multiple reflections after the sound source has stopped.

2. The sound that persists in an enclosed space, as a result of repeated reflection or scattering, after the source of the sound has stopped.

reverse thrust   *Thrust* applied to a moving object in a direction to oppose the object's motion.

reverse transition   Change from a *turbulent* to a *laminar boundary layer* as a result of flow acceleration (laminarization).

revetment   A wall of concrete, earth, sandbags, or the like installed for protection, as against the blast of exploding fuel during a *rocket abort.*

revolution   1. Motion of a *celestial body* in its *orbit*; circular motion about an *axis* usually external to the body.

In some contexts, the terms revolution and rotation are used interchangeably, but, with reference to the motions of a celestial body, revolution refers to motion in an orbit or about an axis external to the body, whereas rotation refers to motion about an axis within the body. Thus, the *Earth*

revolves about the *Sun* annually and rotates about its axis daily.

2. One complete cycle of the movement of a *spacecraft* or celestial body in its orbit, or of a body about an external axis, as a revolution of the Earth about the Sun; or "revolution 17" of a *Space Shuttle flight.*

revolve   To move in a path about an *axis,* usually external to the body accomplishing the motion, as in the *planets* revolve about the *Sun.* Therefore, *revolution.*
See: rotate, orbit

Reynolds number [Symbol: Re]   A *nondimensional number* used in establishing and predicting changes in *fluid* flow regimes (i.e., *laminar* flow, transition flow, *turbulent* flow). It is defined as the ratio of *inertia forces* (momentum forces) to *viscous* forces in fluid flow. That is,

$$\text{Re} = \frac{\text{inertia force}}{\text{viscous force}} = \frac{\rho Vl}{\mu}$$

where ρ is the (mass) density, V is the fluid velocity, l is the characteristic length, μ is the absolute viscosity.

rib   A fore-and-aft structural member of an *airfoil* used for maintaining the correct covering contour and also for bearing stress.
See: Orbiter structure

ribbon parachute   A type of parachute having a canopy consisting of an arrangement of closely spaced tapes. This parachute has high porosity with attendant stability and slight opening shock.

rich   Of a combustible mixture; having a relatively high proportion of *fuel* to oxidizer; more precisely, having a value greater than *stoichiometric.*

Richardson number [Symbol: N_{Ri}]   A *nondimensional number* arising in the investigation of *shearing* flows of a stratified fluid (e.g., atmospheric shear). It is defined as the ratio of *buoyant* forces to *turbulent* force.

$$Nri = \frac{\text{buoyant force}}{\text{turbulent force}} = \frac{gl \,\Delta \rho}{2q}$$

where $\Delta \rho$ is the density difference, g is the acceleration of gravity, l is the characteristic length, q is the dynamic pressure ($\rho V^2/2$) where V is the velocity.

ring seal   Piston-ring type of seal that assumes its sealing position under the pressure of the *fluid* to be sealed.

rise off   Term denoting that a given event occurs only as a result of vehicle vertical motion from the *launch pad.*

riser   One or more straps by which a parachute harness is attached to a harness holding a person or hardware.

rise time   The time required for the leading edge of a *pulse* to rise from one-tenth of its final value to nine-tenths of its final value. Rise time is proportional to time constant.

rocket engine   The portion of the chemical propulsion system in which combustible materials (propellants) are supplied to a chamber and burned under specified conditions and the thermal energy is converted into *kinetic energy,* or *thrust,* to propel the vehicle to which the engine is attached. The term "rocket engine" usually is applied to a machine that burns *liquid propellants* and therefore requires rather complex systems of *tanks, ducts,* pumps, flow-control devices, etc.; the term "*rocket motor*" customarily is applied to a machine that burns *solid propellants* and therefore is relatively simple, requiring basically only the solid propellant grain within a case, an igniter, and a *nozzle.* In general, the term "rocket engine" is used to refer to chemical-propulsion—system engines as a class.

rocket motor
See: rocket engine

Rockwell hardness   Indentation hardness (of metals and plastics) determined by measuring surface indentation or penetration by a diamond cone or steel ball under a specified load.

rod   A relatively long, slender body of material used in or in conjunction with a *nuclear reactor.* It may contain fuel, absorber, or material in which *activation* or *transmutation* is desired.
See: control rod

roentgen [abbr. r]   A unit of exposure to *ionizing radiation.* It is that amount of *gamma* or *X-rays* required to produce ions carrying 1 electrostatic unit of electrical charge (either positive or negative) in 1 cubic centimeter of dry air under standard conditions. Named after Wilhelm Roent-

gen, German scientist who discovered X-rays in 1895. Compare this term with *curie, rad,* and *rem.*

roentgen equivalent, man
See: rem

roentgen-equivalent-physical
See: rep

roll   Rotational or oscillatory movement about the longitudinal (lengthwise) axis of a vehicle.
See: attitude

rolling element   Ball, needle, or tapered roller in a rolling-element bearing.

rolling-element bearing
See: bearing

rollout   That portion of landing an *aerodynamic* vehicle following *touchdown.*

root-mean-square error [Symbol: σ]   In statistics, the square root of the *arithmetic mean* of the squares of the deviations of the various items from the arithmetic mean of the whole. Also called *standard deviation.*

rotary seal   Mechanical seal that rotates with the shaft and is used with a stationary mating ring.

rotate   To turn about an internal *axis.* Said especially of *celestial bodies.* Hence rotation. Compare this term with *revolve.*

rotating service structure (RSS)   An environmentally controlled facility at the *launch pad* that is used for inserting *payloads* vertically into the *Orbiter cargo bay.*

rotation   1. Turning of a body about an *axis* within the body, as the daily rotation of the *Earth.*

2. One turn of a body about an internal axis, as a rotation of the Earth.

rotational speed   *Revolutions* per unit time.

rotation and translation hand controllers
See: Orbiter structure

rotor   Turbopump shaft plus all attachments that rotate with it.

rounding error   In computations, the error resulting from deleting the less significant *digits* of a quantity and applying some rule of correction to the part retained. Also called round-off error.

round off   To adjust or delete less significant *digits* from a number and possibly apply some rule of correction to the part retained.

routine   A set of (machine) instructions arranged in proper sequence enabling a *computer* to perform a desired operation, such as the solution of a mathematical problem.

row The horizontal *vector* of a *matrix.*

rumble 1. A form of *combustion instability,* especially in a *liquid-propellant rocket engine,* characterized by a low-pitched, low-frequency rumbling noise.
2. The noise made in this kind of combustion.

runaway regulator Regulator that has failed in the fully open position, the result being uncontrolled downstream pressure.

run in Period of initial operation during which the wear rate is greatest and the contact surface of mating components is developed.

S

sabot A device fitted around or in back of a *projectile* in a gun barrel or launching tube to support or protect the projectile or to prevent the escape of gas ahead of it. The sabot separates from the projectile after launching.

safe-and-arm device A built-in safety mechanism in the *Space Shuttle Solid Rocket Booster* that prevents the propellant from igniting prematurely even if the NASA standard initiators are inadvertently fired.
See: Solid Rocket Booster

safe/arm (S/A) system Mechanism in a solid-propellant igniter that in the SAFE condition physically prevents the initiating charge from propagating to the energy release system.
See: safe-and-arm device

safe operating pressure Maximum operating *pressure* allowable without using shields to protect personnel and associated hardware.

safety critical Facility, support, test, and flight systems containing:

a. Pressurized vessels, lines, and components.

b. Propellants, including cryogenics.

c. Hydraulics and pneumatics.

d. High voltages.

e. Radiation sources.

f. Ordnance and explosive devices or devices used for ordnance and explosive checkout.

g. Flammable, toxic, cryogenic, or reactive elements or compounds.

h. High temperatures.

i. Electrical equipment that operates in the area where flammable fluids or solids are located.

j. Equipment used for handling program hardware.

k. Equipment used for personnel walking and work platforms.

Used as an adjective to describe something (e.g., a function, data, equipment) that may affect the safety of ground and flight personnel, the *Space Shuttle system,* the *Orbiter, payloads,* the general public and public or private property.

safety factor
See: factor of safety

safety rod A standby *control rod* used to shut down a *nuclear reactor* rapidly in emergencies.
See: scram

safety training Instructions that alert a trainee to conditions or operations that could be significantly dangerous to the operator or to other hazards that could damage equipment or property.

safing 1. An action taken to retreat from an armed condition.
2. Actions taken to eliminate or control hazards.

saltwater switch An electric switch in which salt water acts as the conductor between two metal pins.
See: Solid Rocket Booster

Salyut A series of Soviet space stations placed in Earth orbit beginning in April 1971. The stations, smaller than spacelab, carried on a wide range of experiments of the Earth and its atmosphere the Sun and the effects of space on humans.

satellite See specific types, e.g. artificial satellite

Satellite Power System [abbr. SPS] A very large system deployed in space and designed to provide useful energy to a

Fig. 1   The basic Satellite Power System concept.   (Drawing courtesy of the Department of Energy and NASA.)

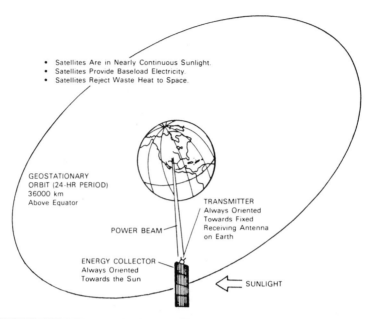

- Satellites Are in Nearly Continuous Sunlight.
- Satellites Provide Baseload Electricity.
- Satellites Reject Waste Heat to Space.

GEOSTATIONARY
ORBIT (24-HR PERIOD)
36000 km
Above Equator

TRANSMITTER
Always Oriented
Towards Fixed
Receiving Antenna
on Earth

POWER BEAM

ENERGY COLLECTOR
Always Oriented
Towards the Sun

SUNLIGHT

Fig. 3   Typical terrestrial antenna array, measuring some 8 km by 12 km (5 mi by 7.5 mi), which would transform the SPS-beamed microwaves into electricity. Microwave levels outside the antenna area and even beneath the antenna would be kept below environmental safety limits. The land under the antenna could, perhaps, also be used for agriculture.   (Photograph courtesy of the Boeing Aerospace Co.)

Fig. 6   An SPS reference system (NASA/DOE-1980).   (Drawing courtesy of NASA/DOE.)

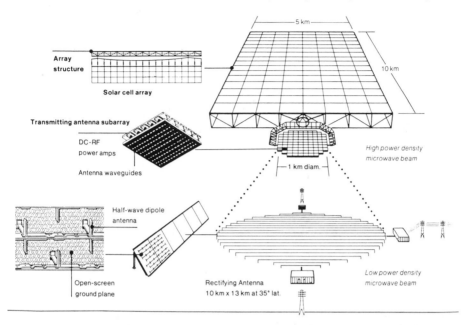

Array structure

Solar cell array

Transmitting antenna subarray

DC-RF power amps

Antenna waveguides

5 km

10 km

High power density microwave beam

1 km diam.

Half-wave dipole antenna

Open-screen ground plane

Rectifying Antenna
10 km x 13 km at 35° lat.

Low power density microwave beam

Fig. 5   Construction of a Satellite Power System assembly in low-Earth-orbit. A Space Shuttle Orbiter (upper right) docks at the facility's assembly bay.   (Photograph courtesy of the Boeing Aerospace Co.)

power grid on Earth. Each SPS is placed in geosynchronous orbit some 36,000 km above the Earth's equator where it experiences sunlight over 99% of the time and is in continuous line-of-sight contact with its ground receiving station. Electrical power is produced by photovoltaic or heat energy conversion of sunlight. This power is then converted to radio frequency energy at high efficiency, and formed into a focused beam precisely aimed at the ground station. The ground station receiving antenna reconverts the radio frequency energy into electricity for distribution in a terrestrial power grid. Construction of the SPS could use extraterrestrial materials assembled by workers at a spacebase or by using terrestrial materials placed in low-Earth orbit.

satelloid   A *vehicle* that revolves about the *Earth* or other *celestial body,* but at such *altitudes* as to require sustaining *thrust* to balance *aerodynamic drag.*

Saturn   The sixth planet from the Sun and, with a diameter of 120,000 km, the second largest in the system. Composed mainly of helium and hydrogen, the planet is so light that it would float on water if there were an ocean large enough to hold it. It takes nearly 29.5 "Earth-years" for the planet to complete a single orbit around the Sun. But a "day" lasts only 10 hours and 39 minutes. The planet is surrounded by perhaps a thousand ringlets, grouped in a series of rings. These rings are composed of countless low-density particles orbiting individually around the equator at progressive distances from the cloud tops. Analysis indicates the rings are mostly ice and frosted rock. In theory, the rings resulted either from the shattering of a moon or passing body which ventured too close to Saturn's great tidal forces, or from the incomplete formation of primordial planetary material. Saturn has 10 major satellites: Janus, Mimas, Enceladus, Tethys, Dione, Rhea, Titan, Hyperion, Iapetus and Phoebe. The Voyager probes found at least three new satellites.

S-band   A *frequency* band extending approximately from 1.55 to 5.2 *gigahertz.*

scalar   Any physical quantity whose field can be described by a single numerical value at each point in space. A scalar quantity is distinguished from a *vector* quantity by the fact that a scalar quantity possesses only *magnitude,* whereas a vector quantity possesses both magnitude and direction.

scalar product   A *scalar* equal to the product of the magnitudes of any two *vectors* and the cosine of the angle θ between their positive directions.

For two vectors A and B, read A dot B, and occasionally as (AB). If the vectors A and B have the components A_x, B_x, A_y, B_y and A_z, B_z along rectangular Cartesian x, y, and z axes, respectively, then

$$A \cdot B = A_x B_x + A_y B_2 = |A| |B| \cos \theta = AB \cos \theta$$

If a scalar product is zero, one of the vectors is zero or else the two are perpendicular. Also called dot product.

scaler   An electronic device that produces an output *pulse* whenever a prescribed number of input pulses have been received. Also called scaling circuit.

The number of input pulses per output pulse of a scaler is termed the scaling factor. A binary scaler is a scaler whose scaling factor is 2. A decade scaler is a scaler whose scaling factor is 10. It is frequently used for rapid counting of radiation-induced pulses from *geiger counters* or other radiation detectors.

scanner   A *radar* mechanism incorporating a rotatable antenna, or radiator, motor drives, mounting, etc., for directing a searching *radar beam* through space and imparting *target* information to an indicator.

scattered light   Light striking fine *aerosol* particles is reflected (scattered) in all directions. Similarly, *photons* of *resonance*-line *wavelength* are absorbed by *atoms* and reemitted in all directions.

scattering   1. *(particle)* A process that changes a particle's *trajectory.* Scattering is caused by particle collisions with *atoms, nuclei,* and other particles or by interactions with fields of *magnetic force.* If the scattered particle's *internal energy* (as contrasted with its *kinetic energy*) is unchanged by the collision, elastic scattering prevails; if there is a change in the internal energy, the process is called inelastic scattering.
2. *(photon)* Scattering may also be viewed

as the process by which small particles suspended in a medium of a different refractive index diffuse a portion of the incident *radiation* in all directions. In scattering, no energy transformation results, only a change in the spatial distribution of the radiation. Along with *absorption,* scattering is a major cause of the attenuation of radiation by the *atmosphere.* Scattering varies as a function of the ratio of the particle diameter to the *wavelength* of the radiation. When this ratio is less than about one-tenth, Rayleigh scattering occurs in which the scattering coefficient varies inversely as the fourth power of the wavelength. At larger values of the ratio of particle diameter to wavelength, the scattering varies in a complex fashion described by the Mie theory; at a ratio of the order of 10, the laws of geometric optics begin to apply.

scavenging   In chemistry, the use of a nonspecific precipitate to remove one or more undesirable *radionuclides* from solution by *absorption* or coprecipitation. In atmospheric physics, the removal of radionuclides from the *atmosphere* by the action of rain, snow or dew.

schlieren (German, streaks, striae)   1. Regions of different density in a *fluid,* especially as shown by special apparatus.
2. Pertaining to a method or apparatus for visualizing or photographing regions of varying density in a field of flow.
See: schlieren photography

schlieren photography.   A method of photography for flow patterns that takes advantage of the fact that light passing through a density gradient in a *gas* is *refracted* as though it were passing through a *prism.* Compare this term with *shadowgraph.*

scientific airlock   An opening in a manned *spacecraft* or *space station* from which *experiment* equipment can be extended outside (into outer space) while the interior of the vehicle retains its atmospheric integrity (i.e., remains pressurized).

scintillation counter   An instrument that detects and measures *ionizing radiation* by counting the light flashes (scintillations) caused by *radiation* impinging on certain materials (phosphors).

scram   The sudden shutdown of a *nuclear reactor,* usually by rapid insertion of the *safety rods.* Emergencies or deviations from normal reactor operation cause the reactor operator or automatic control equipment to scram the reactor.

screaming   A form of *combustion instability,* especially in a *liquid-propellant rocket engine,* of relatively high *frequency* and characterized by a high-pitched noise.

screeching   A form of *combustion instability,* especially in an *afterburner,* of relatively high *frequency* and characterized by a harsh, shrill noise.

search for extraterrestrial intelligence   The objective of the contemporary search is the detection of the existence of an intelligent species, which, like humans, radiates electromagnetic energy into intersteller space in a manner that enables scientists to recognize it as an artifact against the background of natural radiation. Consequently, scientists look for radio signals on a narrow band. These are markedly different from known natural sources which occupy a wide bandwidth. Communication with a species is not an immediate concern.

scrub   To cancel a scheduled rocket firing, either before or during *countdown.*

sealant   Liquid/solid mixture installed at joints and junctions of components to prevent leakage of *fluid* (esp. *gas*) from the joint or junction.

sealed cabin   The occupied space of an *aircraft, aerospace vehicle* or *spacecraft* characterized by walls that do not allow any gaseous exchange between the cabin (inner) *atmosphere* and its surroundings and containing its own mechanisms for maintenance of the cabin atmosphere.

sea-level engine   *Rocket engine* designed to operate at sea level; i.e., the *nozzle* flows full at sea-level pressure.

seal extrusion   Permanent displacement, under the action of *fluid pressure,* of part of a seal into a gap provided for such displacement.

sealing rings   Mechanical devices designed to fit together tightly when two *spacecraft* or a spacecraft and a *space station* are docked so that their cabin atmospheres will not leak out.

secondary leakage   Leakage from a *fluid-*system component to the exterior.

secondary radiation *Electromagnetic* or particulate *radiation* resulting from absorption of other radiation in matter. For example, radiation caused by *cosmic rays* impacting any material and removing electrons and *nuclei* from it.

secondary seal Seal intended to limit *secondary leakage;* e.g. a seal on a valve shaft.

second law of thermodynamics An inequality asserting that it is impossible to transfer thermal energy (*heat*) from a colder to a warmer system without the occurrence of other simultaneous changes in the two systems or in the environment.

It follows from this law that during an *adiabatic* process, *entropy* cannot decrease. For reversible adiabatic processes entropy remains constant, and for irreversible adiabatic processes it increases.

Another equivalent formulation of the law is that it is impossible to convert the heat of a system into work without the occurrence of other simultaneous changes in the system or its environment. This version, which requires an engine to have a cold sink as well as a hot source, is particularly useful in engineering applications.

Another formulation of the second law is that the change in entropy (ΔS) for an *isolated system* is greater than or equal to zero. Mathematically

$$(\Delta S)_{iso} \geq 0$$

See: first law of thermodynamics

section One of the cross-section parts that a *rocket vehicle* is divided into, each adjoining another at one or both of its ends. Usually described by a designating word, as in nose section, aft section, center section, tail section, thrust section, tank section, etc. *See:* Orbiter structure

secular Pertaining to long periods of time on typically the order of a century, as secular perturbations, secular terms.

secular perturbations Changes in the *orbit* of a *planet* or *satellite* that operate in extremely long cycles; long term *perturbations.*

secular terms In the mathematical expression of an *orbit,* terms for very long period *perturbations,* in contrast to periodic terms, terms of short period.

Seebeck effect The establishment of an *electric potential difference* tending to produce a flow of *current* in a *circuit* of two dissimilar metals the junctions of which are at different temperatures.

It is the phenomenon involved in the operation of a *thermocouple* and is named for the German scientist Thomas Seebeck, who first observed the phenomenon in 1822.

selective absorption *Absorption* that varies with the *wavelength* of *radiation* incident upon the absorbing substance.

A substance which absorbs in such fashion is called a selective absorber and is to be contrasted with an ideal *black body, white body,* or *gray body.*

selective scattering Scattering that varies with the *wavelength* of radiation incident upon the scattering *particles.* In general, the largest and most complex degree of selectivity is found for *wavelengths* nearly equal to the diameter of the scattering particles.

selenocentric 1. Relating to the center of the *Moon;* referring to the Moon as a center.
2. Orbiting about the Moon as a central body.

selenographic 1. Of or pertaining to the physical geography of the *Moon.*
2. Specifically, referring to positions on the Moon measured in *latitude* from the Moon's *equator* and in *longitude* from a reference *meridian.*

selenoid A *satellite* of the *Earth's Moon;* a *spacecraft* in *lunar* orbit.

selenology That branch of astronomy that treats of the *Moon,* its surface, motion, constitution and the like. Selene is Greek for moon.

self-cooled Term applied to a *combustion chamber* or *nozzle* in which temperature is controlled or limited by methods that do not involve flow within the wall of coolant supplied from an external source.

self-pressurization Increase of *ullage* pressure by *vaporization* or *boiloff* of contained *fluid* without the aid of additional *pressurant.*

semicircular canals Structures of the inner ear, the primary function of which is to register movement of the body (in space). They respond to change in the rate of movement, and play a key role in the mechanism of balance and orientation.

semiconductor   An electronic *conductor,* with *resistivity* in the range between metals and insulators, in which the electrical charge carrier concentration increases with increasing temperature over some temperature range. Certain semiconductors possess two types of carriers, namely, negative *electrons* and positive *holes.*

sensible atmosphere   That part of an atmosphere that offers significant resistance to a body passing through it.

sensitivity   1. The ability of electronic equipment to amplify a *signal* measured by the minimum stength of signal input capable of causing a desired value of output. The lower the input signal for a given output, the higher the sensitivity.
2. In measurements, the derivative representing the change in the variable being measured.
3. A measure of the relative susceptibility of a *propellant* to *deflagnation* or *detonation* under specified conditions.

sensor   The component of an instrument that converts an input signal into a quantity that is measured by another part of the instrument. Also called "sensing element."

separation   1. The action of a *fallaway section* or *companion body* as it casts off from the remaining body of a *vehicle,* or the action of the remaining body as it leaves a fallaway section behind it.
2. The moment of this action.
See: Solid Rocket Booster, External Tank

separation velocity   The *velocity* at which an *aerospace vehicle* or *space vehicle* is moving when some part or section is separated from it; specifically, the velocity of a space probe or *satellite* at the time of separation from the *launch vehicle.*

sequencer   A mechanical or electronic device that may be set to initiate a series of events and to make the events follow in a given sequence.

serial impact cost   The cost of removal and installation of (Shuttle) flight kits assessed when the time required for installation and removal exceeds 24 hours during the *Orbiter turnaround.*

servo   Control, usually by hydraulic means, of a large *moment of inertia* by the applica-

tion of a relatively small moment of inertia.

set   1. To place a *storage* device in a prescribed state.
2. To place a *binary cell* in the one state.

shadowgraph   1. A picture or image in which steep density *gradients* in the flow about a body are made visible, the body itself being presented in silhouette.
2. The optical method or technique by which this is done.

A shadowgraph differs from a *schlieren photograph* in that the schlieren method depends on the first derivative of the *refractive index* while the shadow method depends on the second derivative. Interference measurements give the refractive index directly.

shaft   A bar (almost always cylindrical) used to support rotating pieces or to transmit power or motion by rotation.

shaft riding elements   Components such as collars and sleeves that are attached to the surface of the pump shaft and are not an integral part of the shaft.

shaker   An electromagnet device capable of imparting known *vibratory acceleration* to a given object.

shake table   Device for subjecting components or assemblies to vibration in order to reveal vibrational *mode* patterns; also called "shaker."

shake table test   A laboratory test for vibration tolerance, in which the device to be tested is placed in a vibrator or *shake table.*

shear wave   A *wave* in an elastic medium that causes an element of the medium to change its shape without a change of volume. Mathematically, a shear wave is one whose *velocity* field has zero divergence. Also called rotational wave.

shelf life   Storage time during which an item remains serviceable, i.e. will operate satisfactorily when put in use.

shell   One of a series of concentric spheres, or orbits, at various distances from the *nucleus,* in which, according to atomic theory, *electrons* move around the nucleus of an *atom.* The shells are designated, in the order of increasing distance from the nucleus, as the k, l, m, n, o, p, and q shells. The number of electrons which each shell can contain is limited. Electrons in each shell have the same energy level and are further grouped into subshells.

See: electron capture, K-capture

shield   A body of material used to prevent or reduce the passage of particles or radiation.

A shield may be designated according to what it is intended to absorb, as a *gamma-ray* shield or *neutron* shield, or according to the kind of protection it is intended to give, as a background, biological, or thermal shield. The shield of a *nuclear reactor* is a body of material designed to prevent the escape of neutrons and radiation into a protected area, which frequently is the entire space external to the reactor. It may be required for the safety of personnel or to reduce radiation sufficiently to allow use of counting instruments.

shielding   The arrangement of *shields* used for any particular circumstances; the use of shields.

shim rod   A *nuclear reactor control rod* used in making infrequent coarse adjustments in *reactivity,* as in startup or shutdown. Compare this term with *regulating rod.*

shirt-sleeve environment   A *space station*'s or *vehicle*'s cabin atmosphere that is similar to that of the *Earth*'s surface, i.e., not requiring a *pressure suit.*

shock   1. *shock wave*

2. A blow, impact, collision, or violent jar.

3. A sudden agitation of the mental or emotional state or an event causing it.

4. The sudden stimulation caused by an electrical discharge on the animal or human organism (e.g., electric shock).

shock front   1. A shock wave regarded as the forward surface of a *fluid* region having characteristics different from those of the region ahead of the wave.

2. The front side of a shock wave.

shock isolator   A resilient support that tends to isolate a system from applied *shock.* Also called shock mount.

shock tube   A relatively long tube or pipe in which very brief high-speed *gas* flows are produced by the sudden release of gas at very high pressure into a low-pressure portion of the tube; the high-speed flow moves into the region of low pressure behind a shock wave.

shooting star   *meteor*

shot   A colloquial *aerospace* term describing the act or instance of firing a *rocket.*

shroud   1. Short extension of the outer wall of a plug nozzle downstream of the throat.

2. Continuous covering of the outer surfaces of (a) an *impeller* or other rotative component; (b) or of a payload being carried by a *launch vehicle.*

shroud line   Any one of the cords attaching a *parachute*'s load to the canopy; also called rigging line.

shutdown   1. The process of decreasing *rocket engine thrust* to zero.

2. To reduce the power level of a *nuclear reactor* to zero (i.e., to stop the *neutron chain reaction*).

shutoff valve   Valve that terminates the flow of *fluid;* usually a two-way valve that is either fully open or fully closed.

Shuttle   The prime element of the U.S. *Space Transportation System* for space research and applications in future decades. Carrying *payloads* of up to 29,500 kg (65,000 lb$_m$), the Shuttle will replace most expendable launch vehicles currently being used and will also provide the first system capable of launching deep space missions into their initial low Earth orbit. The Space Shuttle is also capable of returning payloads from *orbit* on a routine basis. In fact on 14 April 1981 the *Orbiter* "Columbia" became the first "spaceship" (as far as we know!) to land on Earth—culminating a historic and highly successful maiden flight of the Space Shuttle. *Shuttle crews* will be able to retrieve *spacecraft* from Earth orbit and either repair and redeploy them or return them to Earth for refurbishment and eventual reuse. The Shuttle can also be used to conduct missions in which scientists and engineers perform innovative experiments while in orbit around the Earth.

The Space Shuttle is consequently a truly versatile *aerospace* vehicle. It takes off like a rocket, maneuvers in orbit like a spacecraft and lands back on Earth like an airplane. The Shuttle is designed to carry heavy payloads into low Earth orbit. While other launch vehicles have done this, they could only be used just once. Unlike these expendable launch vehicles, the Space Shuttle Orbiter can be reused more than 100 times!

The versatility of the Shuttle permits the checkout and repair of unmanned satellites in orbit or their return to Earth for repairs that cannot be accomplished in space. The Shuttle era provides the potential for considerable savings in spacecraft costs and revolutionary design changes in space systems of the 1980s, 1990s and beyond. Some of the advanced Earth *satellites* that the Shuttle can transport to low Earth orbit and eventually maintain include those involved in environmental protection, energy resource development, meteorology, communications, navigation, agriculture, mapping, oceanography and many other fields—all of which can make life better for people here on Earth!

Interplanetary spacecraft can also be placed in Earth orbit by the Shuttle together with a propulsive upper stage, such as the *Inertial Upper Stage* (IUS), which is being developed by the U.S. *Department of Defense.* After the IUS and its spacecraft

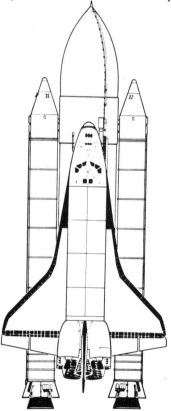

SPACE TRANSPORTATION SYSTEM
The Space Shuttle flight system, consisting of the Orbiter vehicle, the External Tank and two Solid Rocket Boosters.   (Drawing courtesy of NASA.)

payload have been removed from the Orbiter's *cargo bay* and checked out, the IUS is ignited to accelerate the spacecraft to its ultimate destination in *cislunar* or deep space. Upper propulsive stages, like the IUS, will therefore routinely be used to boost satellites to operational Earth orbits higher than the Shuttle's maximum altitude, which is approximately 1,000 km (or about 600 nm).

Unmanned space platforms, such as the *Space Telescope,* which can multiply man's view of the universe, and the *Long Duration Exposure Facility,* which can demonstrate the effects on materials of long exposure to the space environment, can be placed in orbit, erected and returned to Earth by the Shuttle.

The operational Space Shuttle will also have a short *turnaround time.* As the system matures the Shuttle will be refurbished and ready for another journey into space within weeks after landing. Thus, the Shuttle will quickly provide a vantage point in space for direct manned observation of interesting but transient astronomical or geophysical events as well as meteorological, agricultural or environmental crises on Earth. Information from such timely Shuttle observations would contribute to sound decisions for dealing with such urgent matters or to full investigation of such rapidly occurring phenomena.

The Shuttle also permits routine manned access to space. For example it will be used to transport a complete scientific laboratory, called *Spacelab,* into near-Earth orbit. Developed by the *European Space Agency,* Spacelab is designed to operate in *zero gravity* (i.e., *weightlessness*). Spacelab provides scientific facilities for as many as four *payload specialists* (non-career scientific astronauts) to conduct in-orbit experiments in such fields as *astrophysics, planetary sciences,* medicine, *space-based manufacturing, solar physics* and a host of other equally exciting disciplines. Spacelab will remain attached in the Orbiter's cargo bay throughout the mission. Upon its return to Earth, the Spacelab is removed from the Orbiter and prepared for its next mission. Spacelab can be reused approximately 50 times.

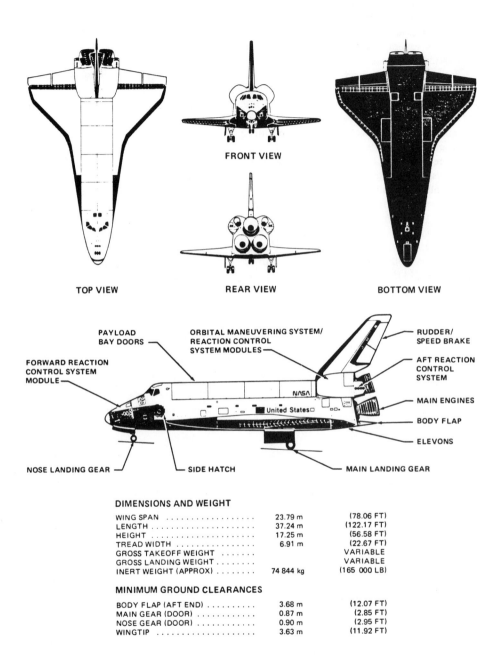

FRONT VIEW

TOP VIEW   REAR VIEW   BOTTOM VIEW

PAYLOAD
BAY DOORS

ORBITAL MANEUVERING SYSTEM/
REACTION CONTROL
SYSTEM MODULES

RUDDER/
SPEED BRAKE

FORWARD REACTION
CONTROL SYSTEM
MODULE

AFT REACTION
CONTROL
SYSTEM

MAIN ENGINES

BODY FLAP

ELEVONS

NOSE LANDING GEAR   SIDE HATCH   MAIN LANDING GEAR

DIMENSIONS AND WEIGHT

WING SPAN	23.79 m	(78.06 FT)
LENGTH	37.24 m	(122.17 FT)
HEIGHT	17.25 m	(56.58 FT)
TREAD WIDTH	6.91 m	(22.67 FT)
GROSS TAKEOFF WEIGHT		VARIABLE
GROSS LANDING WEIGHT		VARIABLE
INERT WEIGHT (APPROX)	74 844 kg	(165 000 LB)

MINIMUM GROUND CLEARANCES

BODY FLAP (AFT END)	3.68 m	(12.07 FT)
MAIN GEAR (DOOR)	0.87 m	(2.85 FT)
NOSE GEAR (DOOR)	0.90 m	(2.95 FT)
WINGTIP	3.63 m	(11.92 FT)

Fig. 2   The Space Shuttle Orbiter.   (Drawing courtesy of NASA.)

The Space Shuttle will bring within reach space projects that many considered impractical not too long ago. For example, the Shuttle can carry into low Earth orbit the modular units for self-sustaining human habitats and settlements. The inhabitants of such *"space stations"* could be employed in the manufacture of special pharmaceuticals, metals, electronic components, crystals and optical components. Manufacturing in the weightless, high-vacuum conditions of outer space has great potential for creating new alloys, producing new medicines and glasses of unusual purity and enabling the growth of very large crystals. Only time and the market forces of the next two decades can really identify the full impact that the Shuttle will have on human activities, both in space and on Earth. What **can** be said at this point is that the first successful flight of the Space Shuttle "Columbia" (12–14 April 1981) has initiated a new age of space exploitation. (See Figs. 1 and 2) The creative use of the Shuttle will trigger a terrestrial renaissance that will favorably affect not only man's scientific pursuits but also his social, cultural and economic activities! With the advent of the operational Space Shuttle, space has the potential to become mankind's major pathway into the next millennium.

Shuttle components   The *Space Shuttle* flight system has three main components, or units: the *Orbiter,* the *External Tank* and two *Solid Rocket Boosters.* (See Fig. 1.)

The Orbiter is the crew- and *payload*-carrying unit of the Shuttle flight system. It is 37 meters (121 ft) long, has a wingspan of 24 m (79 ft) and weighs approximately 68,000 kg (150,000 lb$_m$) without fuel. It is about the size and weight of a DC-9 commercial airplane. (See Fig. 2.) The Orbiter can transport a payload of up to 29,500 kg (65,000 lb$_m$) into low Earth orbit. This payload is carried in a cavernous *cargo (or payload) bay,* which is 18.3 m (60 ft) long and 4.6 m (15 ft) in diameter. The cargo bay is flexible enough to provide accommodations for unmanned *spacecraft* in a variety of shapes and sizes and for fully equipped scientific laboratories, such as *Spacelab.*

The Orbiter's three liquid rocket engines, called the *Space Shuttle Main Engines,* each have a *thrust* of 2,100,000 newtons (470,000 lb$_f$). They are fed propellants from the External Tank, which is 47 m (154 ft) long and 8.7 m (28.6 ft) in diameter. At *lift-off* this tank holds some 703,000 kg (1,550,000 lb$_m$) of propellant, consisting of *liquid hydrogen* (the fuel) and *liquid oxygen* (the oxidizer), which are kept in separate pressurized compartments of the External Tank. Just before reaching orbital velocity the Orbiter *jettisons* the External Tank, and it enters the atmosphere and breaks up over a remote ocean area. The External Tank is the only part of the Shuttle flight system that is not reusable.

Each Solid Rocket Booster has a sea level thrust of 11,600,000 newtons (2,600,000 lb$_f$). These solid boosters ignite simultaneously with the Orbiter's main liquid engines to lift the Shuttle off its *launch pad.* At a predetermined point in the flight, the two Solid Rocket Boosters separate from the Shuttle flight vehicle and parachute to sea, where they are recovered for reuse.

Shuttle crew   The *Orbiter* crew consists of the *commander* and *pilot.* Additional crew members who may be required to conduct Orbiter and *payload* operations are a *mission specialist* and from one to four *payload specialists.* The *mission specialist* is responsible for managing Shuttle equipment and resources supporting payloads during the flight, while payload specialists are in charge of specific payload equipment. A commander plus a pilot or pilot-qualified mission specialist are alway required to operate the Orbiter. Makeup of the remainder of the crew depends on the mission requirements, complexity and duration. Detailed responsibilities of the mission specialist and payload specialist(s) are tailored to meet the requirements of each individual flight. The commander, pilot and mission specialist are NASA astronauts and are assigned by NASA. Payload specialists may or may not be "career" astronauts. They are nominated by the *payload sponsor* and certified for flight by NASA.

In general *Space Transportation System* (STS) crew members (i.e., the commander, pilot and mission specialist) are responsible for operation and management of all STS systems, including payload support systems that are attached either to the Orbiter or to *standard payload carriers.* The payload specialist, on the other hand, is responsible for payload operations and management and the attainment of payload objectives.

The responsibility for in-orbit management of Orbiter systems and attached payload support systems as well as for *extravehicular activity* and payload handling with the *remote manipulator system* rests with the basic crew, because extensive training is required for safe and efficient operation of these systems. Assignment of these functions within the basic crew will vary to meet the requiremens of each flight. In general, the commander and pilot will manage Orbiter systems and standard payload support systems, such as *Spacelab* and the *Inertial Upper Stage.* The mission specialist and/or payload specialists will manage payload support systems that are mission dependent and very directly connected with the payload, such as instrument pointing sybsystems. Table 1 shows the typical breakdown of functions within a Shuttle crew.
See: commander, pilot, mission specialist, payload specialist, payload specialist training
Table 1 Typical Shuttle crew functions.

Shuttle Crew Functions

COMMANDER (career astronaut)
- Responsible for overall space vehicle operations
- Proficient in spacelab systems operations and management
- May support experiment operation at experiment sponsor's discretion and as time allows

PILOT (career astronaut)
- Proficient in spacelab systems operations and management
- Operates remote manipulator system (RMS)
- Second crewman for EVA operations

MISSION SPECIALIST (scientist and career astronaut)

- Proficient in payload operations
- Knowledgeable in spacelab systems
- Prime crewman for EVA operations
- Responsible for spacelab systems/payload coordination
- May perform RMS operations

PAYLOAD SPECIALIST (scientist)
- Proficient in experiment operations
- Responsible for attainment of experiment objectives
- Responsible for management of experiment operations
- Knowledgeable in operation of caution and warning, hatches, tunnel, and life support

Shuttle crew equipment The normal complement of crew equipment will provide supplies for a standard *Shuttle crew* of four for a mission of 7 days. The equipment is designed to be used in the *Orbiter shirtsleeve environment* by 90 percent of the male/female population (the 5th to 95th percentile). All the equipment with the exception of the flashlight is designed for use only inside the *pressurized crew compartment* of the Orbiter. The flashlight is also designed to operate while on extravehicular activity. The survival kit and lifevest are designed to remain completely functional following an *abort* landing.

The Orbiter crew clothing will be issued on a standard sizing schedule to fit the male/female crew. All clothes with the exception of underwear are common to both sexes. Cotton-blended tube socks are provided for all crewmembers. Clothing for Shuttle flights is listed in Table (1) All clothing will be made of commercially available fabrics. With routine maintenance and repair, the useful life of the clothing is expected to be 30 operational missions based on a nominal mission duration of seven days. All clothing will be recycled between missions for use by crewmembers of equivalent size. A 7-day mission clothing set for one crewmember will not exceed 10 kilograms (22 pounds).

The personal hygiene kit is designed to support a crewmember's personal hygiene requirements for a standard mission lasting 7 days. The kit contains a razor, shaving cream, a styptic pencil, skin emollient, stick deodorant, nail clippers, comb and brush dental floss, toothbrush, toothpaste, anti-chap lip balm, and soap. Standard products are expected to be used in the kits, but each crewmember will be allowed to make a limited personal selection of alternate commercial preparations and optional kit components (limited to readily available off-the-shelf items). [See also: Orbiter crew accommodations]

The crew will also have both paper and cloth towels suitable for general-purpose dry utility wipes, or, when wet, suitable for washcloths for shaving, cleansing, and other similar tasks. Trash containers will be provided for temporary stowage of waste materials. [See: Orbiter crew accommodations] The crew will be provided with a Swiss Army-type pocketknife, scissors, sunglasses, a chronograph, a sleeping mask, sleeping earplugs, general-purpose adhesive tape, a *Velcro* kit, in flight restraining devices, a *Mylar* mirror with Velcro attach points, and a small portable desk assembly.

Space Shuttle Main Engine(s) (SSME)
SPACE SHUTTLE SHIRT-SLEEVE CLOTHING

Article of clothing	Quantity per crewmember
Early flights	
Shorts	1 pair per day
T-shirts	1 per day
IVA gloves	1 pair per flight
Shoes	1 pair per flight
Constant-wear garment	3 per flight
Jacket	1 per flight
Athletic supporter	4 per flight
Trousers	1 pair per 7 days plus 1 spare per flight
Handkerchiefs	
Shirt	1 per 3 days
Socks	
Operational flights	
Underwear	1 set per day
One-g footwear	1 pair per flight
Jacket	1 per flight
Trousers	1 pair per 7 days plus 1 spare per flight
Shirt	1 per 3 days
Gloves	1 pair per flight
IVA footwear	1 pair per flight
Brassiere	1 per day

Table 1   Space Shuttle Crew Shirt-Sleeve clothing.   (Table courtesy of NASA.)

A cluster of three Space Shuttle Main Engines provides the main propulsion for the *Orbiter* vehicle. The *liquid hydrogen/ liquid oxygen* engine is a reusable high-performance rocket engine capable of various *thrust* levels. (See Fig. 1.) Ignited on the ground prior to launch, the main engine cluster operates in tandem with the *Solid Rocket Boosters* during the initial ascent. After the boosters separate the main engines continue to *burn* with a nominal operating time of approximately 8.5 minutes.

The main engines develop thrust by using high-energy *propellants* in a staged combustion cycle. The propellants are partially combusted in dual preburners to produce the high-pressure hot gas that drives the *turbopumps*. Combustion is completed in the main combustion chamber. (See Fig. 2.) The cycle ensures maximum performance because it eliminates *parasitic losses*.

Each Space Shuttle Main Engine operates at a liquid oxygen/liquid hydrogen mixture ratio of 6 to 1 to produce a sea level thrust of 1,668 kilonewtons (370,000 lb$_f$) and a vacuum thrust of 2,091 kilonewtons (470,000 lb$_f$). The engines can be throttled over a thrust range of 65% to 109%. This provides a high thrust level during *lift-off* and the initial ascent phase but allows the thrust to be reduced to limit acceleration to 3 *g* during the final ascent stage. Thus, *Shuttle crew* members and passengers will experience only 3 g during launch and less than 1.5 g during a typical *re-entry*. (See Fig. 3.) These acceleration loads are about one-third the levels experienced on previous U.S. manned space flights. The engines are *gimbaled* to provide *pitch, yaw and roll* control during the Orbiter boost phase.

Modified airline maintenance procedures are being used to service the engine without removing it from the vehicle between *Shuttle* flights. Most engine components can be replaced in the field as line replaceable units without extensive engine recalibration or *hot-fire testing*. These procedures result in an economical and efficient vehicle *turnaround*.

Operation of the Space Shuttle Main Engines

The flow of liquid hydrogen and liquid

oxygen from the *External Tank* is restrained from entering the engine by prevalves located in the Orbiter above the low-pressure turbopumps. (See Fig. 4, Nos. 1 and 11.) Before firing, the prevalves are opened to allow propellants to flow through the low-pressure turbopumps and the high-pressure turbopumps (Fig. 4, Nos. 2 and 12) and then to the main propellant valves (Fig. 4, Nos. 3 and 13). On the liquid oxygen side, the system also fills two preburner valves (Fig. 4, Nos. 7 and 14). The *cryogenic propellants* are held in the duct long enough to

chill the engine and liquefy in the respective propellant systems. The chilling process is aided by bleedlines (not illustrated in Fig. 4) that allow circulation of the propellants.

In the start sequence the hydrogen and oxygen sides operate almost simultaneously. On the hydrogen, or *fuel*, side the ignition command from the Orbiter opens the main fuel valve (Fig. 4, No. 3). This permits hydrogen to flow into the coolant loop, through the nozzle tubes (Fig. 4, No. 5) and through channels in the main combustion chamber (Fig. 4, No. 6). A portion

Space Shuttle Main Engines

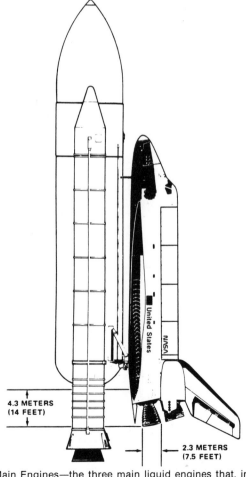

THRUST
Sea level: 1670 kilonewtons (375 000 pounds)

Vacuum: 2100 kilonewtons (470 000 pounds)

(Note: Thrust given at rated or 100-percent power level.)

THROTTLING ABILITY
65 to 109 percent of rated power level

SPECIFIC IMPULSE
Sea level: $356.2 \dfrac{N/s}{kg} \left(363.2 \dfrac{lbf/s}{lbm} \right)$

Vacuum: $4464 \dfrac{N/s}{kg} \left(455.2 \dfrac{lbf/s}{lbm} \right)$

(Given in newtons per second to kilograms of propellant and pounds-force per second to pounds-mass of propellant)

CHAMBER PRESSURE
20 480 kN/m² (2970 psia)

MIXTURE RATIO
6 parts liquid oxygen to 1 part liquid hydrogen (by weight)

AREA RATIO
Nozzle exit to throat area 77.5 to 1

WEIGHT
Approximately 3000 kilograms (6700 pounds)

LIFE
7.5 hours, 55 starts

4.3 METERS
(14 FEET)

2.3 METERS
(7.5 FEET)

United States

NASA

Fig. 1   A data summary for the Space Shuttle Main Engines—the three main liquid engines that, in conjunction with the Solid Rocket Boosters, provide the thrust to lift the Orbiter off the ground for the initial ascent.   (Drawing courtesy of NASA.)

Fig. 2   Major components of the Space Shuttle Main Engine.   (Drawing courtesy of NASA.)

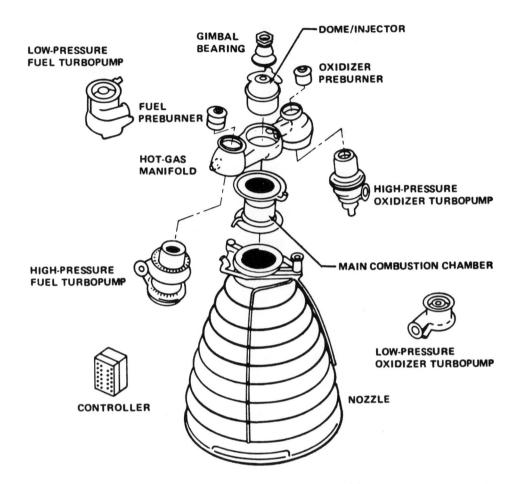

of this coolant loop flow is diverted by the coolant control valve (Fig. 4, No. 4) to the preburners (Fig. 4, Nos. 8 and 15). Some of the hydrogen used in the coolant loop is warmed in the process to virtually ambient conditions and is tapped off at the main combustion chamber and routed back to the low-pressure turbopump to drive the *turbine* to that pump. This flow passes through the turbine and is returned to the walls of the two preburners where it cools the preburners, the hot-gas manifold (Fig. 4, No. 9) and the main injector (Fig. 4, No. 10).

On the oxygen, or *oxidizer,* side the ignition command opens the main oxidizer valve (Fig. 4, No. 13). The liquid oxygen flows through the two turbopumps (Fig. 4, Nos. 11 and 12) to the main injector and

also (through valves Nos. 7 and 14 in Fig. 4) to the two preburners. Oxygen tapped off downstream of the high-pressure oxidizer turbopump (Fig. 4, No. 12) is routed to the low-pressure turbopump (Fig. 4, No. 11) to drive the liquid turbine for that pump. This flow continues through the low-pressure oxidizer turbopump, thus re-entering the circuit.

Spark igniters located in the dome of both preburners and the main combustion chamber initiate combustion. The two preburners are operated at mixture ratios of less than one part oxygen to one part hydrogen to produce hot gas, or hydrogen-rich steam. The hot gas, or steam, is used to drive the turbines of the two high-pressure turbopumps before it enters the hot-gas manifold. This hydrogen-rich steam is transferred by the hot-gas manifold from the turbines to the main injector where it is mixed with additional liquid oxygen from the high-pressure oxidizer turbopump for combustion. This combustion process is completed at a mixture ratio of six parts oxygen to one part hydrogen.

The *pogo suppressor* (Fig. 4, No. 16) is provided to absorb any closed-loop longitudinal dynamic oscillations (i.e., vibration) that might be generated between vehicle structural components during the engine combustion process. A suppressor is not required on the hydrogen side of the engine because the low *density* of that fluid has been shown to be insufficient to transmit any appreciable vibration.

Another major component of the engine is the controller, which operates all engine controls. Mounted on the engine (see Fig. 2), the controller includes a computer to integrate commands received from the Orbiter with data input from sensors located on the engine. The controller monitors the engine before ignition, controls *purges* before and during operation of the engine, manages the engine's redundancy features, receives and transmits data to the Orbiter for either storage or transmission to the ground and operates the engine control valves. The five control valves (Fig. 4, Nos. 3, 4, 7, 13 and 14) effectively control the entire engine operation.

Combustion Devices in the Space Shuttle Main Engine

Combustion devices are located in those parts of the main engine where controlled combustion, or burning, of the liquid oxygen and liquid hydrogen occurs. The five major components in this group are: the ignition system, the preburners, the main injector, the main combustion chamber and the nozzle assembly.

The ignition system starts the combustion process. There are three ignition units, one for the main chamber injector and one for each of the two preburner injectors. Each ignition unit, located in the center of its respective injector, includes a small combustion chamber, two spark igniters (which are similar to spark plugs) and propellant supply lines. At engine start all six spark

igniters are activated, igniting the propellants as they enter the igniter combustion chamber and thus providing an ignition source for propellants entering the preburners and the main combustion chamber. The ignition unit remains active for the duration of engine operation, but the spark igniters are turned off after ignition is complete.

Each main engine has fuel and oxidizer preburners that provide hydrogen-rich hot gases at approximately 1,030° K (760°C or 1,400°F). These gases drive the fuel and oxidizer high-pressure turbopumps. The preburner gases pass through turbines and are directed through a hot-gas manifold to the main injector where they are injected into the main combustion chamber, together with liquid oxygen, and burn at approximately 3,590° K (3,315°C or 6,000°F).

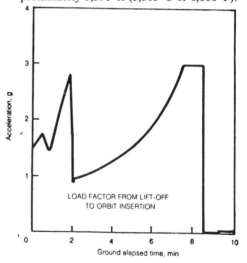

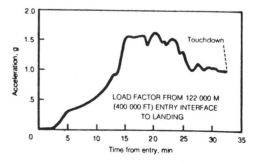

Fig. 3   Typical acceleration levels experienced by Shuttle crew members and passengers during ascent to and descent from orbit.   (Drawing courtesy of NASA.)

Fig. 4   Space Shuttle Main Engine propellant flow.   (Drawing courtesy of NASA.)

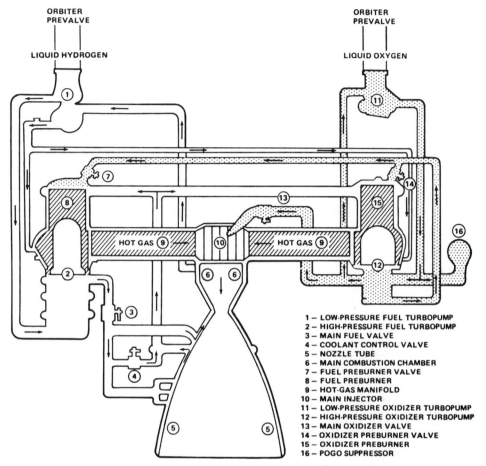

1 — LOW-PRESSURE FUEL TURBOPUMP
2 — HIGH-PRESSURE FUEL TURBOPUMP
3 — MAIN FUEL VALVE
4 — COOLANT CONTROL VALVE
5 — NOZZLE TUBE
6 — MAIN COMBUSTION CHAMBER
7 — FUEL PREBURNER VALVE
8 — FUEL PREBURNER
9 — HOT-GAS MANIFOLD
10 — MAIN INJECTOR
11 — LOW-PRESSURE OXIDIZER TURBOPUMP
12 — HIGH-PRESSURE OXIDIZER TURBOPUMP
13 — MAIN OXIDIZER VALVE
14 — OXIDIZER PREBURNER VALVE
15 — OXIDIZER PREBURNER
16 — POGO SUPPRESSOR

The design of the two preburners is similar. Each consists of fuel and oxidizer supply manifolds, an injector, stability devices, a cylindrical combustion zone and an ignition unit. The supply manifolds ensure uniform propellant distribution so that each injector element receives the correct amount of oxygen and hydrogen. The preburner injectors consist of many individual injection elements that introduce the propellants in concentric streams. Each oxygen stream is surrounded by its companion hydrogen stream. The injector contains *baffles* to help maintain stable combustion in the preburners and thus suppress disturbances that might occur in the combustion process. Gaseous hydrogen flows through passages in each baffle for cooling and is then discharged into the combustion chamber.

The cylindrical combustion zone consists of a structural shell and a thin inner liner. The liner is cooled by passing gaseous hydrogen between it and the structural wall.

The main injector performs the vital function of finally mixing all the liquid oxygen and liquid hydrogen together as thoroughly and uniformly as possible to produce efficient combustion.

The main injector is an intricately fabricated component, consisting of a thrust cone, an oxidizer supply manifold, two fuel cavities, 600 injection elements and an ignition unit. The thrust cone transmits the total thrust of the engine through the gimbal bearing to the Orbiter vehicle. The oxidizer supply manifold receives oxygen from the high-pressure turbopump and distributes it evenly to the 600 injection elements. One of the two fuel cavities supplies the fuel-rich hot gases that originate in the preburners and are used to run the high-pressure turbines. The other fuel cavity supplies gaseous hydrogen from the hot-gas manifold cooling circuit.

The 600 main injection elements have the same basic design as the preburner injection elements, that is, an outer fuel *shroud* that surrounds a central oxidizer stream. The propellants are thoroughly mixed as they are introduced into the main combustion zone for burning at approximately 3,590° K (3,315°C or 6,000°F).

Seventy-five of the injection elements also form baffles that divide the injector into six compartments. The baffles are designed to suppress any *pressure* disturbances that might occur during the combustion process.

The main combustion chamber is a double-walled cylinder between the hot-gas manifold and the nozzle assembly. Its primary function is to receive the mixed propellants from the main injector, accelerate the hot combusted gases to *sonic velocity* through the *throat* and expand them supersonically through the nozzle. The main chamber operating pressure at rated power level is approximately 20,700 kN/m² (3,000 psi). The main combustion chamber consists of a coolant liner, a high-strength structural jacket, coolant inlet and outlet manifolds and actuator struts.

The internal contour of the coolant liner forms the typical contraction-throat-expansion shape common to conventional rocket engine combustion chambers. The contraction area ratio (the ratio of the area at the injector face to the throat area) is 2.96 to 1. The expansion area ratio (the ratio of the area at the aft end of the combustion chamber to the throat area) is 5 to 1. The contraction contour is shaped to minimize the transfer of *heat* from the combustion gases to the coolant liner. The expansion contour accelerates the combustion gases to the 5-to-1 expansion ratio with minimal energy loss.

The coolant liner passes hydrogen coolant (fuel) through 390 channels. Approximately 25% of the total hydrogen flow is used to cool the liner. The chamber jacket goes around the outside of the liner to provide structural strength. Inlet and outlet coolant manifolds are welded to the jacket

and the liner. Two actuator struts are bolted to the chamber and are used, in conjunction with *hydraulic actuators,* to gimbal the engine during flight when it is necessary to change the direction of the thrust.

To provide maximum thrust efficiency the nozzle assembly (see Fig. 5) allows continued expansion of the combustion gases coming from the main combustion chamber. It is designed for a 77.5-to-1 thrust chamber expansion ratio for thrust efficiency at high altitudes. The nozzle assembly is the largest component of the engine, measuring approximately 3 meters (10 ft) in length and 2.4 meters (8 ft) in diameter at the base. The nozzle assembly consists of a forward manifold subassembly and a stacked tube nozzle subassembly.

The forward manifold subassembly provides the attachment to the main combustion chamber. It also distributes hydrogen to the main chamber and nozzle cooling circuits and to both the fuel and oxidizer preburners.

Engine Systems

The hot-gas manifold is a double-walled, hydrogen-gas-cooled structural support and fluid manifold. It is the structural backbone of the engine and interconnects and supports the preburners, high-pressure turbopumps, main combustion chamber and main injector.

The hot-gas manifold conducts hot gas (hydrogen-rich steam) from the turbines to the main chamber injector. The area between the wall and the liner provides a coolant flow path for the hydrogen gas that is exhausted from the low-pressure fuel turbopump turbine. This protects the outer wall and liner against the temperature effects of the hot gas from the preburners. After cooling the manifold the hydrogen also serves as coolant for the primary faceplate, the secondary faceplate and the main combustion chamber acoustic cavities. The high-pressure turbopumps are stud-mounted to the canted flanges on each side of the hot-gas manifold. The preburners are welded to the upper end of each side of the hot-gas manifold above the high-pressure turbopumps.

The heat exchanger is a single-pass coil

Fig. 5   The nozzle assembly of the Space Shuttle Main Engine.   (Drawing courtesy of NASA.)

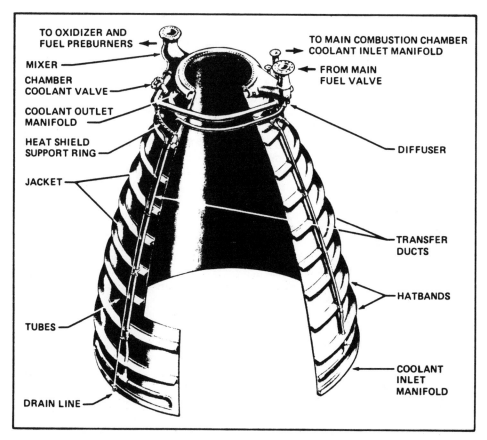

pack installed in the oxidizer side of the hot-gas manifold. It converts liquid oxygen to gaseous oxygen for vehicle oxygen tank and pogo-system accumulator pressurization. The heat exchanger consists of a helically wound small tube, approximately 0.8 meter (2.6 ft) long, in series with two parallel larger tubes, each approximately 7.9 meters (25.8 ft) long. The tubes are attached to supports welded to the inner wall of the hot-gas manifold coolant jacket. The hot turbine exhaust gases from the high-pressure oxidizer turbopump heat the liquid oxygen to a gas. Liquid oxygen, tapped off the discharge side of the high-pressure oxidizer turbopump, is supplied to the inlet of the heat exchanger through an anti-flood valve.

The oxygen is heated to a gas in the small tube (first stage) and to the final outlet temperature in the two larger tubes (second stage). A bypass line with openings around

the heat exchanger injects an unheated portion (approximately 30%) of the total oxygen flow into the outlet of the heat exchanger for control of temperature and flow rate. Openings in the heat exchanger bypass line and in the vehicle control the heat exchanger flow rate.

The gimbal bearing assembly is a spherical low-friction universal joint that has ball-and-socket bearing surfaces. The bearing assembly provides the mechanical interface with the vehicle for transmitting thrust loads and permits angulation of the actual thrust vector (force) about each of two vector control axes. The gimbal bearing is attached to the engine main injector by bolts that allow lateral positioning of the bearing. The gimbal bearing position is established by optical alignment during engine buildup to ensure that the actual thrust vector is within 30 minutes of arc to the engine center line and 1.5 cm (0.6 in.) of the gimbal center. Cycle life is obtained by low-friction anti-galling bearing surfaces that operate under high loads.

The pneumatic control assembly provides for the following: (1) control of ground-supplied gaseous nitrogen used for engine prestart *purges* and of vehicle-supplied helium for operational purge, (2) control of oxidizer and fuel bleed valves and (3) emergency shutdown control of the main propellant valves in the event of electrical power loss to the engine. The pneumatic control assembly consists of a manifold with ports to which *solenoid* valves and pressure-actuated valves are attached.

The oxidizer and fuel bleed valves are opened by pneumatic pressure from the pneumatic control assembly during engine-start preparation. This provides a recirculation flow for propellants through the engine to ensure that the propellants are at the required temperatures for engine start. At engine start these valves are closed by venting the actuation pressure.

The pneumatic control system purge check valves are spring-loaded, normally closed *poppet valves* that isolate propellants from the pneumatic systems. These check valves are opened by pressure actuation.

Turbopumps

The SSME propellant feed system includes four turbopumps, two of which are low pressure and two high pressure. There is one of each for the liquid hydrogen fuel and liquid oxygen oxidizer. All four turbopumps are line replaceable units for maintenance purposes.

The low-pressure fuel turbopump is an axial-flow (in-line) pump driven by a two-stage turbine. It raises the pressure of the fluid being applied to the high-pressure fuel pump to prevent *cavitation*. (Cavitation is the formation of partial vacuums in a flowing liquid.) The rotor assembly is supported on three ball bearings, which are cooled internally by liquid hydrogen. The low-pressure fuel turbopump nominally operates at a speed of 14,700 revolutions per minute (rpm), develops 1,790 kilowatts (2,400 brake horsepower) of power and increases the pump pressure from 207

kN/m^2 to 1,600 kN/m^2 (30 psia to 232 psia) at a flow rate of 67 kg/sec (147 lb_m/sec). The turbine is driven by gaseous hydrogen at a nominal inlet pressure of 29,434 kN/m^2 (4,269 psia).

The high-pressure fuel turbopump is a three-stage *centrifugal* pump driven directly by a two-stage turbine. The latter, in turn, is driven by hot gas that is supplied by the fuel preburner. Fuel flows in series through the three *impellers* from the pump inlet to the pump outlet, and the flow is redirected between impellers by interstage *diffusers*. Two double sets of ball bearings support the rotating assembly. A thrust bearing at the pump end of the rotating assembly provides axial rotor thrust control during startup and shutdown, while a dynamic self-compensating balance system distributes axial forces during main stage operation. The bearings are cooled internally with liquid hydrogen, and two dynamic seals are used to prevent turbine-to-pump leakage.

The high-pressure fuel turbopump is a high-speed, high-power device that operates at a nominal speed of 35,000 rpm and develops 46,435 kilowatts (62,270 brake horsepower) of power. It increases the pressure from 1,213 kN/m^2 to 42,817 kN/m^2 (176 psia to 6,210 psia) at a flow rate of 67 kg/sec (147 lb_m/sec). The nominal turbine inlet pressure and temperature are 35,605 kN/m^2 (5,164 psia) and 961°K (688°C or 1,271°F), respectively.

The low-pressure oxidizer turbopump is an axial-flow pump that is driven by a six-stage turbine and powered by oxidizer propellant. Because the pump and turbine propellants are both liquid oxygen, the requirements for dynamic seals, purges and drains have been eliminated. The primary function of the low-pressure oxidizer pump is to maintain sufficient inlet pressure to the high-pressure oxidizer pump to prevent cavitation. The rotor assembly is supported by two ball bearings, which are cooled internally with oxidizer. Turbine-drive fluid at 30,944 kN/m^2 (4,488 psia) is provided from the high-pressure oxidizer pump discharge. The low-pressure oxidizer pump nominally operates at a speed of 5,150 rpm, develops 1,096 kilowatts (1,470 brake horsepower) and increases the pump pressure from 690 to 2,861 kN/m^2 (100 psia to 415 psia) at a flow rate of 401 kg/sec (883 lb$_m$/sec).

The high-pressure oxidizer turbopump consists of a main pump, which provides liquid oxygen to the main injector, and a boost pump, which supplies liquid oxygen to the preburners. The main pump has a single inlet, and flow is split to a double-entry impeller with a common discharge. Two double sets of ball bearings support the rotor assembly and are cooled internally with liquid oxygen. Dynamic seals within the turbopump prevent the mixing of liquid oxygen and turbine gases. The turbopump rotor axial thrust is balanced by a self-compensating balance piston.

The high-pressure oxidizer turbine is powered by hot gases generated by the oxidizer preburner. These gases pass through the turbine blades and nozzles and discharge into the hot-gas manifold. The turbine housing is cooled by gaseous hydrogen supplied by the oxidizer preburner coolant jacket.

The high-pressure oxidizer turbopump is a high-speed, high-power device that operates at a nominal speed of 29,057 rpm with a turbine inlet pressure and temperature of 36,046 kN/m^2 (5,228 psia) and 817 K (544 C or 1,011 F), respectively. The main oxidizer pump develops 15,643 kilowatts (20,977 brake horsepower) of power with a pump pressure increase from 2,482 kN/m^2 to 31,937 kN/m^2 (360 psia to 4,632 psia) at a flow rate of 484 kg/sec (1,066 lb$_m$/sec). The preburner pump pressure increases from 30,592 kN/m^2 to 52,642 kN/m^2 (4,437 psia to 7,635 psia) at a flow rate of 39 kg/sec (86 lb$_m$/sec with 1,098 kilowatts (1,472 brake horsepower) of power.

Main Valves

The main propellant valves consist of the main oxidizer valve, the main fuel valve, the oxidizer preburner oxidizer valve, the fuel preburner oxidizer valve and the chamber coolant valve. All except the chamber coolant valve are *ball-type valves* and have two major moving components. These are the integral ball-shaft cams and the ball-seal retracting mechanism. The ball inlet seal is a machined plastic, bellows-loaded, closed seal. Redundant shaft seals, with an over-

board drain cavity between them, prevent leakage along the shaft (actuator end) during engine operation. Inlet and outlet sleeves align the flow to minimize turbulence and the resultant pressure loss. Ball seal wear is minimized by cams and a cam-follower assembly that moves the seal away from the ball when the valve is being opened.

All valves are operated by a *hydraulic servoactuator* mounted to the valve housing and receive electrical control signals from the engine controller.

The main oxidizer valve controls oxidizer flow to the main chamber liquid oxygen dome and the main chamber augmented spark igniter. The main fuel valve controls the flow of fuel (hydrogen) to the thrust chamber coolant circuits, the low-pressure fuel turbopump turbine, the hot-gas manifold coolant circuit, the oxidizer preburner, the fuel preburner and the three augmented spark igniters. The oxidizer preburner oxidizer valve controls the flow of oxidizer (oxygen) to the oxidizer preburner and the oxidizer preburner augmented spark igniter. During main stage operation the valve is modulated to control engine thrust between minimum and full power levels. The fuel preburner oxidizer valve controls the flow of oxidizer to the fuel preburner and the fuel preburner augmented spark igniter. During main stage operation the valve is modulated to maintain the desired engine mixture ratio. The chamber coolant valve is a *gate-type valve* that serves as a throttling control to maintain proper fuel flow through the main combustion chamber and nozzle coolant circuits.

Hydraulic power is provided by the Orbiter for the operation of the five valves in the propellant feed system. Servoactuators mounted to the propellant valves convert vehicle-supplied hydraulic fluid pressure to the rotary motion of the actuator shaft by electrical input command. Two servovalves, which are integral with each servoactuator, convert the electrical command signal from the engine controller to hydraulic flow that positions the valve actuator. All actuators, except for the chamber coolant valve, have an emergency shutdown system to pneumatically close the propellant valves in the proper order.

The checkout, start, in-flight operation and shutdown of the SSMEs are managed by dual-redundant 16-*bit digital computers* and their input and output electronics. This electronics package, mounted on the engine, is called a controller.

The controller interfaces with the hydraulic actuators and their position feedback mechanisms, spark igniters, solenoids and sensors to provide closed-loop control of the thrust and propellant mixture ratio, while monitoring the performance of critical components on the engine and providing the necessary redundancy management to ensure the highest probability of proper and continued engine performance. These monitoring and control tasks are repeated every 20 milliseconds (50 times per second). Critical engine operation parameters (temperature, pressure and speed) are monitored for exceeding predetermined values, which would indicate an impending engine malfunction. If any of these critical parameters are exceeded, the controller would perform a safe engine shutdown. Status information is reported to the Orbiter vehicle for proper action and postflight evaluation.

The controller receives commands from the Orbiter's guidance and navigation computers for the checkout, start, thrust-level requirements (throttling) and shutdown. The controller, in turn, performs the necessary functions to start, change from one thrust level to another within one-percent accuracy and shut down as defined by the commands from the vehicle. In addition to controlling the engine, the controller performs self-tests and switches to the backup computer channel and its associated electronics in the event of a computer failure. Similar tests are performed on the components that interface with the controller, and necessary actions are taken to remove faulty components from the active control loop. One failure in any of the electronic components can be tolerated, and normal engine operation will continue. Some second failures can also be tolerated, if the only result is degraded performance of the engine. However, in all cases the controller will perform an engine shutdown when all methods for engine electrical monitoring and control

are exhausted.

The controller is packaged in a sealed, pressurized chassis, with cooling provided by *convection heat transfer,* through pin fins. It is functionally divided into five subsystems: input electronics, output electronics, computer interface electronics, the digital computer and power supply electronics. Each of the five subsystems is duplicated to provide dual-redundant capability.

The input electronics subsystem receives data from the engine sensors, conditions the signals and converts them to digital form for computer use. The sensors for engine control and critical parameter monitoring (i.e., *redlines*) are dual redundant, while the sensors for data only are non-redundant.

The output electronics subsystem converts the computer digital control commands into voltages suitable for powering the engine spark igniters, the solenoids and the propellant valve actuators.

The computer interface electronics subsystem controls the flow of data within the controller, the input data to the computer and the computer output commands to the output electronics. It also provides the controller interface with the Orbiter vehicle for receiving engine commands (via triple-redundant channels) from the vehicle and for transmission of engine status and data (via dual-redundant channels) to the vehicle.

The digital computer subsystem is an internally stored general-purpose digital computer that provides the computational capability necessary to perform all engine control and monitoring functions. The computer memory has a program storage capacity of 16,384 words.

The power supply electronics subsystem converts the 115-volt, three-phase, 400-hertz vehicle power to the individual voltages required to operate the computers, input and output electronics, computer interface electronics and other engine electrical components.

The controller *software* is an on-line, *real time,* process control program. The program will process inputs from the engine sensors; control the operation of actuators, solenoids and spark igniters; accept and process vehicle commands; provide and transmit data to the vehicle; and provide test/checkout and monitoring capabilities.

Shuttle mission categories An evaluation of various possible space *missions* has identified four principal Shuttle mission categories for the next few years: (1) *payload* deployment, (2) sortie or attached payload, (3) *retrieval* and (4) assembly. More advanced missions, such as crew transport and resupply missions between terrestrial bases and permanently *manned space platforms,* must await a national or international commitment to construct and operate a *space station* in *low Earth orbit.*

In a payload deployment mission the payload is placed in orbit by the Shuttle *Orbiter* using the *remote manipulator system* (RMS) to remove mission equipment (e.g., a *spacecraft* with or without an orbital transfer vehicle) from the *cargo bay* and place it in space. This "free-flying" payload is then committed (i.e., turned over) to ground control for further testing, transport to final mission orbit and operation. For such deployment missions basic crew and *payload specialist* functions and tasks are essentially limited to flight readiness and cargo bay removal activities.

These functions typically include: (1) attaching the RMS to the payload, (2) releasing the retaining clamps, (3) removing the payload from the cargo bay, (4) releasing the payload outside the Orbiter and (5) turning the payload over to ground control. Special instrumentation in the Orbiter's aft flight deck may also be used to run evaluation checks on the deployed payload.

In a sortie mission the Orbiter itself serves as a spacecraft, providing stabilization, power, thermal control, telemetry, etc., to attached payloads. Sortie missions lasting between 7 and 30 days are anticipated. For Shuttle missions involving captive or attached payloads, payload specialist and

Fig. 1 Artist's conception of the Space Shuttle with pressurized Spacelab module in a low-Earth-orbit sortie mission. At the aft end of the cargo bay are two Spacelab pallet sections on which are mounted various scientific instruments, including telescopes, sensors and antennas. (Photograph courtesy of NASA.)

basic crew functions and tasks require a greater degree of technical knowledge and skill than payload deployment missions. The "brassboard" protoflights of new space systems and equipment represent one of the greatest potential applications of the Shuttle and its ability to fly scientists and engineers as payload specialists. In a sortie mission the payload specialist is required to function as operator, analyst and even repairman for the payload. Typical activities for the scientist or engineer in orbit include: (1) initial activation of the experiment or protoflight equipment; (2) calibration; (3) target location; (4) optimization of operations; (5) real-time evaluation and analysis of data; (6) coordination with investigators on Earth; (7) powering down and stowing for re-entry; and (8) maintenance, repair and modification of the equipment as appropriate. (See Fig. 1.)

A retrieval mission involves recovering an orbiting space system and requires a rendez-vous and docking operation. The "retrieved" system may either be captured and returned to Earth or repaired, refurbished and serviced in orbit and then returned to operation. Each retrieval mission involves similar tasks and functions, which include: (1) rendezvous and docking maneuvers, (2) acquisition with the RMS, (3) installation in the payload bay, (4) inspection and testing, (5) repair and refurbishment, (6) preoperational activities and (7) operational deployment or stowing for return to Earth. Such retrieval activities would involve all *Shuttle crew* members and could include *extravehicular activities* (EVA).

An assembly mission would be used for a payload that is too large or complex to be launched as a complete unit. Using manipulators, automatic devices and/or EVA, crew members would assemble or construct such a payload in orbit and then transfer it to its operational location in space. The crew functions and tasks needed for assembly are

Fig. 2   Artist's conception of the manufacture of large beams in low Earth orbit, as a prelude to the assembly of very large structures in space.   (Photograph courtesy of NASA.)

similar to those involved in a payload retrieval mission and would include: (1) EVA, (2) the use of powered and hand tools and (3) inspection and preoperational testing. (See Fig. 2.)

Shuttle mission profile   A typical *Space Shuttle mission* may be divided into four basic phases: (1) the boost phase, which occurs from *lift-off* to orbital insertion; (2) the in-*orbit* operations phase; (3) the de-orbit and re-entry phase; and (4) the ground servicing and *turnaround operations* phase, during which the vehicle is prepared for its next journey into space. The Shuttle mission begins at lift-off with the simultaneous ignition of the *Orbiter's* three main liquid rocket engines and the two *Solid Rocket Boosters* (SRB's). When the *Shuttle flight system* has cleared the launch area, it performs a roll maneuver and achieves the desired *launch azimuth.* Some fifty seconds after lift-off, *maximum dynamic pressure* is encountered. The vehicle assembly continues to rise, and the SRB's burn out and separate about 120 seconds into the flight. After the Solid Rocket Boosters have performed their propulsive function and separate, they parachute to the ocean. Special recovery vessels then recover the spent boosters and tow them to shore for refurbishment and reuse.

Meanwhile, the Orbiter's three main liquid engines continue to *burn,* being fed *liquid hydrogen* and *liquid oxygen* from the huge, dirigible-like *External Tank* (ET). About eight minutes after launch the main engines cut off (*MECO*), and the External Tank is *jettisoned.* The External Tank falls back to Earth to break up re-entering the atmosphere, its surviving fragments falling into a remote ocean area. The ET is the only part of the Space Shuttle flight system that is not reused. Throughout the ascent phase the *thrust profile* is tailored to keep acceleration levels at or below 3 *g.*

Free of the External Tank the Orbiter, after coasting for a short time, fires its two small *orbital maneuvering system* (OMS) engines, which are fed from internal fuel tanks. A typical burn of some 105 seconds allows the Orbiter to achieve *orbital velocity* (7,847 meters per second, or 17,500 mph). The initial elliptical orbit ranges from 110 km (60 nm) at its lowest point, called *perigee,* to 280 km (150 nm) at *apogee.* A second OMS firing of some 95 seconds, halfway around the world from the launch site, reshapes the egg-shaped flight path to a circular orbit, and the Space Shuttle is ready to perform its orbital mission. Depending on mission requirements the Orbiter will spend from 1 to 30 days in orbit and then prepare for its return journey to Earth. About halfway around the world from the landing site, the small reaction control system thrusters are fired in short bursts to turn the Orbiter "tail-first." Then the OMS engines are fired for approximately 2 minutes to slow the vehicle down and to lower its flight path in a slow curve toward Earth. Half an hour later, about 150 km up and with the Orbiter again flying "nose first," the crew begins to feel the drag of the thin top layer of the *atmosphere.* Now begins one of the most critical and demanding parts of the Shuttle mission.

During the de-orbit and re-entry phase, the Orbiter must change from a *spacecraft* to an aircraft, while slowing down from its orbital speed of approximately 8,000 meters per second (18,000 mph) to approximately 100 meters per second (225 mph) for landing. Above the atmosphere vehicle maneuvering is done by firing the small reaction control system rockets in the nose and tail; deep in the atmosphere attitude and direction are controlled by a conventional aircraft rudder and flaps. However, at the middle speeds and altitudes experienced during re-entry, rocket and aerodynamic controls must be skillfully blended. The commander and pilot are assisted in this tricky task by five onboard computers.

Edging into the atmosphere the crew uses the reaction control system thrusters to angle the nose up so that the aerospace craft pushes into the thickening blanket of air at about a 40° angle of attack. A special insulation that sheds heat so readily that one side is cool enough to hold in bare hands while the other side is red hot serves as the Orbiter heat shield. This insulation survives temperatures up to 1,533 K (1,260 C or 2,300 F) for up to 100 flights with little or no refurbishment. As *aerodynamic heating* raises the Orbiter's heavily insulated

Fig. 1   Typical Space Shuttle mission profile.   (Drawing courtesy of NASA.)

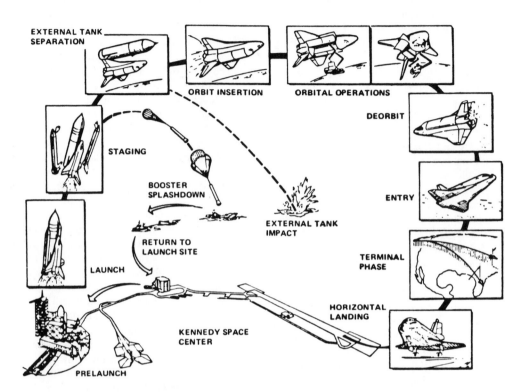

underside to more than 1,500 K, *ionization* of the surrounding atmosphere causes a *blackout* of communications with the ground for some time.

At about 93 km (50.3 nm) altitude, the air becomes dense enough so that *aerodynamic* controls take hold, and the Orbiter vehicle becomes a heavy glider. Without fuel for its main engines, the Orbiter can't go around for a second landing approach. However, it can maneuver to the right or left of its entry path as much as 2,034 km (1,100 nm) to make an "emergency" landing at any of several airports or military air bases. In such a case, it would be ferried back to its home base on the back of a specially fitted NASA 747 air transport. The Orbiter touches down like an airplane on a runway at the Kennedy Space Center in Florida or Vandenberg Air Force Base in California. After landing the Orbiter is towed away for servicing to prepare it for its next journey into space.

The Shuttle will be launched from the NASA Kennedy Space Center for missions requiring equatorial (east-west) orbits and from Vandenberg Air Force Base for missions requiring *polar* (north-south) orbits.

Shuttle mission simulator (SMS)   A computer-controlled training device with fully functional *Orbiter* forward and aft crew stations. It is a standard hardware item for mission-independent training.

Shuttle names   The names of the first four *Space Shuttle Orbiters* intended for operational space flight have been selected from sea vessels used in world exploration. These are "Columbia," Orbiter vehicle (OV) 102, "Discovery," Orbiter vehicle 103, "Challenger," Orbiter vehicle 99, and "Atlantis," Orbiter vehicle 104. The first Orbiter vehicle to come off the assembly line, "Enterprise" OV 101, was used in the *Approach and Landing Test* flights at NASA's *Dryden Flight Research Center* in California. It was named by popular request after the "Starship Enterprise" in the "Star Trek" motion picture and television series.

Shuttle procedures simulator (SPS)   The training hardware used to establish Shuttle crew flight procedures. It is a standard hardware item for mission-independent training.

Shuttle propulsion systems   These *Space Shuttle* systems consist of the three *main engines*, the *Solid Rocket Boosters* and *External Tank*, and the *orbital maneuvering system* and *reaction control system*. (See Fig. 1.) The main engines and boosters provide the thrust for the launch phase of a Shuttle *mission*. The orbital maneuvering system *thrusts* the *Orbiter* vehicle into *orbit* and provides the thrust for transfer from one orbit to another, for *rendezvous* with another *spacecraft* and for de-orbit. The reaction control system provides the thrust needed to change *velocity* in orbit and to change the *attitude* (pitch, yaw or roll) of the Orbiter when the vehicle is above 21,000 meters (70,000 ft).

sideband   1. Either of the two *frequency bands* on both sides of the *carrier frequency* within which fall the frequencies of the wave produced by the process of *modulation*.
2. The wave components lying within such a band.

side hatch   A hatch through which the *Space Shuttle* crew enters the *Orbiter* crew compartment. It also serves as the primary

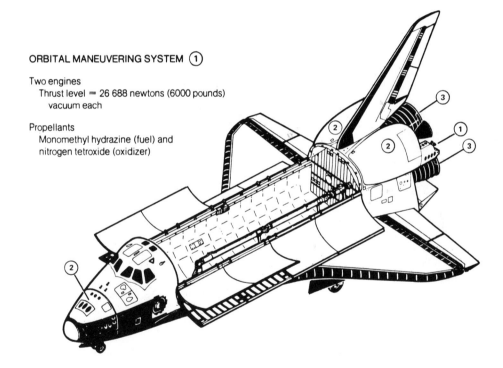

ORBITAL MANEUVERING SYSTEM ①

Two engines
    Thrust level = 26 688 newtons (6000 pounds) vacuum each

Propellants
    Monomethyl hydrazine (fuel) and nitrogen tetroxide (oxidizer)

REACTION CONTROL SYSTEM ②

One forward module, two aft pods

38 primary thrusters (14 forward, 12 per aft pod)
    Thrust level = 3870 newtons (870 pounds)

Six vernier thrusters (two forward, four aft)
    Thrust level = 111.2 newtons (25 pounds)

Propellants
    Monomethyl hydrazine (fuel) and nitrogen tetroxide (oxidizer)

MAIN PROPULSION ③

Three engines
    Thrust level = 2 100 000 newtons (470 000 pounds) vacuum each

Propellants
    Liquid hydrogen (fuel) and liquid oxygen (oxidizer)

Fig. 1   Shuttle propulsion systems.   (Drawing courtesy of NASA.)

emergency escape route.

See: Orbiter structure

signal 1. A visible, audible, or other indication used to convey information.

2. The information to be conveyed over a communication system.

3. Any carrier of information; opposed to *noise.*

signal-to-noise ratio [abbr. S/N] A ratio that measures the comprehensibility of a data source or transmission link, usually expressed as the root-mean-square *signal amplitude* divided by the root-mean-square *noise* amplitude. The higher the S/N ratio, the less the interference with reception.

simple harmonic motion [abbr. SHM] A motion such that the displacement is a *sinusoidal* function of time.

simple harmonic quantity A *periodic* quantity that is a *sinusoidal* function of the *independent variable.* Thus,

$$y = A \sin (wx + \varphi)$$

where y is the simple harmonic quantity; A is the amplitude; w is the angular frequency; x is the independent variable; and φ is the phase of the oscillation.

simulator A heavily computer-dependent training facility that imitates flight hardware responses and can be used for flight practice.

sine wave A *wave* that can be expressed as the sine of a linear function of time, or space, or both.

single-degree-of-freedom system A mechanical *system* for which only one *coordinate* is required to define completely the configuration of the system at any instant.

See: degree of freedom

single-entry compressor A *centrifugal compressor* that takes in air or *fluid* on only one side of the *impeller,* the impeller being faced with *vanes* only on that side.

single failure point [abbr. SFP] A single element of *hardware,* the failure of which would lead directly to loss of life, vehicle, or mission. Where safety considerations dictate that abort be initiated when a redundant element fails, that element is also considered a single failure point.

single-stage compressor A *centrifugal compressor* having a single *impeller* wheel, with vanes either on one or on both sides of

the wheel; also, an *axial-flow compressor* with one row of *rotor* blades and one row of *stator* blades. Axial-flow compressors are normally multistage.

single-stage rocket A *rocket vehicle* provided with a single rocket propulsion system.

See: stage

single-stage turbine A *turbine* having one set of *stator* blades followed by a set of *rotor blades.*

sink 1. In the mathematical representation of fluid flow, a hypothetical point or place at which the *fluid* is absorbed.

2. A *heat sink.* Compare this term with *source.*

sintered ceramic A *ceramic* body or coating prepared by heating a ceramic powder below its melting point but at a sufficiently high temperature to cause interdiffusion of *ions* between contacting *particles* and subsequent adherence at the points of contact.

sintering The bonding of adjacent surfaces of *particles* in a mass of powders, usually metal, by heating.

skin The covering of the body, of whatever material, such as the covering of a *fuselage,* of a wing, of a hull, of an entire *aircraft,* etc.; a body shell, as of a *rocket;* the surface of a body.

skin temperature The outer surface *temperature* of a body.

skirt The lower outer part of a *rocket vehicle.*

skirt fog The cloud of steam and water that surrounds the engines of a *rocket* being launched from a *wet emplacement.*

slave Device that follows an order given by a *master* through remote control.

slenderness ratio A dimensionless number expressing the ratio of a *rocket vehicle* length to its diameter.

slew To change the position of an *antenna* assembly by injecting a signal into the positioning *servo-mechanism.*

slewing 1. Of a *gyro,* the rotation of the *spin axis* caused by applying *torque* about the axis of rotation.

2. In *radar,* changing the scale on the display.

slip coefficient Index to the ability of an *impeller* with a finite number of *blades* to impart the same tangential whirl to the *fluid* as an impeller with an infinite number of blades.

slip flow Flow in the transition regime of gas dynamics, wherein the *mean free path* of the gas *molecules* is of the same order of magnitude as the thickness of the boundary layer. The gas in contact with a body surface immersed in the flow is no longer at rest with respect to the surface.

slipstream Flow of *fluid* around a structure that is moving through the fluid.

sliver Portion of *solid-propellant grain* remaining at the time of *web* burnout.

slosh baffles Assemblies within the *liquid oxygen* tank of the *Space Shuttle*'s *External Tank*.
See: External tank

slow neutron A *thermal neutron.*

slurry A *suspension* of fine solid *particles* in a liquid.

slurry fuel A *fuel* consisting of a *suspension* of fine solid *particles* in a liquid.

small self-contained payload A research and development *payload* that is small—

less than 91 kg (200) lb$_m$) or 0.14 m^3 (5 ft^3)—requires no *Space Shuttle* services and can be flown on a space-available basis.
See: get-away special

snubber A device used to increase the stiffness of an elastic system, usually by a large factor, whenever the displacement becomes larger than a specified amount.

soft landing The act of landing on the surface of a *planet* without damage to any portion of the vehicle or *payload* except possibly the landing gear.

soft radiation *Ionizing radiation* that has a low penetrating capability. Frequently "soft" is applied to *X-rays* of relatively low *frequency* (long wavelengths). Compare this term with *hard radiation.*
See: soft X-rays

soft sealing surface Surface fabricated of material (plastic or *elastomer*) that can yield or deform to provide the sealing action.

soft x-rays *Photons* with energy from 100 *electron volts* to 10 kiloelectron volts.

sol The *Sun.*

solar 1. Of or pertaining to the *Sun* or

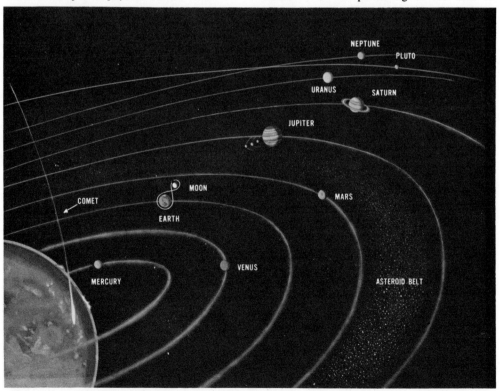

Fig. 1   The Solar System and its major celestial components.   (Photograph courtesy of NASA.)

caused by the Sun, as solar radiation, solar atmospheric tide.

2. Relative to the Sun as a *datum* or reference, as solar time.

solar activity   Any type of variation in the appearance or energy output of the *Sun*.

solar eclipse   The abscuration of the light of the *Sun* by the *Moon*.

A solar eclipse is partial if the Sun is partly obscured, total if the entire surface is obscured, or annular if a thin ring of the Sun's surface appears around the obscuring body.

solar limb   The extreme edges of the apparent solar disk.
See: Sun

solar protons   *Protons* emitted by the *Sun*, especially during *solar* flares.

Solar System   The nine known planets of the Solar System can be divided into two categories: the Jovian planets and the terrestrial planets. Jupiter, Saturn, Uranus and Neptune, the Jovian planets, are believed to consist of large cores of solid hydrogen and heavier elements surrounded by extensive atmospheres of heavy gases. As a group, the Jovian planets are less dense than the terrestrial planets and have many more satellites. Mercury, Venus, Earth, Mars and Pluto, the terrestrial planets, are dense, solid bodies without extensive atmospheres. Between the orbits of Mars and Jupiter, the majority of asteroids, of which over 1,600 are presently known, are situated. These diminutive objects are thought to be remnants of a large planet which disintegrated. The planets travel in elliptical orbits of relatively low eccentricity around the Sun.

solenoid   Helically wound coil of insulated wire that when conducting electricity generates a *magnetic field* that actuates a movable core.

solenoid valve   Valve actuated by an integrally mounted *solenoid actuator.*

solid angle [Symbol: Ω]   A portion of the whole of space about a given point, bounded by a conical surface with its vertex at that point and measured by the area cut by the bounding surface from the surface of a sphere of unit radius centered at that point.

solid lubricant   Dry film lubricant.

solid propellant rocket   A rocket propelled by a chemical mixture or compound containing a fuel and oxidizer that burn to produce very hot gasses at high pressure. The compounds burn without the introduction of outside oxygen. Solid rocket propellant systems are more reliable than rockets using liquid propellants. They have very few moving parts and can be fired at a moment's notice.

Solid Rocket Booster (SRB)   Two Solid Rocket Boosters operate in parallel to augment the *thrust* of the *Space Shuttle Main Engines* (SSMEs) from the *launch pad* through the first two minutes of powered *flight*. These boosters also assist in guiding the entire vehicle during the initial ascent. Following separation they are recovered for refurbishment and reuse. In addition to its basic component, the solid rocket motor (SRM), each booster contains several subsystems: the structural, thrust vector control (TVC), separation, recovery, and electrical and instrumentation. (See Fig. 1)

The heart of the SRB is the solid rocket motor, the largest *solid propellant* motor ever developed for space flight and the first built to be used on a manned launch vehicle. (Larger solid motors have been test-fired, however, but have not been carried through the complete development cycle to flight.) The huge solid rocket motor is composed of a segmented motor case loaded with solid propellants, an ignition system, a movable *nozzle* and the necessary instrumentation and *integration* hardware. (See Fig. 2.)

Each motor case is made of 11 individual weld-free steel segments, each heat-treated, hardened and machined to the exact dimensions required. The 11 segments are held together by 177 high-strength steel pins at each case segment joint. The clevis-type joints are wrapped with reinforced fiberglass tape and sealed with a rubber seal band that is bonded to the case with adhesives.

These segments consist of the forward dome segment, six cylindrical segments, the aft *External Tank* (ET) attach ring segment, two stiffener segments and the aft dome segment. From the 11 segments four subassemblies, or "casting segments," are preassembled before loading the propellants. These four subassemblies are called the forward casting segment, two center casting segments and the aft casting segment. The assembled case has an overall length of 35.3 meters (115.7 ft) and a diameter of 3.7 meters (12.2 ft).

Insulation inside the motor case is designed to protect the case so it can be used 20 times. The propellant inside the motor burns at a temperature of 3,475 K (3,204 C or 5,800 F) for about two minutes. Approximately 11.3 metric tons (12.5 tons)

of insulation are applied inside the motor. The thickness of the insulation varies from 0.25 cm to 12.7 cm (0.1 in. to 5.0 in.), depending on the time of exposure to the hot gases. Most of this insulation consists of a material called nitrile butadiene rubber (NBR), which has been used in previous rocket motors.

The insulation is applied in sheets that stick together and are laid down in such a way that the insulation adheres to an adhe-

Solid Rocket Boosters

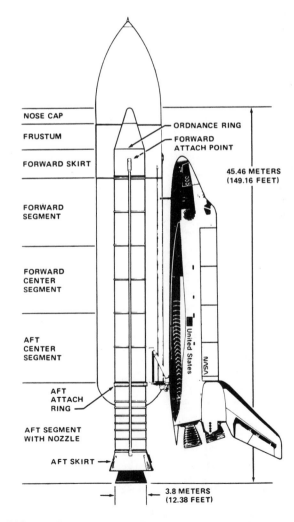

STATISTICS FOR EACH BOOSTER

THRUST AT LIFT-OFF
11 790 kilonewtons (2 650 000 pounds)

PROPELLANT
Atomized aluminum powder
(fuel), 16 percent

Ammonium perchlorate
(oxidizer), 69.83 percent

Iron oxide powder
(catalyst), 0.17 percent (varies)

Polybutadiene acrylic acid
acrylonitrile (binder), 12 percent
Epoxy curing agent, 2 percent

WEIGHT
Empty: 87 550 kilograms
              (193 000 pounds)
Propellant: 502 125 kilograms
              (1 107 000 pounds)
Gross: 589 670 kilograms
              (1 300 000 pounds)

THRUST OF BOTH BOOSTERS
AT LIFT-OFF
23 575 kilonewtons (5 300 000 pounds)

GROSS WEIGHT OF BOTH BOOSTERS
AT LIFT-OFF
1 179 340 kilograms (2 600 000 pounds)

NOSE CAP

FRUSTUM

FORWARD SKIRT

FORWARD
SEGMENT

FORWARD
CENTER
SEGMENT

AFT
CENTER
SEGMENT

AFT
ATTACH
RING

AFT SEGMENT
WITH NOZZLE

AFT SKIRT

ORDNANCE RING

FORWARD
ATTACH POINT

45.46 METERS
(149.16 FEET)

United States

NASA

3.8 METERS
(12.38 FEET)

Fig. 1   A summary of Space Shuttle Solid Rocket Booster data.   (Drawing courtesy of NASA.)

sive applied to the inside of the case walls. The insulated casting segment is placed in an autoclave (similar to a pressure cooker) to cure the insulation by vulcanization at a temperature of 422° K (149°C or 300° F) for 2 to 2.5 hours. After cooling, the insulation becomes a solid material firmly bonded to the case wall. In the dome of the aft segment, carbon fiber-filled ethylene propylene diene monomer is applied over the NBR. The final step in protecting the motor case from the extreme temperature is to spray a thick liner material over the insulation to form a bond between it and the propellant. The liner material, an asbestos-filled carboxyl terminated polybutadiene polymer, is compatible with the insulation and propellant. After curing for 44 hours at 330°K (57°C or 135°F), the lined casting segments are placed vertically into casting pits that are 6 meters (20 ft) square and 12 meters (40 ft) deep.

The propellant used in the solid rocket motor has been thoroughly proven in previous programs. It has excellent safety char-

acteristics, which have been demonstrated by standard Department of Defense hazard classification tests. The propellant type is known as PBAN, which means polybutadiene acrylic acid acrylonitrile terpolymer. In addition to the PBAN (which serves as a binder), the propellant consists of approximately 70 percent ammonium perchlorate (an *oxidizer*), 16 percent powdered aluminum (a *fuel*) and a trace of iron oxide to control the burning rate. Cured propellant looks and feels like a hard rubber eraser. The combined polymer binder and its curing agent is a synthetic rubber. Flexibility of the propellant is controlled by the ratio of binder to curing agent and the solid ingredients, namely oxidizer and aluminum.

Each solid rocket motor contains more than 450,000 kg (1 million lb_m) of propellant, which requires an extensive mixing and casting operation. The propellant is mixed in 2,271-liter (600 gallon) bowls and then taken to special casting buildings and poured into casting segments. A core *mandrel* is positioned in the casting segment prior to the casting operation. The propellant is poured under vacuum into the segments around the mandrels. After the pouring the vacuum is released, and a cover is placed over the casting pit. The segments are then cured for four days at a temperature of 330 K (57 C or 135 F). Following this the mandrel is removed, creating the burning cavity of the motor.

The high thrust level achieved by the *Shuttle flight vehicle* during *lift-off* results from the fact that the propellant in the forward segment has the configuration of an 11-point star. After lift-off, at 62 seconds into the flight, thrust is reduced by the total *burnout* of the star points, which limits *flight dynamic pressure*. Thrust then gradually increases, because of the design of the burning cavity. When the flame surface of the burning propellant reaches the liner surface, the thrust again starts to decay (i.e., decrease) and continues to decay until burnout (some 10 seconds later).

The solid rocket motor ignition system is located in the forward dome segment of the SRM. (See fig. 3.) The ignition sequence is fast moving and begins when two devices known as NASA standard initiators are

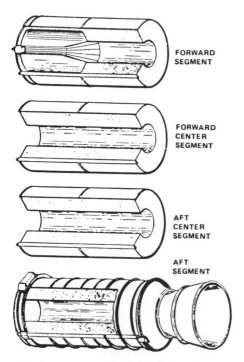

FORWARD SEGMENT

FORWARD CENTER SEGMENT

AFT CENTER SEGMENT

AFT SEGMENT

Fig. 2   The solid rocket motor, heart of the Solid Rocket Booster.   (Drawing courtesy of NASA.)

Fig. 3   Solid rocket motor igniter.   (Drawing courtesy of NASA.)

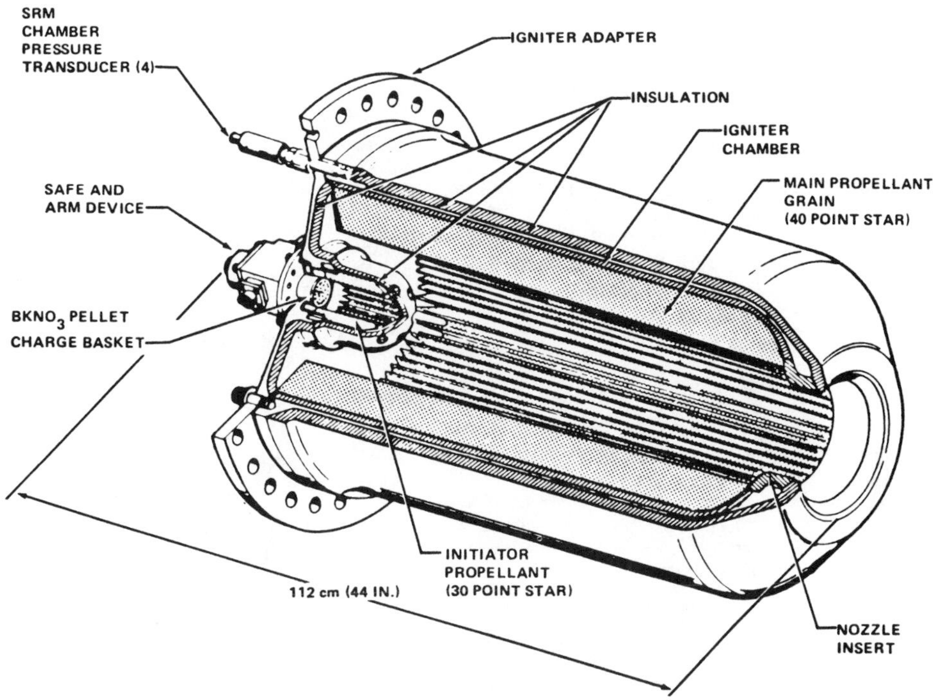

fired, igniting a booster charge of boron potassium nitrate pellets. These pellets start a small rocket motor, called the pyrogen igniter, which is simply a motor within a motor. The first motor is the igniter initiator. It is approximately 18 cm (7 in.) long and 13 cm (5 in.) in diameter. The second motor is the main igniter. It is approximately 91 cm (36 in.) long and 53 cm (21 in.) in diameter. The igniter motor flame reaches 3,172 K (2,899 C or 5,250 F) to start the solid rocket motor propellant burning. Once the SRM propellant begins to burn, flame-spreading occurs in approximately 0.15 second, and the motor reaches full operating *pressure* in less than 0.5 second.

A built-in safety mechanism, the safe-and-arm device, prevents the propellant from igniting prematurely, even if the NASA standard initiators are inadvertently fired. The entire ignition system, including the safe-and-arm device, is 112 cm (44 in.) long and weighs almost 318 kg (700 lb$_m$), of which nearly 64 kg (140 lb$_m$) is igniter propellant.

The huge nozzle is 4.19 meters (13.75 ft)

The huge nozzle is 4.19 meters (13.75 ft) long and weighs more than 9,950 kg (22,000 lb$_m$). The nozzle throat is 137 cm (54 in.) in diameter, and the exit cone is 376 cm (148 in.) in diameter. (See fig. 4.) To survive a temperature of 3,474° K (3,204° C or 5,800° F) for two minutes, materials that restrict and *ablate* rather than absorb heat are used to make a liner that is attached to the metal shell of the nozzle.

As the solid rocket motor propellant burns, huge quantities of hot gases are formed and forced through the nozzle. It restricts the flow of these gases, providing pressure and producing thrust. The gaseous products accelerate rapidly as they expand past the throat (narrow part) of the nozzle, which causes the gases to speed up to approximately 9,700 km/h (6,000 mph) by the time they leave the exit cone.

A flexible bearing allows the nozzle to move, or *gimbal*, to control the direction of the rapidly moving gases. During the SRB recovery sequence, a linear-shaped charge separates most of the nozzle exit cone, which is not recovered. This is done to pre-

Fig. 4   Space Shuttle solid rocket motor nozzle.   (Drawing courtesy of NASA.)

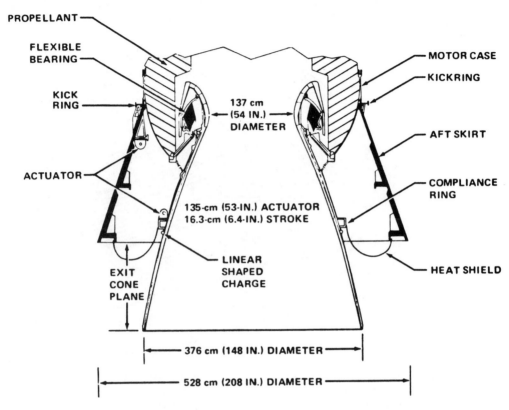

vent excessive stress on the boosters when they hit the water.

The spent (used) solid rocket motors are returned to the manufacturing plant the same way they were delivered, that is, separated into four casting segments. The first stop for the segments after arrival at the plant is a washout facility, where the insulation and any remaining propellant are removed. To accomplish this the casting segment is positioned on a tilt table and raised to a 30° angle. Streams of water at pressures up to 41,370 kN/m^2 (6,000 psi) are used to remove the residual propellant and insulation materials. The casting segments are then disassembled into the original 11 smaller case segments and sent through a degreasing and gritblasting process. From there the segments undergo magnetic particle inspection to determine whether any cracks or defects exist. Next the case segments are filled with oil and hydroproof tested, during which process the oil pressure is raised to 7,612 kN/m^2 (1,104 psig). A

second magnetic particle inspection is performed to determine if the hydroproof test resulted in any damage. The refurbished case segments are then reassembled into casting segments, repainted and prepared again for flight.

The SRB structural subsystem (see fig. 5) provides structural support for the Shuttle vehicle on the launch pad; transfers thrust loads to the External Tank/*Orbiter* configuration; and provides the housing, structural support and bracketry needed for the recovery system, the electrical components, the separation motors and the thrust vector control system.

The SRB forward assembly consists of the nose cap, the frustum, the ordnance ring and the forward skirt.

The nose cap bears the *aerodynamic* load and houses the pilot and *drogue parachutes*. It is made of lightweight stiffened aluminum. The 145-kg (320-lb$_m$) cap is 190.5 cm (75 in.) in overall length and has a base diameter of 172.11 cm (67.76 in.).

The 1,606-kg (3540-lb$_m$) frustum houses the three main parachutes of the recovery system, the altitude switch and frustum location aids and the flotation devices. The frustum also provides structural support for a cluster of four booster separation motors. It is a truncated cone 320 cm (126 in.) long, 370.8 cm (146 in.) in diameter at the base (this is identical to the SRB diameter) and 172.11 cm (67.76 in.) at the top where it joins the nose cap.

The ordnance ring connects the frustum with the forward skirt and contains a linear-shaped pyrotechnic charge that cuts the frustum and forward skirt apart. The 145-kg (320-lb$_m$) ring is 15.2 cm (6 in.) wide and 5.1 cm (2 in.) thick with a diameter of 370.8 cm (146 in.).

The forward skirt houses flight *avionics*, rate gyro assemblies, *range safety system* panels and systems tunnel components. The structure also contains a towing pendant assembly that is deployed from a parachute riser after splashdown. A forward *bulkhead* seals the skirt from seawater intrusion and provides additional buoyancy. The forward skirt is 317.5 cm (125 in.) long and 370.8 cm (146 in.) in diameter and weighs 2,919 kg (6,435 lb$_m$). The structure is a welded cylinder of individual aluminum skin panels, varying in thickness from approximately 1.3 cm to 5 cm (0.5 in. to 2 in.), and contains a welded aluminum thrust post that absorbs axial thrust loads from the External Tank.

The aft skirt is a truncated cone 229.9 cm (90.5 in.) long, 370.8 cm (146 in.) in diameter at the top (identical to the SRB diameter) and 538.5 cm (212 in.) at the base. Weighing 5,443 kg (12,000 lb$_m$), it is manufactured out of high-strength 1.3 cm to 5 cm (0.5 in. to 2 in.) thick aluminum stiffened with integrally machined aluminum longerons. The skin sections are rolled to the *skirt's* contour and welded together to provide the structural capability of supporting the entire 2,041,200 kg (4,500,000 lb$_m$) weight of the Space Shuttle on the *mobile launch platform* until launch, and of absorbing and transferring side loads during SRM nozzle gimbaling while in flight. The SRM nozzle gimbals a nominal 4.7° in all directions and up to 6.65° under certain conditions. The aft skirt also provides mounting for the thrust vector control system and structural support for the aft cluster of four booster separation motors.

The weight of the entire Space Shuttle is borne by holddown post assemblies that provide rigid physical links between the mobile launch platform and the two aft skirts. The uppermost components of the holddown post assemblies are four forged aluminum posts welded to the skirt's exterior. Each post has a rectangular base 50.8 cm by 30.5 cm (20 in. by 12 in.) and tapers into the contour of the aft skirt; and each is designed to withstand compression loads in a 344,700 kN/m^2 to 413,700 kN/m^2 (50,000 psi to 60,000 psi) range to support 255,150

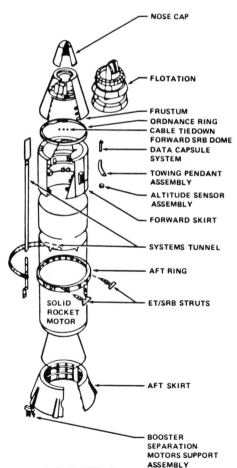

NOSE CAP

FLOTATION

FRUSTUM
ORDNANCE RING
CABLE TIEDOWN
FORWARD SRB DOME
DATA CAPSULE
SYSTEM

TOWING PENDANT
ASSEMBLY

ALTITUDE SENSOR
ASSEMBLY

FORWARD SKIRT

SYSTEMS TUNNEL

AFT RING

ET/SRB STRUTS

SOLID
ROCKET
MOTOR

AFT SKIRT

BOOSTER
SEPARATION
MOTORS SUPPORT
ASSEMBLY

Fig. 5   The structural system of the Solid Rocket Booster.   (Drawing courtesy of NASA.)

kg (562,500 lb$_m$), or one-eighth of the weight of the flight-ready Space Shuttle. These posts rest on aft skirt shoes, which provide the interface between the post and the launch platform.

At lift-off an electrical signal is sent to 16 detonators, 2 in each of 8 frangible (i.e., easily broken) nuts holding the Solid Rocket Boosters to the launch pedestal. Their detonation cracks open the nuts, releasing their grip on the holddown posts and permitting lift-off.

The External Tank is attached to each SRB in two locations: the thrust post of the forward skirt at the forward end and three aft attach struts mounted to the ET attach ring at the aft end. A single pyrotechnic bolt joins the thrust post of the forward skirt and the ET attach fitting. It is designed to carry the 899 kN (202,000 lb$_f$) tension load that occurs after SRM thrust has dropped to zero. The three aft attach struts are designed to react to lateral loads induced by SRB/ET movements, both on the mobile launch platform (because of *cryogenic* loading) and after lift-off (because of dynamic loads associated with ascent). These struts also provide the separation joint for SRB/ET separation. Each strut is designed to carry a maximum, 1,753 kN (394,000 lb$_f$) tension or compression load. Embedded in each strut

is a single pyrotechnic bolt. The ET attach ring, located at the top of the aft motor casting segment, provides the structure on which the three aft attach struts are mounted.

When pressure *transducers* sense a pressure drop at thrust tail-off, SRB separation from the External Tank is electrically initiated. The single forward separation bolt is broken when pressure cartridges force tandem pistons to function, causing the bolt housing to give way. The same technique is used to break the separation bolts in the three aft attach struts.

Each Solid Rocket Booster has a systems tunnel that provides protection and mechanical support for the cables associated with the electrical and instrumentation subsystem and the linear-shaped explosive charge of the range safety system. The tunnel extends along almost the entire length of the booster, being approximately 41 meters (133 ft) long, 25 cm (10 in.) wide and 13 cm (5 in.) thick. It is constructed of an aluminum alloy and weighs 457 kg (1,008 lb$_m$).

The thrust vector control system is the assembly located in the aft skirt that gimbals the SRM nozzle and thus helps to steer the entire Shuttle vehicle. Rate gyros continuously measure the rate of SRB *attitude*

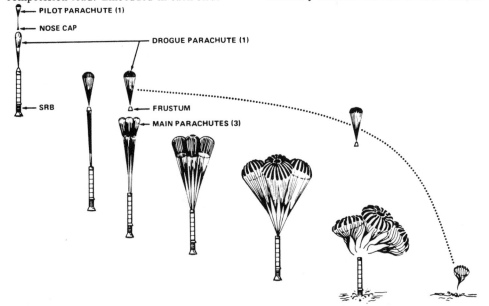

Fig. 6   Solid Rocket Booster parachute deployment sequence.   (Drawing courtesy of NASA.)

Fig. 7   Solid Rocket Booster recovery system.   (Drawing courtesy of NASA.)

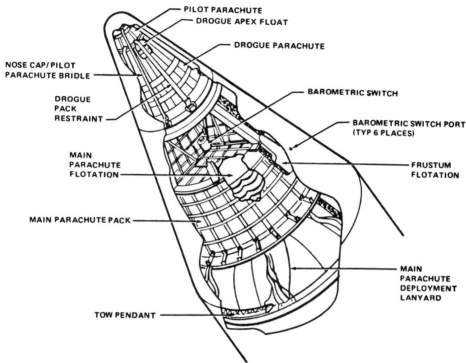

deviation, and the Orbiter computer signals the TVC electromechanical *servoactuators* to impart to the nozzle the force to create *yaw, pitch and roll* vehicle movements. The normal gimbal range is 4.7° in all directions and up to a maximum of 6.65°.

The thrust vector control system in each SRB is composed of two power modules or hydraulic power units and two servoactuators. Each power unit normally provides the power to drive a single actuator; however, both units are interconnected to both actuators, enabling either to drive both actuators (at a slightly reduced response rate). Each hydraulic power unit has an auxiliary power unit, a fuel supply module, a fuel isolation valve, a hydraulic fluid reservoir, a hydraulic pump and a hydraulic manifold. Early in the countdown the hydraulic power units are leak-tested with helium and pressurized with gaseous nitrogen, and the fuel supply module is loaded with liquid hydrazine. At 20 seconds before takeoff (a *T-time* of T-20 seconds), signals originating in the launch processing system are sent to the TVC system through a *multiplexer-demultiplexer* in the SRB aft integrated

electronics assembly. Thereafter, the TVC system is on internal command. A heat shield (thermal curtain) insulates the TVC system and its servoactuators from high heat rates due to radiated thermal energy from gases escaping from the SRM nozzle and the Space Shuttle Main Engines. Not only must the TVC components be protected during flight and preserved for subsequent reuse, but the liquid hydrazine stored in the TVC system must also be kept cool enough to prevent it from starting to burn spontaneously. (Liquid hydrazine can burn above 394 K—121 C or 250 F.)

After separation the Solid Rocket Boosters coast upward and then fall to Earth in a *ballistic trajectory* for almost four minutes. (See fig. 6.) The boosters attain a maximum speed of approximately 4,650 km/h (2,890 mph) during this trajectory before being slowed by atmospheric drag.

The parachutes of the recovery system have canopies of concentric nylon ribbons, spaced like a venetian blind. The ribbon construction adds tensile strength for high-velocity deployment. The pilot parachute, stored in the nose cap (see Fig 7), is 3.5

meters (11.5 ft) in diameter. It is designed for a maximum load of 6,584 kg (14,515 lb$_m$) during drogue parachute deployment. To begin deployment at an *altitude* of 4,694 meters (15,400 ft), a *barometric switch* actuates three thrusters on the frustum that eject the nose cap. As the nose cap moves away from the vehicle, the pilot parachute is deployed. As soon as the pilot parachute inflates, cutters release the drogue parachute pack. The drogue parachute is 16.5 meters (54 ft) in diameter and is designed to sustain a maximum load of 122,470 kg (270,000 lb$_m$). Initially it inflates approximately 60 percent, and then a reefing line is cut to allow 80 percent inflation. At an altitude of 2,835 meters (9,300 ft), a second reefing line is cut to allow full canopy inflation.

A second signal from the barometric switch activates an ordnance train at an altitude of 2,012 meters (6,600 ft.) detonating a circular charge around the ordnance ring and separating the frustum from the booster. The three main parachutes then deploy out of the base of the frustum, and the frustum continues its descent attached to the drogue parachute.

The three main parachutes, each 35 meters (115 ft) in canopy diameter with a combined loading capacity of 242,200 kg (534,000 lb$_m$), begin lowering the SRB on 52-meter (172-ft) lines at an initial descent rate of 376 km/h (233 mph). The main canopies undergo a double disreefing (i.e., the canopies partly inflate, then two sets of reefing lines are cut to permit greater, then complete, inflation). When fully inflated at an altitude of 670 meters (2,200 ft), the main parachutes *decelerate* the vehicle to 111 km/h (69 mph). Atmospheric pressure further slows the descent to a velocity of approximately 95 km/h (60 mph) at impact. Upon splashdown an impact switch activates an ordnance train that causes segmented nuts to release and separate the main parachutes. The SRBs are then recovered for refurbishment and reuse.

The electrical system (see Fig. 8) distributes power to, from and within the Solid Rocket Boosters for operation during ascent and descent and range safety. It uses

Fig. 8   Solid Rocket Booster electrical system. (Drawing courtesy of NASA.)

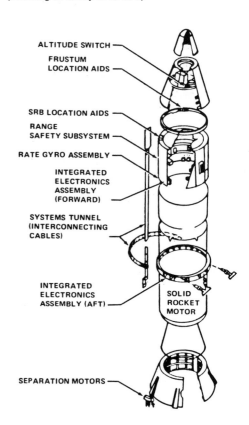

ALTITUDE SWITCH
FRUSTUM LOCATION AIDS
SRB LOCATION AIDS
RANGE SAFETY SUBSYSTEM
RATE GYRO ASSEMBLY
INTEGRATED ELECTRONICS ASSEMBLY (FORWARD)
SYSTEMS TUNNEL (INTERCONNECTING CABLES)
INTEGRATED ELECTRONICS ASSEMBLY (AFT)
SOLID ROCKET MOTOR
SEPARATION MOTORS

two power sources, the Orbiter *fuel cells* and the SRB (recovery) batteries. The range safety system has an independent, redundant electrical circuitry that is activated during the final *countdown* hour and shut down at SRB/ET separation. Power from the fuel cells and from the recovery battery (one for each booster) is routed through two integrated electronics assemblies. During ascent each SRB is controlled by commands from the Orbiter, which are processed through the electronic assemblies. This system also allows data to be received.

All commands are automatic and, for all critical functions, are routed by redundant *solid-state* components through redundant *buses* over redundant channels. In addition the *Shuttle crew* can initiate SRB/ET separation whether or not the automatic separation signal is received.

All information relating to mission safety—ignition of the solid rocket motors, performance of the thrust vector control system, separation—is conveyed over redundant hard lines. All other commands and questions are routed through the two redundant multiplexer-demultiplexers of the integrated electronics assemblies. These devices receive and send coded messages over a single pair of wires and therefore save considerable weight.

The SRB location aids consist of a radio-frequency beacon (radio transmitter) with a 17 km (9 nm) range and a flashing white strobe light with a 9 km (5 nm) range, pulsating at 500 watts intensity (similar to the output of an aircraft beacon). These aids are actuated by the altitude switch and powered from the recovery battery.

The frustum location aids consist of a radiofrequency beacon and a strobe light of the same range, output and configuration as the SRB location aids. These aids are powered by internal batteries and are activated by the closing of a "saltwater switch," in which salt water acts as the conductor between two metal pins.

All major Solid Rocket Booster functions except steering—from launch pad release through SRB/ET separation to recovery of the SRB segments—depend on electrically initiated pyrotechnics. The boosters use the following five types of *explosive-actuated devices:*

1. NASA standard initiator cartridges that contain compressed explosives.
2. Frangible (easily broken) nuts with built-in weak points that break open from the shock imparted by the cartridges.
3. Thrusters with pressure-producing pyrotechnic charges that impart velocity to a released component to move it away from another component.
4. Separation bolts that incorporate tandem pistons exerting pressure on one another, causing the outer housing to stretch and break at a weakened separation groove.
5. Linear-shaped charges that compress explosive powder into rigid, chevron-shaped channels and, upon detonation, cut apart large structures by essentially the same technique as that used by the steel industry to cut bridge girders.

All but the frangible nuts and separation bolts use "ordnance trains," which consist of NASA standard detonators connected to a metal-sheathed fiberglass-wrapped fuse (called a confined detonating fuse) through a fuse manifold. The pencil-lead-thin fuse differs from industrial mild detonating fuses only by being "confined" in fiberglass. All SRB pyrotechnics are triggered by electrical signals from the pyrotechnic initiator controllers.

Small solid-fueled booster separation motors (BSMs) "translate," or move, the Solid Rocket Boosters away from the Orbiter's still-thrusting main engines and *External Tank.* Four booster separation motors are clustered in each SRB frustum; another cluster of four is mounted on each SRB aft skirt. Both clusters are mounted on the SRB sides closest to the External Tank. The thrust of the clusters moves the SRBs away from the Orbiter.

Each of the 16 booster separation motors on the two SRBs is 79 cm (31.1 in.) long and 32.64 cm (12.85 in.) in diameter and weighs 69 kg (152 lb_m). Each BSM has a *specific impulse* of 250 at vacuum and develops a nominal 97,860 newtons (22,000 lb_f) of thrust. The nozzles of the booster separation motors are protected against accidental ignition from *aerodynamic* and radiant (SSME plume) heating. The aft clusters have aluminum nozzle covers 19 cm (7.5 in.) in diameter. The ignition blast fractures these covers at predetermined notches, and the exhaust plumes carry them away from the Orbiter and External Tank. The forward separation motor clusters, located within the frustum except for protruding nozzles, are close enough to the Orbiter that blow-away covers might strike the vehicle. Therefore, these nozzle exits are protected by stainless steel covers 19 cm (7.5 in.) in diameter that are merely blown open like doors when the booster separation motors are fired and thus remain attached to the motors. Detonators ignite the booster separation motors, which burn a nominal 0.66 second (maximum 1.05 seconds) to push the SRBs away from the Orbiter/ET.

The exterior surfaces of the Solid Rocket Boosters, exposed to thermal loads, are insulated by ablative materials to withstand friction-induced heat, which varies between SRB lift-off and SRB splashdown. Maximum temperatures of approximately 1,533 K (1,260 C or 2,300 F) are encountered at the time of SRB/ET separation, when the boosters are exposed to the plumes of the three Space Shuttle Main Engines. Maximum descent temperatures are in the range of 578 K to 589 K (304 C to 315 C or 580 F to 600 F). Splashdown temperatures are limited to 344 K (71 C or 160 F). Various types of insulation, each with special characteristics, are used to protect the SRBs. Cork and a sprayable ablative material are the primary insulating materials. Modified fiberglass in also used in protruding areas, which are subjected to especially high heating conditions.

Solid Rocket Booster retrieval and processing Associated with *Space Shuttle* operations at the Kennedy Space Center is the retrieval and refurbishment of the two reusable *Solid Rocket Boosters* that are burned in parallel with the Orbiter's three *main engines* to place the Shuttle in orbit. Solid Rocket Booster burnout occurs at T + 2 minutes [*See:* T-time] and separation occurs 4 seconds later at an *altitude* of 44.3 kilometers (27.5 miles). The boosters begin a 70-second coast to an *apogee* of 67 kilometers (41.6 miles) and then *freefall* to an impact zone in the Atlantic Ocean approximately 258 kilometers (160 miles) downrange from Launch Complex 39. The SRB *splashdown* "footprint" measures approximately 11 to 17 kilometers (7 by 10 miles).

The Solid Rocket Boosters freefall until they reach an altitude of 4700 meters (15400 feet). At this point, the nose cap is *jettisoned* and the SRB pilot parachute is deployed. This pilot parachute deploys the 16.5-meter (54-foot) diameter, 499-kilogram (1100-pound) drogue parachute, which stabilizes and slows the fall of the booster. At an altitude of 2000 meters (6600 feet), the *frustum* is separated from the SRB forward skirt, deploying the three main parachutes. Each of the main parachutes is 35 meters (115 feet) in diameter and weighs approximately 680 kilograms (1500 pounds) dry and almost twice that when wet. The 74840-kilogram (165000-pound) boosters impact

in the predetermined splashdown area 6 minutes 54 seconds after *lift-off* at a velocity of 95 km/h (59 mph).

Waiting to recover the boosters and their various components are two specially designed and constructed retrieval vessels. The two boats—designated the UTC *Liberty* and the UTC *Freedom*—are 54 meters (176 feet) long, have beams 11.3 meters (37 feet) long, have a depth of 4.6 meters (15 feet), and draw 2.7 meters (9 feet) of water. Of molded steel hull construction, the recovery vessels have sophisticated electronic communications and navigation equipment, including a satellite navigation system, search radars, collision-avoidance sonars with transponders, radars, loran C, vhf and single-sideband high-frequency radio systems, direction finders, fathometers, and gyro compasses. Each vessel has a displacement of 955 metric tons (1052 tons). At sea, propulsion will be provided by twin diesel engines with a combined power output of 2163 kilowatts (2900 horsepower). Maneuvering ability will be provided by a diesel-driven 317-kilowatt (425-horsepower) bow thruster.

The Hangar AF Disassembly Facility is located on the eastern shore of the Banana River, a shallow arm of the sea where many manatees or sea cows, an endangered species, make their homes. Propulsion in the Banana River will be provided by a 317-kilowatt (425-horsepower) waterjet stern thruster, eliminating the danger of propellers killing or maiming these marine animals. The vessels will have a sustained speed capability of 24 km/h (13 knots), a range of 11100 kilometers (6900 statute miles) and a complement of 24 (12 operating crewmembers and 12 retrieval specialists).

Both ships will be used on each Shuttle mission; each ship will recover one SRB casing, three main parachutes, and a frustum/drogue combination. [See: Solid Rocket Booster] To help locate scattered components, tracking devices have been placed on each one. The SRB casings and each frustum/drogue parachute combination are equipped with radio devices that will emit signals picked up by the recovery vessels' radio direction finder. The three main parachutes will be located by sonar. Recovery begins with retrieval of the main parachutes. Each recovery vessel has four large deck reels—1.7 meters (5.5 feet) across—each capable of holding one parachute. The parachute's winch lines are fed onto the spool and the parachutes are wound around them like line on a fishing reel. Retrieval of the frustum/drogue parachute begins in the same way. The drogue parachute is wound around one of the large reels until the 2270-kilogram (5000-pound) frustum is approximately 30 meters (100 feet) from the ship. The drogue parachute shroud lines are then rolled in until the frustum can be hoisted out of the water by a 9-metric-ton (10-ton) crane.

Recovery of the two spent solid booster casings, the last phase of the recovery mission, will be accomplished by using the nozzle plug. The nozzle plug is 4.4 meters (14.5 feet) tall and 2.3 meters (7.5 feet) in diameter and weighs 1590 kilograms (3500 pounds). Because it is remotely controlled, the plug can move in any direction. It is powered by six hydraulically driven propellor-type thruster motors. The four thrusters that give the plug its horizontal movement are positioned around the middle of the plug. The other two thrusters

are located near the bottom and move the plug up and down. The thrusters propel the plug with a horizontal velocity of 0.9 m/s (3 ft/s), a downward vertical velocity of 0.85 m/s (2.8 ft/s), and an upward vertical velocity of 1.4 m/s (4.5 ft/s).

After the plug is launched overboard, it is moved out to the bobbing solid booster casing tethered by 183 meters (600 feet) of *umbilical* cable. Although it is possible to maneuver the plug at a distance of 122 to 137 meters (400 to 450 feet) from the ship, the practical operating range is between 30 and 60 meters (100 and 200 feet). At the top of the plug is a video camera that allows the operator to view the position of the plug in relation to the casing on a shipboard television monitor. Once it is in position, the plug is lowered to a depth of approximately 44 meters (145 feet). As it descends, the television camera is used to inspect the booster casing for any damage incurred during launch, reentry, or impact. Once it reaches the bottom of the booster, it is inserted and three 0.9-meter (3-foot) metal arms are extended, locking the plug into the booster's throat.

Docking is verified by sensors on the plug's shock mitigation units, located just above the four horizontal thruster motors.

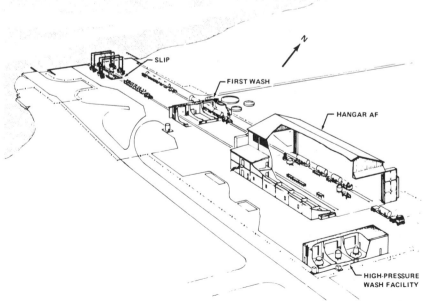

Fig. 1   SRB Disassembly Facility at Cape Canaveral Air Force Station   (Drawing courtesy of NASA.)

Compressed air is then pumped into the water-filled cavity through the umbilical cord at a pressure of up to 517 kilopascals (75 psi). As the water is forced out, the booster will begin its rotation from the vertical to the horizontal. When the water level recedes to a certain level, an inner-tube-type bag is inflated, sealing off the throat of the booster. A 3-meter (10-foot) long dewatering hose is then deployed from the nozzle plug and the remaining water is forced out through it. The booster is then towed back to the KSC Solid Rocket Booster Disassembly Facility.

The Solid Rocket Booster Disassembly Facility is located in Hangar AF at the *Cape Canaveral Air Force Station* on the eastern shore of the Banana River [See: Fig. 1]. Access to the Atlantic Ocean, from which the boosters are retrieved by ship after *jettison* during the Shuttle launch phase, is provided by the locks at Port Canaveral. A tributary channel from the Disassembly Facility ties in with the main channel on the Banana River to KSC. Recovery vessels tow the expended boosters in the horizontal position into the Disassembly Facility's offloading area where they are properly centered in a hoisting slip. Mobile gantry cranes on the hoisting slip lift the booster onto a standard-gage tracked dolly for safing and preliminary washing. The *nose cone frustums* and *parachutes* are offloaded for processing at other facilities. After safing and washing, the SRB casings are moved into the Disassembly Facility for disassembly to the level of major elements, consisting of four SRB segments, the aft skirt assembly, and the forward skirt assembly. [See: SRB] The segments then undergo final cleaning and stripping before they are shipped to the *Vehicle Assembly Building* by truck. From there, the segments will be shipped by rail to the prime contractor in Utah for final refurbishing and loading with *propellant*.

The Parachute Refurbishment Facility is located in the KSC Industrial Area to the south of the Operations and Checkout Building. Parachute systems recovered concurrently with SRB casing retrieval are delivered to the refurbishment facility on reels provided on the retrieval vessels for that purpose. The parachutes for the Solid Rocket Boosters are washed, dried, refurbished, assembled and stored in this facility. New parachutes and hardware from manufacturers also are delivered to the Parachute Refurbishment Facility.

solid rocket motor (SRM)   The heart of the *Space Shuttle Solid Rocket Booster*. This huge solid rocket motor, the largest *solid propellant* motor ever developed for space flight, is composed of a segmented motor case loaded with solid propellants, an ignition system, a movable nozzle and the necessary instrumentation and integration hardware.
See: Solid Rocket Booster

solid-state device   A device that uses the electric, *magnetic,* and *photic* (of or pertaining to light) properties of solid materials.

solidus temperature   Temperature at which melting starts.

somatic effects of radiation   Effects of *radiation* limited to the exposed individual, as distinguished from genetic effects (which also affect subsequent, unexposed generations). Large radiation doses can be fatal. Smaller doses may make the individual noticeably ill, may merely produce temporary changes in blood-cell levels detectable only in the laboratory, or may produce no detectable effects whatever. Also called physiological effects of radiation.

sonar   (From sound, navigation, and ranging.) A method or system, analogous to radar used under water, in which high-frequency *sound waves* are emitted so as to be reflected back from objects; used to detect the underwater objects of interest.

sonic   1. In *aerodynamics,* of or pertaining to the speed of sound; that which moves at *acoustic velocity* as in sonic flow; designed to operate or perform at the speed of sound, as in sonic leading edge.
2. Of or pertaining to sound, as in sonic amplifier.
In sense 2, accoustic is preferred to sonic.

sonic boom   A noise caused by a *shock wave* that emanates from an *aircraft* or *aerospace vehicle* traveling at or above *sonic velocity.*

A shock wave is a pressure disturbance and is received by the ear as a noise or clap.

sonic speed   *Acoustic velocity;* by extension, the speed of a body traveling at a *Mach* number of 1.

sorbent   The material that takes up *gas* by *sorption.*

sorption   The taking up of *gas* by *absorption, adsorption, chemisorption,* or any combination of these processes.
See: absorption

Sortie Support System (SSS)   One of the primary systems for use with U.S. Department of Defense *Shuttle sortie missions.* It is now being developed for the *Space Test Program, U.S. Air Force, Space Division.* (See Fig. 1.) The SSS consists of the following: (1) the sortie support equipment (SSE) required for mechanical support, electrical power, *thermal control,* data handling, communications, *experiment* orientation, computer software and *flight crew* interfaces; (2) the support and test equipment necessary to test, support and maintain the SSE; and (3) the astronaut training equipment needed to train flight crews and ground support personnel in the operation of the SSE.

The Sortie Support System permits operation and control of *attached payloads* by the *payload specialist* from the *Orbiter's aft flight deck.* Although this system is being developed for Shuttle sortie missions, it is also applicable to *deployed payloads.*
See: Shuttle mission categories, Shuttle crew

sound   1. An *oscillation* in *pressure, stress,* particle displacement, particle velocity, etc., in a medium with internal *forces* (e.g., elastic, *viscous*), or the superposition of such propagated oscillations.
2. A sensation evoked by the oscillation described above in the human ear.

The term sound wave or elastic wave may be used for concept 1 and the term sound sensation for concept 2. Not all sound waves can evoke an auditory sensation, e.g., ultrasound.

sounding   1. In *geophysics,* any penetration of the natural environment for scientific observation.
2. In *meteorology,* same as upper-air observation. However, a common connotation is that of a single complete *radiosonde* observation.

sound waves   Pressure oscillations in the air. Specifically, a mechanical disturbance

advancing with infinite *velocity* through an elastic medium and consisting of longitudinal displacements of the medium, i.e., consisting of compressional and *rarefactional* displacements parallel to the direction of advance of the disturbance; a longitudinal wave. Sound waves are small-amplitude *adiabatic* oscillations. The wave equation governing the motion of sound waves has the form

$$\nabla^2 \varphi = \frac{1}{a^2} \frac{\partial^2 \phi}{\partial t^2}$$

where ∇^2 is the Laplace operator, ϕ is the velocity potential, a is the speed of sound, and t is the time; the density variations and velocities are small. As so defined, this includes waves outside the frequency limits of human hearing, which limits customarily define sound.

Gases, liquids and *solids* transmit sound waves, and the propagation velocity is characteristic of the nature and physical state of each of these media. In those cases where a steadily vibrating sound generator acts as a source of waves, one may speak of a uniform wave train; but in other cases (explosions, lightning discharges) a violent initial disturbance sends out a principal wave, followed by waves of more or less rapidly diminishing amplitude. Also called acoustic wave, sonic wave.
See: ultrasonic, infrasonic, pressure wave

Soyuz   A series of over 30 manned spacecraft set in Earth orbit by the Soviet Union. The craft was composed of three sections: an orbital module, a propulsion and instrumentation section and a reentry module. The first craft was launched in April 1967.

space   1. Specifically, the part of the universe lying outside the limits of the *Earth's atmosphere.*
2. More generally, the volume in which all *celestial bodies,* including the Earth, move.

space-air vehicle   A vehicle operable either within or above the *sensible atmosphere.* Also called *aerospace vehicle.*

space base   A largescale permanent manned facility in space. It would support a crew of up to 60 and could be expanded if necessary. It would operate in low-Earth-orbit approx. 500 km at typical inclinations of 28.5°, 55° or 97°. Geosynchronous orbit

Fig. 1   A space construction base for a crew of 19 (circa 1987).   (Drawing courtesy of NASA.)

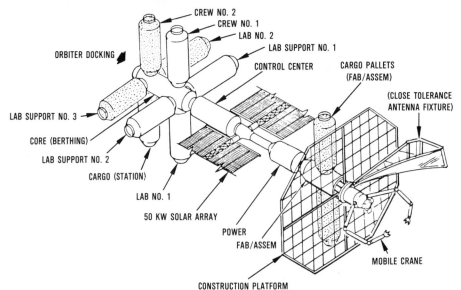

Fig. 3   A 60-person, modularized space base. This facility is powered by twin-110 kW electric nuclear reactors (far right of picture) and features telescoping sections and modular laboratories. (Photograph courtesy of NASA).

is the preferred location for a number of Earth application and scientific sensor operations. NASA has examined bases of a variety of sizes from 8 persons to 50. It has also investigated manned orbiting lunar stations and lunar surface bases.

space capsule A container used for carrying out an experiment or operation in *space*.

A capsule is usually assumed to carry a living organism or equipment.

The "Pioneer I" spacecraft, launched by the United States on 11 October, 1958 by a ThorAble I rocket at the Eastern Test Range. (Photograph courtesy of NASA).

spacecraft In general a manned or unmanned platform that is designed to be placed into an *orbit* about the Earth or into a *trajectory* to another *celestial body*. The spacecraft is essentially a combination of hardware that forms a space platform, such

as the NASA *Multimission Modular Spacecraft*. It provides structure, wiring, thermal control and subsystem functions, such as attitude control, command, data handling and power. Spacecraft come in all shapes and sizes, each frequently tailored to meet the needs of a specific *mission* in space.

spacecraft computer
See Orbiter avionics system; Orbiter displays and controls

spacecraft imaging systems
See: images from space

space defense
See: U.S. Air Force role in space

Spacelab (SL) Spacelab is a flexible laboratory system that features several interchangeable elements that can be put together in various configurations to meet the particular needs of a given flight. The major elements are a habitable module in which scientists can work in a *shirt-sleeve environment,* and platforms called pallets which can be placed in the Shuttle cargo bay behind the module. These pallets will hold instruments requiring direct exposure to space. [See Fig. 1]

Various configurations of these Spacelab elements can be fitted into the Shuttle to meet the needs of any particular mission. The habitable module can be carried alone; or with one or more pallets. The pallets can also be carried into space on a mission that

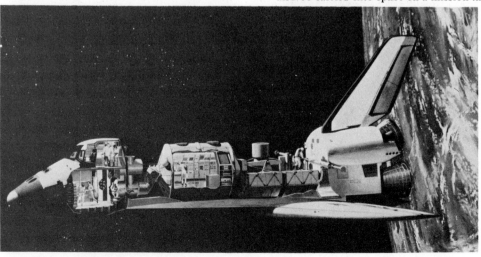

Fig. 1　A key Shuttle payload is Spacelab- a multipurpose laboratory that permits scientists to conduct experiments in the gravity free environment of space. (Photograph courtesy of NASA.)

Fig. 2   The Spacelab components.   (Drawing courtesy of NASA.)

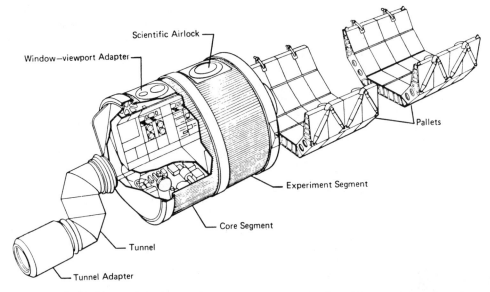

uses no habitable module. Even when no habitable module is carried, the facility is still called Spacelab. (See Fig. 1)

The module comes in two four meter (13.1-ft.) diameter segments. The "core segment" houses data processing equipment and utilities for both the pressurized module and pallets when flown together. It also has laboratory fixtures such as airconditioned experiment racks and a work bench. The second section, called the experiment racks and a work bench. The second section, called the experiment segment, provides more pressurized work space, additional experiment racks and work benches if needed. The core segment can be flown by itself (called the "short module" configuration), or coupled in tandem with the experiment segment (called the "long module" configuration). The short module, which consists of the core segment and two cone-shaped end sections, measures 4.26 meters (15.4 ft.) in length. When the two segments are assembled, the maximum outside length, including end cones, is 7 meters (23 ft.)

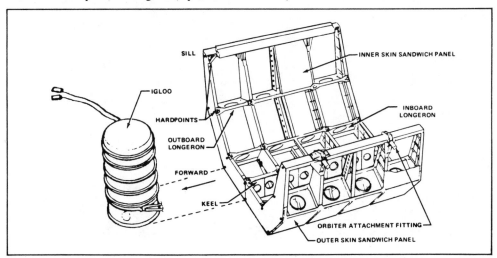

Fig. 3   The pallet segment and igloo of Spacelab.   (Drawing courtesy of NASA.)

The pallets are uniform. Each is a u-shaped platform four meters (13.1 ft.) wide and 3 meters (10 ft.) long. Heavy items, such as *telescopes,* are attached to reinforced "hard points" on the pallet's main structure. Lightweight payload equipment can be mounted on the aluminum honeycomb panels that cover the inside of the pallet. [See Fig. 3]

Up to five pallets may be flown on a single mission. When pallets are flown without a habitable module aboard, subsystems needed for equipment operation (which would normally be housed in the core segment of the module) are placed instead in a pressurized cylinder mounted to the front frame of the first pallet. This cylinder is called the "igloo".

Racks for experiment equipment that goes into the habitable module are also standardized. The 48-centimeter (19-inch-wide) racks can be arranged in single and double assemblies. Normally, the racks and floor are first put together outside the module, checked out as a unit, then slid into

the module where connections are made between the rack-mounted experiment equipment, the subsystems in the core segment, and the primary structure. [See Fig. 4]

Spacelab has other important elements. When a habitable module is flown an access tunnel about one meter (3.3 ft.) in diameter connects the module with the mid-deck level of the *Orbiter cabin.* The *hatch* between the cabin and tunnel is left open during a mission, so the Shuttle cabin, tunnel and module all share the same pressure and common air. The tunnel has lighting and handrails to allow easy passage between Spacelab and the Orbiter mid-deck, where the Spacelab crew will sleep and take their meals. The length of the tunnel varies with the number of pallets carried behind the habitable module. The more pallets aboard, the further forward the module is carried, hence the shorter the tunnel. Its length ranges from 1.4 meters (4.5 ft.) to 5.7 meters (18.9 ft.). [See Fig. 5]

A high quality window-view-port assem-

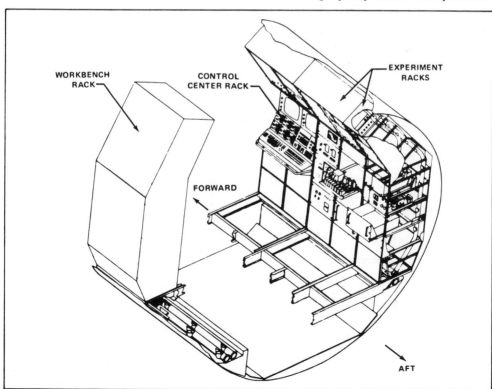

WORKBENCH RACK

CONTROL CENTER RACK

EXPERIMENT RACKS

FORWARD

AFT

Fig. 4   Cutaway view of the core segment.   (Drawing courtesy of NASA.)

bly is normally carried in the "ceiling" of either the core segment or experiment segment of the module. A small viewport is permanently mounted in the aft end cone of the module. A scientific airlock may be installed in the overhead area of the experiment segment for use in exposing small instruments to the space environment outside.

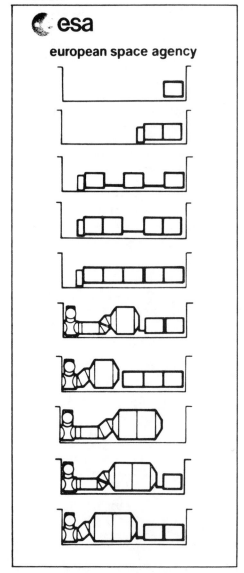

Fig. 5   Developed   by the   European Space Agency,   Spacelab can be configured in several combinations of habitable module and pallets to best support   individual mission   requirements. (Drawing courtesy of NASA.)

The major subsystems of Spacelab include: structure, environment control, electrical power and distribution, command and data management, and common payload support equipment.

The *European Space Agency* is developing Spacelab as an essential part of the United States' *Space Transportation System.* Ten European nations are involved: Germany, Belgium, Denmark, Spain, France, United Kingdom, Italy, The Netherlands, Switzerland, and, as an observer state, Austria. The European Space Agency is performing all development work and providing the first flight unit without charge to NASA. The United States is funding the purchase of the second flight unit.

The European Space Agency's prime contractor is ERNO at Bremen, Germany, but industrial firms in all ten countries involved take part in the project, Some 50 firms funnel part to Bremen for assembly and integration. [See Fig. 6]

Since Spacelab is to be a part of a NASA system, the space agency is heavily involved in its development. The Marshall Space Flight Center's Spacelab Program Office has responsibility for the technical and programmatic monitoring of Spacelab design and development activities in Europe. Marshall also has overall responsibility for management of the first three Spacelab missions.

Modules for all flight configurations contain the same basic internal arrangement of subsystem equipment; the main difference is the volume available for experiment equipment installation. Although subsystem equipment is located in the core segment, about 60 percent of the volume is available for experiments.

The interior design provides flexibility to the user. The floor, designed to carry racks with installed equipment, is in segments. The floor itself consists of a load-carrying beam structure and is covered by panels on the main walking surface. Except for the center floor plates, the panels are hinged to allow underfloor access, both in orbit and on the ground.

Fig. 1   The full scale soft mock up of the Spacelab in the environmental tent in Bremen, Germany.   (Photograph courtesy of NASA).

Fig. 1   The full scale soft mock up of the Spacelab in the environmental tent in Bremen, Germany.   (Photograph courtesy of NASA).

Mission-dependent experiment racks are available for experiments, experiment switching panels, remote acquisition units, intercom stations, and similar equipment. The standard 483-millimeter (19-inch) racks can accommodate laboratory equipment. As many as two double and two single racks can be installed in the core segment, four double and two single in the experiment segment. If experiment racks are replaced by stand-alone experiment equipment, the same attachment points must be used.

The center-aisle volume is the maximum envelope available for payload equipment mounted on the floor without effect on crew habitability and safety requirements. The underfloor volume is available for payload use only in the experiment segment. If the racks are not used, the volume they would occupy is available for other experiment equipment. The module interior is sized and shaped to allow optimum task performance by crewmembers in a *weightless* environment. The module can accommodate as many as three payload specialists working a 12-hour shift. [See Fig. 7] For shift overlap, as many as four can be accommodated for an hour. The cabin air temperature is maintained between (291 and 300 K) 64° and 81° F. The airflow is directionally controllable and is between 16.5 and (0.084 and 0.201 m/sec) 39.6 ft/min.

Foot restraints, handholds, and mobility aids are provided throughout the Spacelab so that crew-members can perform all tasks safely, efficiently, and in the most favorable body position. The basic foot restraint system is identical in Orbiter and Spacelab. Fixed handrails and handholds are distributed throughout the habitable area, such as along the standard racks and along the overhead utilities/storage support structure.

In addition to handrails, the overhead structure contains lights and air ducts. The

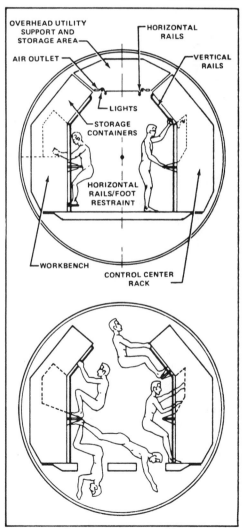

Fig. 7   Spacelab crew working area (looking forward) and examples of the many working positions.   (Drawing courtesy of NASA.)

nominal illumination level in the module is 200 to 300 *lumens*/m^2. At the workbench, it is 400 to 600 lumens/m^2. Individual lights can be turned off as necessary. A workbench in the core segment is intended to support general work activities rather than those associated with a unique experiment. One electrical outlet (28 volts dc, 100 watts) is available to support experiment equipment. Also associated with the workbench are such items as wipes for housekeeping tasks, writing instruments and paper, and a stowage pouch with individual compartments, allowing easy removal of single items without disturbing other stowed equipment. Stowage containers at the workbench, the racks, and in the ceiling provide storage space for experiment hardware, spare parts, consumables, and other loose equipment. Each ceiling container has a volume of 0.12 cubic meter (4.24 cubic feet) and a loading capacity of 36 kilograms (79.4 pounds). Workbench and rack containers are smaller, having a volume of 0.075 cubic meter (2.65 cubic feet) and a load capacity of 22.4 kilograms (49.4 pounds). A grid pattern of mounting holes inside the container is provided for attachment of internal restraints. The container door can be fastened open. Two workbench containers will be available for payload use on most missions; eight ceiling and four rack containers are exclusively for experiment use. In a long-module configuration, there is room for 14 ceiling containers, but only 8 are provided as baseline hardware. The optical window takes the space of at least five (depending on space required for payload activities) ceiling containers. The airlock requires the space of six containers.

Rack and ceiling containers can be converted to accommodate as much as 150 kilograms (331 pounds) of film in plastic foam protective carriers. A range of standard equipment (such as tools and contingency maintenance items), stowed in the workbench containers, is used to support Spacelab activities but will also be available for experiment use. In addition, a tool and maintenance assembly, with a utility box for stowage, will include off-the-shelf tools, specially designed tools, and maintenance items. Utility straps and *bungees* are provided so that crewmembers working on orbit can temporarily restrain equipment at various locations throughout Spacelab.

The transfer tunnel connecting the Spacelab module and the Orbiter enables crew and equipment transfer in a shirtsleeve environment. Mobility aids are installed in the tunnel. The lighted tunnel has the same internal atmosphere as the Spacelab module. It has a minimum of about 1 meter (3.28 feet) clear diameter, sufficient for a box with dimensions of 0.56 by 0.56 by 1.27 meters (1.84 by 1.84 by 4.17 feet) and a crewmember (including one equipped for *EVA*) 1.95 meters (76.77 inches) tall and with a maximum elbow width of 0.75 meter (2.46 feet). The tunnel can also be used for ground access to the Spacelab while it is still horizontal.

For extravehicular activity, Spacelab provisions allow a crew-member in a *pressure suit* to move through the EVA hatch (the Orbiter personnel *airlock* in the Spacelab tunnel adapter, or the *docking module,* depending on the mission configuration), up the end cone of the module, over the module, down the aft cone, and along the pallet. The size of the *airlock* and associated hatches limits the external dimensions of a package that can be transferred to *payloads* to 0.9 meter (2.95 feet) in diameter and 1.4 meters (4.59 feet) long.

For experiments requiring specific operating conditions, certain elements of the payload support equipment can be flown. These items are listed in Table 1.

Functional interfaces between the Orbiter, module and pallet are provided by suitable utility connections. Access to the module from the Orbiter cabin is by means of a tunnel. The latter is attached to the EVA adapter which in turn is linked to the Orbiter cabin itself. The variable length of the tunnel permits some freedom in placing the Spacelab elements in the cargo bay to provide better viewing conditions and/or to satisfy the centre-of-gravity constraints placed on Spacelab and its payload by the Orbiter.

A relatively mild environment is anticipated for Spacelab experiments. Some of the principal parameters are given in Table 2.

Table 1 Spacelab payload support items.

Airlock	One airlock (1 m diameter, 1 m long) available, to be used in top opening of experiment module; allows direct access to space from module. Supports up to 100 kg
High Quality Window	Skylab-type window permits observations from inside module for high-quality viewing in the visible and near-infrared parts of the spectrum; size 41 × 55 cm
Viewport	Two viewports (30 cm diameter) available, one in the aft end cone*, the other one in a top opening of the module
Experiment Vent Assembly	Permits venting of gases from experiment chambers etc, located in the module
Feed-through*	Permits passage of lines (fluids, signals, etc) peculiar to experiments between module and pallet

*always flown

Credit: NASA/ESA

Table 2 Principal Spacelab environment parameters

Parameter	Approximate values
Acceleration	Maximum 3 g linear acceleration during ascent and descent; typically 10^{-4}g on orbit
Vibration	145 dB acoustic noise in cargo bay (launch only); 136 dB acoustic noise inside module (launch only); typically 4g RMS random vibration input to equipment in racks in the module
Thermal	Inside the module: equipment cooling through forced air in the range 20 to 40°C, cabin air in the range 18 to 27°C (adjustable). One cold plate (10 to 40°C) and 4kW heat exchanger available On pallet: equipment cooling by cold plates with temperatures in the range 10 to 40°C
Contamination	Arises from Orbiter, Spacelab and experiment equipment. Precautions will be taken in design to reduce level as far as practically possible. Dumps can be programmed

Credit: NASA/ESA

The Spacelab concept possesses considerable flexibility in its application to a variety of missions. This very important characteristic arises from two sources.

First, the Space Shuttle flight parameters may be varied so that the *orbit inclination,* orbit altitude (200 to 900 km) and resulting ground coverage may be selected for mission compatibility. During the first few years of Shuttle operation, the East Coast launch site at Kennedy Space Center will be used so that the possible range of inclinations is 28.5 to 57°. Later, the West Coast site at Vandenberg Air Force Base will become available, thereby ensuring orbit inclinations up to 104°. Also the Orbiter orientation (all directions with an accuracy up to 0.5° per axis) and flight duration (initially up to 12, eventually up to 30 days) can be adjusted as required.

In the second place, Spacelab mission flexibility results from the modular approach adopted in the design. The module and pallet can be varied in size by selecting from the available Spacelab elements in such a way that the resulting configuration fits the needs of the mission in question. Three basic configurations are apparent— module only, module plus pallet, pallet only. Further flexibility is introduced by the payload support equipment available. The subsystem elements are also modularised so that certain components (eg, cold plates, equipment racks, recorder) may be used or removed as required. This feature means that on a particular Spacelab flight, only those mission-dependent elements required by the experimenter are flown, permitting additional payload weight to be substituted for unnecessary equipment.

The Orbiter/Spacelab subsystems provide basic services for running Spacelab itself and for the payload. These services are available to the experiments via standard interfaces and have been designed to ensure that near-laboratory-type equipment may be used in the module. The actual resources available to the payload are a function of the configuration being flown. Table 3 summarises the services provided in the case of four typical Spacelab configurations. It is the task of the mission planner to ensure that the sums of the requirements of all experiments in the payload do not exceed the total resources available. The services provided include the use of an instrument pointing subsystem.

An instrument pointing subsystem (IPS) can provide precision pointing for payloads that require greater pointing accuracy and stability than is provided by the Orbiter. The IPS can accommodate a wide range of payload instruments of different sizes and weights. The *gimbal* system is attached to the payload when on orbit, and performs the control maneuvers required by the observation program.

During ascent and descent, the payload is physically separated from the IPS to avoid imposing flight loads from the IPS to the payload. The payload is supported by the payload clamp assembly, which distributes the flight loads of the payload into the pallet hardpoints. The payload clamp assembly is capable of mounting and distributing the load of a nominal (2000-kilogram) 4410-pound payload and the IPS into a single unmodified pallet without exceeding safe loading conditions.

The IPS provides three-axis attitude control and stabilization for experiments. The typical payload characteristics are shown in Table 4.

During orbital operations, the IPS is capable of continuous operation in full solar illumination or in a completely shadowed configuration.

Overall control of the IPS during normal operations is exercised from the Spacelab control console using the keyboard and display of the command and data management subsystem. The flight operating software is capable of interfacing, through the Spacelab subsystem, with the Orbiter data-handling system. The IPS control system uses this Spacelab subsystem for all normal operations. Emergency retraction or *jettison* is exercised from a separate IPS control panel located on the Orbiter *aft flight deck.*

Spacelab offers numerous advantages over the experimentation capability of the *Apollo* and *Skylab programs.* Obviously, much more room, time and facilities will be offered by Spacelab than were available in Apollo. Skylab had more room, but the cost of getting crews to and from the orbiting space station

Table 3 Spacelab services for users

Spacelab configuration	Short module + 9 m pallet	Long module	15 m pallet	Independently suspended pallet
Payload weight (kg)[1]	5000–6000	4800–6200	7700–8300	8800–9400
Volume for experiment equipment:				
Inside module (m³)	8	22	—	—
On pallet (m³)	100	—	160	100
Pallet mounting area (m²)	51	—	85	51
Electrical power (28 V DC 115/200 V at 400 Hz AC)[2]:				
Average (kW)[1]	2.5–4.0	2.5–4.5	4.5–5.5	4.5–5.5
Peak (kW)[1]	8	8	9	9
Energy (kWh)[3]	~250	~300	~550	~550
Experiment-support computer with central processing unit and data acquisition system	64 K core memory of 16 bit words, 350,000 operations per second, 15 K core available to users			
Data handling: Transmission through Orbiter Storage digital data	Up to 50 Mbps Up to 30 Mbps total of 3×10^{10} bits	Up to 50 Mbps Up to 30 Mbps total of 3×10^{10} bits	Up to 50 Mbps Up to 1 Mbps total of 3×10^9 bits[4]	Up to 50 Mbps Up to 1 Mbps total of 3×10^9 bits[4]
Instrument pointing subsystem IPS	Mounted on pallet, will provide arc second pointing for payloads up to 3000 kg			

1. Depends on the amount of mission-dependent equipment flown
2. Depends on the power and energy consumed by mission-dependent equipment and the degree of usage by switchable subsystem equipment
3. Energy can be increased by the addition of payload-chargeable kits, each providing 840 kWh and weighing approximately 350 kg (at landing)
4. Using Orbiter Payload Recorder.

Credit: NASA/ESA

Table 4 Characteristics of nominal payloads with the IPS.

Parameter	Large payload	Small payload
Mass, lb (kg)	4410 (2000)	440 (200)
Dimensions, ft (m)	(2 by 4)	(1 by 1.5)
Moment of inertia about payload center of gravity, kgm²		
About axis perpendicular to line of sight	1200	20
About line-of-sight axis	1000	25
Center of gravity offset from center of rotation of gimbal axes, ft (m)		
Along line of sight	6.6 (2.5)	4.9 (1.50)
Perpendicular to line of sight	1 (0.30)	0.33 (0.10)

was great. Also, experiments could not be added or removed between missions, and the experiment *hardware* was not returned to *Earth.*

Spacelab flights will be shorter, at much less cost. Also, non-career astronaut scientist, called *payload specialists,* in reasonably good health can actually go into Earth orbit to carry out investigations, instead of relying on a NASA astronaut as in the past.

The new orbiting laboratory will be capable of supporting almost any kind of experiment, including many which would be extremely difficult or impossible on the ground.

The absence of gravity in orbit allows scientists to conduct experiments that Earth's gravity interferes with on the ground. For example, many elements do not combine uniformly on Earth because gravity causes the heavier molecules to "settle" to the bottom. In space they mix evenly. Crystals can be allowed to form in a purer state while "floating" in space, uncontaminated (and unrestrained) by containers necessary to hold them under gravity conditions. Lubricants, when weightless, spread over surfaces more evenly and medical substances can be mixed more uniformly and produced with higher degrees of purity in the absence of gravity.

Spacelab will fly above most of the Earth's obscuring, filtering atmosphere. So, any telescopes aboard will be able to see farther into space with greater clarity, and its sensors can study *gamma-rays* and other radiation from space, *radiation* which is essentially blocked by the *atmosphere.*

The reusable laboratory will be used to conduct a wide variety of experiments in such fields as life sciences, *plasma* physics, astronomy, high-energy *astrophysics,* solar physics, atmospheric physics, materials sciences and Earth observations.

Earth observations, for example, will provide data for land-use studies, pinpoint pollution, spot crop and forest diseases, provide leads in the search for minerals, waters and other natural resources, enable more accurate mapping and make information available for many other services.

Spacelab's potential benefits to mankind are almost limitless. Using Spacelab, we will

acquire important fundamental knowledge of man: what controls his biological systems, knowledge of his environment and what processes are at work in it, and what control he has over himself and his world.

Scientists are convinced that the more man knows about himself and his surroundings the better he will be able to make changes for the betterment of the human race.

Spacelab will be used by scientists from countries around the world. Its use is open to research institutes, scientific laboratories, industrial companies, government agencies and individuals. While many missions are government sponsored, Spacelab is also intended to provide services to commercial customers.

Each experiment accepted has a *"principal investigator"* assigned as the single point of contact for that particular scientific project. The principal investigators for all experiments on a given mission form what is called the Investigators Working Group. This group coordinates scientific activities in preparation for, and during the actual flight. The investigators prepare the equipment for their experiments in accordance with size, weight, power and other limitations established for the particular mission. Responsibility for experiment design, development, operational procedures and crew training rests with the investigator. Only after it is completed and checked out is the equipment shipped to the Kennedy Space Center for installation on Spacelab.

Each mission has a "mission scientist," a NASA scientist who, as chairman of the Investigators Working Group, serves as the interface between the science-technology community and NASA's payload management people. Through the mission scientist the science-technology needs of the mission and the investigators' goals are injected into the decision making process.

The usual crew for a Shuttle/Spacelab mission consists of: two *pilot astronauts* who operate the Shuttle and carry out the flight plan; one or two *mission specialist* astronauts, who operate Orbiter/Spacelab systems and support payload operations; and up to four "payload specialists," the individual who actually operate the scientific equipment and conduct the experiments. A total of seven persons may fly on any one mission. [See: Shuttle crew]

Payload specialists are not career astronauts. They are scientists or technicians of average physical ability who accompany experiments into orbit. A payload specialist may be selected to fly on more than one mission, but NASA will not create a permanent corps of payload specialists. [See: payload specialist] Payload specialists' training includes visits with the investigators to observe and discuss all aspects of experiment development and to practice performing the experiments using the actual equipment that will be flown in the Spacelab later. The payload specialists also provide feedback to the investigators on how difficult the equipment is to operate. [See: payload specialist training] This helps refine equipment design and procedures and avoid crew operational problems later.

The Marshall Spaceflight Center is responsible for the overall management of the first three Spacelab missions. The experiments to be flown have been defined and the payload specialists for the first two missions selected. Selection of the final experiment complement and the science crew for the third mission is in progress.

The first mission, sponsored jointly and shared equally by NASA and the European Space Agency, is currently scheduled for late 1983. It will feature about 75 separate investigations during seven days in orbit. The Spacelab configuration will consist of the two-segment, long module and one pallet. International in nature, Spacelab Mission 1 will have experiments from the United States, Japan, the Netherlands, United Kingdom, Belgium, France, Germany, Italy and Switzerland. The disciplines represented by these experiments are: atmospheric physics; Earth observations; space plasma physics; materials sciences technology; astronomy; solar physics; and life sciences.

The science crew *for Spacelab 1* will consist of two payload specialists, who will operate the experiments with support provided by two mission specialists. Five payload specialists—two Americans and three Europeans—are now in training for the mission.

Prior to the mission, one American and one European will be selected to actually fly the mission. The other three will provide support from the *Payload Operations Control Center* on the ground. The mission specialists will be astronauts Dr. Owen Garriott and Dr. Robert Parker. The payload specialists selected for Spacelab Mission 1 are: Dr. Michael L. Lampton, a member of the research staff at the Space Sciences Laboratory, University of California at Berkeley. His main fields include astronomy, space physics and optical and electronics engineering. Dr. Byron K. Lichtenberg of the Massachusetts Institute of Technology, who specializes in biomedical engineering. Dr. Ulf Merbold, staff member at the Max-Planck Institute, Stuttgart, West Germany. His main interests are the study of crystal lattice defects and low-temperature physics. Dr. Wubbo Ockels, of Groningen University, the Netherlands. He has been involved deeply in nuclear physics experimental investigations. Dr. Claude Nicollier, of Vevey, Switzerland, a scientist at the Astronomy Division of the European Space Agency's Space Science Department in the Netherlands.

Thirteen major experiments are to be carried out on the second Spacelab mission in the fields of life sciences, plasma physics, infrared astronomy, high energy physics, solar physics, atmospheric physics and technology research. This mission is sponsored solely by the United States.

The configuration for Spacelab 2 is three pallets with igloo and a special fixture at the rear for a large cosmic ray experiment. It will also feature an Instrument Pointing System, a special item of Spacelab equipment designed to support experiments. This mission is currently scheduled for late 1983-1984.

Four payload specialists are now in training for the mission. As on Spacelab Mission 1, only two of these payload specialists will actually fly in space. The other two will provide ground support. The four are: Dr. George W. Simon, of Sacramento Peak Observatory, Sunspot, NM. He specializes in problems concerning velocity, magnetic and intensity fields in the solar convection zone, photosphere, chromosphere, transition zone and corona of the *Sun*. Dr. Dianne K. Prinz, of the Naval Research Laboratory, Washington, D.C. She is a research physicist specializing in solar-terrestrial relationships and the design of optical instrumentation. Dr. John-David F. Bartoe, also of Naval Research Laboratory. He is co-investigator on two Spacelab 2 experiments on the mission, experiments to study the *ultra-violet* radiation from the *Sun*. Dr. Loren W. Acton, of Space Sciences Laboratory, Lockheed Missiles and Space Co., Palo Alto, Calif. His principle duties include conducting studies of the Sun and other *celestial objects* using advanced space instruments.

The Spacelab 3 mission will be the first operational flight of Spacelab. Experiments will be conducted in the applications, science and technology disciplines with emphasis on material processing. This requires the low gravity environment of Earth orbit with minimum vehicle attitude maneuvers. The Spacelab 3 mission will last for seven days duration and is scheduled for launch from the Kennedy Space Center in 1984-85. On-orbit activities for the crew of six will be conducted 24 hours a day in a 370-kilometer (200 nautical mile) high orbit that is inclined to the equator at 57 degrees. A single orientation, appropriate for sustaining low gravity, will be maintained for the full duration of the mission.

As manager of the first three flights, the NASA Marshall Center's Spacelab Payload Project Office has responsibility for insuring that planned experiments meet agreed-upon criteria; that equipment is shipped to the launch site on schedule; that *integration* is accomplished, checkout is completed; and that the mission is carried out as planned.

Each Spacelab flight has a Marshall mission manager who is responsible for managing all activities necessary for the integration of the scientific instruments into a compatible payload within Shuttle capabilities. He is the single point of authority for the entire payload on his assigned mission; the "broker" between individual experiments and Shuttle. He represents the users in negotiating the use of Shuttle and Space-lab accommodations. The mission manager is also responsible for program control, science and technology management, payload definition and interfaces with Shuttle, integrated payload mission planning, control of payload integration, payload specialist training, assuring *launch site* payload integration, defining payload mission operations and data handling requirements, conducting payload operations, and directing payload specialist activity during flight operations.

During the actual mission, the Marshall Center mission manager oversees payload operations from a special control center located adjacent to Shuttle Mission Control at the Johnson Space Center in Houston. This complex, called the *Payload Operations Control Center,* is the hub of experiment activity. From there the mission manager, principal investigators and backup payload specialists will control and direct experiment operations in the orbiting Spacelab.

Spacelab users will, therefore, be drawn from the various disciplines of science, applications and technology. Investigations have shown that, at least, the following fields are likely to obtain benefits from the utilization of Spacelab:

—high-energy astrophysics
—ultraviolet, optical, infared and X-ray stellar, planetary and solar astronomy
—atmospheric, ionospheric (plasma) and magnetospheric physics
—life sciences (including biology, biomedicine, behavior)
—remote earth-sensing (meteorology, land-use planning, resources, pollution control, etc)
—material sciences (eg., crystal growth, pure metals and alloys, composite materials) and fluid physics
—processing and manufacturing in space (eg., electrophoresis, high-strength materials)
—communications and navigation
—advanced technology in all disciplines

These fields are cited as typical and additional areas that could benefit from using Spacelab will be identified as the program matures. It is foreseen that Spacelab will play an important role in the various development phases of those disciplines, which include pure research, instrument R&D, experimental processes and the execution of operational programs.

Spacelab provides a capability for two modes of experimentation—man-tended activities or automated observations. The choice of mode is left to the experimenters who may prefer to have an operator in attendance who can improve the overall efficiency of the planned experimentation and fully exploit unexpected events. On the other hand, automated operation of the equipment may be preferable.

In the Life Sciences Flight Experiment Program, for example, NASA is studying the use of Spacelab to conduct research in the null gravity and altered environments (radiation, acceleration, light, magnetic fields, etc.) of space. The Shuttle/Spacelab presents a unique capability to perform numerous experiments in all fields of life sciences; i.e., biomedicine, vertebrates, man/system integration, invertebrates, environment control, plants, cells, tissues, bacteria, and viruses. Broad objectives are to use the space environment to further knowledge in medicine and biology for application to terrestrial needs, as well as to ensure human well-being and performance in space.

These efforts are structured for the widest participation from the the public and private sectors and is characterized by low-cost approaches, many flight opportunities, short experiment turnaround times, provi-

sions for qualified investigators to fly with their experiments, and maximum use of existing or modified off-the-shelf hardware.

Initially, 7-day missions are scheduled with the Spacelab dedicated to life sciences experiments. Eventually, dedicated flights as long as 30 days are envisioned. The carrier will generally consist of the pressurized Spacelab module with racks and payload support equipment; pallets may or may not be used. The Spacelab will be outfitted with selected common operational research equipment [Table 5] especially acquired to support life science activities. [See Fig. 8]. For a given flight, 10 to 20 life sciences experiments will be carefully selected and developed that can be operated by one or two onboard payload specialists. These specialists may be supported by other onboard crewmembers and may receive real-time and off-line support from ground-based scientists and engineers via air-to-ground data, television, and voice communications links to the POCC and to remote monitoring areas. Several dedicated life sciences Spacelab missions are currently being planned.

A representative life sciences laboratory layout is shown in Fig. 8. This and other typical Spacelab configurations will support medical / biological / technological applications experiments on subjects of man, primates, small vertebrates, plants, cells, and tissues, and will provide null gravity research capability for advanced life support systems.

Function	Equipment
Blood sample	Freezer, clinical centrifuge, hematocrit reader, radiation detector
Urinanalysis	Total solids refractometer, pH meter, freezer, refrigerator
Cardiovascular function	Limb plethysmogram, cardiotachometer, phonocardiogram, oscilloscope, biomedical recorder
Small vertebrate physiology	Glove box, surgical table, mass measurement device, veterinary kit, small vertebrate holding facility
Primate monitoring	Telemetry receiver, biomedical recorder, oscilloscope, primate holding unit
Cell, tissue growth, morphogenesis	Microscope, still camera, colony chamber, biological specimen holding facility incubator
Vestibular function	Rotating chair, electrocardiograph, electroencephalograph, electromyograph, electro-oculograph, cardiotachometer, biomedical recorder

Table 5   Typical Spacelab core operational research equipment.   (Table courtesy of NASA.)

Other payload configurations will consist of carry-on laboratories and minilaboratories, which will be designed to fit either in the Orbiter mid deck or in the Spacelab. Carry-on laboratories will generally be designed to require a minimum of inflight crew time for operation or maintenance and will minimize inflight control and data transmission. Shared Spacelab missions (with mini-laboratories) will be similar in all respects to the dedicated Skylab missions except that the life sciences experiments will make up only a part of the payload. One or more racks will usually contain not only the necessary common operational research equipment and subsystems, but also the experiments they support.

In addition to the basic services provided by Spacelab, certain experiment facilities (eg, furnaces, *telescopes*, high-power *lasers*) will be available for certain missions. These 'facilities' are not provided as part of the Spacelab Program, but will be supplied from 'user' sources, and experimenters from a wide variety of disciplines may take advantage of them for exploring their particular problem areas. In this way it will be possible to attract users of Spacelab from all levels of the scientific and technical communities, be it small university groups or large government agencies. Thus, the participation of an experimenter in Spacelab activities may take four basic forms:—by provision of a complete experiment unit, ie, facility plus detectors or samples;—by supplying experiments for use with a common facility, eg, 'behind the focus' type experimentation;—by provision of an independent experiment which does not utilise a facility; and—by use of the data generated during a Spacelab mission without the provision of any equipment itself, as in the case of certain earth-observation data.

In planning experiments that require attendance, it must be stressed that payload specialists who fly in Spacelab will be scientists and technicians rather than professional astronauts, and no rigorous, long-duration, pre-flight training is anticipated. In some cases it may be desirable to involve the user community on the ground. This involvement can be achieved by communicating experiment data to the ground, in realtime, via the *Tracking and Data Relay Satellite System*. Conversely, commands may be transmitted from the ground to Spacelab.

Spacelab and Shuttle activities are repeated from flight to flight but with different payload complement. The overall responsibility for these operations rests with NASA. Spacelabs and their payloads may be decoupled from the Shuttle turn-around cycle, thereby permitting more time for off-line payload preparation and integration. It is intended that the experimenter be given an active role both on the ground and during the flight itself. The various phases envisaged for experiment *integration* are described as follows:

—Level IV: integration and checkout of experiment equipment with individual experiment mounting elements (eg, racks and pallet segments)—activities that will be possible at the user's home facility.

—Level III: combination, integration and checkout of all experiment mounting elements (eg. racks, rack sets and pallet segments) with experiment equipment already installed, and of experiment and Spacelabe software, ie, payload integration normally carried out at Kennedy Space Center (KSC).

—Level II: integration and checkout of the combined experiment equipment and experiment mounting elements with the flight subsystem support elements (ie, core segment, igloo) and experiment segments when applicable—activities normally performed at KSC.

—Level I: Integration and checkout of the Spacelab and its payload with the Shuttle Orbiter, including the necessary pre-installation testing with simulated interfaces—this procedure is carried out at the actual launch site.

The Level II integration procedure for module-located experiments is facilitated by the roll-out design concept adopted for Spacelab. The payload is contained in the rack and floor combination, which is literally rolled into the Spacelab shell by a

Fig. 8   Typical Spacelab equipment dedicated to on-orbit life sciences investigations.   (Drawing courtesy of NASA.)

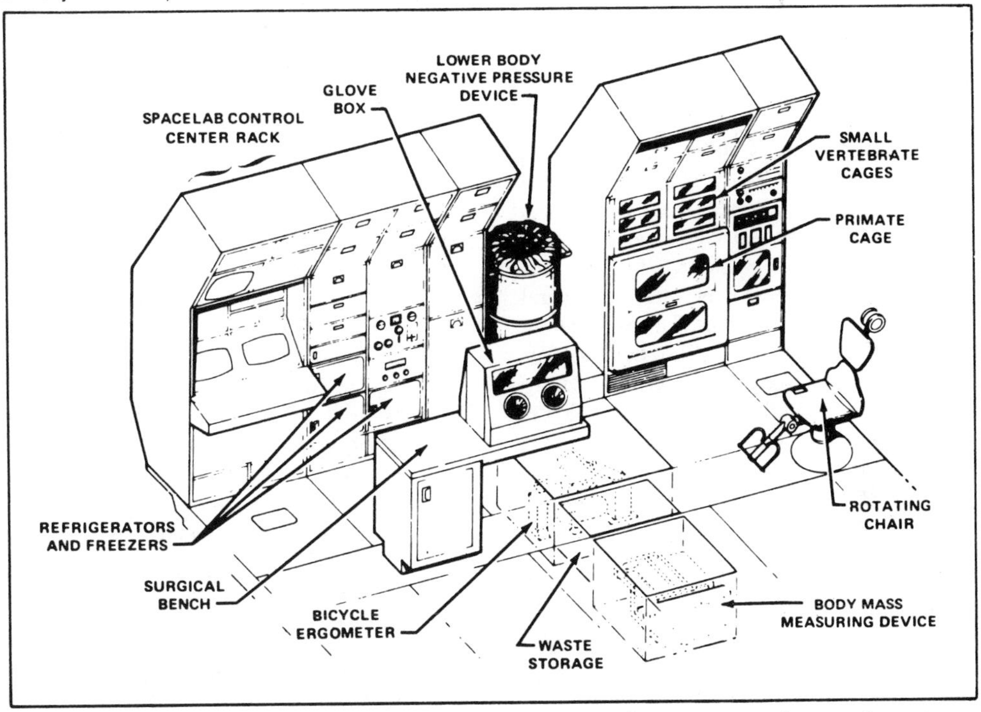

roller-rail system. Special organizations have been set up in Europe and the USA to ensure that the relevant integration phases are effectively executed. These organizations also ensure that adequate support is given (including the necessary ground support equipment) to the experimenter during the equipment development phase.

Spacelab, which the European Space Agency built, is one of the most exciting payloads to be flown by the Space Shuttle. Through its creative utilization by international teams of scientists, man will initiate the "humanization of space" and all the universal terrestrial benefits that such a significant sociotechnical process implies.

space medicine   A branch of *aerospace medicine* concerned specifically with the health of persons who make, or expect to make, flights into *space* beyond the *sensible atmosphere.*

space motion   Motion of a *celestial body* through *space.*

That component perpendicular to the line of sight is termed proper motion and that component in the direction of the line of sight, radial motion.

space nuclear power systems   Systems employing nuclear energy to provide electrical power for space missions. Two types of systems are used: the radioisotope powered generator and the nuclear reactor. Radioisotope systems are used for missions requiring power in the 500 to 2000 watt range. Beyond that level of power a nuclear reactor is the technology of choice. These power systems are used on missions where continuous, reliable power is needed over a long period and where low weight per power output and compact structure are important. They have been used in 18 missions including the Pioneer, Viking and Voyager flights.

space nuclear propulsion   The use of nuclear reactors to propel a space vehicle 1) by generating electrical power for an electric propulsion unit or 2) as a source of thermal energy to heat a propellant to extremely high temperatures for subsequent expulsion out a nozzle. In the latter application of nuclear fission energy, the system is called a "nuclear rocket."

In a nuclear rocket, chemical combustion is not required. The propellant, typically hydrogen, is a separate substance heated by the energy released in the fission process—which occurs in the reactor's core.

Space Operations Center [abbr. SOC]

NASA is currently studying the concept of a manned space station in low-Earth orbit. The SOC would provide a base for extended orbital missions. It is a self-contained orbital facility built of several Shuttle-launched modules. The SOC would contain six core elements. Two identical service module halves would form a central spine with multiple docking ports. These would provide the station with power, communications, etc. Two modules with living quarters would be attached to the service module. Each would contain a command center capable of controlling the entire base. The two modules are joined by a tunnel. A logistics module would store provisions and equipment. The SOC would also contain construction facilities (for building large space systems) and a flight support facility (for launching multistage rocket vehicles). A control cab and Shuttle-type manipulator on a movable boom serve both operations. The station would also contain a manned orbital transfer vehicle for sortie missions to geosynchronous orbit.

space rescue

See: extravehicular mobility unit

space settlement

A large space habitat for perhaps 1,000 to 100,000 persons. The inhabitants would be members of a space manufacturing complex work force, possibly constructing and assembling satellite power systems. Fig. 1 is a NASA artist's concept of one such settlement viewed from an approaching space vehicle. Depicted are a 32 km (20 mi.) long 6.4 km (4 mi.) diameter cylinder and its twin. In this concept, each cylinder would rotate around its axis once every 114 seconds to create Earth-like gravity. Solar energy would be the source of power and lunar or asteroid raw materials would be used for construction and manufacturing. The cylindrical

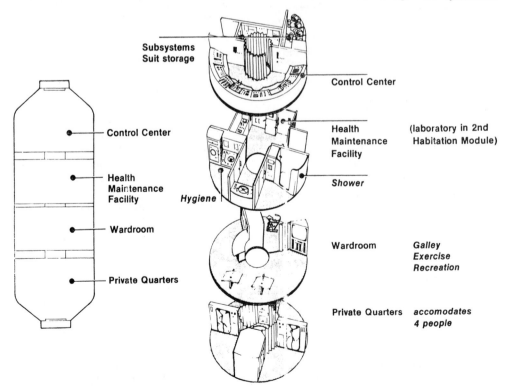

Fig. 2   Habitation module of the Space Operations Center.   (Drawing courtesy of NASA.)

Fig. 1 An artist's concept of a 1.6 km diameter, spherical space settlement. (Photograph courtesy of NASA.)

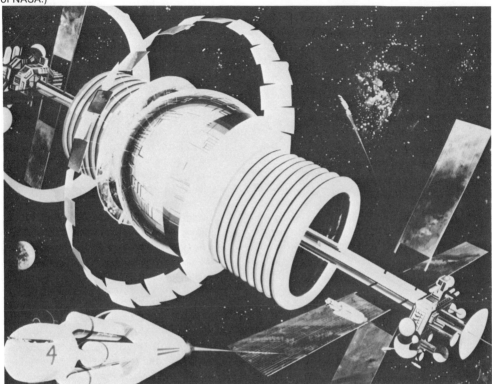

portion is the living area and its interior could be fashioned to resemble a terrestrial landscape. The "teacup" shaped containers ringing the cylinder are agricultural stations and the cylinder is capped by a manufacturing and power station. Large moveable, rectangular mirrors on the sides of the cylinders, hinged at the lower end, would direct sunlight into the interior, regulate the seasons and control the day-night cycle.

space simulator A device which stimulates some condition or conditions existing in *space* and used for testing equipment, or in training programs.

space station A permanently orbiting, large spacecraft which is designed to accommodate long-term human habitation in space. Fig. 1 depicts an early space station concept, a large 50-person "satellite laboratory," which is rotated to provide artificial gravity. The more recent "modularized" space station concept is shown in Fig. 2. Here, transfer vehicles can be seen going between three such stations in low-Earth orbit.

space suit
See: extravehicular mobility unit
space suit assembly [abbr. SSA]
See: extravehicular mobility unit
space system A system of equipment consisting of *launch vehicle(s), spacecraft, ground support equipment,* and test hardware, used in ground testing launching, operating, and maintaining space vehicles or spacecraft.

Space Telescope [abbr. ST] NASA's Space Telescope is a multi-purpose optical *telescope* planned for launch into orbit by the *Space Shuttle.* It will enable man to piece together the puzzle of the *Universe*: how it began, how it grew how it is changing and how those changes will affect *Earth.* With its unique capability for sharply defined imagery, large auxiliary equipment payload, and efficient operation, it will be the most powerful telescope ever built. The Space Telescope will enable scientists to gaze seven times farther into space than ever before, possibly to the edges of the visible *Universe!* With the Space Telescope,

Fig. 1   An artist concept of the NASA Space Telescope in orbit with the Tracking and Data Relay Satellite System and ground tracking station. Delivered by the Space Shuttle, the Space Telescope will be controlled remotely by ground stations. Data acquired by the telescope will be sent electronically to a computer and converted into formats suitable for scientific analysis. (Photograph courtesy of NASA.)

astronomers will   look at *celestial* sources such as *quasars, galaxies, gaseous nebulae,* and *Cepheid variable stars* which are 50 times fainter than those seen by the most powerful telescopes on the ground. [See Fig. 1]

The Space Telescope will be a long-lifetime general-purpose telescope capable of utilizing a wide variety of different scientific instruments at its focal plane. It will weigh about 11,000 kilograms (24,000 pounds) and will have a length of 13.1 meters (43 feet) and a diameter of 4.26 meters (14 feet). The major elements of the Space Telescope will be an Optical Telescope Assembly, a Support Systems Module and the Scientific Instruments.

The Optical Telescope Assembly [See Fig. 2] will mount a (2.4-meter) 94-inch reflecting *cassegrain*-type *telescope*. A *meteoroid* shield and sunshade will protect the optics. The telescope itself will have a Ritchey-Chretien folded optical system with the secondary mirror inside of the prime focus. The primary mirror will be made of ultra-low expansion glass. The mirror will be heated during operation to about optical shop temperatures 21°C (70°F) to minimize variations from its original accuracy. Additional heat required to maintain the 21°C (70°F) temperature will come from electrical strip heaters which will radiate to the back of the mirror.

Figure 3 is a schematic showing how starlight enters the open front end of the telescope [See Fig. 4] and is projected from the primary mirror to the secondary mirror, and then directed to a focus inside the scienfic instruments at the rear. The light baffles preclude unwanted light, which may have been reflected from some part of the telescope, from reaching the image formed within the scientific instruments.

The open front end of the Space Telescope will be similar to most Earth-bound telescopes and will admit light to the primary mirror in the rear of the telescope. The primary mirror will project the image to a smaller secondary mirror in front. The beam of light will then be reflected back through a hole in the primary mirror to the scientific instruments in the rear.

The pointing and stabilization control system can point the telescope to an accuracy of 0.01 *arc-second* and can hold onto a *target* for extended periods within 0.007 arc-second. (This angle is only slightly larger than that made by a dime when viewed at a distance from Washington, D.C., to Boston.) The resolving power of two point images will be about 0.1 arc-second. The references used for pointing stability will be accomplished using precision *gyros* and bright field stars or "guide" stars.

The scientific instruments will provide the means of converting the telescope images to useful scientific data. The instruments, and their sensors, located directly behind the telescope, will communicate images in a variety of ways. The modular instruments will fit behind the focal plane and will contain imaging systems, *spectrum* analyzers (to find out about atomic structure and material content of objects observed), and light intensity and *polarization* calibrators. Devices for exact control of temperature, direction and stability, and the equipment to generate power will be located in similar modular packages.

The apertures of the Scientific Instruments are located at the principal focus. Since stray light suppression is extremely important in reaching faint light levels, the forward end of the telescope is enclosed and well baffled, with the aperture door serving as a *Sun* shield.

The Scientific Instruments include two cameras, two *spectrometers* and a *photometer*.

Fig. 2   Components of the Space Telescope.   (Drawing courtesy of NASA.)

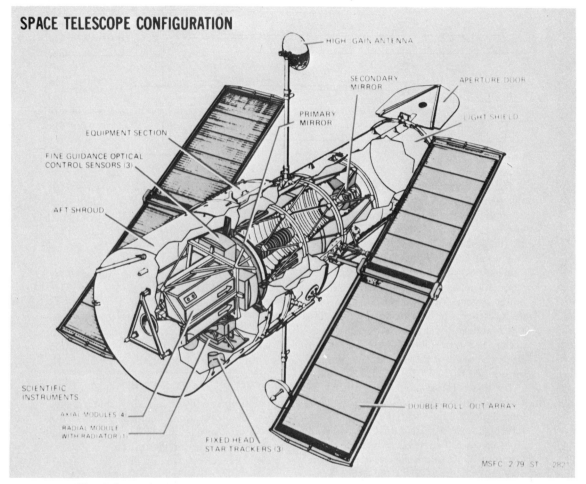

SPACE TELESCOPE CONFIGURATION

The Faint Object Camera, provided by the *European Space Agency,* and the Wide Field Planetary Camera are distinguished by their *fields of view,* spatial resolution and *wavelength range.* Both instruments cover the *ultraviolet* and blue regions of the spectrum.

The Wide Field Planetary Camera covers the red and *near-infrared* regions as well. The Faint Object Camera has a very small field of view, but can use the highest spatial resolution which the Space Telescope optics can deliver. The Wide Field Planetary Camera covers a field at least 40 times larger, but with a resolution degraded by a factor of two to four.

The two *spectrographs,* the High Resolution Spectrograph and the Faint Object Spectrograph, provide a wide range of spectral resolutions which would be impossible to cover in a single instrument. Both instruments will record ultraviolet radiation. Only the Faint Object Spectrograph covers the visible and red regions of the spectrum.

The fifth instrument, the High Speed Photometer, is a relatively simple device capable of measuring brightness variability over time intervals as short as 0.0001 second. It can also be used to measure ultraviolet polarization and to calibrate other instruments.

Since it can be serviced by the Space Shuttle, it is expected that competition for new instruments may be opened every few years to guarantee that the Space Telescope instrument payload represents the best possible configuration.

Fig. 3   Space Telescope light path.   (Drawing courtesy of NASA.)

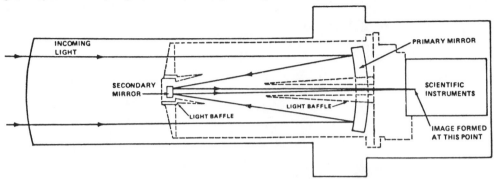

The Support Systems Module will enclose the Optical Telescope Assembly and Scientific Instruments and also provide all interfaces with the Shuttle Orbiter. The module will contain a very precise pointing and stabilization control system, the communications system, thermal control system, data management system and electric power system. Electrical power to operate the telescope will be provided by batteries which are charged by the two *solar panels* during the Sun side of its Earth *orbit*. Images received by the telescope will be transmitted to Earth by *telemetry*.

The Space Telescope is a new concept in *space-borne astronomical observatories*. Its long design lifetime and the capability of refurbishment and scientific instrument replacement will enable it to operate at the forefront of astronomical research for two decades or more. Due to the Space Telescope's unique scientific importance, its operations are designed for maximum flexibility while at the same time insuring *spacecraft* safety and operational efficiency. The Space Telescope will orbit the Earth at a nominal altitude of approximately (500 kilometers) 310 miles with an orbital *inclination* of 28.8 degrees.

The Space Shuttle will launch Space Telescope into orbit, and also serve as a base from which astronauts may make repairs by replacing modular components, including new instrument packages. [See Fig. 5] Each modular package can be replaced in orbit without affecting the overall system. The Shuttle can also bring the telescope back to Earth, if necessary, for extensive maintenance or overhaul, and later relaunch it. The plan is to update the

scientific instruments, on a selective basis, in orbit after about two-and-a-half years and to return the Space Telescope to Earth via the Shuttle after five years for major refurbishment.

Although in orbit, the Space Telescope will be operated in a manner similar to that of ground-based telescopes. The five instruments can be selected as appropriate for specific observations and the telescopes can be pointed to targets of interest. Plans for orbital operations include a limited amount of interaction between the observer and the telescope during the observation.

Principal Shuttle/Space Telescope mission operations phases will include: launch, orbital operation, deployment, checkout, on-orbit service, and re-entry. After orbital insertion and circularization, the *Orbiter* will be maneuvered into the proper position and the Space Telescope will be raised in preparation for deployment. After a preliminary checkout, the Space Telescope will be positioned in space by the Orbiter *remote manipulator system.*

As the Orbiter remains nearby for help if needed, the Space Telescope will receive an initial checkout from ground-based operators to make sure that all systems are operating properly. For the Orbiter to retrieve the telescope, the steps will be reversed. During re-visit, the Space Telescope will be captured and positioned in the Shuttle's *pay-load bay* by the Orbiter remote manipulator system. A crew in space suits will provide necessary service, making repairs and replacing equipment. This capability will make the Space Telescope a practical long-term reality. The Space Telescope in the Orbiter's cargo bay will be isolated from the intense heat generated during *re-entry*. After it enters the Earth's *atmosphere*, the Orbiter will land like a conventional airplane. Servicing and repairs will then prepare the Space Telescope for its next mission.

NASA's Office of Space Science, *NASA Headquarters,* Washington D.C., is responsible for overall direction of the Space Telescope program.

—The *Marshall Space Flight Center,* Huntsville, Ala., is NASA's lead center, responsible for overall project management.

—The *Goddard Space Flight Center,* Greenbelt, Md. is responsible for the scientific instruments, mission operations and data reduction.

—The *Johnson Space Center* is responsible for the Space Shuttle and flight crew operations.

—*The Kennedy Space Center* is responsible for the Space Shuttle launch operations.

—*European Space Agency* members will provide one of the scientific instruments, as well as the solar arrays, and will participate in flight operations.

The Space Telescope prime contractors are Lockheed Missiles and Space Company, Sunnyvale, Cal., responsible for the Support Systems Module and systems engineering, and the Perkin-Elmer Corp., Danbury, Conn., responsible for the Optical Telescope Assembly.

NASA will oversee the operation of an independent Science Institute responsible for conducting an overall Space Telescope science program to implement NASA policy, and detailed science planning and routine operations. The Institute will carry out NASA science objectives, solicit and select observational proposals and coordinate research and international participation. It will determine general viewing schedules, target sequence and target availability within the context of spacecraft constraints. The Institute will also reduce and analyze data, conduct basic research, evaluate and disseminate the science data and store the data in archives.

The Space Telescope's unique capability for high angular resolution imagery, large auxiliary equipment payload and efficiency of operation will make it the most powerful telescope ever built, enabling man to gaze seven times farther into space than now possible—perhaps to the outer edges of the universe.

The largest Earth-based telescopes in operation today can see an estimated two billion *light years,* into space. [See Table 1] The Space Telescope will be able to see much deeper—14 billion light-years. Some scientists believe the universe was formed nearly 14 billion years ago—so the Space Telescope might provide views of galaxies at the time they were formed.

Studying the stars isn't merely a matter of distance, however; it is also one of clarity. All Earthbound seeing devices have distorted vision because the Earth's atmosphere blurs the view and smears the light. The clearer images provided by the Space Telescope will enable scientists to evaluate the mass, size, shape, age and evolution of the universe more comprehensively.

With a telescope in Earth orbit, long time exposure images more than 10 times sharper than those from the ground can be achieved. Another great advantage of orbital observation is the absence of atmospheric material that absorbs the ultraviolet and infrared radiation from stars. The crisper images of the telescope, combined with the darker sky background, will also permit much fainter objects to be detected. By concentrating the starlight into a smaller area, the contrast with the

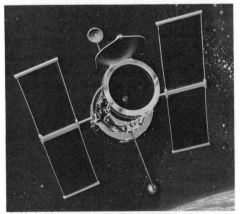

Fig. 4   The open end of the Space Telescope. This 10-ton unmanned orbiting telescope will enable scientists to see deep into space—seven times farther than is now possible, perhaps even to the outer edges of the Universe.   (Photograph courtesy of NASA.)

background (which is lower, due to the absence of scattered light and *airglow* emission) is improved, while the concentration means that the exposure times to reach a given brightness level will be reduced.

For faint object photography, the Space Telescope should be able to go perhaps 50 times fainter than the same detection system on the ground. One of its scientific instruments will use *solid-state* imaging-type sensors of much higher sensitivity, in some respects, than photographic films. An additional fringe benefit of these sensors is the freedom from the limited storage range of the photographic emulsion, thus permitting very long periods of observation and even fainter magnitude limits.

Table 1 World's Largest Optical Telescopes

6.0m	236″	Special Astronomical Observatory, USSR
5.0m	200″	Hale Telescope; Mt. Palomar, Hale Observatories
4.0m	158″	Cerro Tololo Inter-American Observatory, Chile
4.0m	158″	Mayall Telescope; Kitt Peak National Observatory, Arizona
3.9m	153″	Anglo-Australian Telescope; Siding Spring, Australia
3.8m	150″	United Kingdom Infrared Telescope; Mauna Kea, Hawaii
3.6m	142″	European Southern Observatory, Chile
3.6m	141″	Canada-France-Hawaii Telescope; Mauna Kea, Hawaii
3.5m	140″	National Astronomical Observatory, Italy
3.1m	122″	NASA Infrared Telescope; Mauna Kea, Hawaii
3.0m	120″	C. Donald Shane Telescope; Lick Observatory, California
2.7m	107″	McDonald Observatory, Texas
2.6m	102″	Byurakan Observatory, USSR
2.6m	102″	Crimean Observatory, USSR
2.5m	100″	Hooker Telescope; Mt. Wilson, Hale Observatories
2.5m	100″	Irenee du Pont Telescope; Hale-Carnegie Institution, Chile
2.5m	98″	Isaac Newton Telescope; Royal Greenwich Observatory, UK
2.4m	94″	NASA Space Telescope to be launched in the 1980's
2.3m	92″	Wyoming Infrared Telescope; Jelm Mt., Wyoming
2.3m	90″	Steward Observatory, Arizona
2.2m	88″	University of Hawaii, Mauna Kea
2.2m	88″	Max Planck Institute, Calar Alto Mt.
2.2m	88″	Max Planck Institute, SW Africa
2.16m	85″	University of Mexico, Mexico
2.15m	84″	Kitt Peak National Observatory
2.15m	84″	La Plata Observatory, Argentina
2.1m	82″	Otto Struve Telescope; McDonald Observatory, Texas

MMT Six 1.8m mirrors functioning as one 4.4 mirror, Mt. Hopkins Observatory, Arizona
Credit: courtesy Perkin-Elmer

A further advantage for observational programs lies in the accessibility of all of the sky and almost 24 hours of observing conditions. With ground-based observatories, most optical observations are made only during twilight and dark hours and even then only when it is reasonably clear. With the Space Telescope, it will be possible to make some observations even in sunlight (although not to the faintest levels) and realize about 4,500 hours of observation per year. (Excellent ground-based observatories obtain about 2,000 hours per year.)

The Space Telescope is expected to contribute a great deal to the study of little-understood energy processes in celestial objects, the early stages of star and solar-system formation, such highly-evolved objects as *supernova* remnants and *white dwarf stars,* and the origin of the universe.

With the Space Telescope, scientists can look at galaxies so far away that they will be seen as they were billions of years ago. They should have much to reveal about the birth and growth of cosmic structures like our galaxy.

The Space Telescope may be able to search for planets that orbit other stars in the same way the Earth orbits the Sun. Data in this area will tell us about basic physical processes in the universe and indicate the chances for the existence of other life-supporting planets. It will provide a new perspective on the neighboring planets, giving continuing information about their physical conditions and atmosphere—the kind of information needed to build and equip spacecraft for further exploration.

Astronomical observations in the past have often suggested potential solutions to problems on Earth. For example, the first confirmation of nuclear *fusion* was from the study of the Sun. If the quasars are as intensely luminous as they appear, they have the concentrated power of millions of stars. The Sun would be pale in comparison. Are quasars really so powerful? Can their energy principle be used on Earth? For the first time, with the Space Telescope, quasars can be seen well enough to investigate these possibilities.

Space Tracking and Data Network [abbr. STDN]
See: STS tracking and communications network

space tug   An upper stage installed in the *Orbiter's* cargo bay for *payload* launch or recovery and landing (on Earth via the *Space Shuttle*). The space tug will be developed specifically with the capability for delivery, retrieval and servicing of payloads in *orbits* and *trajectories* beyond the (altitude) capability of the Orbiter vehicle alone. It will be retrievable for both refurbishment and multiple reuse.

space-walk communications
See: extravehicular mobility unit

Space Watch
See: U.S. Air Force role in space

spalling   Flaking off of particles and chunks from the surface of a material as a result of localized stresses

span   1. The dimension of a craft measured between lateral extremities; the measure of this dimension.
2. Specifically, the dimension of an *airfoil*

from tip to tip measured in a straight line. Span is not usually applied to vertical airfoils.

spandex   Any of various synthetic textile elastic fibers.

spark chamber   An instrument for detecting and measuring the paths of *elementary particles*. It is analogous to the cloud chamber and bubble chamber. It consists of numerous electrically charged metal plates mounted in a parallel array, the spaces between the plates being filled with an inert gas. Any *ionizing* event causes sparks to jump between the plates along the *radiation* path through the chamber. Compare this term with *bubble chamber, cloud chamber.*

spatial   Pertaining to *space*

special microwave imagery sensor [abbr. SSMI*]
See: Defense Meteorological Satellite Program
*SSMI = special sensor, microwave imagery

special nuclear material [abbr. SNM]
Term referring to *plutonium*-239, *uranium*-233, uranium containing more than the natural abundance of uranium-235, or any material artificially enriched in any of these substances.

Special (or Restricted) Theory of Relativity   A theory developed by Albert Einstein in 1905 that is of great importance in atomic and nuclear physics. It is especially useful in studies of objects moving with speeds approaching the speed of light. Two of the results of the theory with specific application in nuclear physics are statements (a) that the *mass* of an object increases with its *velocity* and (b) that mass and *energy* are equivalent.
See: mass-energy equation

species   A particular kind of atomic *nucleus, atom, molecule* or *ion*; a *nuclide.*

specific activity   The *radioactivity* of a *radioisotope* of an *element* per unit *mass* of the element in a sample. The activity per unit mass of a pure radionuclide. The activity per unit mass of any sample of radioactive material.

specific gas constant [Symbol R]   This can be shown by the following equation:

$$R \equiv \frac{R_u}{MW}$$

where R_u is the *universal gas constant* and MW is the *molecular weight* (molar mass) of the gas. It should be noted that the specific gas constant is a function of the molecular weight of a substance and is therefore different for each gas. For example, the specific gas constant for air (assuming a molecular weight of 28.97) is 287.04 J/kg-K.
See: ideal gas, universal gas constant

specific heat [Symbol: c]   The ratio of the *heat* absorbed (or released) by unit mass of a system to the corresponding *temperature* rise (or fall). If this ratio varies with temperature, it must be defined as a differential quotient dQ/dT, where dQ is the infinitesimal increment of heat per unit mass and dT is the infinitesimal increment of temperature.

For *ideal gases* the *thermodynamic* process must be specified; two specific heats are defined, one being the specific heat in a constant-pressure process

$$c_p = (dQ/dT)_p = (\partial h/\partial T)_p$$

where h = enthalpy; and the other, the specific heat in a constant-volume process

$$c_v = (dQ/dT)_p = (\partial u/\partial T)_v$$

where u = internal energy.

specific humidity   The ratio of the mass of vapor to the mass of non-condensable gas; more specifically in reference to air-water vapor mixtures, the specific humidity can be defined as the ratio of the mass of water vapor to the mass of air in the mixture. Also called "humidity ratio."
See: absolute humidity, relative humidity

specific impulse [Symbol: I$_{sp}$]   Performance index for rocket propellants, equal to the thrust produced by propellant combustion divided by the mass flowrate.

$$I_{sp} \equiv \frac{thrust}{mass\ flowrate}$$

specific ionization   The number of *ion pairs* formed per unit of distance along the track of an *ion* passing through matter.
See: ionization, ionizing radiation

specific power   The energy delivered per unit mass of fuel in a *nuclear reactor* or in a

radioisotope power source. Typically expressed in kilowatts of thermal energy per kilogram or watts per gram.

specific propellant consumption   The reciprocal of the *specific impulse,* i.e., the required *propellant* flow to produce one *newton* (pound-force) of *thrust* in an equivalent *rocket.*

specific volume [Symbol: v]   *Volume* per unit mass of a substance. The reciprocal of *density.*

spectral   Of or pertaining to a *spectrum.*

spectral line   A bright, or dark, line found in the *spectrum* of some radiant source. Bright lines indicate emission, dark lines indicate absorption.
See: absorption line, emission line

spectrograph   An optical instrument for resolving light or other *electromagnetic radiation* into its component *wavelengths.*

spectrography   The method of remotely analyzing stellar objects or other light sources by means of their radiated *spectrum.*

spectroheliograph   1. An image of the *Sun* in a particular *spectral* region.
2. A device for producing spectroheliographs.

spectroheliometer   A device for measuring the *spectrum* of the *Sun.*

spectrometer   An instrument that spreads light into a *spectrum* and measures the intensity at different *wavelengths.*

specular reflection   *Reflection* in which the reflected radiation is not diffused; reflection as from a mirror. The angle between the normal to the surface and the incident beam is equal to the angle between the normal to the surface and the reflected beam. Any surface irregularities on a specular reflector must be small compared to the wavelength of the incident radiation.

speed of light [Symbol: c]   The speed of propagation of *electromagnetic radiation* through a perfect vacuum; a universal dimensional constant equal to 299,792.5 kilometers per second. Also called velocity of light.

spent (depleted) fuel   *Nuclear reactor fuel* that has been irradiated (used) to the extent that it can no longer effectively sustain a *chain reaction.*

spherical coordinates   A system of *coordinates* defining a point on a sphere or spheroid by its angular distances from a *primary great circle* and from a reference secondary great circle, as *latitude* and *longitude.*

spin rocket A small *rocket* that imparts spin to a larger *rocket vehicle* or *spacecraft.*

spin stabilization Directional stability of a *spacecraft* obtained by the action of gyroscopic forces which result from spinning the body about its axis of symmetry.
See: gyro

spin table A flat platform on which human and animal subjects can be placed in various positions and rapidly rotated, much as on a phonograph record, in order to simulate and study the effects of prolonged tumbling at high rates. Complex types of tumbling can be simulated by mounting the spin table on the arm of a centrifuge.

spiral arms The part of the *spiral galaxy* where the *gas,* dust and young *stars* are concentrated, distributed in a spiral pattern throughout the galaxy.

spoiler A plate, series of plates, tube, bar, or other device that projects into the *airstream* about a body to break up or spoil the smoothness of the flow, especially such a device that projects from the upper surface of an *airfoil,* giving an increased *drag* and a decreased *lift.*

Spoilers are normally movable and consist of two basic types; the flap spoiler, that is hinged along one edge and lies flush with the airfoil or body when not in use, and the retractable spoiler, which retracts edgewise into the body.

spontaneous emission The decay of an *atom* or *ion* in an *excited energy state* E_j to a lower state E_i without the influence of any external perturbation. This process results in the emission of a *photon* of energy

$$h\mu = E_j - E_i$$

where h is the *Planck constant* and μ is the frequency.

spontaneous fission *Fission* that occurs without an external stimulus. Several heavy isotopes decay mainly in this manner; examples: californium-252 and californium-254. The process occurs occasionally in all *fissionable* materials, including uranium-235.

sporadic meteor A *meteor* that is not associated with one of the regularly recurring *meteor showers* or streams.

"Sputnik I" On 4 October 1957 the Soviet Union launched "Sputnik I," the first man-made object to be placed in *orbit* around the *Earth.* A 29-m (96-ft) rocket boosted this *artificial satellite* into orbit. This 83.5-kg (184-lb$_m$) satellite provided scientists with information on atmospheric and *electron* densities. "Sputnik I" also transmitted temperature data for 22 days before its batteries ran down. It entered the Earth's *atmosphere* and burned up on 4 January 1958.

sputtering Dislocation of surface *atoms* of a material from bombardment by high-energy atomic *particles.*

squeeze-film damping *Friction damping* produced by *pressure* and flow *forces* in a thin film of *fluid* subjected to high load and *shear.*

squib Term for an *electroexplosive device.*

SRB thermal protection system *Ablative* materials used to insulate the exterior surfaces of the *Space Shuttle Solid Rocket Booster* (SRB), which are exposed to thermal loads, to enable them to withstand friction-induced heat.
See: Solid Rocket Booster

stability 1. The property of a body, as an *aircraft, aerospace vehicle* or *rocket* to maintain its *attitude* or to resist displacement; if displaced, to develop *forces* and *moments* tending to restore the original condition.
2. Of a *fuel,* the capability of a fuel to retain its characteristics in an adverse environment, e.g., extreme temperature.

stable Incapable of spontaneous change. For example, not radioactive.

stable isotope An isotope that does not undergo *radioactive decay.* Compare this term with *radioisotope.*

stacked tube nozzle subassembly A component of the nozzle assembly in the *Space Shuttle Main Engine.*
See: Space Shuttle Main Engine

stacking Assembling the coolant tubes of a *liquid rocket thrust chamber* vertically on a *mandrel* that simulates the chamber/nozzle contour; this procedure facilitates fitting and adjusting the tubes to the required contour prior to brazing.

stage 1. A separable, self-contained, self-propelled section of a *launch vehicle*.
2. A set of rotor blades and stator vanes in a *tubine* or an *axial-flow pump,* or one set of *impeller* and associated flow passages in a *centrifugal-flow pump.*
3. The degree of polymerization of a synthetic resin.

stage-and-a-half A *liquid rocket propulsion* unit of which only part falls away from the *rocket* vehicle during flight, as in the case of *booster rockets* falling away to leave the *sustainer engine* to consume remaining *fuel.*

staged combustion *Rocket engine* cycle in which propellants are partially burned in a preburner prior to being burned in the *combustion chamber.*

staging 1. Separating a stage or set of stages from a spent stage of a *launch vehicle.*
2. Incorporating two or more stages in a *pump* or *turbine.*
3. Increasing the *molecular weight* of a resin without effecting a cure.
See: stage.

stagnation Condition in which flowing *fluid* is brought to rest *isentropically.*

stagnation point Point in a flow field about a body immersed in a flowing *fluid* at which the fluid particles have zero *velocity* with respect to the body.

stagnation pressure 1. *Pressure* that a flowing *fluid* would attain if brought to rest *isentropically.*
2. Pressure of a flowing fluid at a point of zero fluid *velocity* in a body around which the fluid flows.

stagnation region The region in the vicinity of a *stagnation point* in a flow field about a body where the *fluid velocity* is negligible.

stagnation temperature *Temperature* that a flowing *fluid* would attain if the fluid were brought to rest *isentropically* from a given flow velocity (same as total temperature); for an *ideal gas,* the process need only be *adiabatic.*

stall 1. In a valve, condition wherein the actuation force is equal to the dynamic force plus the friction force, the valving element thus being stopped in a partially open position.
2. In a *pump,* loss of capability when flow separation in the *rotor* or *strator* flow passages progresses to the point at which *headrise* drops abruptly.

standard 1. An exact value, or a concept, that has been established by authority or agreement, to serve as a model or rule in the measurement of a quantity or in the establishment of a practice or procedure.
2. A document that establishes engineering and technical limitations and applications for items, materials, processes, methods, design or engineering practices.

standard deviation [Symbol: σ] A measure of the dispersion of data points around their *mean value.* It is the positive square root of the *arithmetic mean* of the squares of the deviation from the arithmetic mean of the *population*:

$$\sigma = \sqrt{\ \Sigma d^2 m/n}$$

where m is the arithmetic mean; d is deviation from the arithmetic mean; and n is number of points.

standard temperature and pressure [abbr. STP] Usually a temperature of 0° C but also used to designate a temperature of 15° C and 1 standard *atmosphere.*

standing wave A *periodic wave* having a fixed distribution in space which is the result of *interference* of progressive waves of the same frequency and kind. Such waves are characterized by the existence of *nodes* or partial nodes and *antinodes* that are fixed in space.

star A self-luminous celestial body composed of gases. It generates energy by means of nuclear fusion reactions in its core.

star grain A hollow *rocket-propellant* grain with the cross section of the hole having a multipointed shape.
See: solid-propellant rocket

Starlab A NASA planned Spacelab facility designed to investigate the visual, ultra-violet and near-infrared properties of extra-galactic space, the Milky Way galaxy and the Solar System with both high spatial resolution and wide-angle imaging spectroscopy. The Starlab is a Shuttle-based telescope that will use both film and telemetry data to record scientific data. The instruments used could be controlled by onboard investigators or ground station scientists. The payload will be returned to Earth following each mission.

The Starlab telescope and instrument assembly.   (Drawing courtesy of NASA.)

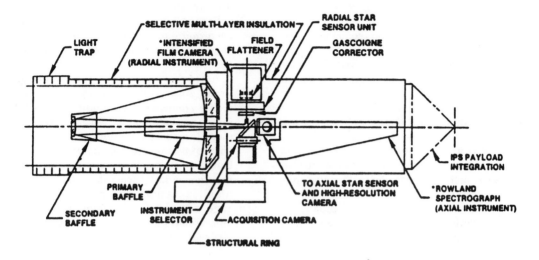

starting pressure   In rocketry, the minimum *chamber pressure* required to establish shock-free flow in the exit plane of a supersonic *nozzle*.

starting torque   Turning or twisting force required to initiate rotary motion.

star tracker(s) [ST]
See: Orbiter guidance, navigation and control system

state variable   Any independent variable in a problem that must be specific to define a condition of state, as for example a component of position.

state vector   Ground-generated *spacecraft* position, *velocity,* and timing information *uplinked* to the spacecraft computer for use as a navigational reference by the (Shuttle) crew.
See: Orbiter avionics system

static   1. Involving no variation with time.
2. Involving no movement, as in static test.
3. Any radio *interference* detectable as *noise* in the audio stage of a receiver.

static conversion   *Energy* conversion in which no moving parts or equipment are utilized.

static firing   The firing of a *rocket engine* in a hold-down position to measure *thrust* and accomplish other tests.

static pressure   1. The *pressure* with respect

to a stationary surface tangent to the *mass-flow velocity vector.*
2. The *pressure* with respect to a surface at rest in relation to the surrounding *fluid.*

static seal   Device used to prevent leakage of fluid through a mechanical joint in which there is no relative motion of the mating surfaces other than that induced by changes in the operating environment.

static testing   The testing of a *rocket* or other device in a stationary or hold-down position, either to verify structural design criteria, structural integrity, and the effects of limit loads or to measure the *thrust* of a *rocket engine.*

stationary wave   A *standing wave* in which the net energy *flux* is zero at all points.

Station Conferencing and Monitoring Arrangement [abbr. SCAMA]
See: STS tracking and communications network.

station keeping   The sequence of maneuvers that maintains a space vehicle in a predetermined *orbit.*

stator   Part of a *turbopump* assembly that remains fixed or stationary relative to a rotating or moving part of the assembly.

statute mile   5280 feet = 1,6093 *kilometers* = .0869 *nautical mile.*

stau time   Average length of time spent

within the *combustion chamber* by each gas *molecule* or *atom* involved in the combustion process; also called residence time.

STDN tracking stations
See: STS tracking and communications network

steady flow   A flow whose *velocity vector* components at any point in the *fluid* do not vary with time.
See: streamline flow

steady state   Condition of a physical system in which parameters of importance (*fluid velocity, temperature, pressure,* etc.) do not vary significantly with time; in particular, the condition or state of *rocket engine* operation in which *mass, momentum,* volume and pressure of the combustion products in the *thrust chamber* do not vary significantly with time.

steerhorn   A lug or boss on an *engine bell* to which a steering control rod is attached.

step rocket
See: multistage rocket

steradian [Symbol: sr]   The unit *solid angle* that cuts unit area from the surface of a sphere of unit radius centered at the vertex of the solid angle. There are 4π steradians in a sphere. The steradian is the supplementary *SI unit* of solid angle.

stiffness   Resistance to deflection.

stochastic process   An ordered set of observations in one or more dimensions, each being considered as a sample of one item from a *probability* distribution.

stoichiometric   Of a mixture of chemicals, having the exact proportions required for complete combination, applied especially to combustible mixtures used as *propellants.*

stoichiometric combustion   The burning of *fuel* and *oxidizer* in precisely the right proportions required for complete reaction with no excess of either reactant.

stopping point   The end of the highly luminous path of a visual *meteor.* Also called Hemmungspunkt.

stopping power   A measure of the effect of a substance upon the *kinetic energy* of a *charged particle* passing through it. Compare this term with *cross section.*
See: absorption

storable   Of a *liquid:* subject to being placed and kept in a tank without benefit of special measures for *temperature* or *pressure* control, as in storable *propellant.*

storable propellant   *Rocket propellant* (usually liquid) capable of being stored for prolonged periods of time without special *temperature* or *pressure* controls.

storage   1. The act of storing information.
See: store
2. Any device in which information can be stored. Also called a memory device.
3. In a *computer,* a section used primarily for storing information. Such a section is sometimes called a memory or a store. The physical means of storing information may be electrostatic, ferroelectric, *magnetic, acoustic,* optical, chemical, electronic, electrical, mechanical, etc., in nature.

storage capacity   The amount of information, usually expressed in *bits,* that can be retained in *storage.*

stowing   The process of placing a *payload* in a retained, or attached, position in the *Orbiter's cargo bay* for ascent to or return from *orbit.*

strain gauge   Instrument used to measure the strain or distortion in a structural member or test specimen subjected to a load.

strange particles   A class of very short-lived *elementary particles* that decay more slowly than they are formed, indicating that the production process and decay process result from different fundamental reactions. They include *K-mesons* and *hyperons.*

stratospheric fallout
See: fallout

stream   A group of *meteoroids* with nearly identical orbits, also called meteor stream.

streamline   A line whose tangent at any point in a *fluid* is parallel to the instantaneous *velocity vector* of the fluid at that point.
In *steady-state* flow, the streamlines coincide with the *trajectories* of the fluid *particles;* otherwise, the streamline pattern changes with time.

stress concentration   Localized increase in the stress in a structural member.

stress corrosion   Corrosion of a metallic surface enhanced (i.e., increased in rate) by the existence of localized stresses, whether applied or residual.

stress-corrosion cracking   Delayed fracture of a material as a result of the combined action of static tensile stress and a corrosive (agressive) environment.

stretchout   An action whereby the time for completing an action, especially a contract, is extended beyond the time originally programmed or contracted for.

stringer   A slender, lightweight, lengthwise fill-in structural member in a *rocket* body serving to reinforce and give shape to the *skin.*

strongback   A heavy beam, bar or *truss* structure for taking a strain.

structural test motor [abbr. STM]   Motor model constructed for the purpose of evaluating the structural integrity of a *solid rocket motor* design.

STS-1, The First Space Shuttle Mission
"Today our friends and adversaries are reminded that we are a free people capable of great deeds. We are a free people in search of progress for mankind."—President Reagan welcoming the crew of *Columbia,* April 14, 1981.

A new era in space promising countless benefits for people everywhere opened at 1:21 p.m. EST, April 14, 1981. At that time, the crew of the *Space Shuttle orbiter Columbia*—John W. Young, *commander,* and Robert L. Crippen, *pilot*—made a perfect landing on the hardpacked bed of Rogers Dry Lake in California's Mojave Desert, after a nearly flawless voyage in space.

This was the first airplane-like landing of a craft from *orbit.* Moreover, Columbia appeared hardly the worse for wear after its searing atmospheric entry when temperature soared perhaps as high as 1650° C (3000° F).

Its appearance was not deceptive. After a careful inspection of Columbia, NASA technicians reported that its condition after its historic flight was excellent and that *Columbia* should be capable of making at least a hundred round trips between *Earth* and Earth orbit.

Reusability is one of the goals of the Space Shuttle. The Shuttle includes three major units: the *Orbiter,* of which *Columbia* is an example; two *solid rocket boosters,* which also are recovered and reused; and the orbiter's *external fuel tank* for which there are no plans for reuse at present. [*See:* External Tank; Solid Rocket Booster]

The solid rockets that helped launch Columbia were recovered in the Atlantic off Daytona Beach, Florida. They were determined to be suitable for refurbishment at a fraction of the cost of buying new rockets.

Reusability of the Orbiter and the solid rocket boosters is one of the keys to significant cost reductions in space operations that the Shuttle is expected to make possible. The Shuttle is also designed to facilitate space operations and to open space to ordinary people of all nations who have important work to do there. The Shuttle is the kingpin of NASA's *Space Transportation System* (STS) which will include many more facilities to improve and lower the cost of space operations.

The solid rocket boosters used in STS-1 marked other firsts: the first time solid rockets have been used to launch a manned spacecraft and the first recovery of boosters for reuse.

The STS-1 mission was also the first time that any American *spacecraft* has been put into orbit without prior unmanned testing. As a result, the mission was conservatively planned in the interest of safety. STS-1 mission objectives were a safe ascent and safe return of *Columbia* and its crew.

Another first was the launch of an airplane-like craft, also called an *aerospace vehicle,* into space with both wings and landing gear. STS-1 also marked the return of Americans to space after an absence of nearly six years.

STS-1 is the first of four planned orbital test flights leading to an operational capability late in 1982. The flights are designed to prove and improve the flight system as well as refurbishment capability, turnaround time, payload capability, and tracking and data acquisition.

Columbia will be used for the flight tests. When the STS is operational, Columbia will be joined by *Challenger,* which is in production, and later the planned *Discovery* and *Atlantis* and possibly a fifth orbiter, giving the United States the world's first fleet of manned *aerospace vehicles.*

(The Launch)

At 7 a.m. EST, April 12, 1981, Columbia's three *main liquid-hydrogen-fueled rocket engines* and two solid rocket boosters generated nearly 28.6 million newtons (6.5 million pounds) of thrust to lift the approximately 2-million-kilogram (4½-million-pound) Space Shuttle from Launch Pad 39A at Kennedy Space Center, Fla. Rising on a pillar of orange flame and white steam, the Shuttle cleared its 106-meter (348-foot) high launch tower in six seconds and reached Earth orbit in about 12 minutes. The solid rocket boosters and external fuel tank had been shed prior to orbit.

"Man, that was one fantastic ride," exclaimed Crippen, who was on his first space flight, as his heartbeat rose from 60 to 130 per minute. Young, a veteran of four space flights including an Apollo Moon landing, had a heartbeat rise of from only 60 to 85. Later, the 50-year-old Young said he was excited too and jocularly added: "But I just can't make it go any faster."

Young and Crippen changed their orbit from its original eliptical 106 by 245 kilometers (65 by 152 statute miles) to a circular orbit of 245 kilometers by firing their *Orbital Maneuvering System* at *apogee* (orbit high point). Later, they raised their orbit to nearly 277 kilometers (172 statute miles). They found *Columbia* easy to control.

They tried out all systems and conducted many engineering tests. They checked the computers, the jet thrusters used in orienting *Columbia,* and the opening and closing of the cavernous *cargo bay* doors. Aside from the obvious reason of being able to release *satellites* into and retrieve satellites from space, opening the cargo bay doors is critical to deploy the *radiators* that release the heat that builds up in the crew compartment. Closing them is necessary for return to Earth. (See: Orbiter Structure)

Young and Crippen also documented their flight in still and motion pictures. One view of the cargo bay which was telecast to Earth indicated that all or part of 16 heat-shielding tiles were lost, probably due to stresses of launch, from two pods on the tail section that house Columbia's OMS rockets. The loss was not considered critical as these pods are not subjected to intense heat. However, there are areas on the space-

craft's underbelly, nose, wings, and tail where the frictional heat generated by entry into the atmosphere could reach 1650°C (3000°F). As it turned out, the more than 30,000 tiles all adhered.

The tiles are made of a material that sheds heat so readily that it can be red hot on one side and cool enough to touch with the bare hand on the other. Moreover, unlike other heat-shielding materials used on previous spacecraft they are not burned away by high temperatures. [See: Orbiter thermal protection system]

Young and Crippen wore ordinary coveralls while in orbit. They kept *Columbia* in a tail-forward position and upside down relative to Earth. The upside-down position provided a better view of Earth and its horizon for orientation.

About 12:22 p.m. EST, April 14, Young and Crippen fired their OMS rockets for approximately 2 minutes and 27 seconds to reduce speed from their *orbital velocity* of around 28,000 kilometers (17,500 miles) per hour. At the time, they were over the Indian Ocean and began an hour-long descent to their landing field in California. They fired their attitude control (orientation) *thrusters* to turn *Columbia* right side up relative to Earth and nose forward. They fired thrusters again to pitch *Columbia's* nose up 40° so that the brunt of atmospheric entry pressures and temperatures would be taken by *Columbia's* broad well protected underside. About 12.48 p.m. EST, while over the western Pacific Ocean, *Columbia* began atmosphere entry.

After completing the fiery entry into the atmosphere, *Columbia's* computers transitioned from the steering rockets to the *rudder* and *elevons* (a combination of *ailerons* and *elevators* commonly used on delta-winged craft) to pilot *Columbia* through the atmosphere. They found the 89,000-kilogram (98-ton) *Columbia* relatively easy to control.

Columbia continued its descent like a powerless glider. Air *drag* caused it to steadily lose speed as its altitude dropped.

According to plan, Young and Crippen guided *Columbia* over their landing strip on the bed of Rogers Dry Lake in the Mojave Desert of California, banked sharply left, and looped back. They touched down at a speed of some 346 kilometers per hour (215 mph), which is about twice that at which commercial jetliners ordinarily land. The touchdown marked the successful conclusion of STS-1—2 days, 6 hours, 20 minutes and 52 seconds after lift-off from Florida.

As soon as Colombia stopped, it was surrounded by a convoy of vehicles carrying specialists who took measures to remove dangerous concentrations of explosive or poisonous gases in *Columbia's* cargo bay or in space surrounding its engines. They ventilated the entire craft, and withdrew residual fuel from engines. It took about an hour to assure that *Columbia's* vicinity was safe. The crew was then permitted to leave *Columbia* and to go to a waiting medical van. Young exited first. Before going to the van, he carefully inspected *Columbia's* exterior. His inspection completed, he smiled broadly and gave a thumbs-up sign.

The age of routine manned space exploitation had begun!

STS associated payload   A specific complement of instruments, space equipment and support hardware carried into space by the *Shuttle* to accomplish a *mission* or discrete activity.

STS flight control team   Personnel in the NASA *Mission Control Center* on duty to provide real-time support for the duration of each *Space Shuttle flight*. Their support is a *standard service*.

STS flight operations center   The NASA center (Johnson Space Center) responsible for *Space Transportation system* (STS) *flight planning* (utilization planning, flight design, crew activities planning, operations planning and associated activities) and STS flight control activities in the execution of a *Shuttle flight*. This center is sometimes referred to as the "STS flight operator."

STS landing operations   When the *Space Shuttle Orbiter* returns to *Earth* from its mission in space, it will land at Kennedy Space Center on one of the world's most impressive runways. The concrete runway of the Shuttle Landing Facility (SLF) is located northwest of the *Vehicle Assembly Building* on a northwest/southeast align-

ment (330° northwest/150° southeast) [*See:* Fig. 1]. The runway is 4572 meters (15,000 feet) long and 91 meters (300 feet) wide with a 305-meter (1,000-foot) safety overrun at each end. A paved runway at the *Dryden Flight Research Center* at the *Edwards Air Force Base* in California matches the KSC runway in length and width and has an overrun of 8 kilometers (5 miles) that extends into a dry lake bed. The Shuttle Landing Facility runway in 40.6 centimeters (16 inches) thick in the center with the thickness diminishing to 38 centimeters (15 inches) on the sides. The runway is not perfectly flat but has a slope of 61 centimeters (24 inches) from centerline to edge. Underlaying the concrete paving, completed in late 1975, is a 15.2-centimeter (6-inch) thick base of soil cement. The runway is grooved to prevent hydroplaning in wet weather. The Orbiter's wheels will run across 6.35-millimeter (0.25-inch) wide grooves that cross the runway at intervals of 2.86 centimeters (1.125 inches). The grooves, together with the slope of the runway from centerline to edge, provide rapid drainoff of any water from a heavy Florida rain to help combat hydroplaning. The total length of the grooves is 13,600 kilometers (8450 miles).

In general terms, the SLF runway is roughly twice as long and twice as wide as average commercial landing facilities, although a number of domestic and foreign airports have landing strips far exceeding average dimensions. The SLF includes a 168- by 150-meter (550- by 490-foot) *aircraft* parking apron, or ramp, located near the southeastern end of the runway. The landing facility is linked with the *Orbiter Processing Facility* by a 3.2-kilometer (2-mile) towway.

Located on the northeast corner of the SLF ramp area is the mate/demate device (M/DD) used to raise and lower the Orbiter from its 747 carrier aircraft during ferry operations. The device is an open-truss steel structure with the hoists and adapters required to mate the Orbiter to and demate it from the 747. Such a mating or demating operation must be performed before and after each ferry flight. Movable platforms for access to certain use in the open air and is designed to withstand winds of up to 200 km/h (125 mph). Lightning protection is included. The M/DD is 46 meters (150 feet) long, 28 meters (93 feet) wide, and 32 meters (105 feet) high. A similar device is in use at the NASA Dryden Flight Research Center in California and a third will be erected at the Vandenberg Air Force Base in California in support of Space Shuttle operations there. A portable stiff-leg derrick system was designed and built by KSC for use at the Marshall Space Flight Center in Alabama. The portable system is also available for use in retrieving an Orbiter compelled to make a landing at a number of contingency sites located around the world. Also located at the SLF is a landing aids control building near the aircraft parking apron.

The Orbiters will be guided automatically to safe landings by a sophisticated microwave scanning beam landing system (MSBLS). At deorbit, the Space Shuttle Orbiter is committed to a landing at a precise point on a relatively small target. Unlike conventional aircraft, it lacks propulsion and the high-speed glide to landing must be accomplished perfectly the first time. The energy management procedures to do this require the most precise and up-to-date *vectors* that can be derived from available navigation systems. Landings may be made on the runway from the northwest to the southeast (Runway 15) or from the southeast to the northwest (Runway 33), and MSBLS ground stations are duplicated to permit an approach from either direction. A decision on which runway is to be used must be made before the Orbiter's *altitude* drops below 41,150 meters (135,000 feet) and its velocity falls below 8,230 km/h (5,115 mph).

Each of the two MSBLS ground stations consists of two sites adjacent to the landing runway. An elevation site is located approximately 1,067 meters (3,500 feet) in from the runway threshold and an *azimuth/distance-measuring* equipment site is located approximately 396 meters (1,300 feet) beyond the stop end of the runway.

The ground stations are located 94 meters (308 feet) to the west of the runway's centerline. In addition, equipment is provided for remote control and monitoring and for remote display of maintenance and monitor information. The azimuth/distance-measuring sites on the far end of the runway send signals that sweep 15° on each side of the landing path to provide directional and distance data. Signals from the companion shelter near the touchdown point sweep the landing path to provide elevation data up to 30°. Equipment onboard the Orbiter receives these data and automatically makes any needed adjustments to the glide slope.

When all facilities and systems are ready for landing, the deorbit will be initiated at approximately 60 minutes before landing and Orbiter entry will occur at approximately 30 minutes before touchdown [*See:* Fig. 1a]. The landing approach begins with retrofire and continues with the Orbiter engaged in terminal-area-energy-management maneuvers after it is in the Earth's atmosphere. These maneuvers are designed to place the Orbiter in a favorable position to intercept the landing approach corridor at the correct altitude and speed. Final landing guidance with microwave scanning beam landing system (*See:* Fig. 1b) begins at an altitude of 4,074 meters (13,365 feet), a range from the runway threshold of about 12 kilometers (7.5 miles), and a velocity of about 682 km/h (424 mph). The time from *touchdown* will be 86 seconds. This phase of the flight will be fully automatic with the crew monitoring, supervising, and backing up the MSBLS. The final approach will have an initial glide slope of 22° (more than 7 times as steep as the 3° slope of a commercial airliner on a straight-in approach) and flare or pull-up maneuvers will be required to bring the Orbiter to the final glide slope of 1.5° and touchdown speed of 346 km/h (215 mph). The crew will initiate landing gear deployment 22 seconds before touchdown at an altitude of 91.4 meters (300 feet) at a distance of 1,695 meters (5,560 feet) from the runway threshold. The targeted touchdown point is 841 meters (2,760 feet) past the threshold of the runway.

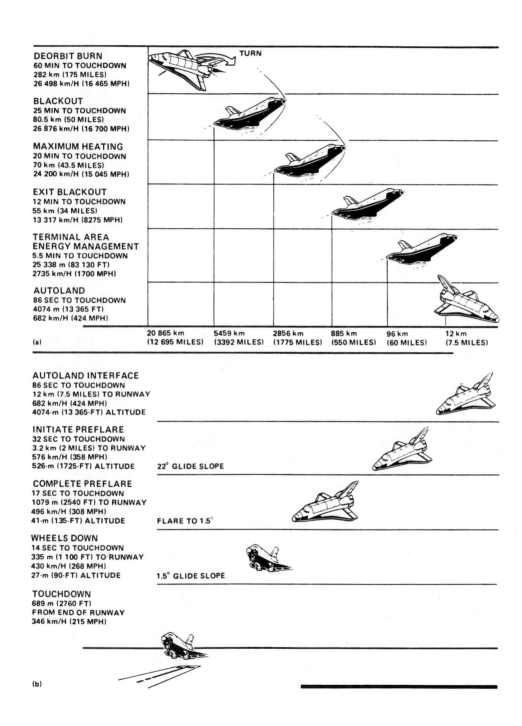

DEORBIT BURN
60 MIN TO TOUCHDOWN
282 km (175 MILES)
26 498 km/H (16 465 MPH)

TURN

BLACKOUT
25 MIN TO TOUCHDOWN
80.5 km (50 MILES)
26 876 km/H (16 700 MPH)

MAXIMUM HEATING
20 MIN TO TOUCHDOWN
70 km (43.5 MILES)
24 200 km/H (15 045 MPH)

EXIT BLACKOUT
12 MIN TO TOUCHDOWN
55 km (34 MILES)
13 317 km/H (8275 MPH)

**TERMINAL AREA
ENERGY MANAGEMENT**
5.5 MIN TO TOUCHDOWN
25 338 m (83 130 FT)
2735 km/H (1700 MPH)

AUTOLAND
86 SEC TO TOUCHDOWN
4074 m (13 365 FT)
682 km/H (424 MPH)

| 20 865 km (12 695 MILES) | 5459 km (3392 MILES) | 2856 km (1775 MILES) | 885 km (550 MILES) | 96 km (60 MILES) | 12 km (7.5 MILES) |

(a)

AUTOLAND INTERFACE
86 SEC TO TOUCHDOWN
12 km (7.5 MILES) TO RUNWAY
682 km/H (424 MPH)
4074-m (13 365-FT) ALTITUDE

INITIATE PREFLARE
32 SEC TO TOUCHDOWN
3.2 km (2 MILES) TO RUNWAY
576 km/H (358 MPH)
526-m (1725-FT) ALTITUDE   22° GLIDE SLOPE

COMPLETE PREFLARE
17 SEC TO TOUCHDOWN
1079 m (2540 FT) TO RUNWAY
496 km/H (308 MPH)
41-m (135-FT) ALTITUDE   FLARE TO 1.5°

WHEELS DOWN
14 SEC TO TOUCHDOWN
335 m (1 100 FT) TO RUNWAY
430 km/H (268 MPH)
27-m (90-FT) ALTITUDE   1.5° GLIDE SLOPE

TOUCHDOWN
689 m (2760 FT)
FROM END OF RUNWAY
346 km/H (215 MPH)

(b)

Fig. 4   Reentry burn cycly for the Space Shuttle.   (Chart courtesy of NASA.)

STS launch facilities   The NASA John F. Kennedy Space Center (KSC) is responsible for prelaunch checkout, for launch of the *Space Shuttle* and its *payloads,* and for ground *turnaround* and support operations. This responsibility extend to Space Transportation System (STS) operations capability development, including the construction and maintenance of STS payload and flight element processing facilities and the development of ground operations management systems and plans, processing schedules, and logistics systems and their use in support of the STS and payloads. It also extends to the establishment of NASA requirements for facilities and ground operations support at Vandenberg Air Force Base (VAFB) in California and designated contingency landing sites. The Kennedy Space Center also supports the Department of Defense in the development and verification of requirements for ground operations at Vandenberg and will maintain NASA facilities and ground support equipment at the California launch site.

The launch facilities of *Launch Complex 39* (LC-39) and the technical support base of KSC's Industrial Area were carved out of virgin savanna and marsh in the early 1960's as the departure point for the *Apollo Pro-gram* manned explorations of the Moon. Twelve manned and unmanned Saturn V/Apollo missions were launched from Complex 39 between November 9, 1967, and December 7, 1972. Minimal modifications to LC-39 permitted the launch of both the smaller Saturn IB and the much larger Saturn V. The *Skylab* space station was placed in a 435-kilometer (270-mile) high circular *orbit* by a Saturn V launch vehicle on May 14, 1973, and three Apollo spacecraft with three-man crews were launched by Saturn IB's on May 26, July 28, and November 16, 1973. The Saturn/Apollo era ended on July 15, 1975, with the launch of a Saturn IB/Apollo and its three-man crew as the U.S. contribution to a joint mission with the Soviet Union. [*See:* Apollo-Soyuz Test Project]

In reshaping the Kennedy Space Center for the Space Shuttle, planners took maximum advantage of existing buildings and structures that could be modified, scheduling new ones only when a unique requirement existed. The only major totally new facilities required to support the Space Shuttle were the Shuttle Landing Facility and the Orbiter Processing Facility.

Kennedy Space Center is located at *lati-*

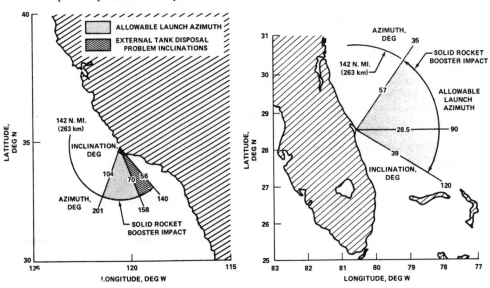

Fig. 1   Shuttle orbit inclinations and launch azimuth from KSC and VAFB.   (Drawing courtesy of NASA.)

tude 28.5° N and *longitude* 80.5° W. It occupies an area of approximately 56,700 hectares (140,000 acres) of land and water. This area, with the adjoining water bodies, provides sufficient expanse to afford adequate safety to the surrounding communities for Shuttle launch and landing activities. Space Shuttle launches will be conducted either from KSC or from Vandenberg Air Force Base in California; the initial launches will be made from the Florida facility. Fig. 1 shows the various orbital inclinations and the related launch *azimuths* for each site. Together, these capabilities will satisfy all known future space launch requirements.

Payloads as heavy as 29,500 kilograms (65,000 pounds) can be launched due east from KSC into an orbit of 28.5° inclination. Payloads of as much as 14,500 kilograms (32,000 pounds) can be launched from VAFB into the highest inclination orbit of 104°. *Polar orbit* capabilities up to 18,000 kilograms (40,000 pounds) can be achieved from Vandenberg.

Shuttle operations at KSC will continue and improve upon the mobile launch concept introduced for the Apollo Program. Using the mobile concept, the space vehicle is thoroughly checked out in an enclosed building before it is moved to the pad for final preparations. This method provides greater protection of flight hardware from the elements, more systematic checkout processing using computer techniques, and minimal pad time. Operations in facilities at Launch Complex-39 (LC-39), at the KSC Industrial Area, and at the Cape Canaveral Air Force Station (CCAFS) will be conducted with a smoothness of flow that will permit rapid Orbiter landing-to-launch processing when the Space Transportation System becomes operational in the 1980's. Space-Shuttle-related facilities at LC-39, and elsewhere at KSC and CCAFS, and their operational functions are described here.

Space Shuttle Orbiters will be processed between missions in a structure analogous to a sophisticated aircraft hangar—the

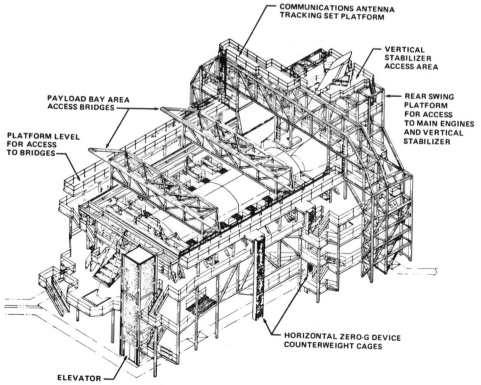

Fig. 2   Inside the Orbiter Processing Facility.   (Drawing courtesy of NASA.)

Fig. 3   The Vehicle Assembly Building at KSC.   (Photograph courtesy of NASA.)

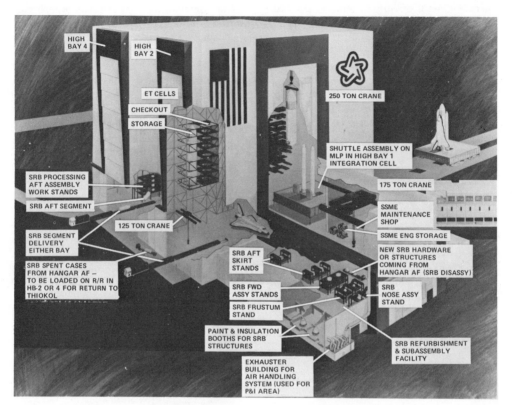

Orbiter Processing Facility (OPF). Once the Space Transportation System becomes fully operational, the Orbiter Processing Facility will be capable of handling two Orbiters simultaneously. The OPF is located on the west side of the Vehicle Assembly Building (VAB) in order to minimize Orbiter towing distance as the processing flow continues.

The Orbiter Processing Facility consists of two identical high bays connected by a low bay. Each high bay is 60 meters (197 feet) long, 46 meters (150 feet) wide, and 29 meters (95 feet) high. Each bay has an area of 2,700 square meters (29,000 square feet) and is equipped with two 27-metric-ton (30-ton) bridge cranes with a hook height of approximately 20 meters (66 feet). The low bay separating the two high bays is 71 meters (233 feet) long, 30 meters (97 feet) wide, and 7.5 meters (24.6 feet) high. A 930-square-meter (10,000-square-foot) annex is located on the north side of the facility. [Fig. 2]

In the high bays, an underfloor trench

system contains electrical, electronic, communication, instrumentation, and control cabling; hydraulic supply and return piping; gaseous nitrogen, oxygen, and helium piping; and compressed air distribution piping. Gaseous nitrogen, helium, and compressed air are supplied from the system in the Vehicle Assembly Building. The low bay houses areas for electronic equipment, mechanical and electrical equipment shops, thermal protection system repair, and self-contained atmospheric pressure ensemble (SCAPE) suiting. It also includes provisions for a communications room, offices, and supervisory control rooms. The high-bay area has an emergency exhaust system in case of a hypergolic fuel spill. Fire protection systems are provided in both high bays and in the low bay.

The Vehicle Assembly Building [Fig. 3] built for the vertical assembly of the Saturn vehicles used in the Apollo, Skylab, and Apollo-Soyuz programs—is the heart of Launch Complex 39. With modifications, it will be used in assembling the Shuttle vehicle. One of the largest buildings in the world, the Vehicle Assembly Building covers a ground area of 3.3 hectares (8 acres) and has a volume of 3,665,000 cubic meters (129,428,000 cubic feet). It is 160 meters (525 feet) tall, 218 meters (716 feet) long, and 158 meters (518 feet) wide. The building is divided into a high-bay area 160 meters (525 feet) tall and low-bay area with a height of 64 meters (210 feet). A transfer aisle running north and south connects and transects the two bays, permitting the easy movement of vehicle stages. The high-bay area is divided into four separate bays. The two on the west side of the structure—Bays 2 and 4—are used for processing Solid Rocket Booster (SRB) motors and the External Tank (ET). The two bays facing east—Bays 1 and 3—are used for the vertical assembly of Space Shuttle vehicles atop Mobile Launcher Platforms.

Extendable platforms, modified to fit the Shuttle configuration, will move in around the vehicle to provide access for *integration* and final testing. When checkout is

Fig. 4   The Space Shuttle leaving the Vehicle Assembly Building on top of the Mobile Launcher Platform.   (Photograph courtesy of NASA.)

complete, the platforms will move back and the doors will be opened to permit the Crawler-Transporter to move the Mobile Launcher Platform (MLP) and assembled Shuttle vehicle to the launch pad. The high-bay door openings are 139 meters (456 feet) high. The lower door opening is 46 meters (152 feet) wide and 35 meters (114 feet) high with four door leaves that move horizontally. The upper door opening is 104 meters (342 feet) high and 23 meters (76 feet) wide with seven door leaves that move vertically. [*See:* Fig. 4]

The low bay will be the site for a Refurbishment and Subassembly Facility (RSF) for SRB segments. Existing pneumatic, environmental control, light, and water systems have been modified in both bays. The north doors to the VAB transfer aisle have also been widened 12 meters (40 feet) to permit entry of the towed Orbiter from the Orbiter Processing Facility. The doors are slotted at the center to accommodate the *Orbiter's vertical stabilizer.* [*See:* Orbiter structure] The Vehicle Assembly Building has more than 70 lifting devices, including two 227-metric-ton (250-ton) bridge cranes.

The solid rocket motor segments and associated hardware will be shipped to KSC by rail from the contractor facility in Utah. The segments will be transported horizontally and will have transportation covers. End rings will provide segment handling points, environmental protection, and protection of the *solid grain propellant* and the outer edge of each segment from potential impact damage. On arrival at KSC, the segments will first be moved into High Bay 4 for transportation cover removal and offloading. They are then placed vertically into buildup workstands or in-process areas. The SRB segments, forward and aft closures, *nozzle* assemblies, nozzle extensions, and associated *hardware* will be given receiving inspections. *Inert* SRB elements such as forward *skirts, frustums, nose caps, recovery systems,* electronics and instrumentation components, and elements of the *thrust vector control system* will be received in the VAB Refurbishment and Subassembly Facility (RSF) located in the low bay.

The structural assemblies and components required to build up the nose assembly, the frustum assembly, the forward and aft skirts, and the *External Tank* attachment hardware will be shipped to KSC and put into storage or routed to the VAB low bay for buildup. As the STS program matures, recovered hardware will be routed to the Refurbishment and Subassembly Facility for reuse. Assembly and checkout of the SRB forward skirt and nose assembly and the aft skirt assembly will also be performed in the RSF. When completed, the aft skirt assemblies will be transferred to High Bay 4 for assembly with the aft solid rocket motor segments.

The stacking of the SRB major assemblies will begin after the buildup of the aft booster assemblies in High Bay 4, the assembly and checkout of the forward nose skirt assemblies in the RSF, and the alignment of the Mobile Launcher Platform (MLP) support posts. The stacking operation will be accomplished in the following sequence:

1. The aft booster assemblies will be transferred from the buildup area in High Bay 4 to the High Bay 1 or 3 integration cells and attached to the MLP support posts.

2. Continuing serially, the aft, forward center, and forward rocket motor segments will be stacked to form complete solid rocket motor assemblies. An alignment check of the complete flight set of SRB assemblies will be performed after the stacking operations are completed. Integrated and automated systems testing of the assembled *Solid Rocket Boosters* will be accomplished on the Mobile Launcher Platform, using the launch processing system (LPS) to supply ET/Orbiter simulation.

The External Tank will be transported to KSC by barge from the Michoud Assembly Facility at New Orleans, Louisiana. On arrival at KSC, the *tank* and the associated hardware will be offloaded. The External Tank will be transported horizontally to the Vehicle Assembly Building where it will be transferred to a vertical storage or checkout cell. In the Shuttle's operational phase, High Bays 2 and 4 will each contain one storage and one checkout cell. Initial operations will be in High Bay 4. The storage cells provide only the minimum access and equipment required to secure the External

Tank in position. After transfer to the checkout cell, both permanent and mobile platforms will be positioned to provide access to inspect the tank for possible damage in transit and to remove hoisting equipment. The *liquid oxygen* and *liquid hydrogen* tanks will then be sampled and will receive a blanket pressure of gaseous nitrogen and gaseous helium, respectively, in preparation for a normal checkout.

The ET subsystems checkout includes an inspection of the external insulation and connection of ground support equipment (including the launch processing system) to the appropriate interfaces. Electrical, instrumentation, and mechanical function checks and tank and line leak checks will be performed in parallel. After satisfactory checkout of the ET subsystems, ground support and LPS equipment will be removed and stored and ET closeout will be initiated. Forward hoisting equipment will be attached and work platforms stored—or opened—in preparation for transfer to the Mobile Launcher Platform. The External Tank will be hoisted vertically from the checkout cell with the 227-metric-ton (250-ton) high-bay crane and transferred to a Mobile Launcher Platform in High Bay 1 or 3 for mating with the Space Shuttle vehicle's twin Solid Rocket Boosters. After ET/SRB mating, the integration cell ground support equipment will be connected and intertank work platforms will be installed.

The *Orbiter* will be towed into the transfer aisle through the north door of the VAB. Once in position, the lifting beams will be installed and the erection slings attached. The Orbiter is then lifted, with the landing gear retracted. The Orbiter will be rotated from the horizontal to the vertical position using the 227-and 159-metric-ton (250- and 175-ton) cranes. It is then transferred to the Shuttle assembly area in High Bay 1 or 3 and lowered and mated to the External Tank, which has previously been mated with the Solid Rocket Boosters on the Mobile Launcher Platform. After mating is complete, the erection slings and the load beams will be removed from the Orbiter and the platforms and stands will be positioned for Orbiter/ET/SRB access. The Orbiter will be mated in the VAB with its fin toward the transfer aisle (toward the south at the pad).

After the Orbiter has been mated to the ET/SRB assembly and all *umbilicals* have been connected, an electrical and mechanical verification of the mated interfaces will be performed to verify Shuttle vehicle interface compatibility. A shuttle interface test will be performed using the launch processing system to verify Shuttle vehicle interfaces and Shuttle vehicle-to-ground interfaces. The launch processing system will be used to control and monitor Orbiter systems as required in the Shuttle vehicle assembly and checkout station. After interface testing is complete, ordnance is installed and connected before the Orbiter is transferred to the *pad.* Complete external access to the Shuttle vehicle will be provided in the VAB. Payload access will be limited to access provided internal to the Orbiter through the crew compartment; the payload bay doors will not be opened in the Vehicle Assembly Building.

If the Vehicle Assembly Building is the heart of Launch Complex-39 (LC-39), the Launch Control Center (LCC) is its brain. The Launch Control Center is a four-story structure located on the southeast side of the VAB and connected to it by an enclosed utilities bridge. No changes will be made to the exterior of the Launch Control Center to adapt it for Shuttle operations. On the inside, Firing Rooms 1 and 2 have been equipped with the highly automated launch processing system designed for Shuttle checkout and launch. Firing Rooms 3 and 4 will not be directly concerned with Shuttle operations at this time. Compared to a firing room manned for a Saturn/Apollo launch, a Space Shuttle firing room will be a lonely place: launch with the Shuttle launch processing system will require approximately 45 operational personnel— one-tenth of the 450 needed for an Apollo launch.

The launch processing system will automatically control and perform much of the Space Shuttle vehicle checkout while the vehicle components are being prepared for launch. It will also provide the capability for work order control and scheduling and will conduct countdown and launch opera-

tions. The final *countdown* will require approximately 2.5 hours, compared to 28 hours for a Saturn/Apollo countdown. During systems-to-vehicle integration, the launch processing system will interface with the Solid Rocket Booster, the External Tank, the Space Shuttle Main Engines (SSME's), and the Orbiter systems. LPS hardware interface modules will be located in areas such as the Orbiter Processing Facility, the VAB high bays, the Hypergol Maintenance Facility, and various other sites that support Shuttle maintenance and checkout.

The launch processing system has been divided into the central data subsystem (CDS) and the checkout, control, and monitor subsystem (CCMS). The central data subsystem consists of two large-scale computers (Honeywell H-6680) that store test procedures, vehicle processing data, a master program library, historical data, pre- and post-test data analyses, and other data. The CDS is located on the second floor of the Launch Control Center. The checkout, control, and monitor subsystem consists of consoles, minicomputers, a large mass storage unit, and related equipment and is located on the third floor in the firing rooms that will be used to actually process and launch the vehicle. Vehicle checkout, countdown, and launch will be conducted with

Fig. 5   The STS Mobile Launcher Platform at KSC. (Photograph courtesy of NASA.)

the support of the information sorted in the CDS. Automatic checkout from the firing rooms will be accomplished by using computer programs to monitor and record the prelaunch performance of all electrical and mechanical systems. Command signals from the LPS computer will be sent to the various components and test circuits. While a component is functioning, a sensor will measure its performance and send these data back to the LPS for comparison with the checkout limits stored in the system's computer memory. (Certain test requirements and limits are established for each component and stored in the LPS memory.) When the checkout program is complete, a signal will indicate whether or not its performance has been satisfactory. If unsatisfactory, the LPS computer will then provide data that support isolation of the fault. This process will continue through vehicle checkout.

The Mobile Launcher Platform is a transportable launch base for the Space Shuttle. Two platforms are now available for Launch Complex 39 and the capability exists to add a third. The MLP's are mobile launchers used for Saturn/Apollo missions that have undergone major changes to adapt them for Shuttle. The most striking visual change is the removal of the 121-meter (398-foot) high tower and its hammerhead crane that soared up from the launch platform. The addition of a permanent launch tower (the fixed service structure) at each of the two launch pads has eliminated the need for towers on the Mobile Launcher Platforms.

The Mobile Launcher Platform [*See:* Fig. 5] is a two-story steel structure 7.6 meters (25 feet) high, 48.8 meters (160 feet) long, and 41.1 meters (135 feet) wide. The platform is constructed of welded steel up to 15.2 centimeters (6 inches) thick. At their park site north of the VAB, in the VAB high bays, and at the launch pad, the Mobile Launcher Platforms rest on six 6.7-meter (22-foot) tall pedestals. The single 13.7-meter (45-foot) square opening in the center of the Apollo mobile launchers that allowed hot exhausts from the Saturn V/IB to escape into the flame trench during lift-off has been replaced by three openings in the

MLP—two for SRB exhaust and one for SSME exhaust [*See:* Fig. 6]. The SRB exhaust holes are 12.8 meters (42 feet) long and 6 meters (20 feet) wide. The SSME exhaust opening is 10.4 meters (34 feet) long and 9.5 meters (31 feet) wide. Inside the platform are two levels with rooms and compartments housing LPS hardware interface modules, system test sets, *propellant* loading equipment, and electrical racks.

Unloaded, the Mobile Launcher Platform weighs 3,733,000 kilograms (8.23 million pounds). The total weight with an unfueled Space Shuttle aboard is 4,989,500 kilograms (11 million pounds); the total weight with the propellant-laden Shuttle aboard is 5,761,000 kilograms (12.7 million pounds). The Space Shuttle vehicle is supported and restrained on the Mobile Launcher Platform during assembly, transit, and pad checkout by the SRB support/holdown system. Four conical hollow supports for each Solid Rocket Booster are located in each SRB exhaust well. The supports are 1.5 meters (5 feet) high and have a base diameter of 1.2 meters (4 feet).

Two tail service masts (TSM's), one located on each side of the SSME exhaust hole, support the *fluid,* gas, and electrical requirements of the Orbiter's liquid oxygen and liquid hydrogen aft umbilicals. The TSM assembly also protects the ground half of those umbilicals from the harsh launch environment. At launch, the SRB ignition command fires an explosive link allowing a

Fig. 6   An exterior view of the Mobile Launcher Platform.   (Photograph courtesy of NASA.)

Fig. 7 The Space Shuttle Columbia on the pad Complex 39 (KSC) during the 20-second flight-readiness firing on February 20, 1981. (Photograph courtesy of NASA.)

9072-kilogram (20,000-pound) counter-weight to fall, pulling the ground half of the umbilicals away from the Space Shuttle vehicle and causing the mast to rotate into a blastproof structure. As it rotates backward, the mast triggers a compressed gas *thruster,* causing a protective hood to move into place and completely seal the structure from the SSME exhaust.

Other MLP systems include the hydrogen burnoff system and the post-shutdown engine deluge system. During the engine start sequence for the *Orbiter's main engines,* the liquid hydrogen fuel begins to flow first and the engines are started in a hydrogen-rich condition. The preignition flow of liquid hydrogen is expected to create a small gaseous hydrogen cloud in the vicinity of the Orbiter's main engines, posing the possibility of overpressures that might damage the heat shields or other components in the aft section of the Orbiter when the cloud is detonated by igniting the engines. To eliminate this possibility, a 1.5-meter (5-foot) long boom is suspended from each of the tail service masts such that they rest below and alongside the nozzles of the number two and three engines. Each boom holds four flare-like devices called radially outward firing igniters. These devices, which are ignited at T—5 seconds [*See:* T-time] and burn for 10 seconds, will burn off excess hydrogen from the nozzles as it

emerges and thus prevent the buildup of a potentially hazardous cloud of explosive gas.

The post-shutdown engine deluge system consists of 22 water nozzles spaced around the SSME exhaust hole in the MLP. The system is designed to cool the aft end of the Orbiter following the *flight-readiness firing* of the main engines before the first Space Shuttle flight [*See:* Fig. 7]. It will also be available in the event of an on-pad *abort.* Experience obtained during main engine testing has shown that after the engines are shut down, there can still be some residual hydrogen remaining in the vicinity of the nozzles that can burn for a appreciable period of time. The system is fed by a 15-centimeter (6-inch) diameter supply line and can provide a flow of up to 9,460 liters/min (2,500 gal/min) directed at the Orbiter's engine nozzle/boattail area to control hydrogen afterburning and provide cooling water.

Work platforms used in conjunction with the MLP provide access to the *SSME nozzles* and to the *Solid Rocket Boosters* after erection in the Vehicle Assembly Building or while the Space Shuttle is undergoing checkout at the pad. The Orbiter engine service platform is positioned beneath the MLP and raised by a winch mechanism through the exhaust hole to a position directly beneath the three engines. An elevator platform with a cutout may then be extended upward around the *engine bells.* The Orbiter engine service platform is 10.4 meters (34 feet) long and 9.4 meters (31 feet) wide. Its retracted height is 3.7 meters (12 feet) and the extended height is 5.5 meters (18 feet). The weight is 27,200 kilograms (60,000 pounds). Two SRB service platforms provide access to the SRB nozzles after the vehicle has been erected on the Mobile Launcher Platform. The platforms are raised from storage beneath the MLP into the SRB exhaust holes and hung from brackets by a turnbuckle arrangement. The SRB platforms are 1.2 meters (4 feet) high, 6 meters (20 feet) long, and 6 meters (20 feet) wide. Each weighs 4,500 kilograms (10,000 pounds). The Orbiter and SRB service platforms are moved down the pad ramp to a position outside the exhaust area before launch.

The two Crawler-Transporters used to move the Space Shuttle are the same tracked vehicles previously used to move Saturn/Apollo/Skylab flight hardware between the facilities of Launch Complex 39. They have been refurbished to carry the assembled Space Shuttle on its Mobile Launcher Platform between the VAB and LC-39's two launch pads. Modifications consisted primarily of replacing outdated electronic and electrical equipment. Both massive vehicles are expected to perform their Shuttle transportation functions throughout the duration of the program.

The transporters [*See:* Fig. 8] are 39.9 meters (131 feet) long and 34.7 meters (114 feet) wide. They move on four double-tracked crawlers, each 3 meters (10 feet) high and 12.5 meters (41 feet) long. Each shoe on the crawler track weighs 0.9 metric ton (2,000 pounds). The maximum speed unloaded is 3.2 km/h (2 mph); loaded speed is 1.6 km/h (1 mph). The unloaded weight is 2,721,000 kilograms (6 million pounds). The transporters have a leveling system designed to keep the top of the space vehicle vertical within ±10 minutes of arc—about the dimensions of a basketball. This system also provides the leveling operations required to negotiate the 5-percent ramp leading to the launch pads and to keep the load level when

it is raised and lowered on pedestals at the pad and within the Vehicle Assembly Building.

The overall height of the transporter is 6.1 meters (20 feet) from ground level to the top deck on which the Mobile Launcher Platform is mated for transportation. The deck is flat and about the size of a baseball diamond—27.4 meters (90 feet) square. Each transporter is powered by two 2,050-kilowatt (2,750-horsepower) diesel engines. The engines drive four 1,000-kilowatt generators that provide electrical power to 16 traction motors. Through gears, the traction motors turn the four double-tracked crawlers spaced 27.4 meters (90 feet) apart at each corner of the transporter.

The Crawler-Transporters move on a roadway 40 meters (130 feet) wide, almost as broad as an eight-lane turnpike. The crawlerway consists of two 12.2-meter (40-foot) wide lanes separated by a 15.2-meter (50-foot) wide median strip and provides a traveling surface for the transporters between the VAB and Launch Pads A and B. The distance from the VAB to Pad A is 5.5 kilometers (3.4 miles). The distance to Pad B is 6.8 kilometers (4.2 miles). The roadway is built in three layers with an average depth of 2.1 meters (7 feet). The top surface on which the transporters operate is river gravel. The thickness of the gravel is

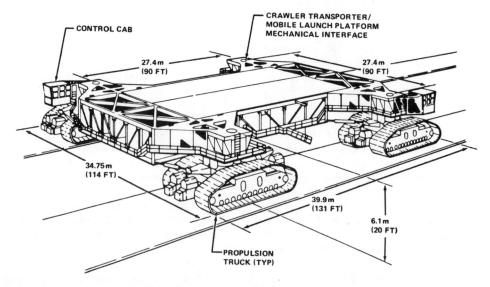

Fig. 8   The crawler-transport at the Kennedy Space Center.   (Drawing courtesy of NASA.)

Fig. 9   The surface arrangement at Launch Complex 39-Pad A, KSC.   (Drawing courtesy of NASA.)

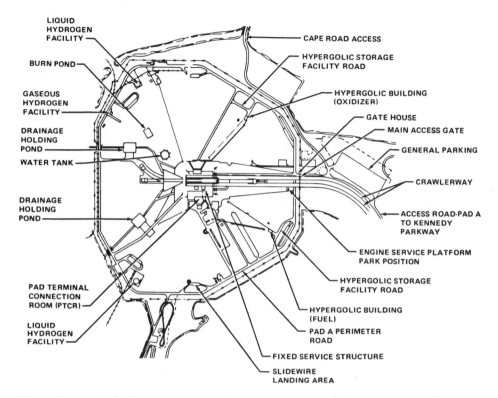

20.3 centimeters (8 inches) on curves and 10.2 centimeters (4 inches) on the straightaway sections.

The Launch Complex 39 *pads* are roughly octagonal in shape [*See:* Fig. 9]; each contains about 67 hectares (one-fourth square mile) of land. The pads are elevated above the surrounding terrain: Pad A is 14.6 meters (48 feet) above sea level and Pad B is 16.8 meters (55 feet) above sea level. The hardstand area at the top of each pad measures 119 by 99 meters (390 by 325 feet). Pad A construction was completed in mid-1978 and construction at Pad B is scheduled for completion in 1982. Major changes include construction of new *hypergolic fuel* and *oxidizer* support areas at the southwest and southeast corners, respectively, of the pad; removal of the *RP-1* support area; erection of a new fixed service structure (FSS); addition of a rotating service structure (RSS); and replacement of the Saturn *flame deflectors* with three new flame deflectors. The upper portion of the *umbilical* tower removed from each Mobile

Launcher Platform during modification has been installed at each pad to serve as a fixed service structure.

Access to the Space Shuttle on the pads [*See:* Fig. 10] will be provided through the FSS vent arm for electrical power and for venting hydrogen from the External Tank; through the FSS Orbiter access arm for crew and passenger *ingress*; through the FSS External Tank gaseous oxygen vent arm for preventing icing on the External Tank and for venting oxygen vapors and nitrogen away from the vehicle; through the MLP tail service masts for propellant loading and electrical power; through the RSS mid-body umbilical unit for *fuel-cell* servicing and life-support functions; through the RSS facilities for loading and offloading *payloads;* and through the RSS hypergolic umbilical system for servicing the *orbital maneuvering* and *reaction control systems* with *fluids* and gases.

The fixed service structure, located on the west side of the pad, is a square cross-section steel structure that provides access to the Shuttle Orbiter and to the rotating service structure. The FSS is essentially an open-framework structure 12.2 meters (40 feet) square and is permanently fixed to the pad surface. It incorporates several sections of the Saturn V umbilical towers removed from the Apollo mobile launchers in their conversion to Mobile Launcher Platforms. The FSS tower supports the hinge about which the rotary bridge supporting the RSS pivots as it moves between the Orbiter checkout position and the retracted position. A *hammerhead crane* situated atop the FSS provides hoisting services as required in pad operations. FSS work levels are at 6.1-meter (20-foot) intervals beginning at 8.2 meters (27 feet) above the surface of the pad. The height of the FSS from the pad surface to the top of the tower is 75.3 meters (247 feet). The height to the top of the hammerhead crane is 80.8 meters (265 feet), and

the top of the lightning mast is 105.8 meters (347 feet) above the pad surface.

The FSS has three service arms: an access arm and two vent arms. The Orbiter access arm (OAA) swings out to the Orbiter *crew compartment* hatch to provide personnel access to the forward compartments of the Orbiter. The outer end of the access arm ends in an environmental chamber that mates with the Orbiter and will hold six persons. The arm remains in the extended position until 2 minutes before launch to provide emergency *egress* for the crew. The Orbiter access arm is extended and retracted by two rotating actuators that rotate it through an arc of 70° in approximately 30 seconds. In its retracted position, the arm is latched to the FSS. The OAA is located 44.8 meters (147 feet) above the pad. It is 19.8 meters (65 feet) long, 1.5 meters (5 feet) wide, and 2.4 meters (8 feet) high and

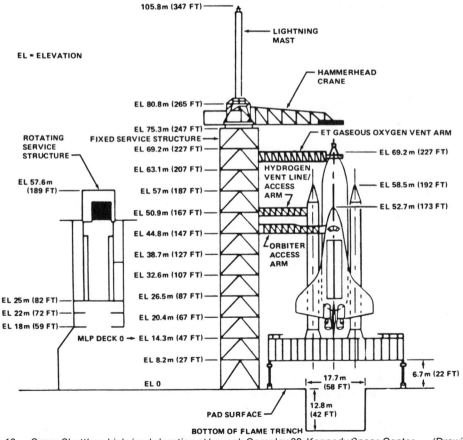

Fig. 10   Space Shuttle vehicle/pad elevation at Launch Complex 39, Kennedy Space Center.   (Drawing courtesy of NASA.)

weighs 23,600 kilograms (52,000 pounds).

The External Tank hydrogen vent line and access arm consists of a retractable access arm and a fixed supporting structure. This arm allows mating of the ET umbilicals and contingency access to the intertank interior while protecting sensitive components of the system from the launch environment. The vent arm supports small helium and nitrogen lines and electrical cables, all mounted on a 20.3-centimeter (8-inch) inside-diameter hydrogen vent line. At SRB ignition, the umbilical is released from the Shuttle vehicle and retracted 84 centimeters (33 inches) into its latched position by a system of counterweights. The service lines rise approximately 46 centimeters (18 inches), pivot, and drop to a vertical position on the fixed structure where they are protected from the launch environment. All this activity occurs in approximately 4 seconds. The vent arm itself rotates through 210° of arc to its stowed position in about 3 minutes. The fixed structure is mounted on the northeast corner of the FSS 50.9 meters (167 feet) above the surface of the pad. The vent arm is 14.6 meters (48 feet) long and weighs 6800 kilograms (15,000 pounds).

The External Tank gaseous oxygen vent arm is attached to the fixed service structure between the 69-meter (227-foot) and 63-meter (207-foot) levels. The arm suspends a hood or cap that will be lowered over the top of the External Tank and sealed by means of an inflatable collar. Heated gaseous nitrogen is introduced into the hood to warm the inflatable vent seals around the two gaseous oxygen louvers at the top of the External Tank. This prevents vapors from the liquid oxygen vent system at the top of the tank from condensing into ice, which could possibly become dislodged during lift-off and damage the Orbiter's *thermal protection system tiles*. The system also provides exhaust ducts to carry the oxygen vapors and the nitrogen away from the vehicle and serves as a rain shield for the top of the tank. The vent system arm is 24.4 meters (80 feet) long, 2.4 meters (8 feet) high, and 1.5 meters (5 feet) wide. The diameter of the vent hood is 4 meters (13 feet). The weight of the arm and hood is 16,329 kilograms

(36,000 pounds).

The vent system arm is extended at approximately T—12 hours and the vent seals are inflated to an operating *pressure* of 3.4 to 6.9 kN/m^2 (0.5 to 1 psi) Prior to *cryogenic* loading, which begins at approximately T—5 hours. The vent system supports launch preparation until the ET vent valve closes at T—2 minutes 55 seconds. Approximately 30 seconds later, the seals are deflated and the hood is retracted (retraction time approximately 30 seconds). After the hood has been stowed, the arm is retracted (retraction time approximately 1 minute). The arm will be in the retracted position at approximately T—45 seconds and is held there hydraulically until T—0, when it is latched to the FSS for lift-off.

The emergency exit system or slide wire [*See:* Fig. 11] provides an escape route for personnel onboard the Shuttle and on the Orbiter access arm of the fixed service structure until ignition of the Solid Rocket Boosters. Five slide wires extend from the level of the Orbiter access arm to the ground on the west side of the pad. A single stainless-steel basket enclosed with Nomex webbing 2.5 centimeters (1 inch) wide and 3 millimeters (one-eighth inch) thick is suspended by two trolleys from each wire (19 millimeters (0.75 inch) diameter) and positioned on the FSS for ready entry in the event of emergency. Each basket holds a maximum of two persons. When boarded, each basket slides down a 366-meter (1,200-foot) wire to the landing zone area west of the pad. The deceleration system is composed of a catch-net system and drag chains. The descent takes approximately 35 seconds. After the basket has been stopped by the deceleration system, those onboard may run or walk to the bunker area a short distance to the west.

The lightning mast extends above the fixed service structure and surrounding pad equipment and provides protection from lightning strokes. The 24.4-meter (80-foot) tall fiberglass mast is grounded by a cable that starts from a ground anchor 335 meters (1,100 feet) south of the FSS, angles up and over the lightning mast, then extends back down to a second ground anchor 335 meters

(1,100 feet) north of the FSS. The mast functions as an electrical insulator holding the cable away from the FSS and as a mechanical support in rolling contact with the cable. The mast and its support structure extend 30.5 meters (100 feet) above the FSS.

The rotating service structure provides protected access to the Orbiter for change-out and servicing of payloads at the pad. The structure is supported by a rotating bridge that pivots about a vertical axis on the west side of the pad's flame trench. The RSS rotates through 120° (one-third of a circle) on a radius of 49 meters (160 feet). The hinge column rests on the pad surface and is braced to the fixed service structure. Support for the outer end of the bridge is provided by two eight-wheel motor-driven trucks that move along circular twin rails installed flush with the pad surface. The track crosses the flame trench on a new permanent bridge. The rotating service structure is 31 meters (102 feet) long, 15.24 meters (50 feet) wide, and 39.6 meters (130 feet) high. The elevation of the main structure above the surface of the pad extends from 18 to 57.6 meters (59 to 189 feet). The structure has Orbiter access platforms at five levels to provide access to the payload while the Orbiter is being serviced in the RSS. Each platform has independent extensible planks that can be arranged to conform to the individual payload configuration. With the exception of *Spacelab* and other horizontally handled cargoes, payloads may be loaded into the Orbiter from the RSS under environmentally *clean* or *"white room"* conditions.

Payloads will be transported to the rotating service structure in the *payload canister.* The canister restrains and provides environmental protection for the various Shuttle payloads while in transit. It also provides the constant payload envelope required by payload envelope required by payload handling devices and environmental services to the payload itself. The payload canister is 21 meters (69 feet) long, 6.4 meters (21 feet) wide, and 6.4 meters (21 feet) high. It weighs 38,555 kilograms (85,000 pounds). The canister can accommodate payloads up to 18.3 meters (60 feet)

long, 4.6 meters (15 feet) in diameter, and 29,500 kilograms (65,000 pounds) in weight.

Other rotating service structure elements servicing the space vehicle are the Orbiter midbody umbilical unit, the hypergolic umbilical system, and the *orbital maneuvering system* (OMS) pods heated-gas purge system. The Orbiter midbody umbilical unit provides access and services to the mid-fuselage portion of the Orbiter on the pad. Liquid oxygen and liquid hydrogen for the *fuel cells* and gases such as nitrogen and helium are provided through the umbilical unit. The unit is 6.7 meters (22 feet) long, 4 meters (13 feet) wide, and 6 meters (20 feet) high. The Orbiter midbody umbilical unit extends from the rotating service structure at levels ranging from 48.2 to 53.6 meters (158 to 176 feet) above the surface of the pad. The hypergolic umbilical system carries hypergolic fuel and oxidizer, helium, and nitrogen service lines from the fixed service structure to the vehicle. Six umbilical handling units—manually operated and locally controlled—are structurally attached to the rotating service structure. The Orbiter midbody umbilical unit, hypergolic umbilical system, and OMS pod heated-gas purge system connections with the Orbiter are severed when the rotating service structure is prepared for retraction to its park site at approximately the T—6-hour point in the countdown.

The OMS pods heated-gas purge system is suspended from the lower portion of the rotating service structure and consists of heated *purge* covers designed to enclose the OMS pods located on the aft upper portion of the Orbiter. [*See:* Orbiter Structure] The Orbiter's OMS pods have structural skins consisting of layered graphite/epoxy covered by thermal protection system silica tiles and Nomex felt *reusable surface insulation* attached by an adhesive. The graphite/epoxy absorbs moisture from the *atmosphere,* resulting in a degradation of strength under reentry conditions. The heated-gas purge system is designed to reduce the moisture level in the OMS pods to that considered desirable for flight. The OMS pods heated-gas purge covers are *concave* in shape to fit the contours of the pods and have an average height of 4.9 meters (16 feet) with 1.8-meter (6-foot) extensions to cover the reaction control system (RCS) clusters on the aft outboard portions of each pod. The purge covers, fabricated of an aluminum framework covered with an aluminum skin, are located between the 18- and 25-meter (59- and 82-foot) levels of the rotating service structure. A pressurized *herculite* seal between the OMS pods and the covers excludes ambient air. Conditioned air (dried air with a *relative humidity* of 1 to 2 percent) is heated in the pad terminal connection room beneath the pad and flows in pipes attached to the fixed service structure and the rotating service structure before entering the pod covers near their tops. The air flows over the pod structures under a *pressure* of approximately 3.4 kN/m² (0.5 psi). The air leaves the pod covers through vents at the bottom. The temperature of the heated air will be maintained at up to 380°K (107°C or 225°F) for extended periods before the loading of the *hypergolic propellants* for the OMS and RCS systems. After propellant

loading, the purge will be maintained at a temperature of approximately 310°K (38°C or 100°F) for various periods of time until the final *countdown* begins.

Pad structures are insulated from the intense heat of launch by the flame deflector system, which protects the flame trench floor and the pad surface along the top of the flame trench. The flame trench transects the pad's mound at ground level and is 149.4 meters (490 feet) long, 17.7 meters (58 feet) wide, and 12.2 meters (40 feet) high. The Orbiter flame deflector is fixed and is 11.6 meters (38 feet) high, 22 meters (72 feet) long, and 17.6 meters (57.6 feet) wide. The top of the SRB flame deflector abuts with that of the Orbiter flame deflector to form a flattened, inverted V-shaped structure beneath the three MLP exhaust holes. The SRB deflector is 13 meters (42.5 feet) high, 12.8 meters (42 feet) long, and 17.4 meters (57 feet) wide. The deflectors are built of steel and covered with an ablative, or heat shedding, surface with an average thickness of 12.7 centimeters (5

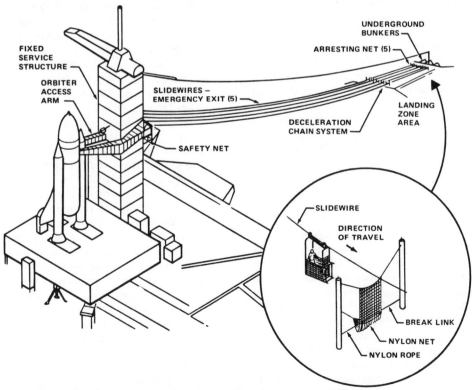

Fig. 11 Space Shuttle emergency exit system. (Drawing courtesy of NASA.)

inches). There are two movable SRB side flame deflectors, one located on each side of the flame trench. They are 6 meters (19.5 feet) high, 13.4 meters (44 feet) long, and 5.3 meters (17.5 feet) wide.

The Shuttle Orbiter with its delicate payloads is much closer to the Mobile Launcher Platform (MLP) surface than was the Apollo spacecraft at the top of a Saturn V/IB *launch vehicle*. A sound-suppression water system has been installed on the pads to protect the Orbiter and its payloads from damage by *acoustical* energy reflected from the Mobile Launcher Platform during launch.

The system includes an elevated water tank with a capacity of 1,135,500 liters (300,000 gallons). The tank is 88.4 meters (290 feet) high and stands on the northeast side of the pad. The water will be released just before ignition of the three Space Shuttle Main Engines and twin Solid Rocket Boosters and will flow through parallel 2.1-meter (7-foot) diameter pipes to the pad area. Water will be pouring from 16 nozzles atop the flame deflectors and from outlets in the SSME exhaust hole in the Mobile Launcher Platform at SSME ignition at T—3.46 seconds [*See:* T-Time]. When SRB ignition and *lift-off* follow at T + 3 seconds, a torrent of water will begin flowing onto the platform from six large quench nozzles or "rainbirds" mounted on its surface. The peak flow from the pre- and post-lift-off systems will be at the rate of 3,406,500 liters/min (900,000 gal/min) 9 seconds after lift-off. The MLP "rainbirds" are 3.7 meters (12 feet) high. The two in the center are 107 centimeters (42 inches) in diameter; the other four have a diameter of 76 centimeters (30 inches).

Acoustical levels reach their peak when the Space Shuttle is approximately 91 meters (300 feet) above the platform and should no longer be a problem when an altitude of 305 meters (1000 feet) is reached. Below the peak level, the rocket exhaust is channeled over the flame deflectors and into the flame french; above the peak level, sound is reflected off the metal plates of the MLP surface. In terms of time, the maximum sound reflection comes approximately 5 seconds after lift-off. The problem ends

after the Shuttle has been airborne for about 10 seconds and has reached an altitude of 305 meters (1000 feet).

Design specifications for Shuttle payloads require the capability to withstand acoustical loads of up to 145 *decibels*. It was anticipated that the sound-suppression system would reduce the acoustical levels within the Orbiter payload bay to about 142 decibels, 3 decibels below the design requirement. The decibel level on the aft heat shield of the Orbiter without water suppression would be approximately 167 or 168; the sound-suppression system was expected to lower the level to 162 or 163. [For comparison, the decibel reading at the base of the Saturn V at lift-off averaged 160.] However, higher than anticipated acoustical environments were experienced in the *STS-1* flight of the Space Shuttle *"Columbia"* and this required some redesign and modification of the sound-suppression system. These modifications were successfully accomplished in time for the *STS-2* flight of Columbia.

The *liquid oxygen* (LO₂ or lox) used as an *oxidizer* by the Orbiter's main engines is stored in a 3,406,500-liter (9,000,000-gallon) storage tank located at the northwest corner of the pad. This ball-shaped vessel is a huge *vacuum,* or *Dewar,* bottle designed to store *cryogenic* (supercold) lox at a temperature lower then 90 K (—183° C or —297° F). Lox is quite heavy—3.8 liters (1 gallon) weighs approximately 4.5 kilograms (10 pounds)—and is transferred to the pad by two main pumps rated at 37,850 liters/min (10,000 gal/min) each.

The *liquid hydrogen* (LH₂) used as a *fuel* by the Orbiter's main engines is stored in a 3,200,000-liter (850,000-gallon) storage tank located at the northeast corner of the pad. This large ball-shaped vessel is also a huge vacuum bottle designed to contain and store liquid hydrogen, a cryogenic fluid much colder than liquid oxygen (lox). Liquid hydrogen vaporizes at temperatures above 20 K (—253°C or —423° F). It is extremely light—3.8 liters (1 gallon) weighs approximately 0.23 kilogram (0.5 pound). Because of its lightness, pumps are not needed for transfer to the pad. Liquid hydrogen vaporizers convert a small portion of the liquid hydrogen stored in the tank into a gas and this gas *pressure* exerted from the top of the tank moves the liquid hydrogen into the transfer lines and to the pad. Vacuum-jacketed transfer lines carry these supercold fluids to the Mobile Launcher Platform where they are fed into the Orbiter through the tail service masts.

The Orbiter's orbital maneuvering system and reaction control system engines use monomethyl hydrazine (MMH) as a fuel and nitrogen tetroxide (N_2O_4) as an oxidizer. These fluids can be stored at ambient temperatures and are hypergolic; that is, they ignite on contact. They are stored in well-separated areas on the southwest (MMH) and southeast (N_2O_4) corners of the pads. These propellants are fed by transfer lines to the pad and through the fixed service structure to the rotating service structure hypergolic umbilical system. Other fluids and gases supplied on the pad include helium, nitrogen, air, *Freon 21*, and ammonia.

The vital links between the launch processing system in the Launch Control Center and the *ground support equipment* and flight *hardware* on the pad are provided by elements located in the pad connection terminal room (PCTR) below the pad's elevated hardstand. All pad launch processing system (LPS) terminals (hardware interface modules) interface with the central data subsystem in the Launch Control Center.

The Hypergol Maintenance and Checkout Facility (HMF) provides all facilities required to process and store the hypergol-fueled modules that make up the Orbiter's forward *reaction control system* (FRCS), orbital maneuvering system (OMS), auxiliary propulsion system (APS), and certain *payload bay kits*. The facility occupies a group of buildings in an isolated section of the KSC Industrial Area approximately 13 kilometers (8 miles) southeast of the Vehicle Assembly Building. Originally built for the Apollo Program, these buildings have been extensively modified to support Space Shuttle operations.

The Launch Equipment Test Facility (LETF) is located in the KSC Industrial Area immediately south of the Operations and Checkout Building. This facility is required to test the operation of launch-critical systems or ground systems that could cause failures in the *space vehicle* if they did not function properly. The facility is able to simulate such launch vehicle events as movement due to the wind, Orbiter engine ignition and lift-off, and the effects of solar heating [i.e. heating by the sun] and cryogenic shrinkage. The ability of the ground systems to react properly to these events must be verified before committing the ground support equipment to a manned launch. The systems tested are the External Tank vent line, the Orbiter access arm, and the ET gaseous oxygen vent arm, all on the fixed service structure, and the tail service masts and the SRB holddown posts on the Mobile Launcher Platform. The test systems include an SRB holddown test stand, a tower simulator, an Orbiter access arm random-motion simulator, and a tail service mast/ET hydrogen vent line random-motion and lift-off simulator. Tests are monitored in a control building on the west side of the LETF complex. The test equipment is not new. Previously located at the NASA Marshall Space Flight Center in Huntsville, Alabama, the equipment was used for similar purposes in the Apollo Program. It was transported to KSC by barge and the simulators were refurbished and modified for the Space Shuttle Program.

Space Shuttle payloads will be processed in a number of facilities at KSC and at the Cape Canaveral Air Force Station. These facilities have been used over the years for Apollo, Skylab, and a wide variety of unmanned spacecraft launched by expendable vehicles. Where necessary, they have been mocified for the Shuttle Program. The Space Shuttle is planned as an economical single-configuration reusable Space Transportation System, and payload processing has been streamlined, standardized, and systemized for maximum efficiency. To shorten *turnaround time,* it is imperative that the payloads be processed expeditiously and be ready for launch simultaneously with the Shuttle vehicle. In addition, funding limitations have demanded that expensive manpower be conserved by a simplified checkout system.

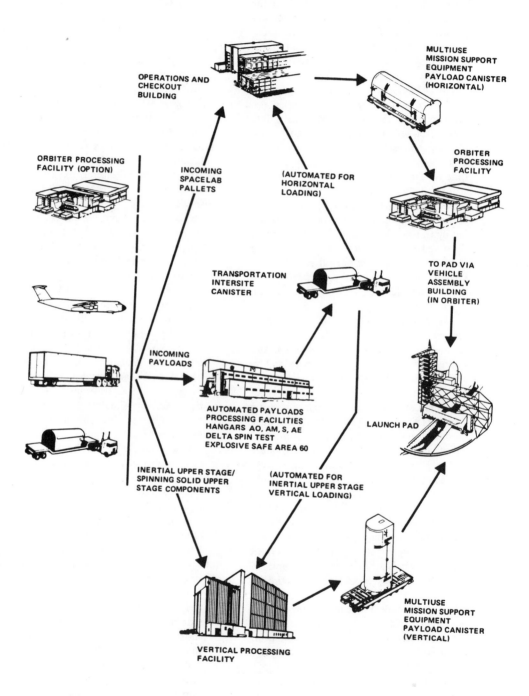

OPERATIONS AND
CHECKOUT
BUILDING

MULTIUSE
MISSION SUPPORT
EQUIPMENT
PAYLOAD CANISTER
(HORIZONTAL)

ORBITER PROCESSING
FACILITY (OPTION)

ORBITER
PROCESSING
FACILITY

INCOMING
SPACELAB
PALLETS

(AUTOMATED FOR
HORIZONTAL
LOADING)

TO PAD VIA
VEHICLE
ASSEMBLY
BUILDING
(IN ORBITER)

TRANSPORTATION
INTERSITE
CANISTER

INCOMING
PAYLOADS

AUTOMATED PAYLOADS
PROCESSING FACILITIES
HANGARS AO, AM, S, AE
DELTA SPIN TEST
EXPLOSIVE SAFE AREA 60

LAUNCH PAD

INERTIAL UPPER STAGE/
SPINNING SOLID UPPER
STAGE COMPONENTS

(AUTOMATED FOR
INERTIAL UPPER STAGE
VERTICAL LOADING)

MULTIUSE
MISSION SUPPORT
EQUIPMENT
PAYLOAD CANISTER
(VERTICAL)

VERTICAL PROCESSING
FACILITY

Fig. 12   STS payload operations at the Kennedy Space Center.   (Drawing courtesy of NASA.)

Payloads will be installed in the Orbiter either horizontally in the Orbiter Processing Facility or vertically at the pad [*See* Fig. 12]. To obtain the shortest possible Shuttle turnaround time, KSC will perform a simulated Orbiter-to-cargo-interface verification of the entire cargo before its installation in the Orbiter. Payloads to be installed horizontally in the Orbiter Processing Facility will be verified in the Operations and Checkout Building. Payloads to be installed vertically at the pad will be verified at the Vertical Processing Facility (VPF). Payloads installed vertically consist primarily of automated spacecraft involving *upper stages* and operations too hazardous to be performed in the Orbiter Processing Facility.

Essential to the processing of payloads at various KSC Shuttle facilities are the multiuse mission support equipment (MMSE): *the payload canister,* the payload canister transporter, and the payload strongback. The payload canister transporter is a 48-wheel self-propelled truck designed to operate between and within Shuttle payload processing facilities. It will be used to transport the *payload canister* and its associated hardware throughout KSC. The transporter is 19.8 meters (65 feet) long and 7 meters (23 feet) wide. Its elevating flatbed has a height of 1.8 meters (6 feet) but can be lowered to 1.6 meters (5 feet 3 inches) or raised to 2.1 meters (7 feet) ± 7.6 centimeters (3 inches). Its wheels are independently steerable and permit the transporter to move forward, backward, or sideways; to "crab" diagonally; or to turn on its own axis like a carousel. It has self-contained braking and stabilization jacking systems. The transporter is driven by a *hydraulic* system powered by a liquid-cooled diesel engine between facilities and by an electric motor using ground power inside the facilities.

The bare transporter weighs 63,500 kilograms (140,000 pounds). With a full load of diesel fuel and with the environmental control system, fluid and gas services, electrical power system, and instrumentation and communication system modules mounted on it, the transporter has a gross weight of 77,300 kilograms (170,500 pounds). The

transporter is steerable from diagonally opposed operator cabs on each end. Its top speed unloaded is 16 km/h (10 mph). The maximum speed of the fully loaded transporter is 8 km/h (5 mph). Because payload handling will require precise movements, the transporter has a "creep mode" that permits it to move as slowly as 0.64 cm/s (0.25 in/s) or 0.023 km/h (0.014 mph). Interfacility drive power is provided by a 298-kilowatt (400-horsepower) diesel engine. Intrafacility drive power comes from an 82-kilowatt (110-horsepower) electric motor. The transporter will carry the payload canister in either the horizontal or vertical position.

The payload strongback provides support for all payload sections during horizontal handling. A major use for the strongback is the handling of *Spacelab* but it will also be used to unload a wide variety of payloads during postflight handling in the Orbiter Processing Facility. Movable attachment points permit the strongback to be used for different payloads. The strongback consists of a rigid steel frame with adjustable beams, brackets, and clamps. It will not induce any bending or twisting loads on payload elements. The strongback is 18.3 meters (60 feet) long, 4.9 meters (16 feet) wide, and 2.7 meters (9 feet) high and weighs 18,150 kilograms (40,000 pounds).

The Operations and Checkout (O&C) Building is a five-story structure containing 55,740 square meters (600,000 square feet) of offices, laboratories, astronaut quarters, and cargo assembly areas. It is located in the industrial Area immediately east of the KSC Headquarters Building. [*See:* Kennedy Space Center.] Horizontally integrated payloads are received, assembled, and integrated into a cargo in the Operations and Checkout Building before being mated with the Orbiter at the Orbiter Processing Facility. The Spacelab and its payloads constitute the majority of horizontal payloads. [*See:* Spacelab.]

The Spacelab assembly and test area in the Operations and Checkout Building was used for the assembly and testing of Apollo spacecraft during the Apollo Program and has received the necessary modifications to adapt it to the Space Shuttle era. The Spacelab assembly and test area is 198 meters (650 feet) long and a uniform 26 meters (85 feet) wide. It is divided into a high-bay area 53 meters (175 feet) long and 32 meters (104 feet) high and a low-bay area 145 meters (475 feet) long and 21 meters (70 feet) high. The assembly and test area is environmentally controlled to 297 K (24° C or 75° F) ±2°. Relative humidity is maintained below 60 percent. The major Spacelab facilities within the O&C Building are two integrated assembly and checkout workstands, an engineering-model workstand, *pallet* staging workstands, a rack/floor workstand, a tunnel maintenance area, and two *end cone* stands. [*See:* Spacelab] The two integrated workstands will be controlled from two automatic test equipment (ATE) control rooms located on the third floor of the O&C Building. The capability exists to switch control of either workstand to either control room. Mechanical and electrical ground support equipment required to support Spacelab assembly and testing are located in and around the workstands. These facilities make it possible to support two independent Spacelab processing flows. To assist in Spacelab/Orbiter interface verification, as Orbiter interface adapter and two racks that simulate the Orbiter's *aft flight deck* are provided at the end of the workstand. Orbiter utility interfaces—electrical, gas, and fluid—will be provided by ground support equipment cables or lines.

The Spacelab ground-operations concept permits a *user* to design and develop an experiment that can be integrated with other individual experiments into a complete Spacelab payload. NASA will provide *racks* and *pallet segments* to *STS users* and *integration* and verification will be possible with minimum Spacelab-unique ground support equipment. The processing of Spacelab payloads will begin with the integration and checkout of experimental equipment with individual-experiment mounting elements—racks for the *Spacelab pressurized module* and pallet segments for the experiments to be exposed to the space environment. Experiments sponsored by the *European Space Agency* with undergo

preliminary integration in Europe before shipment to the United States.

Payload elements will be delivered to the launch site in as near flight-ready condition as is practical. After initial integration is complete, the STS user will be responsible for providing transportation of the payload to the launch site. After delivery of individual experiments and payloads to the Operations and Checkout Building, the *Spacelab "train" of pallets* and racks will be assembled using the pallet and/or rack stands. All work that can be accomplished before processing on the Spacelab integration workstand will be done during this period. Following mechanical buildup of the payload train, the Spacelab elements will be transferred to the Spacelab integration workstand for integration with the *Spacelab module/igloo.*

Spacelab operational hardware will have been undergoing refurbishment and buildup in parallel with the payload buildup. After buildup of the total Spacelab and payload configuration in the workstand, the module aft and cone will be installed, pallets will be positioned, and utilities between the pallets and module will be connected. Upon completion of checkout and test activities, the Spacelab will be hoisted by bridge cranes and the strongback, installed in the payload canister, moved to the Orbiter Processing Facility by the canister transporter. Environmental conditioning, through an air purge and system monitoring, are provided during transport to the OPF.

Payload removal from the canister and installation in the Orbiter will be accomplished in the Orbiter Processing Facility. The cargo to be installed will be hoisted in a horizontal *attitude* from its canister/transporter, positioned over the Orbiter, lowered into the payload bay, and secured in place. The strongback and facility crane will support this operation. After payload installation, the Shuttle payload interfaces will be connected and verified. An integrated test will be conducted to complete the verification of interfaces between the payloads and the Orbiter. Upon completion of testing, the payload bay doors will be closed and latched. The payload bay environment with the doors closed will be maintained by providing a purge of clean air at 294 K (21° C or 70° F) and a maximum relative humidity of 50 percent. At this point, the Orbiter will be powered down for movement to the Vehicle Assembly Building. No power or purge is provided from this point until completion of the Shuttle vehicle assembly in the VAB.

Upon completion of operations in the Orbiter Processing Facility, the Orbiter will be towed to the Vehicle Assembly Building for transition from the horizontal to the vertical position. The Orbiter's cargo is quiescent during VAB operations with no routine access planned. Shuttle power will be available after the MLP and External Tank interfaces have been connected and verified. Following the completion of mating and Shuttle system interface verification checks with the Mobile Launcher Platform, the Space Shuttle vehicle will be rolled out on the Crawler-Transporter to the pad. The Orbiter will be powered down during the movement to the pad and power will not be available to a payload during this time unless it is provided by a self-contained source.

After the Mobile Launcher Platform has been mated hard down on its mounts at the pad and the umbilicals are connected, an interface verification test will be conducted to verify the integrity and serviceability of the pad/Shuttle system interfaces. Access to payloads at the pad is not planned, although the capability exists from the rotating service structure to open the payload bay doors and reach the payload from the extendable platforms of the payload ground handling mechanism.

Vertically integrated payloads are normally received in payload processing facilities in Buildings AE, AO, AM and Hangar S at the Cape Canaveral Air Force Station or in Spacecraft Assembly and Encapsulation Facility-2 (SAEF-2) in the KSC Industrial Area. These payloads consist primarily of automated free-flying satellites or spacecraft using upper stages and their processing normally involves some hazardous operations, which are conducted in explosive safe areas located at the Cape Canaveral Air Force Station. Vertical integration into a complete Shuttle cargo is done in the Vertical Processing Facility (VPF). The VPF highbay area has a ceiling height of 32 meters (105 feet) and a usable floor area of 943 square meters (10,153 square feet).

Spacecraft undergo buildup in the *clean room* environments of the payload processing facility to which they are assigned and where they receive preintegration testing. After functional tests are completed, the spacecraft are moved to an explosive safe area where hazardous operations will be conducted if required. These areas are the Delta Spin Test Facility and Explosive Safe Area 60, both located at the Cape Canaveral Air Force Station. Activities performed in these areas include installation of *solid propellant apogee motors* or *ordnance* separation devices, hydrazine loading, or any other work involving items that are potentially explosive or hazardous. Upon completion of all hazardous operations, the spacecraft is ready for movement to the Vertical Processing Facility.

Shuttle upper stages include the spinning solid upper stage A (SSUS-A), the spinning solid upper stage D (SSUS-D), and the inertial upper stage (IUS). The SSUS-A and SSUS-D are received at the Delta Spin Test Facility where they undergo inspection, assembly, and test. Spacecraft that are scheduled to fly on an SSUS-D are mated with the SSUS-D in the Delta Spin Test Facility and then moved into the Vertical Processing Facility. Spacecraft scheduled to fly on an SSUS-A are mated with the SSUS-A in the Vertical Processing Facility.

The IUS is received at the Solid Motor Assembly Building in the Titan III Complex at the Cape Canaveral Air Force Station. Operations in the Solid Motor Assembly Building are under the management of the U.S. Air Force. When these operations are complete, the IUS is moved to the Vertical Processing Facility of KSC for mating with its spacecraft. Installation in the Vertical Processing Facility and all subsequent operations are conducted under KSC management.

The final integration and testing of vertically integrated payloads is accomplished in the Vertical Processing Facility. Operations in the VPF are conducted under environmentally controlled conditions. The entire VPF environment is temperature controlled

297 K (24°C or 75°F) ± 3°; relative humidity is controlled at 45 ± 5 percent. Spacecraft processing within the VPF will vary depending on the type of upper stage involved. A spacecraft already mated with an SSUS-D is installed directly into one of two workstands after removal from the transporter/container. Spacecraft using an SSUS-A or an IUS upper stage will be mated with the upper stage previously positioned in a workstand in the VPF high bay. Regardless of where the upper stages are mated to their spacecraft, the entire Shuttle cargo will eventually be assembled in a single VPF workstand. Testing of individual payloads (spacecraft mated with an upper stage) will be accomplished before any combined cargo testing or simulated Orbiter to cargo testing.

A major test conducted in the Vertical Processing Facility is the cargo integration test. Cargo integration test equipment (CITE) in the VPF permits—by simulation—the verification of payload/cargo mechanical and functional connections with the Orbiter before they are shipped from the VPF to the launch pad. Similar equipment in the Operations and Checkout Building performs the same functions for payloads that are integrated into the Orbiter horizontally. After completion of testing, the entire Shuttle cargo is inserted into the payload canister by the vertical payload handling device. Environmental conditioning and system monitoring are provided during transport to the pad for insertion in the rotating service structure. Upon arrival at the rotating service structure, the payload is installed as described previously. Installation of the payload into the rotating service structure occurs before transfer of the Space Shuttle vehicle to the launch pad.

There are many variations from the standard flow of horizontally and vertically integrated payloads at KSC. Among the specific flow variations are *small self-contained payloads,* life-sciences payloads, and Department of Defense payloads. Small self-contained or *"get away special"* payloads will be received and built up in the Operations and Checkout Building and installed in the Orbiter in the Orbiter Processing Facility. Because they have limited

Orbiter interfaces, most small self-contained payloads will not require interface verification in the cargo test integration facility and will be built up, attached to a special bridge beam, and installed in the Orbiter much like any other bridge beam.

The flight hardware associated with life-sciences payloads will normally follow the flow for horizontally integrated payloads. (*See:* Spacelab) However, live specimens for these payloads will be received at Hangar L at the Cape Canaveral Air Force Station. Technical activities in Hangar L are managed by *NASA's Ames Research Center;* The Kennedy Space Center will have operations and *maintenance* responsibility. Life-sciences specimens, or live specimens already in their flight containers, are installed at the launch pad by opening the payload bay doors and installing the specimens from a special access platform mounted on the payload ground handling mechanism or through the crew entry hatch with the live specimens in their containers mounted in the Orbiter mid-deck area of the crew cabin.

Department of Defense (DOD) payloads or cargo elements will be shipped from their factory locations by air to be Cape Canaveral Air Force Station. (*See:* U.S. Air Force role in space) Upon arrival, they will be moved to the Solid Motor Assembly Building for assembly and prelaunch testing. The DOD cargo will then be transported in the payload canister to the launch pad, where it will be installed in the rotating service structure for insertion into the Orbiter's payload bay.

The mate/demate stiffleg derrick (MDSD) was developed to meet the single mate/demate cycle of Orbiter 101 "Enterprise" and its 747 carrier aircraft required for Space Shuttle vibration tests at the Marshall Space Flight Center (MSFC). The derrick is portable and also provides the means to recover a Space Shuttle Orbiter compelled to land at any one of a number of contingency landing sites around the world. The derrick may be used to hoist, position, secure, and lower the Orbiter for all Orbiter/747 mate or demate operations. The system consists of a stiffleg derrick with a 91-metric-ton (100-ton) capacity, a 91-

metric-ton (100-ton) capacity mobile crane, a wind restraint system, a communications system, and support accessories.

The mate/demate stiffleg derrick offered the least expensive way of performing the one-time mate/demate cycle at the Marshall Space Flight Center. The stiffleg derrick was removed from a Marshall test stand, refurbished, and modified. It is mounted on a 7.6-meter (25-foot) high triangular steel tower to clear the 747's port wing and permit the boom to reach the Orbiter's aft hoist points. The derrick includes three major subsystems: hoisting, wind restraint, and communications: (1) Hoisting—The stiffleg derrick will hoist the aft end of the Orbiter and the mobile crane will hoist the forward end. (2) Wind restraint—Six plastic-coated steel tagline masts, four nylon taglines, and two trucks provide wind restraint of the Orbiter while it is suspended awaiting further hoisting/lowering operations. (3) Communications—Radio communications, in the form of Handie-Talkies with headsets, are used to coordinate the activities of the mate/demate team. The derrick will be stored at KSC should it be needed at some future date to recover an Orbiter from a contingency landing site. The mobile crane portion of the system would be leased if or when needed.

The towing tractors at the Kennedy Space Center are similar to those that tow aircraft at commercial airports. At KSC, the tractor is used to tow Orbiters, 747 Shuttle carrier aircraft, and other *aircraft* weighing 372,000 kilograms (820,000 pounds) or less between the Shuttle Landing Facility, the Orbital Processing Facility, and the Vehicle Assembly Building. It also tows the External Tank transporter from the VAB dock area to the ET checkout facility in High Bay 4 of the VAB. The tractor is 4.9 meters (16 feet) long, 2.4 meters (8 feet) wide, and 2.2 meters (7.3 feet) high. The basic weight is 29,500 kilograms (65,000 pounds); ballasted weight is 49,900 kilograms (110,000 pounds). Its maximum speed is 32 km/h (20 mph) and its towing speed is 8 km/h (5 mph). The towing tractor is powered by a 191-kilowatt (256-horsepower) diesel engine and its transmission is fully automatic with six forward

speeds and one reverse speed. A diesel-driven ground-power generator provides the Orbiter (or other aircraft being towed) with 115-volt, 400-hertz, three-phase alternating-current electrical power.

STS launch operations (abbr. LO) Operations at the *launch pad* after arrival of the *Space Shuttle vehicle* are controlled from the Launch Control Center. Once in place on the pad support columns, the rotating service structure is extended, the *Mobile Launcher Platform* and the Space Shuttle vehicle are electrically and mechanically mated with the supporting pad facilities and *ground support equipment,* and all *interfaces* are verified. In parallel with these operations, power is applied to the Space Shuttle vehicle and supporting ground support equipment, the launch-readiness test is performed, and *hypergolic* and helium load preparations are completed. Preliminary *Orbiter cabin* closeout will be accomplished as a part of the nonhazardous *countdown* preparations, followed by pad clearance and such hazardous operations as the servicing of hypergolic, high-pressure-gas, *fuel-cell cryogenic* systems, and *payload fluid* servicing, if required. The Space Shuttle is now ready for the loading of *cryogenic propellants* and the boarding of the *flightcrew* for launch.

For operational Space Shuttle flights completion of the hazardous servicing and the rotating service structure retraction will place the vehicle in standby status at approximately T—2 hours in the countdown. The Space Shuttle vehicle will have the capability to be held at this point for up to 24 hours or to proceed with the terminal countdown following a clearance to launch. The final 2 hours of the countdown include final mission *software* update, completion of propellant system purges, propellant line chill, loading of *liquid hydrogen* and *liquid oxygen* aboard the *External Tank,* crew entry, terminal sequence, and *lift-off.*

The launch sequence and ascent profile given in Table 1 is typical for a *Space Shuttle mission* being launched on an *azimuth* of 81.36° (measured clockwise from north) to place the *Orbiter* in a 277-kilometer (172-mile) high circular *orbit* with a 38° inclination. This orbit will be achieved by two *OMS* Orbital Maneuvering System maneuvers following completion of the *SRB* Solid Rocket Boosters and *SSME* Space Shuttle Main Engines burns.

Lift-off will occur 0.3 second after ignition of the Solid Rocket Boosters and release of the Space Shuttle vehicle from the eight holddown posts supporting it on the Mobile Launcher Platform. Tower clearance will occur at T + 6.5 seconds. The Solid Rocket Boosters will be *jettisoned* after 2 minutes of flight and will splash down in the Atlantic Ocean approximately 272 kilometers (169 miles) downrange from *Launch Complex 39*. The External Tank will be jettisoned 8 minutes 50 seconds into the flight on a suborbital *trajectory* that will bring it to impact in the Indian Ocean approximately 58 minutes after lift-off. The first *OMS* burn will place the Orbiter in an orbit with an *apogee* of 277 kilometers (172 miles) and a *perigee* of 96.2 kilometers (59.8 miles). The second OMS burn occurs near orbital apogee and will place the Orbiter into a near-circular orbit of approximately 277 kilometers (172 miles) above the Earth.

STS payload processing facilities
See: STS launch facilities

STS payload support equipment
See: STS launch facilities

STS tracking and communications network *Space Shuttle* missions will be linked to Earth by the "Space Tracking and Data Network" (STDN) operated by the NASA Goddard Space Flight Center (GSFC). Seven new systems have been added to the tracking and communications network to handle the data from Shuttle and Shuttle *payloads*. The major additions to the tracking and communications network for Shuttle missions are as follows:

1. First digital voice system—Voice links from the Shuttle to ground and from the ground to Shuttle are digital. The conversion from *analog* to digital will be made in the tracking stations. After conversion, the voice will be handled the same as for earlier *manned missions*. This new digital system is less immune to "noise" and facilitates voice transmission.

2. "Bent pipe" data flow—All *real-time* data from the Space Shuttle will go directly from the *Orbiter* to the *Mission Control Center* at the NASA Lyndon B. Johnson Space Center (JSC) in Houston. In the bent-pipe mode, all real-time data flows through an electronic "pipeline" directly into Houston without any delay at a tracking site.

3. The *Tracking and Data Relay Satellite System* [*See also:* separate entry] will not be ready for the early Shuttle flights. The work that will ultimately be handled by these satellites will be performed for early Shuttle flights by ground stations. When the system becomes operational in the 1980's, there will be two satellites—or "tracking stations"—in orbit: one over the Atlantic Ocean and one over the Pacific Ocean.

4. *Radar* tracking for reentry flights to Dryden Flight Research Center, California—Five radars provided by *NASA* and the *U.S. Air Force* [*See:* U.S. Air Force role in space] at Edwards and Vandenberg Air Force Bases, and at several other west coast sites, will track the Orbiter during its reentry *flightpath* for landing on the desert. When the first radar "locks" onto Shuttle, the other four "slave" to it and all five perform as one system.

5. Transportable tracking stations—Three special-purpose transportable tracking stations have been added to the basic network for the Space Shuttle. The stations are Ponce de Leon Inlet, Florida (near New Smyrna Beach); Bangor, Maine; and Buckhorn Lake Station, Edwards Air Force Base, California. The station at Ponce de Leon Inlet will back up the Merritt Island station in Florida during powered flight in the event the latter station might experience radio attenuation problems from the solid rocket motor plume. The Bangor site will be used to support high-inclination launchings. The Buckhorn station wil support landings at the Dryden Flight Research Center in California.

6. Domestic communications satellites— For the first time in the history of manned space flight, domestic communications satellites (five in all) will be used to electronically tie Earth stations together. The use of domestic communications satellites will also permit 10 to 20 times more data to

be transmitted. To support the Space Shuttle, ground terminals for domestic communications satellites are situated at Johnson Space Center, Houston, Texas; Kauai, Hawaii; Goldstone, California; Kennedy Space Center, Florida; Dryden Flight Research Center, California; and Goddard Space Flight Center, Greenbelt, Maryland.

7. Dual voice channels—The STS voice communications system has been configured for support of two separate voice conversations upward and downward simultaneously.

The Space Tracking and Data Network consists of 13 tracking stations situated on four continents (North and South America, Australia, and Europe). The network is operated by the Goddard Space Flight Center in Greenbelt, Maryland (a suburb of Washington, D.C.) for the Office of Tracking and Data Systems at NASA Headquarters. The STDN tracking stations are located as follows:

  Ascension Island
  Santiago, Chile
  Bermuda
  Goldstone, California
  Guam
  Kauai, Hawaii
  Madrid, Spain
  Merritt Island, Florida
  Orroral, Australia (near Canberra)
  Quito, Ecuador
  Fairbanks, Alaska
  Rosman, North Carolina
  Winkfield, England

Special-purpose transportable stations are located at Ponce de Leon Inlet, Florida; Bangor, Maine; and Buckhorn Lake Station, Edwards Air Force Base, California. Also supporting the STDN network are several instrumented aircraft. These Air Force aircraft, named *Advanced Range Instrumentation Aircraft* (ARIA), are situated at various spots around the world where ground stations are unable to support missions.

The Space Tracking and Data Network stations are equipped with a variety of *antennas* (9- to 26-meter (30- to 85-feet) diameter), each designed to accomplish a specific task usually in a specific *frequency*

band. Functioning like giant electronic magnifying glasses, the larger antennas absorb radiated electronic signals transmitted by *spacecraft* in a *radio* form called *telemetry*. The center of all STDN activity is the GSFC Communications Center. More than 3 million kilometers (2 million miles) of circuitry connect the 13 remote sites of the NASA Communications Network (NASCOM) and link the Shuttle to the JSC Mission Control Center in Houston. The major switching centers in NASCOM are located at the Goddard Space Flight Center, Greenbelt, Maryland; the Jet Propulsion Laboratory, Pasadena, California; Cape Canaveral, Florida; Canberra, Australia; Madrid, Spain; and London, England. This communications network is composed of telephone, *microwave, radio,* submarine cables, and *communications satellites*. These various systems link data flow through 11 countries of the free world with 15 foreign and domestic carriers and provide the required information between tracking sites and the JSC and GSFC control centers. Special *wide-band* and video circuitry is also utilized as needed. The Goddard Space Flight Center has the largest wide-band system in existence.

The Station Conferencing and Monitoring Arrangement (SCAMA) allows voice traffic managers to "conference" as many as 220 different voice terminals thoughout the United States and abroad with talk/listen capability at the touch of a few buttons. The system is redundant, which accounts for its mission support reliability record of 99.6 percent. All Space Shuttle voice traffic will be routed through SCAMA at the Goddard Space Flight Center. Included in the equipment of the worldwide STDN system are 126 digital computers located at the different stations. The computers at these remote sites control tracking antennas, handle commands, and process data for transmission to the JSC and GSFC control centers. Data from the Shuttle from all the tracking stations around the world are funneled into the main switching computers (Univac 494's) at GSFC where the data are reformatted and transmitted to the Johnson Space Center without delay on special very wide band (1.5 million bits per second) circuits via domestic communications satellites. Commands generated at the Johnson Space Center are transmitted to the main switching computers at GSFC and switched to the proper tracking station for transmission to the Shuttle. The main switching computers at GSFC have been "rebuilt" to increase their switching or throughput capacity to 10 times greater than that of Apollo.

A revolutionary way of tracking will begin in the early 1980's. The new tracking system is called the Tracking and Data Relay Satellite System (TDRSS). The TDRSS will make its maiden flight aboard a Shuttle in the early 1980's. TDRSS is owned by Western Union Telegraph Company with NASA leasing services for a 10-year period. When the TDRSS is deployed by the Shuttle, its assignment will be to track Shuttle flights and the various scientific and applications satellites that will ride into orbit aboard a Shuttle. The new breed of "space trackers" will actually be "tracking stations in the sky." They will ultimately replace more than half the existing ground tracking stations.

The TDRSS will consist of two operational satellites in synchronous or stationary orbit above the *Equator.* The satellites will be spaced approximately 130° apart at longitude 171° W (southwest of Hawaii) and longitude 41° W (northeast corner of Brazil). A TDRSS ground terminal is located at White Sands, New Mexico, for transmitting all tracking and voice data "live" to the control centers at JSC and GSFC. The White Sands Earth Station, managed and owned by Western Union, will include complex electronic equipment, three 18-meter (60-foot) dish antennas, a number of small antennas, and a dual-processor computer system.The TDRSS will provide NASA with greatly improved communications to and from Shuttle and Shuttle-launched payloads at greatly reduced cost. An additional advantage is that, while the worldwide network used for the *Mercury* and *Gemini* Earth-orbital missions could track only about 15 percent of the time, the TDRSS will provide coverage in the 85- to 100-percent range.

To ensure operational reliability, a spare TDRSS will be launched into orbit as a backup and positioned at a synchronous altitude midway between the two operational satellites. Still another TDRSS on Earth will be flight qualified, stored, and ready for launching on short notice. Weighing nearly 2268 kilograms (5000 pounds) in orbit and measuring 17 meters (56 feet) from tip to tip of the solar panels, the TDRSS will be the largest *telecommunications* satellite ever launched. Each satellite will have two spacedeployable antennas 5 meters (16.5 feet) in diameter that unfurl like a giant umbrella. Each antenna weighs approximately 23 kilograms (50 pounds). A ground version of similar size and capability would weigh 2268 kilograms (5000 pounds). To launch the TDRSS, the Space Shuttle Orbiter will climb to an altitude of 200 kilometers (125 miles), then open its payload bay for the subsequent launch of the TDRSS by an inertial upper stage (developed by the U.S. Air Force) into its orbit 35,887 kilometers (22,300 miles) above the Equator.

When the TDRSS becomes operational, it will be able to track and communicate simultaneously with 24 low-orbital *spacecraft,* inlcuding Shuttle. The system will be able to handle up to 300 million *bits* of information per second. Since it takes about 8 bits of information to make one word, this is equivalent to processing 300 14-volume sets of encyclopedias every second. The TDRSS is being developed for shared use by NASA and Western Union. Major contractors to Western Union are TRW for the satellite and Harris Electronic Systems Divison for the satellite-deployable antennas and the ground station in New Mexico. The satellite is designed for a life of 10 years.

stubout   Tubular nipple protruding from a component to which the connecting line is welded or brazed.

stuffing box   Cavity in the housing of a *pump* shaft designed to accept a packing for the purpose of preventing leakage along the shaft.

subassembly   Two or more parts that together form a portion of an assembly or a component replaceable as a whole, but that have a part or parts that are individually replaceable.

subatomic particle Any of the constituent particles of an atom: an *electon, neutron, proton,* etc. Subatomic particles are classified by relative mass into four groups: *leptons, mesons, nucleons* and *hyperons,* from lowest to highest masses, respectively.

subcritical Coined word denoting (1) operation of rotating machinery below a critical speed or (2) *fluid* maintained at *pressures* or *temperatures* below its *critical point.*

subcritical assembly A *nuclear reactor* consisting of a mass of *fissionable* material and *moderator* whose *effective multiplication factor* is less than one and that hence cannot sustain a *chain reaction.* Used primarily for educational purposes.
See: criticality

subcritical mass An amount of fissionable material insufficient in quantity or of improper geometry to sustain a *fission chain reaction.*
See: critical mass, criticality

subcritical reactor A *subcritical assembly.*

sublimation In thermodynamics the direct transition of a material from the solid *phase* to the vapor phase, and vice versa, without passing through the liquid phase.

sublimator
See: extravehicular mobility unit

subliming ablator An *ablation* material characterized by *sublimation* of the material at the heated surface.

submerged nozzle *Nozzle* configuration in which the nozzle entry, throat, and part of all of the nozzle exit cone are cantilevered into the *combustion chamber.*

subsatellite point Intersection of the *local vertical* passing through a *satellite* in *orbit* with the *Earth*'s surface.

subsonic diffuser
See: diffuser

subsonic flow Flow of a *fluid,* as air over an *airfoil,* at speeds less than *acoustic velocity* (speed of sound).

subsystem tank Container of pressurized *fluid* or gas that is mounted internally in a vehicle, is essentially isolated from adverse vehicle loads, and is of *monocoque* design.

subtend To be opposite, as an arc of a circle subtends an angle at the center of the circle, the angle being formed by the radii joining the ends of the arc with the center.

suction specific speed Index to *pump* suction performance relating rotational speed and flowrate to the minimum *net positive suction head* (NPSH) at which the pump will deliver specified performance.

suction surface Convex surface of a pump or turbine blade; along this surface, the *fluid* pressures are lowest.

Sun An incandescent body of gas, by far the largest and most massive object in the Solar System. The sun is a G class star with a mean radius of about 695,000 km. The surface temperature is approximately 5900°K. It is largely a heated plasma which should radiate very nearly as a blackbody. However, the cooler atomic gases of the solar atmosphere and the large temperature gradiant below the surface, coupled with the nonisothermal character of transient surface features, lead to deviations. Scientists agree that nuclear fusion is the source of the Sun's energy. Deep within the interior temperatures are approx. 20 million degrees C and pressures are 70 trillion grams per square centimeter. All of Earth's energy comes from the Sun.

supercritical Coined word denoting (1) operation of rotating machinery above a critical speed or (2) *fluid* maintained at *pressures* or temperatures above its corresponding *critical point.*

supercritical mass A mass of *fuel* whose *effective multiplication factor* is greater than one.
See: critical mass

supercritical reactor A *nuclear reactor* in which the *effective multiplication factor* is greater than one; consequently a reactor that is increasing its power level. If uncontrolled, a supercritical reactor would undergo an *excursion.*
See: criticality

superheating The heating of a vapor, particularly saturated (wet) steam, to a *temperature* much higher than the boiling point at the *exising* pressure. This is done in power plants to improve efficiency and to reduce condensation in the *turbines.*

super-insulation High-efficiency laminated-foil insulator used in low temperature applications; thermal conductivity is 1/10 to 1/150 that of common insulating materials.

superior conjunction   The *conjunction* of a *planet* and the *Sun* when the sun is between the *Earth* and the other planet.

superior planets   The *planets* with *orbits* larger than that of the *Earth: Mars, Jupiter, Saturn, Uranus, Neptune,* and *Pluto.*

supersonic   Of or pertaining to, or dealing with, speeds greater than the *acoustic velocity* (speed of sound).

supersonic compressor   A *compressor* in which a *supersonic velocity* is imparted to the fluid relative to the *rotor* blades, the *stator* blades or to both the rotor and stator blades, producing oblique *shock waves* over the blades to obtain a high pressure rise.

supersonic diffuser   A *diffuser* designed to reduce the *velocity* and increase the *pressure* of *fluid* moving at *supersonic* velocities.

supersonic flow   In aerodynamics, flow of a *fluid* over a body at speeds greater than the *acoustic velocity* (speed of sound) and in which the *shock waves* start at the surface of the body. Compare this term with *hypersonic flow.*

supersonic nozzle   A converging-diverging *nozzle* designed to accelerate a *fluid* to *supersonic* speed.

supersonics   Specifically, the study of *aerodynamics* of *supersonic* speeds.

surface boundary layer   That thin layer of air adjacent to the *Earth*'s surface. Within this layer the wind distribution is determined largely by the vertical temperature gradient and the nature and contours of the underlying surface; shearing stresses are approximately constant.

surface contamination   The deposition and attachment of *radioactive* materials or toxic materials to a surface.

surface tension   The tendency of a liquid that has a large cohesive force to keep its surface as small as possible, forming spherical drops.

surge   A transient rise in power, *pressure,* etc., such as a brief rise in the discharge pressure of a rotary *compressor.*

survey meter   Any portable *radiation detection* instrument especially adapted for surveying or inspecting an area to establish the existence and amount of *radioactive* material present.

Surveyor   The Surveyor Project, begun in 1960, consisted of seven unmanned spacecraft which were launched between May 1966 and January 1968. The craft were used to develop lunar softlanding techniques, to survey potential Apollo landing sites and to improve scientific understanding of the Moon.

suspension   In physical chemistry, a system composed of one substance (suspended phase, suspensoid) dispersed throughout another substance (suspending phase) in a moderately finely divided state, but not so finely divided as to acquire the stability of a *colloidal system.*

Given sufficient time, a suspension will, by definition, separate itself by gravitational action into two visibly distinct portions, whereas a colloidal system, by definition, is stable. Dust in the atmosphere is an example of a suspension of a solid in a gas.

sustainer engine   Auxiliary *booster engine* in a *propulsion system* that provides *thrust* after the main booster engines cease firing.

swaging   Process of tapering or reducing the diameter of a rod or tube by hammering or squeezing.

sweat cooling
See: transpiration cooling

synchrocyclotron   A cyclotron in which the frequency of the accelerating *voltage* is decreased with time so as to match exactly the slowing revolutions of the accelerated *particles.* The decrease in rate of acceleration of the particles results from the increase of mass with energy as predicted by the *Special Theory of Relativity.* Compare this term with *synchrotron.*
See: cyclotron; accelerator

synchronism   The relationship between two or more *periodic* quantities of the same *frequency* when the *phase* difference between them is zero or constant at a predetermined value.

synchronous satellite   An equatorial west-to-east satellite orbiting the *Earth* at an altitude of approximately 35,900 kilometers (22,300 mi.) at which *altitude* it makes one *revolution* in 24 hours, synchronous with the earth's rotation.
See: geosynchronous orbit

synchrotron   An *accelerator* in which *particles* are accelerated around a circular path

Surveyor landing sites on Moon. Surveyors I, III, V and VI all Landed in mare areas ("Apollo zone"). Surveyor VII touched down in the rugged highland region near the Crater Tycho.   (Photograph courtesy of NASA.)

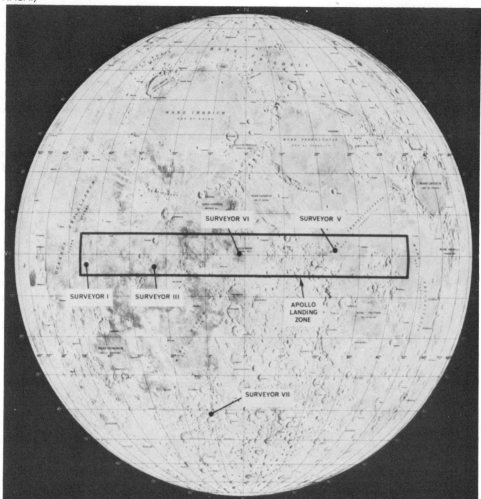

SURVEYOR VI   SURVEYOR V

SURVEYOR I   SURVEYOR III

APOLLO LANDING ZONE

SURVEYOR VII

by *radio-frequency* electric fields. The magnetic guiding and focusing fields are increased synchronously to match the energy gained by the particles so that the orbit radius remains constant. Compare this term with *cyclotron* and *synchrocyclotron*.

synchroton radiation   Radiation produced by high-speed *electrons* or *ions* spiraling around lines of force in a *magnetic field*. The light and x-rays emitted are polarized.

synodic period
    See: orbit

synoptic   Pertaining to or affording an overall view.

system   One of the principal functioning entities comprising the project *hardware* and related operational services within a *project* or flight mission. Ordinarily, a system is the first major subdivision of a project work. Similarly, a subsystem is a major functioning entity within a system. (A system may also be an organized and disciplined approach to accomplish a task; e.g. a failure-reporting system.)

1. Any organized arrangement in which each component part acts, reacts or interacts in accordance with an overall design inherent in the arrangement.

2. Specifically, a major component of a given vehicle such as a propulsion system or a guidance system. Usually called a major system to distinguish it from the systems

subordinate or auxiliary to it.

systematic error   An *error* that is always a function of the magnitude of the quantity observed. When the error is constant it is called a bias error. Systematic errors are often caused by false elements in an instrument.

systems tunnel   A tunnel within each *Space Shuttle Solid Rocket Booster* provides protection and mechanical support for the cables associated with the electrical and instrumentation subsystem and the linear-shaped explosive charge of the *range safety system.*
See: Solid Rocket Booster

szygy   A point of the *orbit* of a *planet* or *satellite* at which it is in *conjunction* or *opposition.*

The term is used chiefly in connection with the *Moon,* when it refers to the points occupied by the Moon at new and full phase.

T

---

tachyon   A hypothetical "faster-than-light" particle. If it exists it might be detected through the emission of *Cerenkov radiation.*

tail   1. The tail surfaces of an aircraft, *aerospace vehicle, missile* or *rocket.*
See: Orbiter
2. The rear portion of a body, as of an aerospace vehicle or rocket.
3. Short for *comet* tail.
4. In nuclear technology *tails.*
See: depleted uranium
2. A container incorporated into the structure of a *nuclear rocket* from which a monopropellant, such as liquid hydrogen, is fed into the nuclear reactor.
3. A ground-based or space-based container for the storage of liquid hydrogen, liquid

oxygen or other liquid propellants until they are transferred to a rocket's tanks or some other receptacle.

tail cone
See: Approach and Landing Test(s)

tailoff   Period of decay in *rocket motor thrust* after the end of effective propellant burning time.

tailoring   Modification of a basic *solid propellant* by adjustment of the propellant properties to meet the requirements of a specific rocket motor.

tail service mast [abbr. TSM]
See: STS launch facilities

takeoff   1. The ascent of a *rocket vehicle* as it departs from the *launch pad* at any angle.
2. The action of an aircraft as it becomes airborne.

Compare this term to *lift-off,* which refers only to the vertical ascent of a rocket or missile from its launch pad.

tandem launch   The launching of two or more *spacecraft* or *satellites* using a single *launch vehicle.*

tank   1. A container incorporated into the structure of a *liquid-propellant rocket* from which a liquid propellant or propellants are fed into the *combustion chamber(s).*
2. A container incorporated into the structure of a *nuclear rocket* from which a *monopropellant,* such as *liquid hydrogen,* is fed into the *nuclear reactor.*
3. A ground-based or space-based container for the storage of liquid hydrogen, *liquid oxygen* or other liquid propellants until they are transferred to a rocket's tanks or some other receptacle.

tankage   The aggregate of the *tanks* carried by a *liquid propellant rocket* or a *nuclear rocket.*
See: external tank

tank components   1. Devices for controlling the behavior of *propellants* (positioning devices, slosh and vortex suspension devices, baffles, standpipes, expulsion devices).

2. Tank insulation.

See: External Tank.

tape mass memories

See: Orbiter avionics system

taper 1. Gradual reduction in or enlargement of coolant-tube diameter.

2. Gradual axial increase in grain perforation area along a section of fixed shape.

taper ratio The ratio of maximum coolant-tube diameter to minimum tube diameter; usually kept below 4 in modern *rocket engine* design.

target 1. Any object, destination, point, etc., toward which something is directed.

2. An object that reflects a sufficient amount of a radiated signal to produce a return, or echo, signal on detection equipment.

3. The *cooperative* (usually passive) or non-cooperative partner in a space *rendezvous* operation.

4. In nuclear technology a material subjected to particle bombardment (as in an *accelerator*) or *neutron* irradiation (as in a *nuclear reactor*) in order to induce a *nuclear reaction.*

Taylor theorem If all the derivatives of a function *f(x)* are continuous in the vicinity of $x = $ a, then *f(x)* can be expressed in an infinite series (the Taylor series):

$$f(x) = f(a) + f'(a)(x - a) + \frac{1}{2!} f''(a)(x - a)^2$$
$$+ \ldots + \frac{1}{n!} f^{(n)}(a)(x - a)^n$$

The case a $=$ 0 is called a Maclaurin series.

Teflon Trade name for synthetic fluorine containing resins used especially for molding articles and for coatings to prevent sticking.

tektite A small glassy body containing no crystals with a high silicon dioxide content (greater than 65%). These objects are found only in certain parts of the world in large areas called "strewn fields" and bear no relationship to the geologic formations in which they occur. Tektites are named as minerals with the suffix "ite," as for example, "australite" (found in Australia), "moldavite" (found in Bohemia and Moravia, Czechoslovakia) and "indochinite" (found in Southeast Asia). Their varied shapes (spheroid to lens-, disc- and pear-shaped) indicate that they may have been

formed from molten material that cooled rapidly. Contemporary theories concerning the origin of tektites include formation by the fusion of materials during the impact of very large *meteorites* and formation via *terrestrial* or *lunar* volcanic activities. The lunar *regolith* has yielded small tektite-like objects.

telecommunications Information transfer by electromagnetic means. Telecommunication systems include radio and television.

telemetry The technique of measuring a quantity or quantities, transmitting the measured value(s) to a distant station, and there interpreting, indicating or recording the quantities measured. The data are transmitted via a *telecommunications* system from the location at which the measurements are made to the location at which they are recorded and/or interpreted.

See: Orbiter communications and data systems

temperature A property that determines the direction of heat (thermal energy) flow.

temperature coefficient of reactivity The change in *nuclear reactor reactivity* (per degree of *temperature*) that occurs when the operating temperature of the reactor changes. This coefficient is said to be positive when an increase in temperature increases reactivity and negative when an increase in temperature decreases reactivity. In reactor design negative temperature coefficients of reactivity are desirable, because they help to prevent power *excursions.*

temperature coefficient of resistance The (usually small) change in the *resistance* of a material with changes in material *temperature.* In general, *semiconductor* and electric insulator materials have a negative coefficient of resistance (i.e., as temperature increases, resistance decreases), while electric conductors (i.e., metals) have a positive coefficient of resistance.

temperature jump Difference in temperature between the *nozzle* wall and the layer of gas *molecules* next to the wall, a result of *rarefaction* of the exhaust gas.

tensor An *array* of functions that obeys certain laws of transformation. A one-row or one-column tensor array is a *vector.*

The motivation for the use of tensors in some branches of physics is that they are invariants not depending on the particular coordinate system employed.

terminal   1. A point at which any *element* in a *circuit* may be directly connected to one or more other elements.
2. Pertaining to a final condition or the last division of something, as *terminal ballistics.*

terminal ballistics   That branch of ballistics dealing with the motion and behavior of *projectiles* at the termination of their flight, or in striking and penetrating a target.

terminator   The boundary line separating the illuminated and dark portions of a non-luminous *celestial* body, such as the *Moon.* Figure 1 provides a dramatic view of *Jupiter's* satellite *Io,* taken by the *"Voyager I"* *spacecraft* on 8 March 1979. A volcanic eruption can be seen on the terminator, where the volcanic cloud is catching the rays of the rising *Sun.*

Terra   The Earth.

terrestrial   Of or pertaining to the Earth. (Compare this term with *extraterrestrial.*)

terrestrial planets   The inner planets, namely, *Mercury, Venus, Earth* and *Mars.* These are similar to Earth in their general properties; that is, they are small, relatively high-density bodies, composed of metals and silicates with shallow atmospheres, as compared to the gaseous *outer planets.*
See: planets, Earth, Mercury, Mars, Venus

tesla [Symbol: T]   The *SI unit* of *magnetic flux density.* It is defined as one *weber* of magnetic flux per meter squared

$$1T \equiv \frac{1 \text{ Wb}}{m^2}$$

Named in honor of Nikola Tesla (1870-1943), a Croatian-American physicist-inventor.

test bed   1. A base, mount or frame within or upon which a piece of equipment, especially an engine, is secured for testing.
2. A flying test bed.

test chamber   A place, section, or room having special characteristics where a person or object is subjected to an experiment, as an altitude chamber.

test firing   The *firing* of a *rocket engine,* either live or static, with the purpose of making controlled observations of the engine or of an engine component.

test flight   A flight to make controlled obser-

vations of the operation or performance of an *aircraft* or rocket, of an *aircraft, aerospace vehicle, rocket* component, of a system, etc.
See: STS-1 Flight

test reactor   A *nuclear reactor* specially designed to test the behavior of materials and components under the *neutron* and *gamma* fluxes and temperature conditions of an operating reactor.

Tethered Satellite System [abbr. TSS]   NASA plans to eventually deploy a sub-satellite from the Space Shuttle, thereby exploiting gravity gradient forces. Because of atmospheric drag, the lifetime of a free-flying satellite at these same altitudes would be very short. This system would have a total mass up to 1500 kg and could be mounted on one pallet. It would consist of a tether reel mechanism with a servo drive motor and control sensors; a boom with docking probe used for initial deployment and subsequent retrieval; a satellite weighing up to 350 kg; up to 100 km of synthetic or metallic tether; a digital control computer and, a control and display panel on the Orbiter aft-flight deck for Shuttle crew operation.

text and graphics hardcopy system (Orbiter)
See: Orbiter communications and data systems

theodolite   An optical instrument that consists of a sighting telescope, mounted so that it is free to rotate around horizontal and vertical axes, and graduated so that the angle of rotation may be measured. The telescope is usually fitted with a right-angle prism so that the observer continues to look horizontally into the eyepiece, whatever the variation of the *elevation angle.*

thermal   1. Of or pertaining to *heat* or *temperature.*
2. A vertical air current caused by differential heating of the terrain.

thermal breeder reactor   A *breeder reactor* in which the *fission chain reaction* is sustained by *thermal neutrons.*

thermal burn   A burn of the skin or other organic material due to radiant heat, such as that produced by the detonation of a *nuclear explosive.*

See: flash burn, radiation burn, radiation illness

thermal-compensating orifice Orifice whose flow area adjusts to compensate for *temperature* changes in the controlled fluid.

thermal conductivity An intrinsic physical property of a substance, describing its ability to conduct *heat* (thermal energy) as a consequence of molecular motion.

thermal cycling Exposure of a component to alternating levels of relatively high and low *temperatures.*

thermal emission The process by which a body emits *electromagnetic radiation* as a consequence of its temperature only.
See: Planck's Law

thermal emissive power The rate of *thermal emission* of *radiant energy* per unit area of emitting surface. Also called emissive power.

thermal equilibrium A condition that exists when *energy* transfer as *heat* between two thermodynamic systems—for example, System One and System Two—is possible but none occurs. We then say that System One and System Two are in thermal equilibrium, and that they have the same temperature.

◼th Law of Thermodynamics

◼duction interval Time period ◼g during which the temperature of the *solid propellant surface* is raised by external heating to the temperature at which chemical reaction rates become significant.

thermal instability The conditions of temperature *gradient, thermal conductivity,* and *viscosity* that lead to the onset of *convection* in a *fluid.* Such gross phenomena as atmospheric winds are an example of this type of instability. In general, if the fluid is conducting as a *plasma,* the application of a *magnetic field* tends to reduce these thermal instabilites.

thermal jet engine A *jet engine* that utilizes *heat* to expand gases for rearward ejection. This is the usual form of *aircraft* jet engine.

thermal (slow) neutron A *neutron* in *thermal equilibrium* with its surrounding medium. Thermal neutrons are those that have been slowed down by a moderator to an average speed of about 2200 meters per second (at room temperature) from the much higher initial speeds they had when

expelled by *fission.* This *velocity* is similar to that of *gas molecules* at ordinary temperatures.

thermal noise The *noise* at *radio frequencies* caused by thermal agitation in a dissipative body.

thermal reactor A nuclear reactor in which the *fission chain reaction* is sustained primarily by *thermal neutrons.* Most reactors are thermal reactors.

thermal shock The development of a steep temperature gradient and accompanying high stresses within a structure.

thermal system
See: multimission modular spacecraft

thermionic Of or pertaining to the emission of *electrons* by *heat.*

thermionic emission Direct ejection of *electrons* as the result of heating the material, which raises electron energy beyond the *binding energy* that holds the electron in the material.

thermistor An *electron device* employing the temperature-dependent change of resistivity of a *semiconductor.*

The thermistor has a very large negative temperature coefficient of resistance and it can be used for temperature measurements or electronic *circuit* control.

thermochemical Pertaining to a chemical change induced by *heat.*

thermochemical cycles
See: hydrogen

thermochemistry A branch of chemistry that treats of the relations of *heat* and chemical changes.

thermodynamic efficiency [Symbol: η_{th}] In *thermodynamics,* the ratio of the *work* done by a heat engine to the total *heat* supplied by the heat source.

$$\eta_{th} \equiv \frac{\text{work}_{out}}{\text{heat}_{in}}$$

Also called thermal efficiency, Carnot efficiency.

thermodynamic temperature scale The *Kelvin temperature scale* or the *Rankine temperature scale.*

thermometer A device for measuring *temperature.*

thermometry Generally considered to be

the measurement of low and moderate temperatures (approximately up to 773 K or 500 C), while *pyrometry* involves measurement of high temperatures.

thermonuclear bomb (device)   A *hydrogen bomb* (device).

thermonuclear reaction   A reaction in which very high temperatures bring about the *fusion* of two light *nuclei* to form the nucleus of a heavier *atom*, releasing a large amount of energy. In a *hydrogen bomb,* the high *temperature* to initiate the thermonuclear reaction is produced by a preliminary *fission* reaction.

thermos flask
See: Dewar flask

Third Law of Thermodynamics   Based upon the work of Nernst, Planck, Boltzmann, Lewis, Einstein and others in the early 1900s, the Third Law of Thermodynamics can be stated as: The *entropy* of any *pure substance* in *thermodynamic equilibrium* approaches zero as the *absolute temperature* approaches zero. This law can be expressed mathematically for a pure substance in equilibrium as:

$$\lim_{T \to 0} S = 0$$

where S is the value of absolute entropy (J/K)

T is the absolute temperature (K)

The Third Law is important in that it furnishes a basis for calculating the absolute entropies of substances, either elements or compounds—data which may then be used in analyzing chemical reactions.
See: entropy, absolute zero

thorium [Symbol: Th]   A naturally *radioactive element* with *atomic number* 90 and, as found in nature, an *atomic weight* of approximately 232. The *fertile* thorium-232 isotope is abundant and can be transmuted to *fissionable* uranium-233 by *neutron* iradiation.
See: fertile material, transmutation

thorium series (sequence)   The series of *nuclides* resulting from the *radioactive decay* of thorium-232. Many man-made nuclides decay into this sequence. The end product of this sequence in nature is lead-208.

three-body problem   The problem in classical *celestial mechanics* that treats the motion of a small body, usually of negligible mass, relative to and under the *gravitational* influence of two other finite point masses.

three-dimensional grain configuration   Grain whose burning surface is described by three-dimensional analytical geometry (one that considers end effects).

three-way valve   Valve having three controlled ports, usually one inlet and two outlet ports.

threshold   Generally, the minimum value of a *signal* that can be detected by the system or *sensor* under consideration.

threshold dose   The minimum dose of *radiation* that will produce a detectable biological effect.
See: absorbed dose.

throat   Portion of a convergent/divergent *nozzle* at which cross-sectional area is minimal, the region of transition from subsonic to supersonic flow of exhaust gases.

throttle valve   Valve to control flowrate of a *fluid* by means of a variable-area flow restriction; this kind of valve may have an infinite number of operating positions as contrasted to a shutoff valve, which is either fully open or fully closed.

throttling   The varying of the rocket engine during powered flight by some technique. Tightening of fuel line changing of *thrust chamber pressure,* pulsed thrust, and variation of *nozzle* expansion are methods to achieve throttling.

through-bulkhead initiator   Contrivance in which a small charge of material is detonated such that the resulting *shock wave* is transmitted through a solid metal interface; the transmitted shock wave detonates an acceptor charge on the internal side of the interface, which then initiates the *deflagration* of heat-producing material that ignites the *propellant.*

thrust   Propulsive force developed by a *rocket engine* during firing.

thrust-balance system   Set or arrangement of devices that provides the (pressure × area) force necessary to balance axial thrust in a *turbopump.*

thrust barrel   Structure in the rocket vehicle designed to accept the thrust load from two or more engines; also called thrust structure.

See: Space Shuttle Engine

thrust chamber   The assembly of *injector*, *combustion chamber*, and *nozzle.*

thrust coefficient   Ratio of *engine thrust* to the product of *nozzle throat* area times nozzle inlet pressure.

thrust collector   Structure attached to a *rocket engine* during testing to transmit the engine *thrust* to thrust-measuring instruments.

thrust cone   A component of the *Space Shuttle Main Engine* that transmits the total *thrust* of the engine through the *gimbal* bearing to the *Orbiter* vehicle.

See: Space Shuttle Main Engine

thrust-time profile   Plot of *thrusts* vs time for the *firing* duration of a *rocket engine.*

thrust vector   Direction of the *thrust force.*

thrust vector control (TCV)   This system, located in the aft skirt of the *Solid Rocket Booster,* is the assembly that *gimbals* the solid rocket motor nozzle and thus helps to steer the entire *Space Shuttle* vehicle.

See: Solid Rocket Booster

tile gap fillers

See: Orbiter thermal protection system

tilt/spin table   The mechanism installed in the *Orbiter's cargo bay* that deploys the Spinning Solid Upper Stage with its *payload.*

time dilation   A consequence of the special theory of *relativity* hypothesizing that if two observers are moving at constant velocity (v) relative to each other (which is close to the *velocity of light* [c], it will appear to each that the other's clock has slowed down. The time intervals will be "dilated" by a factor $[1/\beta]$ where

$$\beta = \sqrt{1 - (v^2/c^2)}$$

See: relativity

time lag   The total time between the application of a signal to a measuring instrument and the full indication of that signal within the uncertainty of the instrument.

time line   The planned schedule for astronauts on a space mission.

time-of-flight spectrometer   A device for separating and sorting *neutrons* (or other *particles*) into categories of similar *energy,* measured by the time it takes the particles to travel a known distance. Compare this term with *mass spectrometer.*

time tic   Markings on *telemetry* records to indicate time intervals.

time tick   A time signal consisting of one or more short audible sounds or beats.

TNT equivalent   A measure of the *energy* release in the detonation of a *nuclear explosive* expressed in terms of the weight of TNT (the chemical explosive trinitrotoluene) which would release the same amount of energy when exploded. It is usually expressed in kilotons or megatons. The TNT equivalence relationship is based on the fact that 1 ton of TNT releases one billion (10^9) calories of energy.

See: yield

tolerance   The allowable variation in measurements within which the dimensions of an item are judged acceptable.

tolerance stackup   Additive effect of all the allowable manufacturing *tolerances* on the final dimensions of the assembly; also called tolerance buildup.

tool caddies

See: extravehicular mobility unit

topside sounder   A *satellite* or *spacecraft* that is designed to measure *ion* concentration in the *ionosphere* from above the *Earth's atmosphere.*

torque box   A structure built of thin walls so as to form a closed box. It is designed to resist *torsional* loads and may consist of one or more completely enclosed compartments called "cells."

See: Orbiter structure

torsion   The state of being twisted.

torsional deflection   Deflection imposed on a flexible joint by applying a *torque* about its longitudinal axis.

total eclipse   An *eclipse* in which the entire source of light is obscured.

total impulse [symbol: I_t]   The integral of the *thrust* (F) over an interval of time t:

$$I_t = \int F \, dt$$

Total impulse is related to *specific impulse* I_{sp} by $I_t = I_{sp}Mdt$, where M = propellant mass flow rate.

total pressure   *Stagnation pressure.*

total radiation   *Radiation* over the entire *spectrum* of emitted *wavelengths.*

total temperature   *Stagnation temperature.*

touchdown   The (moment of) landing of an *aerospace vehicle* or *spacecraft* on the surface of a *planet* or *moon.*

toughness   The ability of a metal to absorb energy and deform plastically before fracturing.

tower simulator
   See: STS launch facilities

tracer, isotopic   An *isotope* of an *element,* a small amount of which may be incorporated into a sample of material (the carrier) in order to follow (trace) the course of the element through a chemical, biological or physical process, and thus may also follow the larger sample. The tracer may be radioactive, in which case the observations are made by measuring the *radioactivity.* If the tracer is *stable, mass spectrometers, density measurement,* or *neutron activation* analysis may be employed to determine *isotopic composition.* Tracers also are called labels or tags, and materials are said to be labeled or tagged when radioactive tracers are incorporated in them.

track   1. The *path* or actual line of movement of an *aircraft, rocket, aerospace vehicle,* etc., over the surface of the *Earth.* It is the projection of the *flightpath* on the surface.
2. To observe or plot the path of something moving, such as an aircraft or rocket, by one means or another, as by telescope or by radar—said of persons or of electronic equipment, as, "the observer, or the radar, tracked the satellite."
3. To follow a desired track.

Tracking and Data Relay Satellite System (TDRSS)   Satellite communication system which provides the principal coverage from *geosynchronous orbit* for all *Space Shuttle flights.* Fig (1) depicts one of four giant telecommunication satellites, built by TRW, which comprise the space segment of the Western Union-owned-and-operated TDRSS. NASA will lease tracking and data relay services for its fleet of Earth-orbiting *spacecraft,* including the Shuttle.

The TDRSS will consist of two geostationary relay satellites 130 degrees apart in longitude (see figures 2 and 3), and a ground terminal located at White Sands, NM. The system will also include two space satellites, one in orbit and one configured for a rapid replacement launch. The purpose of TDRSS is to provide *telecommunication* services between low earth-orbiting user spacecraft and user control and/or data procession facilities. A *real-time,* bent-pipe concept is used in the operation of TDRSS telecommunication services. This bent-pipe concept provides no processing of user data. The system will be capable of transmitting data to, receiving data from, or tracking user spacecraft over a large percentage of the user orbit.

The TDRSS will provide communications and tracking services to approved users. Forward link services include tracking and command data channels at *S-band* and *Ku-band frequencies.* Return link services include a tracking channel and *telemetry* data channels for a wide range of data rates. Return link services are available at S-band and Ku-band frequencies. S-band forward and return link services are provided via the Single Access (SA) service and Multiple Access (MA) service. SA service uses steerable *parabolic antennas* on the Trackng and Data Relay Satellites (TDRS), whereas the MA service uses a TDRS array antenna with ground-implemented phasing. Ku-band forward and return link services are provided only by the SA service.

The Multiple Access service will provide dedicated return link service to low earth-orbiting user spacecraft with real-time, playback and science data rates up to 50 kb/sec. Return link support can be provided simultaneously to each of 20 users during that portion of their orbit which is visible to either TDRS (i.e., 85 percent or more of a user's orbit). Based on current mission model projection, minimal scheduling restrictions will be encountered for return link service. Forward link service is time-shared with a maximum *bit* rate of 10 kb/sec and supports one user per TDRS at a time. The MA service will operate at S-band. Users of this service are called MA users.

The Single Access service will provide a high data rate return link to users with real-time, playback, or science data requirements up to 100 Mb/sec and for users requiring forward link data rates up to 25 Mb/sec. This service will be used on a priority scheduled basis, and normally will not be used for dedicated support to any mission. Forward link service at S-band will provide data rates up to 300 kb/sec, and at Ku-band will provide data rates up to 25 Mb/sec. Return link service at S-band will provide total data rates up to 3.15 Mb/sec, and at Ku-band will provide data rates up to 300 Mb/sec. Users of this service are called SSA or KSA users.

The forward and return link frequency plan is summarized in Fig. 4 for the space-to-space links. The TDRS-to-ground terminal and ground terminal-to-TDRS link channelization is not shown in Fig. 4 since it does not affect the TDRSS/user interface.

train   Anything, such as a luminous *gas* or ionized *particles,* left along the *trajectory* of a *meteor* after the head of the meteor has passed.

trainer   A teaching device or facility that primarily provides a physical representation of flight hardware. The trainer may also have limited computer capabilities. Compare this term with *simulator.*

trajectory   In general, the path traced by any body moving as a result of an externally applied *force,* considered in three dimensions.

Trajectory is sometimes used to mean *flight path* or *orbit,* but orbit usually means a closed path and trajectory, a path which is not closed.

transceiver   A combination *transmitter* and *receiver* in a single housing, with some components being use by both units.
See: transponder

transducer   General term for any device that converts a physical magnitude of one form of *energy* into another form (for example, electical to *acoustic*).

transfer maneuver   *Velocity* change in orbit.

transfer orbit   In *interplanetary* travel, an *elliptical trajectory* tangent to the *orbits* of both the departure *planet* and the target planet.

transfer tunnel and tunnel adapter   Devices that provide for transfer of the *Shuttle crew* and equipment between the *Spacelab* and the *Orbiter crew module.*
See: Orbiter structure

transformation, nuclear   *Transmutation.*

transient   The condition of a physical system in which the parameters of importance (*temperature, pressure, fluid velocity,* etc.) vary significantly with time; in particular, the condition or state of the *rocket engine* operation in which the *mass, momentum,* volume, and *pressure* of the combustion products within the *thrust chamber* vary significantly with time.

transient period   The interval from start or ignition to the time when steady-state conditions are reached, as in a *rocket engine.*

transit   1. The passage of a *celestial body* across a *celestial meridian,* usually called "meridian transit."
2. The *apparent* passage of a celestial body across the face of another celestial body or across any point, area or line.
3. An instrument used by an astronomer to determine the exact instant or meridian transit of a celestial body.

transition point   In *aerodynamics,* the point of change from *laminar* to *turbulent* flow.

translation   Movement in a straight line without rotation.

translunar   Of or pertaining to *space* outside the *Moon's orbit* about the *Earth.*

translunar space   As seen from the *Earth* at any moment, *space* lying beyond the *orbit* of the *Moon.*

transmission grating   A *diffraction grating* in which energy is resolved to *spectral* components on *transmission* through the grating.

transmitter   A device for the generation of *signals* of any type and form which are to be transmitted.

In *radio* and *radar,* it is that portion of the equipment which includes electronic circuits designed to generate, amplify, and shape the *radiofrequency* energy which is delivered to the *antenna* where it is radiated

out into space.
See: receiver

transmutation The transmutation of one element into another by a "nuclear reaction" or series of reactions. Example: the transmutation of uranium-238 into plutonium-239 by absorption of a *neutron.*

transonic Pertaining to the range of speed in which flow patterns change from *subsonic* to *supersonic* or vice versa, about *Mach* 0.8 to 1.2, as in transonic flight, transonic flutter.

transonic flow In *aerodynamics,* flow of a *fluid* over a body in the range just above and just below the *acoustic velocity* (speed of sound).

Transonic flow presents a special problem in aerodynamics in that neither the equations describing *subsonic* flow nor the equations describing *supersonic* flow can be applied in the transonic range.

transonic speed The speed of a body relative to the surrounding *fluid* at which the flow is in some places on the body *subsonic* and in other places *supersonic.*

transparent plasma A *plasma* through which an *electromagnetic wave* can propogate. In general, a plasma is transparent for frequencies higher than the *plasma frequency.*

transpiration The passage of *gas* or *liquid* through a porous solid (usually under conditions of *molecular flow*).

transpiration cooling A form of *mass transfer cooling* that involves the controlled injection of a mass (fluid) through a porous surface and is basically limited by the maximum rate at which this coolant material can be pumped through the surface.
See: mass transfer cooling

transplutonium element An *element* above *plutonium* in the *periodic table,* that is, one with an *atomic number* greater than 94.
See: transuranic element

transponder A combined *receiver* and *transmitter* whose function is to transmit signals automatically when triggered by an *interrogator.*

transuranic element (isotope) An element above *uranium* in the *periodic table,* that is, with an *atomic number* greater than 92. All 11 transuranic elements are produced artifi-

cially and are *radioactive.* They are neptunium, plutonium, americum, curium, berkelium, californium, einstinium, fermium, mendelevium, nobelium and lawrencium.

transuranium element A *transuramic element.*

trap A part of a solid-propellant *rocket engine* used to prevent the loss of unburned *propellant* through the *nozzle.*

trapping The process by which *radiation particles* are caught and held in a *radiation belt.*

traveling plane wave A *plane wave* each of whose *frequency* components has an exponential variation of *amplitude* and a linear variation of *phase* in the direction of propagation.

traveling-wave tube [abbr. TWT] An *electron tube* in which a stream of electrons interacts continuously or repeatedly with a guided *electromagnetic wave* moving substantially in synchronism with it, and in such a way that there is a net transfer of *energy* from the *electron stream* to the wave.

travel time Elapsed time from first motion of the movable element in a flow-control device to full-open or full-closed position.

trim Adjustment of an *aerodynamic vehicle's* controls to achieve *stability* in a desired flight condition.

triple point The point at which the solid, liquid and vapor phases of a substance are in *equilibrium.* For a particular substance the triple point occurs at a unique set of *temperature, pressure* and volume values.

triplet *Injector* orifice pattern consisting of one or more sets of three orifices that produce streams converging to a point; usually *fuel* is injected through outer orifices, and *oxidizer* is injected through the central orifice.

triplexer A *dual-diplexer,* that permits the use of two receivers simultaneously and independently in a radar system by disconnecting the receivers during the transmitted *pulse.*

tritium [Symbol: T] A *radioactive isotope* of *hydrogen* with two *neutrons* and one *proton* in the nucleus. It is manmade and is heavier than *deuterium* (heavy hydrogen).

Tritium is used in industrial thickness gauges, and as a label in experiments in chemistry and biology. Along with deuterium (D) it is also a key fuel for nuclear *fusion* (See D-T reaction in *fusion* section). Its nucleus is a triton. Compare this term with deuterium.
See: hydrogen

triton The nucleus of a *tritium* (^{3}H) atom.

truncation error In computations, the error resulting from the use of only a finite number of terms of an infinite series or from the approximation of operations in the infinitesimal calculus by operations in the calculus of finite differences.

trunnion A pin or pivot on which something can be rotated or tilted.

truss An assemblage of structural members (such as beams) forming a rigid framework.

T-time Any specific time (minus or plus) that is referenced to "launch" time, or "zero," at the end of a *countdown*. T-time is used to refer to times during a countdown sequence that is intended to result in the ignition of a *rocket propulsion* unit or units to launch a *missile* or *rocket vehicle*.

tube crown Portion of the coolant tube that forms the outer wall of the cooling jacket.

tube-wall construction Use of parallel metal tubes that carry coolant to or from the *combustion chamber* or *nozzle* wall.

tumble 1. To rotate end over end—said of a rocket, of an ejection capsule, etc.
2. Of a gyro, to *precess* suddenly and to an extreme extent as a result of exceeding its operating limits of bank or pitch.

tumbling An attitude situation in which the vehicle continues on its flight, but turns end over end about its center of mass.

tumbling system
See: External Tank

tunnel adapter The *Space Transportation System flight kit* used to attach the *Orbiter airlock* to the *Spacelab* tunnel. This is a standard flight kit for Spacelab.
See: Spacelab

turbine Machine consisting of one or more bladed disks (*rotor* or turbine wheel) and one or more sets of fixed *vanes (stator)* inside a casing, so designed that the wheel is turned by incoming *fluid* (usually hot gas) striking the blades.

turbofan A *turbojet engine* that gains pro-

pulsive thrust by extending a portion of the *compressor* or *turbine* blades outside the inner engine case.

The extended blades propel bypass air which flows along the engine axis but between the inner and outer engine casing. This air is not combusted but does provide additional thrust caused by the propulsive effect imparted to it by the extended compressor blading.

turbojet engine A *jet engine* incorporating a *turbine*-driven air *compressor* to take in and compress the *air* for the combustion of fuel. The gases of combustion (or the heated air) are then used to rotate the *turbine* and to create a *thrust-producing* jet. Often called a turbojet.

turbopump An assembly consisting of one or more pumps driven by a hot-gas turbine.
See: Space Shuttle Main Engines

turbopump system An assembly of components (e.g., propellant *pumps, turbine*(s), power source) designed to raise the *pressure* of the *propellants* received from the vehicle *tanks* and deliver them to the main *thrust chamber* at specified pressures and flow-rates.
See: Space Shuttle Main Engines.

turbulence 1. A state of *fluid* flow in which the instantaneous velocities exhibit irregular and apparently random fluctuations so that in practice only statistical properties can be recognized and subjected to analysis.

These fluctuations often constitute major deformations of the flow and are capable of transporting momentum, energy, and suspended matter at rates far in excess of the rate of transport by the molecular processes of *diffusion* and conduction in a nonturbulent or *laminar* flow.

turbulence ring Circumferential protuberance in the gas-side wall of a (gas generator) *combustion chamber* intended to generate *turbulent flow* and thereby enhance the mixing of burning gases.

turbulent boundary layer The layer in which the *Reynolds stresses* are much larger than the *viscous stresses*. When the *Reynolds number* is sufficiently high, there is a turbulent layer adjacent to the *laminar boundary layer*.

turbulent flow Fluid flow in which the *velocity* at a given point fluctuates randomly and irregularly in both magnitude and direction. The opposite of *laminar flow.*

turndown ratio   Ratio of maximum to minimum controlled flowrates of a *throttle valve.*

twenty-four hour satellite
See: geosynchronous

two-body problem   The problem in classical *celestial mechanics* that treats of the relative motion of two point masses under their mutual *gravitational* attraction.

two-dimensional grain configuaration   Solid-propellant grain whose burning surface is described by two-dimensional analytical geometry (cross section is independent of length).
See: Solid Propellant rocket

two-phase flow   Simultaneous flow of gases and solid *particles* (e.g., condensed metal oxide); or of liquid and vapor (gases).

U

UHF   Ultrahigh frequency.

ullage   The amount that a container, such as a *fuel* tank, lacks of being full.

ullage pressure   Pressure in the *ullage* space of a container, either supplied or self-generated.

ultimate load (or pressure)   Load (or pressure) at which catastrophic failure (general collapse or rupture) of a structure occurs.

ultimate stress   Stress at which a material fractures or becomes structurally unstable.

ultra-   A prefix meaning "surpassing a specified limit, scope or range" or "beyond."

ultrahigh frequency (UHF)   A radio frequency in the range 0.3 gigahertz to 3 gigahertz.

ultrasonic   Of or pertaining to frequencies above those that affect the human ear, that is, more than 20,000 hertz (vibrations or cycles per second).

ultrasonics   The technology dealing with *pressure waves* that are similar to sound waves but have frequencies above the audi-

ble (human hearing) limit. Ultrasonics typically involves frequencies of approximately 20 kilohertz and up.

ultraviolet airglow spectrometer   An *ultraviolet spectrometer* especially designed to analyze the faint fluorescent glow of ultraviolet light in a planetary *atmosphere* and thereby permit the study of the gases or *ions* and energy sources that produce this faint light.

ultraviolet (UV) spectrometer   An optical instrument for analyzing the intensity of *ultraviolet* radiation at various *wavelengths.*
See: spectrometer

umbilical   An electrical or fluid servicing line between the ground or a tower and an upright *rocket* vehicle before launch.

umbra   1. The darkest part of a shadow, in which light is completely cut off by an intervening object. A lighter part of the shadow, a region in which light is only partially cut off, is called the "penumbra."
2. The darker central portion of a *sunspot,* which is surrounded by the lighter penumbra.

UMR injector   Injector that produces a *uniform mixture ratio* (UMR) and thus a combustion region with relatively uniform temperature distribution.

underexpansion   A condition in the operation of a *rocket nozzle* in which the exit area of the nozzle is insufficient to permit proper expansion of the propellant gas. Consequently, the exiting gas is at a *pressure* greater than the ambient pressure, and this leads to the formation of external expansion waves.

unidentified flying object (UFO)   An object reportedly seen in the sky by an observer who cannot determine its nature. The vast majority of such "UFO sightings" can, in fact, be explained by known phenomena. These phenomena may, however, be beyond the knowledge or experience level of the observer. Known phenomena that have given rise to UFO reports include: *artificial Earth satellites,* aircraft, high-altitude weather balloons, certain types of clouds and even the *planet Venus.*

There are, nonetheless, some reported sightings that cannot be fully explained on the basis of available data (which may be insufficient or too unreliable scientifically) or on the basis of comparison with known phenomena. It is the investigation of these relatively few puzzling UFO cases that has given rise to popular (although technically unfounded) speculations about "flying saucers," *extraterrestrial* visitations and the arrival of *"little Green men."* While life forms may certainly exist elsewhere in the *Universe,* the reports of relatively haphazard UFO encounters hardly reflect the logical and orderly sequence of contacts one might anticipate from intelligent, star-traveling creatures (should they exist).

Uranus   Seventh major planet with a diameter of 52,400 km and a mass 14.6 times that of the Earth. The principal constituents of Uranus are thought to be hydrogen and helium. Methane has been observed in its atmosphere, which is very clear and deep. No distinctive atmospheric markings such as in the Jovian and Saturnian atmospheres can be discerned. Uranus is believed to have a core of rock and metal surrounded by layers of ice, liquid hydrogen and gaseous hydrogen. There is no clear boundary between the liquid and gaseous hydrogen layers. Uranus has nine faint rings circling the planet at altitudes from about 18,000 to 25,000 km (11,000 to 15,400 mi.) above Uranus's visible cloud tops. The planet has five satellites: Miranda, Ariel, Umbriel, Titania and Oberon.

U.S. Air Force Role In Space   Because there is no distinct boundary between *air* and *space*, for many years the Air Force has seen its responsibility for defense as one without altitude limitations. The development of United States space policy and space operations is a natural evolution of the development of air power. This entry discusses the "space" portion of the Air Force's *"aerospace"* mission—the reasons for operating in space, and some of the current and future space systems that contribute to national defense.

The Air Force has found that some of its functions can be performed better in space than in the *atmosphere* or on *Earth*. In essence, the Air Force operates in space because it makes military and economic sense to be there. The Air Force is there to provide support to United States *terrestrial* forces. The cost of operating space-based systems prohibits "nice to have" operations there. Unlike climbing the mountain "because it's there," the Air Force develops and operates systems in space only when they can provide a cost-effective and significant contribution to the effectiveness of defense combat forces. Such contributions include: improved communications, more precise navigation aids, better and more extensive weather information, the capability for more reliable and more rapid warning of attack, and improved surveillance. President Carter stated, "Photoreconnaissance satellites have become an important stabilizing factor in world affairs. In the monitoring of arms control agreements, they make an immense contribution to the security of all nations. We shall continue to develop them." Space will play a greater role in future military operations; as its military value increases, the importance of space defense will also increase.

The underlying goal of U.S. space policy is to preserve space for peaceful uses. National policy and international treaties prohibit employing weapons of mass destruction in space, but there is a need to insure that no other nation will gain a military advantage through the exploitation of space. As the United States' use of space systems to support its military forces grows, it becomes increasingly important that these systems survive an enemy's attempt to interfere with them. The Soviet Union has developed and tested an anti-satellite (ASAT) interceptor that has an operational capability against our satellites. Former President Carter stated our nation's preference for verifiable limitations on ASAT systems and our opposition to a space weapons race. But in the absence of an agreement, and in the face of proven Soviet cpabilities, the Air Force must work to defend U.S. *satellites,* if necessary. To this end, research and development activities designed to increase satellite survivability against attacks, and to nullify the Soviet space threat are underway. In addition to prototype development of an anti-satellite capability, the Air Force is working on an improved ground-based system to enhance detection and tracking of satellites. It is also working on ways to make our satellites less vulnerable to attack. Finally, command, control and communications are being improved to effectively manage all space defense resources. One vital aspect of space defense is the ability of the United States to detect potential attacks. Currently operational satellites provide early detection of hostile missile launches. Future programs will develop more sophisticated sensors capable of surviving high energy *laser and nuclear weapons* effects.

A new era of space transportation will come into being in the 1980s with the advent of the Space Shuttle and its ability to inexpensively transport a variety of payloads to orbit. It is designed to reduce the cost and increase the effectiveness of using space for defense, commercial, and scientific needs. The Space Shuttle, a national program under National Aeronautics and Space Administration direction, is a breakthrough in the method of placing space systems into orbit. Ultimately, it will be capable of carrying twice as much payload weight and three times as much volume as can the largest present booster, the Titan III. With this extra capability, cheaper but larger *spacecraft* can be built, or additional subsystems or systems for *redundancy* to attain longer satellite life can be added. Satellites will be recovered from low *Earth* orbit, providing the opportunity to refurbish expensive satellites and update them as various technologies mature. Service or repair of satellites while they remain in orbit will extend lifetimes and lower costs.

The Space Shuttle offers great advantages over the current system of using expendable launch vehicles. The new system will make it possible to get into space rapidly and routinely, increase the weight and size of payloads, and do it for less money.

Because the Space Shuttle will orbit at relatively low altitudes—160 to 960 km (100 to 600 miles)—an auxiliary system is being developed by the Air Force to propel payloads farther from earth, out to *geosynchronous orbit* and beyond. This system is the *Inertial Upper Stage* (IUS). [*See:* Fig. 1] The IUS is a *solid-fueled rocket* that will be carried in the Shuttle's payload

bay. After the Shuttle reaches its low Earth orbit, the IUS and its *spacecraft* will be deployed. The IUS rocket will then propel its payload to the desired orbit or onto *interplanetary trajectories.*

One of the satellite systems which will someday be placed in orbit using the Shuttle will be the Navstar Global Positioning System (GPS). Navstar GPS is an excellent example of the unique capability of space to enhance the conduct of traditional military missions. Navstar GPS will consist of 18 satellites placed in 20,200 km (10,900 nautical mile) orbits. Each satellite continually transmits its position and precise time of transmission. Passive receiving sets on the ground, in *aircraft* or on ships receive signals from four different satellites, and this data is processed through an *algorithm* that compares the signals and solves for position and *altitude*. By analyzing this information, a military user can tell his exact position as well as his altitude anywhere on earth to within 16m (50 feet). Six Navstar satellites are now in orbit, having been launched atop Atlas-F boosters. The entire constellation of 18 satellites will be operational in the late 1980s. When the system is fully operational it should provide a solution to night or bad weather precision bombing. Also, receivers are being developed which can pinpoint the position of ships and submarines at sea; manpacks will allow a foot soldier to locate himself in order to target his artillery for first shot accuracy on the enemy. Numerous civilian applications for Navstar GPS exist as well, and the Department of Defense is coordinating its activities relating to radio navigation systems with the Department of Transportation, the Federal Aviation Administration, the Department of Commerce, the State Department and NASA.

The Defense Satellite Communications System, called DSCS, provides for voice and high data communication between virtually anywhere on Earth through satellite relay. The current DSCS orbital configuration is comprised of four DSCS II satellites along with two spares at geosynchronous altitude; this provides almost total global coverage. This capability supports vital national security requirements for worldwide command and control, crises management, intelligence data relay, diplomatic traffic and early warning detection and reporting. In 1981 the Air Force launched the first DSCS III Research and Development satellite. DSCS III satellites are designed for longer life, increased channelization, user flexibility, and enhanced anti-jam protection. The initial capability of the DSCS system using DSCS III production satellites is expected in early 1985. A total of 12 production satellites, including a refurbished qualification satellite, will be acquired to continue DSCS communications into the early 1990s. The satellites can be launched by the Air Force Titan III booster or carried aloft on the Space Shuttle. An example of the value of space systems for communications can be found in the 1975 Mayaguez incident. Communications during that tense situation were provided by the DSCS system, allowing the President to communicate in real time with the Marine commander of the landing party. Also, data transmitted by weather satellites revealed that the primary inflight refueling area for the helicopters was covered with clouds, so the tanker rendezvous point was changed to one that was cloud-free.

The Air Force Satellite Communications System (AFSATCOM) is an *Ultra High Frequency* (UHF), satellite based, communications system. The space segment consists primarily of communications packages, called *transponders*, on the *Fleet Satellite Communications* (FLTSATCOM) System and Satellite Data System spacecraft. The AFSATCOM system provides data communications for command and control of our nuclear capable forces. Terminals procured by the Air Force are installed in command post aircraft, strategic bombers, tanker aircraft and in selected ground command centers. The Army and Navy also have terminals that use the AFSATCOM transponders. The AFSATCOM system provides worldwide direct 100 word per minute teletype communications to and from the National Command Authorities, the military commanders and our nuclear capable forces. The AFSATCOM system became operational in May 1979, with the successful

launch of the second FLTSATCOM satellite. Two more satellites were launched in January and October 1980 to complete the FLTSATCOM portion of the space segment.

The Defense Meteorological Satellite Program [DMSP] is the Department of Defense's single most important source of weather data. Through these satellites, military weather forecasters can detect and observe weather conditions throughout the world. Two satellites are in sun-synchronous orbit, [See: orbit] providing meteorological data to Air Force and Navy ground and shipborne stations. The data can help identify severe weather such as thunderstorms, typhoons and hurricanes, and form three-dimensional cloud analyses which are the basis for computer simulation of weather conditions throughout the world. All of this information is readily available to aid Air Force, Army, Navy or Marine military commanders in performing their mission. While the primary mission of the DMSP satellites is gathering weather data for military users, its information is actually a national resource. For example, data gathered by the satellites are made available to the civilian community through the Commerce Department's National Oceanic and Atmospheric Administration, NOAA.

To provide detection of certain enemy missile launches and keep track of objects in space, two systems have been developed by Air Force Systems Command. One, PAVE PAWS, consists of a pair of solid-state radar systems, one on each coast. The site at Otis AFB, MA, was put into operation by strategic Air Command (SAC) in April 1980. The other site, at Beale AFB, CA, became operational in September of 1980. In the event of a sea-launched *ballistic missile* attack on the continental United States PAVE PAWS provides detection and early warning; it also keeps tabs on satellites in orbit. In its primary role, it would tell the potential numbers and destination of the missiles. This information would then be simultaneously passed to the National Command Authorities in Washington, the North American Air Defense Command in Colorado, and SAC. In its secondary mission, PAVE PAWS feeds display data on the position and velocity of hundreds of satellites orbiting Earth. Using phased array technology, it can track more targets more accurately, and at greater range while, at the same time, using less power. Phased array means that the PAVE PAWS "eyes" are steered electronically, using thousands of small radar antennas coordinated by two large computers. Since no mechanical parts limit the speed of the radar scan, it can more rapidly detect and locate orbiting objects. Its computers also allow it automatically to detect, track and predict the impact point of a missile.

It is very important to the defense community to know exactly what is in space, where it is and where it's going. Using the latest in silicon-chip technology, *Air Force Systems Command* is developing a worldwide network of monitoring stations to improve on our ability to identify and track orbiting objects. It is known as GEODSS, for *G*roundbased *E*lectro-*O*ptical *D*eep *S*pace *S*urveillance, and will consist of five facilties. The first, at White Sands Missile Range in New Mexico, is being tested now. Others will be built in South Korea, Hawaii, and at sites in the Indian Ocean and eastern Atlantic regions. Equipped with powerful telescopes, the installations will relay images from objects in space onto photo-imaging tubes which will convert the picture to electronic impulses which are then fed into computers. The data then fed immediately to NORAD's master computers where all objects in space are monitored constantly. GEODSS will be able to spot an object the size of a soccer ball at 40,000 km (25,000 miles) into space. The system is planned to be completely operational by 1983.

As space activity and importance increase, it has become apparent that centralized management of our operations would be not only more efficient, but also less expensive. The Consolidated Space Operations Center, or CSOC, is being planned to do this. The CSOC will consist of two elements, the Satellite Operations Center and the Shuttle Operations and Planning Center. It will combine in a single facility with the elements necessary to conduct both Space Shuttle and satellite mis-

sion operations. The Satellite Operations Center will perform communications, command and control functions for orbiting U.S. spacecraft. The Space Shuttle Operations and Planning Center will conduct Department of Defense flight planning, readiness and command functions while have direct mission authority of DOD Shuttle missions. Facility construction is scheduled to begin in 1983; it should become operational in 1986. When completed it will be manned by about 300 military, 100 Air Force civilians and 1400 contract personnel.

U.S. Air Force Systems Command [abbr. AFSC] The mission of Air Force Systems Command (AFSC) is to advance *aerospace* science and technology, and develop and acquire the best aeronautical and space systems at the lowest cost. AFSC also meets the major space responsibilities of the Department of Defense (DOD). These include research and development as well as testing and engineering of *satellites, boosters, space probes,* and associated systems needed to support specific National Aeronautics and Space Administration (NASA) projects and programs arising under basic agreements between DOD and NASA.

From headquarters at Andrews AFB, MD, adjacent to the nation's capital, AFSC's commander directs the operations of a number of divisions, ranges, laboratories and development and test centers. The Command has approximately 10,500 officers, 15,600 airmen and 26,600 civilians. Nearly two-thirds of the command's officers hold advanced degrees; six percent have earned their doctorates. AFSC manages approximately $20 billion—nearly one-third of the Air Force's budget. The Command administers over 32,000 contracts having a total face value just under $100 billion.

Located at AFSC Headquarters, the Director of Laboratories provides policy, planning and technical direction to all phases of the Command's research and development laboratories. The directorate monitors the laboratories' programs, operations, and activities to insure they can respond promptly to the ever-changing needs of the Air Force. Research organizations under the Director of Laboratories are:

1 Air Force Office of Scientific Research [AFOSR]

2 Frank J. Seiler Research Laboratory [FJSRL]

3 European Office of Aerospace Research and Development [EOARD]

4 Air Force Wright-Aeronautical Laboratories [AFWAL]

5 Air Force Geophysics Laboratory [AFGL]

6 Air Force Human Resources Laboratory [AFHRL]

7 Air Force Weapons Laboratory [AFWL]

8 Air Force Rocket Propulsion Laboratory [AFRPL]

The AFSC emblem (See Fig. 1) expresses the word "system," and symbolizes the command's mission. The arrow emerging from the unbroken circumference of the circle draws together diverse elements through management, uniting and orienting them in direction and purpose. The lower portion of the arrow denotes various research and development resources upon which the command draws. The middle represents system management procedures that unite those resources. The top of the arrow symbolizes the operational weapon systems that result once the process is complete.

Located at Bolling AFB, District of Columbia, the Air Force Office of Scientific Research exercises Air Force executive management responsibility for Air Force basic research. It awards grants and contracts for research in the basic sciences directly related to Air Force needs. The research supported is selected from unsolicited proposals originating from scientists involved in the search for new knowledge and expansion of scientific principles. AFOSR is also responsible for the activities of the Frank J. Seiler Research Laboratory [FJSRL] and the European Office of Aerospace Research and Development [EOARD].

The Seiler Research Laboratory is located at the USAF Academy, Colorado Springs, CO. This in-house laboratory is engaged in basic research concerned with the physical and engineering sciences. The research usually centers around chemistry,

applied mathematics, and gas dynamics. The laboratory sponsors related research conducted by the faculty and cadets of the USAF Academy. The European Office of Aerospace Research and Development is located in London, England. This unit links the Air Force with the scientific communities in Europe, Africa, and the Near East. It identifies foreign technology, engineering and manufacturing advances which can be applied to meet Air Force requirements.

The Air Force Wright-Aeronautical Laboratories are situated at Wright-Patterson AFB, OH. AFWAL combines common laboratory management and support functions and provides a command and staff element for its four laboratories: Avionics, Aero-Propulsion, Flight Dynamics, and Materials. AFWAL's mission is to plan and carry out Air Force research, exploratory and advanced development, and selected engineering development through its four labs. AFWAL also provides certain technical support to the Department of Defense, and other U.S. government agencies for planning, developing and operating *aerospace* systems.

AVIONICS LABORATORY develops electronic technologies for airborne systems. Research is centered around reconnaissance, weapon delivery, navigation, electronic components, communications, electronic warfare, and *software*. This technology is used to develop *radar* and *electro-optic sensors,* fire control computers, precision *inertial* reference systems, miniaturized digital memories, signal processing warning receivers, adaptive jammers, and information processing systems. The scope of the laboratory's activities runs from concept definition through the building of *hardware* for technology demonstration. The laboratory has a number of facilities at its disposal to support its research programs.

AERO-PROPULSION LABORATORY plans and conducts basic research and development in *air-breathing propulsion,* flight vehicle power, fuel, lubricants, and fire protection. It also supports other AFSC programs to ensure the rapid application of research and technology to advanced weapon systems. To maintain their technical competence, the laboratory's scientists and engineers carry out substantial in-house research and development efforts. They monitor contractor efforts, and compare and trade-off products and technology developed by different contractors. The laboratory works closely with other government, industrial, and academic agencies.

FLIGHT DYNAMICS LABORATORY develops technology for design and fabrication of future aerospace weapon systems. The laboratory also conducts configuration research of advanced vehicles, performs engineering simulations and develops experimental flight vehicles to demonstrate new technologies. The laboratory is the Air Force focal point for nonnuclear survivability/vulnerability, and atmospheric electrical hazards study. It is also the national center for landing gear and tire testing and flight control development. The laboratory has extensive facilities to support its mission.

MATERIALS LABORATORY manages Air Force exploratory development in materials and manufacturing technology programs. It also plans and conducts specific programs in materials research and advance development. These programs are designed to reduce cost, improve *reliability* and performance of *aircraft, missiles, spacecraft* and support equipment. The laboratory provides technical and management support to other agencies involved in aerospace work.

The Air Force Geophysics Laboratory at Hanscom AFB, MA is the Center for research, exploratory, and advanced development in *geophysics*. The laboratory conducts research in *missile* geophysics, *upper atmospheric* and stratospheric operations, *ionospheric* effects, the optical/infrared environment, *meteorology* and the space environment.

The Air Force Weapons Laboratory at Kirkland AFB, NM conducts research and development programs in *laser* and *particle beam* technology, *nuclear weapons* effects and safety, and nuclear survivability/vulnerability. A leader in laser science and engineering since the 1960s, AFWL operates two laser test ranges, an airborne laser

laboratory, and optics development and fabrication facilities. AFWL is the Systems Command focal point for the technical aspects of nuclear safety. This laboratory also assesses the ability of *aircraft, missiles,* and electronics to withstand *nuclear explosion* effects. The laboratory simulates *radiation,* blast and shock effects, and *electromagnetic* pulses from nuclear explosions at its home base and elsewhere. To support its weapons programs, AFWL has the largest scientific computation capability in the Department of Defense.

The Air Force Human Resources Laboratory [AFHRL], located at Brooks AFB, TX, manages and conducts research and advanced development programs in simulation, education, training, and personnel use technology. AFHRL provides an in-house research capability for the Air Force Manpower and Personnel Center. The laboratory's Operational Training Division is located at Williams AFB, AZ, with an operating location at Luke AFB, AZ. Its Logistics and Maintenance Research Division is located at Wright-Patterson AFB, OH, with an operating location at Lowery AFB, CO.

The Air Force Rocket Propulsion Laboratory at Edwards AFB, CA plans and conducts research, exploratory and advanced development to provide timely *rocket propulsion* technology options for Air Force systems. The AFRPL program comprises over half of the national investment in rocket propulsion technology. Doing basic research, AFRPL provides materials, data and analytical tools for exploratory development programs. This area includes synthesis of *propellant* mechanics, combustion, characterization of rocket exhaust plumes and *solid rocket nozzle* material studies. Exploratory development comprises the major portion of the AFRPL program. Its major thrusts include reducing development risks for the M-X Intercontinental Ballistic Missile, providing propulsion options for air launched tactical missiles, and demonstrating higher performance space motors and *satellite thrusters.* Among the many systems which are using AFRPL's technology are missiles such as *Minuteman, Titan,* Sidewinder and the Short Range Attack

Missile and Air Force communications and surveillance *satellites.* Work has also been done for NASA's *Apollo* and *Space Shuttle* programs.

There are other organizations in AFSC which are not assigned to the Director of Laboratories, but nonetheless also contribute to the Systems Command's overall technological base. These organizations include:

1. Aerospace Medical Division [AMD]
2. Armament Division [AD]
3. Aeronautical Systems Division [ASD]
4. Foreign Technology Division [FTD]
5. Electronic Systems Division [ESD]
6. Rome Air Development Center [RADC]
7. Space Division [SD]
8. Air Force Flight Test Center [AFFTC]
9. Arnold Engineering Development Center [AEDC] and
10. Air Force Contract Management Division [AFCMD]

The Aerospace Medical Division at Brooks AFB Texas includes four major organizations which carry out specialized biomedical activities: The Air Force Aerospace Medical Research Laboratory, Wright-Patterson AFB, OH; the USAF School of Aerospace Medicine and the USAF Occupational and Environmental Health Laboratory, both at Brooks AFB, TX; and the Wilford Hall Medical Center, Lackland AFB, TX. Air Force *biotechnology* research and development, world-wide support in epidemiology, flight medicine, environmental health, biomedical education, and comprehensive medical care are activities of these organizations. AMD coordinates medical research and development with AFSC product divisions and other laboratories. It also interfaces with mutually supportive activities in research and development, field support, education, and clinical investigations.

The USAF School of Aerospace Medicine, located at Brooks AFB TX, conducts biotechnology research and development, medical evaluation and education, consultation, and aeromedical support for the U.S. Air Force. Investigations encompass laboratory and clinical studies in all areas of physiological, environmental and dynamic conditions which may affect the health and performance of Air Force personnel under a variety of operational circumstances. It also provides medical evaluations for flying personnel. Consultation service is available to all military services and members of certain allied nations. The school also provides worldwide epidemiological consultation, a reference/referral clinical laboratory service, and several full-time residency programs in *aerospace medicine* for its staff.

The Air Force Aerospace Medical Research Laboratory at Wright-Patterson AFB, OH specializes in theoretical and experimental biotechnology research and development in biodynamics, human engineering, combined aerospace stress effects, and toxic hazards. It is a center of excellence for noise research. Facilities include the Dynamic Environment Simulator, for air combat stress and performance modeling; and a 74 m (250-ft) indoor track that has both a high and low-G deceleration-sled capability for development of personnel impact protection systems.

The Wilford Hall USAF Medical Center is a 1,000-bed hospital and medical complex at Lackland AFB, Texas. This hospital and complex of more that 60 buildings is one of six medical centers in the Air Force and one of the largest in the Department of Defense. A professional staff of more than 3,400 provides care to nearly one million patients a year. Wilford Hall also provides over half of all post-graduate medical education in the Air Force. Residency fellowship training is offered in most medical specialties and subspecialties. The Center also has a clinical research mission. Its research supports clinical investigations with specialized equipment, facilities and assistance not available in the usual hospital organization. There are approximately 100 active research protocols ongoing during any one year. These include investigations in cancer chemotherapy, dentistry, surgery, anesthesiology, internal medicine, dermatology, pediatrics, psychiatry and drugs. Forty-five clinical specialties are available at the center.

The USAF Occupational and Environmental Health Laboratory is located at Brooks AFB, TX. This lab provides professional consultation, specialized laboratory services and operational field support to assist the Air Force in managing occupational, radiological and environmental health programs. The Occupational and Environmental Health Laboratory uses a multidisciplinary approach in providing consultation to Air Force organizations concerned with environmental problems common to all.

The System Command's Armament Division [AD] is located at Eglin AFB, FL. The Division's primary mission is to develop, test, and initially acquire all nonnuclear air armament for the Air Force's tactical and strategic forces. This mission encompasses the spectrum of research technology and development planning to acquisition of armament. Development activities that advance air armament science and technology are conducted in four phases: basic research, exploratory, advance, and engineering development. AD performs the air armament acquisition process from conceptual planning to initial production of military *hardware*. This hardware, developed and produced under the management of AD, fulfills the operational armament needs of the tactical and strategic arms of the Air Force. Among items developed, tested, and initially acquired by AD are air-launched tactical and air defense *missiles,* guided weapons, *aircraft* guns and ammunition, targets, and related armament support equipment.

The Armament Division is extensively involved in the test and evaluation of air armament. Its capabilities in this area include test and evaluation of *electromagnetic* warfare systems, intrustion interdiction systems, and *inertial* navigation systems to name a few. In addition to other AD activities, the Division operates, through its 6585th Test Group at Holloman AFB, NM, the 15,200 m (50,000-foot) precision rocket sled track. AD operates more than 1920 sq km (720 square miles) of land test ranges and facilities and more than 114,000 sq km (44,000 square miles) of test area in the Gulf of Mexico.

The Air Force Armament Laboratory [AFATL] at Eglin AFB, Florida is the principal Air Force laboratory performing research and development of *free fall* scorers. Its efforts provide the future technologi-

cal base for aircraft armaments. These include chemical and fuel-air explosives, energy sources and conversions, electronic and mechanical devices, bombs, dispensers, fuzes, flares, guns and ammunition. In addition to its development responsibilities, AFATL maintains a technical capability and provides consulting services in nonnuclear munition safety, aircraft munition compatibility and analysis, and prediction of weapon effects. The scope of AFATL activities include technical support and consulting services to system developers, government agencies, and to joint international cooperation, standardization, and development efforts. AFATL is organizationally assigned to the Armament Division at Eglin AFB, FL.

The Aeronautical Systems Division [ASD] is located at Wright-Patterson AFB, Ohio. Management control for the development and acquisition of *aeronautical* systems is ASD's mission. Keeping pace with today's rapidly expanding technology, ASD manages funds for *aircraft, missiles,* and related equipment, including research and development efforts covering a multitude of programs and projects. In addition, ASD manages a substantial amount for foreign military sales. Typical of the wide range of systems presently under ASD management are strategic programs, including B-52 offensive *avionics* and integration of the Air Launched Cruise Missile with the B-52; the F-15 Air Superiority fighter; the F-16 (See Fig 2), developed as a multi-role complement to the more sophisticated F-15; the A-10 aircraft; speciically designed for the close air support mission; remotely piloted vehicles (RPVs) to complement manned aircraft by providing low cost systems in a variety of roles; and the Maverick missile, an air-to-surface weapon to knock out enemy tanks, armored vehicles and field fortifications. ASD's many other efforts include developing and acquiring training *simulators,* reconnaissance, strike and electronic warfare systems, air-to-air and air-to-surface missiles, and airlift and tanker aircraft.

The 4950th Test Wing at Wright-Patterson AFB, OH continues the pioneering spirit started by Orville and Wilbur

Wright in the birthplace of aviation; it has 43 aircraft. Many of the Wing's aircraft are flying laboratories and conduct flight testing of electrical countermeasures, environmental research, advanced navigation, satellite communication, new radar, electro-optical, and surface-to-air detection systems, *telemetry* data for missile and *satellite* programs. In addition, the Wing manages the largest precision measurement equipment laboratory within the Air Force and is the major aircraft modification center for AFSC. This expansive organization has the engineers, craftsmen and facilities to design, fabricate and install temporary research and development modifications on test aircraft. Experienced craftsmen build prototype and experimental equipment to support laboratory research and development.

Also at Wright-Patterson AFB, OH is the Foreign Technology Division [FTD]. To prevent possible technological surprise by a potential enemy, FTD acquires, evaluates, analyzes and disseminates foreign aerospace technology to other units. This is accomplished in concert with other divisions, laboratories and centers. Information collected from a wide variety of sources is screened in unique data and laboratory processing equipment. Then scientific and technical specialists analyze it and prepare reports, studies and technical findings as well as assessments of possible hostile technological or operational environments in which Air Force weapon systems can be expected to operate. FTD provides foreign technology support for aerospace and other systems research and development activities.

Located at Hanscom AFB, MA, the Electronic Systems Division [ESD] is responsible for development, acquisition and delivery of electronic systems and equipment for the command, control and communications functions of aerospace forces. These systems take many forms such as a joint U.S.—Canadian network of combined civilian-military radar sites which simultaneously control civil air traffic and insure air sovereignty; a major updating of the underground North American Air Defense Command [NORAD] combat operations center; long-range radars on both the east and west coasts to warn of missile and air-

craft attack; satellite communications terminals for ground, mobile and aircraft use; and a new airborne radar-and-communications post which can give the Air Force an instant air defense and tactical control system anywhere in the world. All of ESD's projects share a common characteristic— helping the modern military commander use his forces effectively.

The Rome Air Development Center [RADC] at Griffiss AFB, New York is the principal organization charged with Air Force research and development programs related to command, control, communications and intelligence. Reporting to Electronic Systems Division [ESD] at Hanscom AFB, RADC is responsible for advancing this technology and also demonstrating selected systems and subsystems in the areas of intelligence, reconnaissance, mapping and charting, and command, control and communications.

Among the Center's unique resources are the nation's most versatile high-power laboratory, a *digital* communications test facility, an experimental cartographic (i.e. map making) facility, and exotic equipment for *reliability* studies.

The Space Division [SD] at Los Angeles Air Force Station, California manages the research, development, acquisition, launch and on-orbit command and control of Department of Defense space systems. SD is also the focal point within DOD for plans and activities associated with *NASA's Space Transportation System* (Space Shuttle) (Fig 3) program. From its Los Angeles headquarters and through its worldwide field units, the Space Division is responsible for:

• Developing the *spacecraft, launch vehicles* and *ground support equipment* to maintain and improve military space capabilites

• Launching, orbiting, commanding and controlling satellites for DOD and other government agencies

• Identifying and developing space systems concepts and technological alternatives to satisfy critical military needs five to ten years in the future

• Managing operations of the Space and Missile Test Organization, the Air Force Satellite Control Facility, and the Manned Space Flight Support Group

Space Division activities are managed by technical program offices for defense meteorological satellites, space navigation systems, technology, space communications, launch vehicles (including the Space Shuttle), space defense systems, the satellite data system, and defense support systems. [See: U.S. Air Force Role In Space]

The Air Force Satellite Control Facility [AFSCF], located at Sunnyvale Air Force Station, California, operates a worldwide network of tracking stations to perform on-orbit tracking, data acquisition, and command and control of DOD *space vehicles.*

The Manned Space Flight Support Group, located at the *Johnson Space Center* near Houston, was activated in June 1979. The group and NASA work together on mission planning for *Space Shuttle flights* that will carry military *payloads.* Air Force personnel are being trained to participate in joint NASA-DOD Shuttle operations.

The Space and Missile Test Organization [SAMTO], at *Vandenberg AFB,* California, provides test management for all DOD-directed *ballistic* and space programs, operates the *Eastern and Western Space and Missile Centers,* and conducts launch operations at both Vandenburg AFB and *Cape Canaveral AFS, FL.* SAMTO activities support Space Division test programs and those of the Ballistic Missile Office, and other military and governmental agencies. The following Centers comprise the Space and Missile Test Organization: Western Space and Missile Center; Eastern Space and Missile Center.

The Western Space and Missile Center [WSMC] at Vandenberg AFB, CA conducts *polar-orbit* space launch operations employing *Titan III* and *Atlas* space boosters (See: Fig. 4). It will also manage Space Transportation System (Space Shuttle) operations scheduled to begin in 1984 at Vandenberg. Launching Minuteman I and III missiles, the Center conducts research and development activities for Air Force

ballistic missile systems and *reentry vehicles.* Tests of the advanced *ICBM* missile, M-X, will be made at Vandenberg AFB. WSMC operates the Western Test Range which stretches halfway around the world from the California coast to the Indian Ocean where it joins the Eastern Test Range. The range supports ballistic, space, and aeronautical tests.

The *Eastern Space and Missile Center* [ESMC], located at Patrick AFB, FL, launches the Titan III space booster and represents DOD interests in Space Shuttle operations from *Kennedy Space Center.* ESMC also operates Patrick AFB which hosts a work force of over 13,000 people and approximately 50 tenants from nearly every major command, the other services, and major governmental agencies. Tenants' missions range from equal opportunity management instruction to geodetic survey from flying training to nuclear test ban monitoring and from weather rocket training to air rescue and recovery. ESMC operates the Eastern Test Range which stretches halfway around the world from the Florida coast into the Indian Ocean where it joins the Western Test Range, forming a worldwide network. Tracking and data-gathering stations are located on the islands of Grand Bahama, Grand Turk, Antigua, and Ascension; other locations support the Range's Pacific operations.

The Ballistic Missile Office [BMO] at Norton AFB, CA is responsible for planning, implementing and managing Air Force ballistic missile systems and subsystems programs. BMO also carries out systems development functions for tri-service (i.e., Army, Navy, and Air Force) advanced ballistic reentry requirements. In addition, BMO provides for alteration of existing intercontinental ballistic missile (ICBM) sites and launch silos and control centers for some Minuteman ICBMs, which it initially developed more than 15 years ago. BMO is managing the development of the M-X system, a new survivable land based ICBM that is expected to be deployed initially in 1986. Current plans call for maintaining survivability for the M-X system by hiding 200 missiles and launchers among 4,600 hardened shelters without revealing actual missile location. The ICBM acquisition approach differs from many other program approaches because BMO acts as "prime contractor" working with more than 25 associated contractors to develop the missile and basing components of the M-X system.

Located at Edwards AFB, CA the Air Force Flight Test Center [AFFTC] plans, conducts, supports and evaluates tests of manned and unmanned aircraft and aerospace research vehicles to include performance, flying, qualities, and subsystem performance, *reliability, maintainability,* and functional capability under climatic extremes.

The Center:
● conducts development testing of advanced and special mission parachutes
● tests and evaluates remotely piloted vehicle (RPV) mid-air recovery systems
● operates the USAF Test Pilot School
● conducts research and development programs for improvement and modernization of its test ranges, facilities, and test techniques
● operates and maintains ranges, instrumentation, and special technical support facilities needed for its mission.

Located approximately 160 km (100 miles) northeast of Los Angeles on the western edge of the Mojave Desert, AFFTC boasts one of the largest ground complexes 1,220 ha (301,000 acres), in the Air Force, and has at its immediate disposal over 39,000 sq km (15,000 sq. mi) of restricted *airspace* in which to conduct its test activities. Within the Edwards reservation are 168 sq km (65 sq. mi) of usable landing area on two dry lake beds with runway lengths up to 12 km (7.5 mi). This complements the 4,600 m (15,000 ft) main runway. Additionally, ideal flying weather and a semi-isolated location combine to afford an environment conducive to safe, highly successful flight test efforts. Major projects currently under evaluation include the Air-Launched Cruise Missile, F-15, F-16, A-10, and B-1 aircraft.

Personnel and facilities of the Air Force Flight Test Center played key roles in the *Approach and Landing Tests* of the Space Shuttle *Orbiter, "Enterprise"* (See: Fig 5).

Edwards AFB is serving as the landing site for the first series of Shuttle orbital flights [*See:* STS-1] and will serve as an alternate landing site for subsequent operational flights.

Collocated at AFFTC are *NASA's Dryden Flight Research Center,* the Air Force Rocket Propulsion Laboratory, the U.S. Army Aviation Engineering Flight Activity, and approximately 60 military tenant and civilian contractor agencies. AFFTC also manages the Utah Test and Training Range [UTTR], a major test facility in northwestern Utah.

The Arnold Engineering Development Center [AEDC] at Arnold AFS, TN operates the most advanced and largest complex of *aerospace flight simulation* test facilities in the world. Its mission is to insure that aerospace *hardware*—aircraft, missiles, spacecraft, *jet* and *rocket propulsion systems,* and other components—will work right, the first time they fly. AEDC operates some 40 test units in which flight conditions can be simulated from sea level to altitudes of 1600 km (1,000 miles) and from *subsonic* velocities to more than 32,000 km/h (20,000 mph). Equipment being tested ranges in size from small-scale models to full-scale vehicles with propulsion systems installed and operating. Some engineering development work for virtually every major U.S. aerospace system has been supported by tests at the Center. In addition, a number of unexpected problems encountered in the operational use of systems have been quickly and economically solved. Tests are conducted for the military, NASA, aerospace industry contractors, and other federal agencies. Construction of the Aeropropulsion Systems Test Facility [ASTF] at Arnold Center began in 1977. Valued at $437 million, ASTF is scheduled to be operational in the 1980s.

The Air Force Contract Management Division [AFCMD] at Kirtland AFB, NM is responsible for DOD contract management activities in those plants assigned to the Air Force under the DOD National Plant Cognizance Program. AFCMD evaluates contractor performance and manages the administration of contractor performance and manages the administration of contracts executed by Air Force, Army, Navy Defense Supply Agency, NASA and other government purchasing agencies when required. Operating from a variety of field locations throughout the United States and Europe, AFCMD has government responsibility for 19 major defense contractors. (See: Fig. 6) It furnishes technical and management direction for contract administration, production and manufacturing improvements, quality assurance, transportation, safety and flight operations.

universal hatches   D-shaped hatches in the *airlock* that allow the airlock to be mounted in the *Orbiter* cabin or in the *cargo bay.*

universal time (UT)   Time defined by the rotation of the *Earth* and determined from the apparent daily motions that reflect this rotation; also called *"Greenwich mean time."*
See: Orbiter displays and controls

unmanned (Orbiter) captive flights
See: Approach and Landing Test(s)

unstable isotope   A *radioisotope.* Compare this term with *stable isotope.*

upcomer   *Nozzle* tube in which *coolant* flows in a direction to that of the exhaust gas flow.

uplink data   Information that is passed from a *ground station* on *Earth* to a *spacecraft, space probe* or *space platform.*

upper air   A term in synoptic meteorology and weather observation that describes the portion of the *atmosphere* above the lower troposphere.
See: atmosphere

upper air observation   A measurement of atmospheric conditions aloft, above the effective range of surface weather observation. Also called "upper air sounding" or "sounding."

upper air sounding
See: upper air observation

upper atmosphere   The general term applied to the outer layers of the *Earth*'s atmosphere above about 30 km (18.6 mi). It includes a portion of the stratosphere, the mesosphere, the thermosphere and the exosphere. Rocket probes and Earth-viewing satellites are frequently used to gather scientific data about this region.

upper stage   1. In general the second, third or later stage in a *multistage rocket.*

2. In the *Space Transportation System* the *Spinning Solid Upper Stage* or *Inertial Upper Stage.* Both of these upper stage systems are designed for launch from the *Orbiter cargo bay* and have propulsion systems capable of delivering *payloads* into *orbits* and *trajectories* beyond the (altitude) capabilities of the Orbiter vehicle alone.

upweight   Launch weight. The term frequently refers specifically to *payloads* and all supporting *items* they require.

uranium [Symbol: U]   A radioactive element with the atomic number 92 and, as found in natural ores, an average *atomic weight* of approximately 238. Its two principal natural isotopes are uranium-235, or ^{235}U (0.7% of *natural uranium*), which is *fissionable,* and uranium-238, or ^{238}U (99.3% of natural uranium), which is *fertile.* Natural uranium also includes a minute amount of uranium-234, or ^{234}U. Uranium is the basic raw material of nuclear energy.

uranium hexafluoride [Symbol: UF$_6$]   A volatile compound of *uranium* and fluorine. UF$_6$ gas is the process fluid in the *gaseous diffusion process.*

uranium series (sequence)   The series of *nuclides* resulting from the *radioactive decay* of uranium-238, also known as the *uranium*-radium series. The end product of the series is lead-206. Many man-made nuclides decay into this sequence.

uranium tetraflouride [Symbol: UF$_4$]   A solid green compound called "green salt." It is an intermediate product in the production of *uranium hexafluoride.*

uranium trioxide [Symbol: UO$_3$]   An intermediate product in the refining of *uranium,* also called "orange oxide."

V

vacuum flask
See: Dewar flask

vacuum gage   An instrument for measuring *pressure* that is lower than *atmospheric pressure.*
See: pressure

vacuum pump A device that supports a flow of gas in a *vacuum system*. Some of the more common types of vacuum pumps are the mechanical pump, the *diffusion* or vapor pump and the *cryopump*.

vacuum system A chamber that has walls capable of withstanding *atmospheric pressure* when evacuated. It has an opening through which the gas within can be removed by means of a pipe or manifold leading to a pumping system.

vacuum tube An *electron tube* evacuated to such an extent that its electrical characteristics are essentially unaffected by the presence of residual gas or *vapor.*

valve Mechanical device by which the flow of *fluid* may be started, stopped, or regulated by a movable part that opens, closes, or partially obstructs a passageway or port in a containing structure, the valve housing.

valve-type disconnect Quick-disconnect coupling that includes valve elements for sealing purposes at separation.

valving element The movable portion of a valving unit that translates or rotates to vary flow or to shut off the flow of liquid.

valving unit The combination of the *valving element* and the valve seat contained in a suitable housing.

Van de Graaff generator (accelerator) An electrostatic machine in which electrically charged particles are sprayed on a moving belt and carried by it to build up a high potential difference on an insulated terminal. Charged particles are then accelerated along a discharge path through a vacuum tube by this potential difference between the insulated terminal and the opposite end of the machine. Named after R.S. Van de Graaff, who invented the device in 1931, a Van de Graaff *accelerator* is often used to inject particles into larger accelerators.

Van der Waal equation (of state) One of the best known of the many laws that have been proposed to describe the *thermodynamic* behavior of real gases and their departures from the *ideal gas law.* It states:

$$[p + (a/v^2)] (v - b) = R_u T$$

where a and b are constants dependent upon the particular gas; p is the *pressure* of the gas; v is its *specific volume* (measured in units of specific volume at normal temperature and pressure); R_u is the *universal gas constant;* and T is the *absolute temperature* (K). This equation (unlike the ideal gas equation of state) allows for the finite size of the *molecules* (the constant b, which compensates for the volume of the molecules) and for the attractive forces between the molecules (the constant a).

Vanguard Early U.S. three-stage rocket, standing 21.6 m (70.8 ft.) high and weighing more than 10,000 kg. (20,000 lbs.). First launched in March 1958, it eventually orbited three satellites.

Vanguard Project Program, launched in 1958, to study the shape, surface and interior of the Earth. Information from the three satellites later revealed the pear shape of the Earth and deepened understanding of field and radiation belts and the space environment around the planet.

vapor A gas whose temperature is below its *critical temperature,* and which can therefore be condensed to the liquid or solid state by the increase of *pressure* alone.

vapor honing Method of eroding (or cleaning) a metal surface by blasting a fine erosive material against the surface, a vaporized fluid being used as a carrier for the material.

vapor pressure 1. The *pressure* exerted by the *molecules* of a given *vapor.* For a *pure substance* confined within a container, it is the vapor's pressure on the walls of its containing vessel; for a vapor mixed with other vapors or gases, it is that particular vapor's contribution to the total pressure (i.e., its *partial pressure*). In *meteorology* the term "vapor pressure" is generally used to denote the partial pressure of water vapor in the *atmosphere.*
2. The sum of the partial pressures of all the vapors in a system.

variable-area exhaust nozzle An exhaust *nozzle* on a *jet engine* or *rocket engine* that has an exhaust opening that can be varied in area by means of some mechanical device, thereby permitting variation in the exhaust *velocity.* Compare this term with *"fixed-area exhaust nozzle."*

variance [Symbol: σ^2] In statistics a measure of variability or spread; the mean-square deviation from the *mean;* that is, the mean of the squares of the differences between individual values of x and mean value μ.

$$\sigma^2 \equiv E[(x-\mu)^2] \equiv E(x^2) - \mu^2$$

where E denotes the expect value. The positive square root (σ) of the variance is called the "standard deviation."

vector   Any quantity, such as *force, velocity,* or *acceleration,* that has both magnitude and direction at each point in space, as opposed to a *scalar,* which has magnitude only. Such a quantity may be represented geometrically by an arrow of length proportional to its magnitude, pointing in the assigned direction. A *unit vector* is a vector of unit length; in particular, the three unit vectors along the positive X-, Y- and Z-axes of rectangular *Cartesian coordinates* are denoted respectively by $\bar{i}$, $\bar{j}$ and $\bar{k}$. Any vector $\bar{A}$ can be represented in terms of its components a_1, a_2 and a_3 along the coordinate axes X, Y and Z, respectively. That is:

$$\bar{A} = a_1\bar{i} + a_2\bar{j} + a_3\bar{k}$$

A vector drawn from a fixed origin to a given point (X, Y, Z) is called a *position vector* and is frequently symbolized by $\bar{r}$. The position vector in rectangular Cartesian coordinates is:

$$\bar{r} = x\bar{i} + yj + z\bar{k}$$

Equations written in vector form are valid in any coordinate system.

vector product   A *vector* whose magnitude is equal to the product of the magnitudes of any two given vectors and the sine of the angle between their positive directions. For two vectors $\bar{A}$ and $\bar{B}$, the vector product is often written $\bar{A} \times \bar{B}$ (which is read "$\bar{A}$ cross $\bar{B}$) and defines a vector perpendicular to both $\bar{A}$ and $\bar{B}$ and so directed that a right-hand rotation about $\bar{A} \times \bar{B}$ through an angle of not more than 180° carries $\bar{A}$ into $\bar{B}$. The magnitude of $\bar{A} \times \bar{B}$ is equal to twice the area of the triangle of which $\bar{A}$ and $\bar{B}$ are coterminous (having a common boundary at some point) sides. If the vector product is zero, one of the vectors is zero, or else the two are parallel. When $\bar{A}$ and $\bar{B}$ are written in terms of their two components along the X-, Y- and Z-axes of the rectangular *Cartesian coordinates,* i.e.,

$$\bar{A} = a_1\bar{i} + a_2\bar{j} + a_3\bar{k}$$
$$\bar{B} = b_1\bar{i} + b_2\bar{j} + b_3\bar{k}$$

then the vector product is the determinant

$$\bar{A} \times \bar{B} = -\bar{B} \times \bar{A} = \begin{vmatrix} \bar{i} & \bar{j} & \bar{k} \\ a_1 & a_2 & a_3 \\ b_1 & b_2 & b_3 \end{vmatrix}$$

The vector product is also called the "cross product."

vector steering   A steering method for *rockets* and *spacecraft* in which one or more *thrust chambers* are gimbal-mounted (gimbaled) so that the direction of the *thrust* (i.e., the thrust *vector*) may be tilted in relation to the *center of gravity* of the vehicle to produce a turning movement.
See: Space Shuttle Main Engines

vehicle   In general an *aerospace* structure, machine or device (such as a *rocket*) that is designed to carry a *payload* through the *atmosphere* and/or space; more specifically a rocket vehicle.
See: launch vehicle, Space Shuttle

Vehicle Building *(*abbr. VAB*)*
See: STS launch facilities

vehicle mass ratio   The ratio of the final mass of a vehicle (M_f), after all the propellant has been expended, to the initial (fully loaded) mass (M_o):

$$\text{Vehicle Mass Ratio} = M_f/M_o$$

The inverse ratio (M_o/M_f) is also sometimes called the mass ratio.

vehicle tank   Tank that serves both as a primary integral structure of a *vehicle* and as a container of *pressurized propellants.*

Velcro   The trade name for a hook-and-pile fastener, generally of nylon, used to replace zippers in some apparel. It is composed of two types of plastic cloth that stick to each other. One type has many fine, hooked spines protruding (sticking out) from the surface; the other type has many narrow holes or crevices in the surface. When the spines are pressed into the holes, they stick there.

velocity   A *vector* quantity that describes the rate of change of position. Velocity has both magnitude (speed) and direction, and it is expressed in terms of units of length per unit of time (e.g., meters per second).

velocity head   Pressure (i.e., head) of a fluid due to the speed of flow, i.e., the difference between *total pressure* and *static pressure*. *See:* pressure.

velocity of light (symbol c)   The velocity of light in a vacuum (free space) is 2.997925×10^8 meters per second.

velocity slip   *Velocity* of the gas *molecules* next to the *nozzle* wall, a result of *rarefaction* of the exhaust gas.

Venera probes   Series of successful Soviet probes to Venus begun in 1961.

vent-and-relief valve   Specialized version of a *relief valve* wherein the assembly acts as an outlet for *ullage* vapor during the filling of a *tank* (e.g., propellant tank) and then performs as a relief valve during operation.

venturi tube   A short tube of having a smaller diameter in the middle than at the ends that is used for measuring the quantity of fluid flowing through a pipe. When a fluid flows through such a tube, the pressure decreases as the diameter becomes smaller, the amount of the decrease being proportional to the speed of flow and the amount of restriction.

vent valve   Pressure-relieving shutoff valve that is operated on external command, as contrasted to a relief valve, which opens automatically when *pressure* reaches a given level.

Venus   Second major planet from the Sun, with a diameter of 12,102 km. Although closest to the Earth, Venus differs starkly from our planet. Its atmosphere, nearly 100 times denser than Earth's, is 97% carbon dioxide. There are three cloud layers. The carbon dioxide and water vapor in the atmosphere create a greenhouse effect that results in a buildup of heat on the surface to a temperature of 455°C (850°F). Venus's extreme heat, poisonous and corrosive atmosphere, crushing surface air pressure, and relative absence of water, make the possibility of finding life on the planet remote.

vernier engine   A *rocket engine* of small *thrust* used primarily to obtain a fine adjustment in the *velocity* and *trajectory* or in the *attitude* of a *rocket* or *aerospace vehicle*. *See:* reaction control system

Vertical Assembly Building (VAB)   The high-bay building near the *Kennedy Space Center launch pad* in which the Shuttle ele-

ments are stacked onto the *mobile launch platform*. The VAB is also used for the vertical storage of the *External Tanks*.

vertical stabilizer   A structural component of an *aerodynamic* vehicle consisting of a fin and rudder assembly. In the case of the *Space Shuttle Orbiter*, the vertical stabilizer consists of a structural fin, the rudder/speed-brake (the rudder splits in half for speed-brake control) and the systems for positioning the rudder/speed-brake control surface. *See:* Orbiter structure

vertigo   The sensation that the outer world is revolving about the patient (objective vertigo) or that he himself is moving in space (subjective vertigo). The word frequently is used erroneously as a synonym for dizziness or giddiness to indicate an unpleasant sensation of disturbed relations to surrounding objects in space.

vestibula   The organ in the ear that provides a sense of "which way is down" (along the force of *gravity*). It is thus responsible for the sense of balance—the "vestibular sense."

vibration   1. Motion due to a continuous change in the magnitude of a given *force* that reverses its direction with time.

Vibration is generally interpreted as the cyclical (symmetrical or nonsymmetrical) fluctuations in the rate at which an object accelerates.

In longitudinal vibration, the direction of motion of the *particles* is the same as the direction of advance of the vibratory motion; in transverce vibration, it is perpendicular to the direction of advance.

2. The motion of an oscillating body during one complete *cycle; oscillations.*

Vickers hardness   Indentation-hardness test in which the indenter is a diamond cone of a specified angle between opposite faces; the load, test duration, and rate of descent of the indenter are specified.

video   Pertaining to the picture signals in a television system or to the information-carrying signals which are eventually presented on the *cathode-ray tubes*.

video display   A television-like display of data (pictures).

Viking, Project   The culmination of a series of missions to explore Mars. Viking was designed to orbit Mars and land and operate on its surface. Two identical spacecraft were launched. Each consisted of an orbiter carrying a sterilized lander. The two landers were the first unmanned spacecraft ever to operate on the surface of another planet for more than a few minutes. Viking 1 reached Mars orbit in June 1976; Viking 2 began orbiting Mars in August 1976. The first landing on Mars occurred in July 1976. At the time of their launches, the Viking spacecraft were the most complex planetary spacecraft developed. Each orbiter has operated far beyond its design lifetime with few anomalies or failures. Viking Orbiter 1 exceeded four years of active flight operations in Mars orbit.

virtual mass   *Mass* of *fluid* near a moving body (e.g., a blade) that moves with the body and thereby increases the effective mass in motion.

viscosity   *Fluid* resistance to flow caused by internal molecular attraction.

viscous flow   The flow of *fluid* through a *duct* under conditions such that the *mean free path* is very small in comparison with the smallest dimension of a transverse section of the duct.

This flow may be either *laminar* or *turbulent*.

viscous fluid   A fluid whose molecular *viscosity* is sufficiently large to make the *viscous forces* a significant part of the total force field in the fluid.

viscous force   The *force* per unit volume or per unit mass arising from the action of tangential *stresses* in a moving *viscous fluid*. This force may then be introduced as a term in the equations of motion.

viscous stresses   The components of the *stress tensor* when the *pressure*, i.e., the mean of the three normal stresses, has been subtracted out from each of the normal stresses.

See: Reynolds stresses

visible radiation   *Electromagnetic radiation* lying within the *wavelength* interval to which the human eye is sensitive, the *spectral* interval from approximately 0.4 to 0.7 micrometers (4000 to 7000 *angstroms*).

The term is without reference to the variable response of the human eye in its reception of radiation.

visible spectrum   That portion of the *electromagnetic spectrum* occupied by the wavelengths of *visible radiation*, roughly 0.4 to 0.7 micrometers (4000 to 70000 angstroms). This portion of the electromagnetic spectrum is bounded on the shortwavelength end by *ultraviolet radiation*, and on the long-wavelength end by *infrared radiation*.

visual magnitude   The *apparent magnitude* of a *star* or other *celestial body* measured by visual observation.

void   Air bubble in a cured *propellant* grain or in a rocket motor insulation.

volumetric efficiency   Measure of the desirability of a given design for an expulsion device ratio of loaded liquid volume to internal tank volume.

vortex   Any flow possessing *vorticity*.

vorticity   A *vector* measure of local *rotation* in a *fluid* flow, defined mathematically as the *curl* of the velocity vector,

$$\Omega = \nabla \times V$$

where Ω is the vorticity; V is the *velocity*; and ∇ is the *del-operator*.

The vorticity component *normal* to a small plane element is the limit of the circulation per unit area as the area of the element approaches zero.

The vorticity of a solid rotation is twice the *angular velocity* vector.

In *meteorology*, the vorticity usually refers to the vertical component of the vorticity as defined above.

vorticity equation   A dynamic equation for the rate of change of the *vorticity* of a parcel, obtained by taking the *curl* of the *vector* equation of motion.

Vostok   Series of six Soviet manned spacecraft. Vostok, launched in April 1961, carried the first man in space.

Voyager   NASA mission to explore the outer planets. Voyagers 1 and 2 completed successful fly-bys of Jupiter in 1979 and Saturn in 1980 and 1981. Voyager 2 will continue to encounter Uranus in January 1986 and Neptune in August 1989. Traveling above the ecliptic plane, Voyager 1 will not encounter another planet.

W

warhead   Originally the part of a *missile* carrying the explosive, chemical or other charge intended to damage the enemy. By extension, the term is sometimes used as synonomous with *payload* or *nose cone.*

warmant passages   Passages provided in cryogenic-valve hydraulic actuators to maintain actuator temperatures above specified operating minimums under extended hold conditions.

water hammer   Literally, the sound of concussion in a conduit when a flowing liquid is suddenly stopped; more generally, the pressure surge in the system that results from such stoppage.

water vapor   Water (H_2O) in gaseous form.

The amount of water vapor present in a given gas sample may be expressed in a number of ways.

wave   A disturbance that is propagated in a medium in such a manner that at any point in the medium the quantity serving as measure of disturbance is a function of the time, while at any instant the *displacement* at a point is a function of the position of the point.

wave equation   The partial differential equation of the form

$$\nabla^2 \varphi = (1/C^2)\,(\partial^2 \varphi / \partial t^2)$$

Where φ is usually a function of the position and time coordinates; ∇^2 is the Laplacian operator; t is the time; and C^2 is a constant.

The general solution to this equation is any function defined over a plane, the *phase front,* moving perpendicular to itself at the speed. Also called equation of wave motion. *See:* wave

waveguide   A system of material boundaries capable of guiding *electromagnetic* waves.

wavelength [Symbol: λ]   In general, the mean distance between maximums (or minimums) of a roughly periodic pattern. Specifically, the least distance between particles moving in the same *phase* of *oscillation* in a *wave* disturbance.

The wavelength is measured along the direction of propagation of the wave, usually from the midpoint of a crest (or trough). to the midpoint of the next crest (or trough). It is related to frequency (μ) and phase speed (c) by

$$\lambda = c/\mu$$

where λ is wavelength. The reciprocal of wavelength is the *wave number.*

wave mechanics   A form of *quantum mechanics.*

wave motion   The oscillatory motion of the particles of a medium caused by the passage of a *wave,* produced by *forces* external to the medium, but propagated through the medium by internal forces. Wave motion per se involves no net translation of the medium.

Various types of *oscillation* are found in natural wave motions. Among the simplest are the linear oscillation parallel to the direction of propagation of a longitudinal wave, the linear oscillation perpendicular to the direction of propagation of a transverse wave, and the orbital motion produced by the passage of a progressive gravity wave.

wave number (symbol v)   The reciprocal of *wavelength;* the number of waves per unit distance of propagation.

The wave number is usually expressed in reciprocal centimeters (cm^{-1}) or reciprocal meters (m^{-1}).

$$\bar{\mu} = 1/\lambda$$

wear rate   Amount of surface wear in a designated time period.

web   Minimum thickness of a solid *propellant grain* from the initial ignition surface to the insulated case wall or to the intersection of another burning surface at the time when the burning surface undergoes a major change; for an end-burning grain, the length of the grain.

weber [Symbol: Wb]   The *SI unit* of *magnetic flux.* It is defined as the *flux* that produces an *electromotive force* of one volt, when, linking a circuit of one turn, the flux is reduced to zero at a uniform rate in one second. Named in honor of Wilhelm Weber, a German physicist (1804-1981) who studied magnetism.

Weber-Fechner law   An approximate psycho-physical law relating the degree of response or sensation of a sense organ and the intensity of the stimulus. The law asserts that equal increments of sensation are associated with equal increments of the logarithm of the stimulus, or that the just noticeable difference in any sensation results from a change in the stimulus which bears a constant ratio to the value of the stimulus.

The Weber-Fechner law is applied to the detection of contrast in the problem of visual range, as well as to many other psychophysical problems. Also called Weber law.

web fraction   Ratio of web to grain outer radius.

weight [Symbol: w]   1. The *force* with which a body is attracted toward the *earth*. 2. The product of the *mass* (m) of a body and the *gravitational acceleration* (g) acting on a body. W = mg

In a dynamic situation, the weight can be a multiple of that under resting conditions. Weight also varies on other planets in accordance with their value of gravitational acceleration.

weight flow rate [Symbol: $\dot{w}$]   Mass flow rate ($\dot{m}$) multiplied by gravity (g)

$$\dot{w} = \dot{m}g$$

welding   Joining two or more pieces of metal by applying *heat, pressure,* or both, with or without filler material to produce a localized union through fusion or recrystallization across the *interface.*

wet   To come in contact with, and flow across (a surface, body or area)—said of a fluid.

wet criticality   *Nuclear reactor criticality* achieved with the coolant present. Compare this term with *dry criticality.*

wet cycle   Operation of a flow-control device with propellant or test fluid in the flow passage.

wet emplacement   A *launch pad* that provides a deluge of water for cooling the *flame bucket,* the *rocket engines,* and other equipment during the launch of a *missile.* Compare this term with *flame deflector* and dry *emplacement.*

wettability   Ease with which a fluid will flow over and adhere to a surface (e.g., molten solder on a heated metal surface).

wetting   The spread of a liquid over a solid surface when adhesive force is larger than cohesive force.

whistler   A *radiofrequency electromagnetic signal* sometimes generated by lightning discharges.

white body   A hypothetical body whose surface absorbs no *electromagnetic radiation* of any *wavelength,* i.e., one which exhibits zero *absorptivity* for all wavelengths; an idealization exactly opposite to that of the *black body.*

In nature, no true white bodies are known. Most white pigments possessing high reflectivity for visible radiation are fairly good absorbers in the infrared; hence, they are not white bodies in the sense of the radiation theory. However, the term white body is used for physical objects with respect to a particular wavelength interval. *See also:* gray body

white room   A clean and dust-free room, used for the assembly and repair of precise spacecraft components and devices, such as *gyros.*

whole body counter   A device used to identify and measure the *radiation* in the body (body burden) of human beings and animals; it uses heavy shielding to keep out background radiation and ultrasensitive *scintillation detectors* and electronic equipment.

wick   A group or braid of thin fibers that "sucks up" a liquid if the adhesive force between the fiber and the liquid is greater than the liquid's cohesive force.

window   Any gap in linear continuum, as "atmospheric windows," ranges of wavelengths in the *electromagnetic spectrum* to which the atmosphere is transparent, or "firing windows," intervals of time during which conditions are favorable for launching a *spacecraft* on a specific mission.

wind tunnel   A tubelike structure or passage, sometimes continuous, together with its adjuncts, in which a high-speed movement of air or other gas is produced, as by a fan, and within which objects such as engines or aircraft, airfoils, rockets (or models of these objects), etc., are placed to investigate the *airflow* about them and the *aerodynamic forces* acting upon them. Tunnels are designated by the means used to produce the gas flow, as hot shot tunnel, arc tunnel, blow down tunnel; by the speed range, as supersonic tunnel, hypersonic tunnel; or by the medium used, as plasma tunnel, light gas tunnel.

wing   The *aerodynamic* lifting surface that provides conventional *lift* and control for the *Orbiter* vehicle. The wing consists of the *wing glove;* the intermediate section (including the main landing gear well); the *torque box;* the forward *spar* for mounting the leading-edge structure thermal protection system; and the wing/elevon interface, the elevon seal panels and the *elevons.*
See: Orbiter structure

wing aft spar
See: Orbiter structure

wing box   A structure built of thin walls so as to form a closed box and designed to resist torsional loads; may consist of one or more completely enclosed compartments called cells; also called a *"torque* box."
See: Orbiter structure

wing glove   The (forward) wing glove is an extension of the basic *Orbiter* wing and aerodynamically blends the wing leading edge into the mid fuselage.
See: Orbiter structure

work hardening   An increase in hardness (stiffness) of a metal brought about by subjecting it, at temperatures below the recrystallization range, to stresses great enough to plastically deform the metal; work hardening may occur during fabrication or during use, but in either event the material loses ductility.

working fluid   A *fluid* (gas or liquid) used as the medium for the transfer of *energy* from one part of a *system* to another part.
See: Carnot cycle, Rankine cycle, Brayton cycle

work station   A facility or functional area where organizational level operations and *maintenance* tasks are performed in direct support of a turnaround cycle, or where intermediate and depot level maintenance tasks on *Shuttle* components (or related ground support equipment) are performed.

X

X-band   A *radio-frequency* band from 5,200 to 10,900 megahertz (5.2 to 10.9 *gigahertz*), designated originally for high-frequency radar; now being used in *spacecraft* radio propagation experiments for space communications.

X ray   A penetrating form of *electromagnetic radiation* of very short wavelength (about 0.1 to 100 angstroms, or 0.01 to 10 nanometers) and high proton energy (about 100 electronvolts to 100 kiloelectronvolts). They are emitted either when the inner orbital *electrons* of an *excited atom* return to their normal state (these are characteristic X rays), or when a metal target is bombarded with high speed electrons (these are *bremsstrahlung*). X rays are always nonnuclear in origin. Compare this term with gamma rays.

X-Ray Observatory [abbr. XRO]   A possible NASA astrophysics mission designed to use X-ray instruments to investigate large galactic phenomena, stellar structure and evolution, the nature of the active galaxies, clusters of galaxies and cosmology. The spacecraft will consist of a large instrument module (over 3,000 kg) plus the spacecraft bus propelled by on-board hydrozine. It will be launched by the Space Shuttle using the remote manipulator system. The mission is scheduled to last two years.

X-Ray Timing Explorer [abbr. XTE]   A planned NASA astrophysics mission designed to study the temporal variability in X-ray emitting objects. A large area counter system will be used to study individual objects over long exposure times while an all-sky monitor will investigate remaining strong X-ray emitters simultaneously over longer timescales with decreased sensitivity. The experiment will be launched by the Space Shuttle, using the remote manipulator system, during the mid-to-late 1980s. It will operate for two years.

Y

Z

yaw   Angular motion of a vehicle about a vertical axis through its midpoint or center of gravity and perpendicular to the longitudinal axis.

yield   The total energy released in *nuclear explosion*. It is usually expressed in equivalent tons of TNT (the quantity of TNT required to produce a corresponding amount of energy). Low yield is generally considered to be less than 20 kilotons; low intermediate yield from 20 to 200 kilotons; intermediate yield from 200 kilotons to 1 megaton. There is no standardized term to cover yield from 1 megaton upward.

yield load   Load that must be applied to a structure to cause permanent deformation of a specified amount.

yoke   Cross member used in mechanical devices (e.g., valves) to connect two movable elements with a single driver.

Young modulus   The ratio of normal *stress* within the proportional limit to the corresponding normal *strain*.

Z   The symbol for *atomic number*.

zenith   That point of the *celestial sphere* vertically overhead. The point 180° from the zenith is called the *nadir*.

zero-g   The condition of free fall and *weightlessness*. When there are no forces on objects in a spacecraft, they are "in zero-g."

zero-power reactor   An experimental reactor operated at such low power levels that a coolant is not really needed and little *radioactivity* is produced. Compare this term with *subcritical assembly*.

Zeroth Law of Thermodynamics   Two systems, each in *thermal equilibrium* with a third system, are in thermal equilibrium with each other. This statement is frequently called the "Zeroth Law of Thermodynamics" and is actually an implicit part of the concept of *temperature*. It is important in the field of *thermometry* and in the establishment of *empirical temperature scales*. *See:* thermodynamics, First and Second Laws of Thermodynamics

Zulu time [Symbol: Z]   Time based on Greenwich Mean Time (GMT).

Appendix I

NASA facilities

A brief description of the program responsibilities of NASA's ten principal field centers and the National Space Technology Laboratories follows: (See Fig [1])

Ames Research Center [ARC], Space environmental physics; simulation techniques; gas dynamics at extreme speeds; configuration, stability, structures, and guidance and control of *aeronautical* and space vehicles; biomedical and biophysical research.

Hugh L. Dryden Flight Research Center [DFRC], General aviation and extremely high-performance *aircraft* and *spacecraft;* flight operations and flight systems; structural characteristics of aeronautical and space vehicles.

Goddard Space Flight Center [GSFC], Scientific research in space with unmanned satellites; research and development of meteorological and communications satellites; tracking and data acquisition operations.

Jet Propulsion Laboratory [JPL], (Operated under contract by the California Institute of Technology): Deep space, lunar and interplanetary spacecraft; operation of related tracking and data acquisition systems.

Lyndon B. Johnson Space Center [JSC], Research and development of manned spacecraft system; development of *astronaut* and crew life support systems; development and intergration of experiments for space flight activities; application of space technology, and supporting scientific, engineering, and medical research.

John F. Kennedy Space Center [KSC], Provision of supporting activities for the major launchings; preparation and integration of space vehicles; collaboration with elements of the Department

of Defense as the *Eastern Test Range* and Corps of Engineers to avoid unnecessary duplication of launch facilities, services, and capabilities.

Langley Research Center [LaRC], Aeronautical and space structures and materials; advanced concepts and technology for future aircraft; aerodynamics of re-entry vehicles; space environmental physics; improved *supersonic* flight capabilities.

Lewis Research Center [LeRC], Power plants and propulsion; high energy propellants; *electric propulsion;* aircraft engine noise reduction; engine pollution reduction; data bank of research information in aerospace safety.

George C. Marshall Space Flight Center, [MSFC], Research and development of *launch vehicles* and systems to launch vehicles and systems to launch manned and unmanned spacecraft; development and integration of *payloads* and experiments for assigned space flight activities; application of space technology and supporting scientific and engineering research.

National Space Technology Laboratories, [NSTL] formerly called Mississippi Test Facility [MTF], Static test firing of large space and launch vehicle engines; also houses certain environmental research and earth resources activities of NASA and other governmental agencies, with emphasis on the use of space technology and associated managerial and technical disciplines.

Wallops Flight Center [WFC], Launch facilities and services for other NASA installations which conduct sub-orbital, orbital, and *space probe* experiments with vehicles ranging from small rockets to the *Scout* four-stage solid fuel rocket. Development of techniques for collection and processing of experimental data.

Appendix II

NASA launch vehicles

Whether it was a manned or unmanned mission, the spacecraft carrying the payload was lifted into space by rocket propulsion systems, commonly referred to as "launch vehicles". A variety of expendable launch vehicles have been developed and used in the U.S. Space Program. Several of these vehicles, such as the Atlas/Agena, the Saturn V, the Saturn IB, and the Titan III-E/Centaur, are no longer in service. Others, such as the Delta, *Atlas/Centaur,* Scout, and *Titan 34D* will remain in service until the *Space Transportation System* becomes fully operational. Then, the Shuttle will replace all expendable launch vehicles currently in use, except the Scout. Consequently, by the mid-1980's the Space Shuttle will carry into space virtually all of America's civilian and military payloads, as well as many international civilian and government payloads. The components, capabilities and applications of the *Space Shuttle* are explored extensively in other entries. The characteristics of the expendable launch vehicles, which helped start the Age of Space, will be summarized here.

The Atlas/Agena stood 36.6 meters (120 feet) high and developed a total thrust at liftoff of approximately 1,725,824 newtons (388,000 pounds-force.) This multi-purpose, two-stage liquid propellant rocket was used to place unmanned spacecraft into earth orbit, or injected payloads into planetary or deep space trajectories. The versatile Atlas/Agena was used in the earlier Mariner probes to Mars and Venus, the Ranger photographic missions to the Moon, the Orbiting Astronomical Observatory (OAO) program, and the early Applications Technology Satellites (ATS) launches. This launch vehicle was also used as the rendezvous target vehicle for the Gemini spacecraft, during a series of two-man orbital missions in 1965-66. In support of the manned lunar landing program, Atlas/Agena launched Lunar Orbiter spacecraft—which pho-

tographed (from an orbit around the moon) possible landing sites. This launch vehicle was last used to launch an Orbiting Geophysical Observatory (OGO) in 1968.

The Saturn V, America's most powerful rocket, developed 34.5 million newtons (7.75 million pounds-force) of thrust at liftoff. When mated to the Apollo spacecraft, this vehicle stood 111 meters (363 feet) tall. The Saturn V "moon rocket" completed its last manned mission on December 7, 1972 when it sent Apollo 17 on the final lunar expedition of the 1970's. This vehicle was also used to lift the unmanned Skylab Space Station into earth orbit on May 14, 1973. Skylab was then used by three separate astronaut crews for a total of 171 days.

The Saturn 1B vehicle when mated to the Apollo spacecraft was 69 meters (223 feet) high and developed 7.1 million newtons (1.6 million pounds-force) of thrust at liftoff. This launch vehicle was used to carry the manned crews of the Skylab Program and the Apollo Soyuz Test Project (ASTP).

The Titan III-E/Centaur had an overall height of 48.8 meters (160 feet) and was a two-stage, liquid propellant rocket to which were attached two large solid rocket motors. At liftoff these solid rockets produced 10.7 million newtons (2.4 million pounds-force) of thrust. The Titan III-E/Centaur was the launch vehicle for two Viking missions to Mars, and two Voyager spacecraft to Jupiter and Saturn.

The Kennedy Space Center (KSC) in Florida launches Delta and Atlas/Centaur rockets for NASA. Delta rockets are also launched from the Western Test Range in California. Scout rockets are launched from the Western Test Range, the Wallops Flight Center (Wallops Island, Virginia) and from the San Marco launch complex (which is off the east coast of Africa).

The Delta launch vehicle is 35.4 meters (116 feet) high and has a liquid-fueled first and second stage. The third stage is a solid-propellant rocket. In one Delta configuration the first stage is augmented by 9 Castor II strap-on solid propellant motors, 6 of which ignite at liftoff. In this configuration, the average first stage thrust with the main liquid engine and six solid motors burning is 2,357,440 newtons (530,000 pounds-force). In another Delta configuration more powerful Castor IV solid rocket boosters are used. With five of these solid motors igniting at liftoff, the Delta has an average first stage thrust of 2,817,000 newtons (632,000 pounds-force). The Delta launch vehicle is called the "workhorse" of the space program. It has successfully lifted over 140 payloads into space. The advanced Delta 3920 launch vehicle will have the capability to insert a 1,270 kg (2,800 lb_m) payload into *geosynchronous* transfer orbit.

The Atlas/Centaur stands approximately 39.9 meters (131 feet) tall. At liftoff, the Atlas booster develops over 1.9 million newtons (431,000 pounds-force) of thrust and the Centaur second stage develops 133,450 newtons (30,000 pounds-force) of thrust in a vacuum. The Centaur was America's first high-energy, liquid hydrogen-liquid oxygen propellant launch vehicle stage. The Atlas/Centaur is NASA's standard launch vehicle for intermediate payloads and has been used to launch earth orbital, earth synchronous (geostationary), and interplanetary missions. As presently configured, the Atlas/Centaur can place 4,300 kg (10,000 pounds-mass) in low earth orbit, 1,905 kg (4,200 lb_m pounds-mass) in synchronous transfer orbit, and 900 kg (2,000 pounds-mass) on an interplanetary trajectory.

The standard Scout launch vehicle is a solid propellant, four stage booster system some 23 meters (75 feet) tall and with a liftoff thrust of 588,240 newtons (132,240 pounds-force). The Scout has been used to place a variety of U.S. and international payloads into inclined, equatorial, and polar orbits in support of various earth orbital, space probe, and reentry missions. The current payload capability of the Scout is 272 kg (600 lb_m).